THE ROLE OF GREEN CHEMISTRY IN BIOMASS PROCESSING AND CONVERSION

THE ROLE OF GREEN CHEMISTRY IN BIOMASS PROCESSING AND CONVERSION

Edited by

Haibo Xie
Nicholas Gathergood

A JOHN WILEY & SONS, INC., PUBLICATION

Published by John Wiley & Sons, Inc., Hoboken, New Jersey.
Published simultaneously in Canada

Wiley also publishes its books in a variety of electronic formats. Some content that appears in print may not be available in electronic formats. For more information about Wiley products, visit our web site at www.wiley.com.

Library of Congress Cataloging-in-Publication Data:

The role of green chemistry in biomass processing and conversion / edited by Haibo Xie, Nicholas Gathergood.
p. cm.
Includes index.
ISBN 978-0-470-64410-2 (cloth)
1. Environmental chemistry–Industrial applications. 2. Biomass energy. I. Xie, Haibo, 1978- II. Gathergood, Nicholas, 1972-
TP155.2.E58R65 2013
333.95′39–dc23

2012017190

Printed in the United States of America

ISBN: 9780470644102

10 9 8 7 6 5 4 3 2 1

CONTENTS

FOREWORD

Many predictions have been made as to when global oil production will reach its maximum, most predicting it to occur in the early 21st century with the demand for oil continuing to rise while production is reducing. When combined with the now very clear fact that remaining oil is difficult to obtain and comes at a very high environmental as well as economic cost, it is inevitable that oil prices will rise probably at a more dramatic rate than we have seen before leading to market and political instabilities. While public and most political attention has focused on the impact of this on energy costs, there is an equally inevitable effect on chemicals derived from petroleum. Indeed, it could be argued that the prospects for chemicals are worse as with energy there are noncarbon alternatives. Clearly, we must quickly seek economically and environmentally sound sustainable alternative feedstocks for the manufacture of key commodity chemicals.

The economics and availability of oil feedstocks is a key factor in the drive to get more sustainable alternatives, but it is not the only driver. Protection of the natural environment is also widely recognized as a key aspect in building a sustainable future. Global warming as a result of CO_2, CH_4, and other emissions; the accumulation of plastics in landfill sites and in the ocean; acid rain; smog in highly industrialized areas; and many other forms of pollution, can all be attributed to the use of oil and other fossil fuels as feedstocks. The challenge for scientists to support a sustainable economy is to produce material products for society which are based on green and sustainable supply chains. We cannot sustainably use resources more quickly than they are produced and we cannot sustainably produce waste more quickly than the planet can process it back into useful resources. We need short-cycle renewable resources.

Biomass offers the only sustainable and practical source of carbon for our chemical and material needs. It is also available for a cycle time measured in years rather than hundreds of millions of years for fossil resources. The concept of a biorefinery is the key to unlocking biomass as a feedstock for the chemical industry. Biorefineries of the future will incorporate the production of fuels, energy, and chemicals, via the processing of biomass.

The move from petroleum to biomass as the carbon feedstock for the chemical industry provides only half the answer. We need to use efficient technologies in the biorefineries and protect the environment: to do this, the concepts outlined by green chemistry must be applied. Green chemistry was originally developed to eliminate the use, or generation, of environmentally harmful and hazardous chemicals as well as reduce waste. Green chemistry today takes a more life cycle point of view and seeks to use clean manufacturing to convert renewable resources into safe products, products that ideally can be recycled at the end of life thus maintaining the principle of "closed loop manufacturing." It offers a tool kit of techniques and underlying principles that any researcher could, and should, apply when developing green and sustainable chemical-product supply chains. This book addresses this challenge by studying in depth how different green chemical technologies can help turn biomass into green and sustainable chemicals. The chapters cover the use of benign solvents, alternative energy technologies, catalytic methods and separation techniques, as well as the basics of biomass, biorefineries, and green chemistry.

After introductory chapters on biorefineries and green chemistry, there are three chapters focusing on how the three most studied alternative reaction media in green chemistry, can be applied to biorefineries. Ionic liquids represent one of the most fascinating of the green chemical technologies – getting around the volatile solvent problem by using nonvolatile liquids that can also be incredibly powerful solvents and even combined catalyst–solvent systems. Ionic liquids are one of the more likely solutions to the problem of often highly intractable biomass. There can be no better solvent from an environmental point of view and in terms of convenience in a biorefinery than water – biomass is inevitably wet anyway and the more we can do processing in water the simpler, safer, and cheaper the biorefinery products are likely to be. Biorefineries will produce a lot of CO_2 and making use of that CO_2 will be an especially important goal; supercritical CO_2 is a rather useful solvent for extractions from biomass and for some downstream chemistry. These "alternative media" chapters are followed by chapters tackling the critical issue of cellulose dissolution for processing – NaOH/urea/water being a very simple and effective medium for dissolving cellulose and then using those solutions, while the organosolv method and especially the organosolv-ethanol process can also be used to help process ligno-cellulosics more generally and even help tackle the problematic issue of lignin valorization.

One of the most popular product types from biomass have been pyrolysis oils that are being seriously considered as partial replacements for petroleum fuels. Chapter 8 addresses this area and includes the vital issue of upgrading since most as-produced pyrolysis oils are not of the required chemical quality for example, they are too acidic, for direct mixing with petroleum. Microwave processing is an alternative to conventional heating as a way to turn biomass into pyrolysis oils as well as for biomass pretreatment and saccrification – some of the topics covered in Chapter 9.

Catalysis is the most important green chemical technology, tackling the fundamental green-chemistry challenges of improved efficiency, better selectivity, and lower energy consumption. Three chapters look at different ways that different catalysts can help make the most out of biomass as a feedstock. Chapter 10 looks at biotransformations and how they can be used to turn biomass into different fuels and chemicals. Heterogeneous catalysts including solid acids and bases and supported metals are often considered to be preferable to homogeneous equivalents as they enable simpler and less wasteful separations at the end of the process and it is appropriate that their use in some biomass conversions are considered here. A particularly interesting and current challenge in biomass conversion is the utilization of glycerol produced in very large quantities as a by-product in the manufacture of biodiesel, one of the most successful biofuels. The use of the glycerol would greatly support biodiesel manufacture and Chapter 12 looks at catalytic ways to help do this.

Green chemistry offers alternatives to conventional reactors and energy sources. Apart from microwaves discussed in Chapter 9, ultrasonics have also proven popular and their use in biorefineries and especially in assisting biofuel production is discussed in Chapter 13. Separations are often the biggest source of waste in a chemical manufacturing process and clever ways to separate complex products in biorefinery processes are essential. In Chapter 14, advanced membrane technologies including the important pervaporation method and different membrane materials including polymers and zeolites are discussed. In the final chapter, the critical issues of ecotoxicity and environmental impact from using biorefineries are addressed including biofuel production and biofuel emissions.

Biomass utilization alone is not the answer to the sustainable production of liquid fuels and organic chemicals but when combined with the best of green chemistry we have the real opportunity to help create a truly sustainable society.

JAMES CLARK

PREFACE

Our high quality of living standards in many parts of the world is largely due to and dependent on the development of fossil-based energy and chemical industries. While the products from these industries have enriched our life, they have also directly or indirectly placed our environment under immense stress. One of most noticeable issues is global warming, caused by the accumulation of "Green House" gases, due to over dependence on nonrenewable fossil-based resources. To counteract this, the concept of green-chemistry was proposed towards the design of products and processes that minimize the use and generation of hazardous substances. The aim is to avoid problems before they occur.

Fossil fuels are considered nonrenewable resources because they take millions of years to form. It is estimated that they will be depleted by the end of this century. Furthermore, the production and use of fossil fuels raises considerable environmental concerns. A global movement toward the generation of energy and chemicals from renewable sources is therefore under way. This will help meet increased energy and chemical-feedstock needs. Biomass has an estimated global production of around 1.0×10^{11} tons per year, through natural photosynthesis using CO_2 as the carbon source. Therefore, the carbon in biomass is regarded as a "carbon neutral" carbon source for the construction of chemicals and materials through biological and chemical approaches. It is estimated that by 2025, up to 30% of raw materials for the chemical industry will be produced from renewable sources. To achieve this goal it will require a major readjustment of the overall techno-economic approach. From a sustainability point of view, and learning from decades of petroleum-refinery process, the introduction and integration of green-chemistry concept into biomass processes and conversion is one of the key issues towards a concept of avoiding problems before they happen.

Biomass can refer to *species biomass*, which is the mass of one or more species, or to *community biomass*, which is the mass of all species in the community. It can include microorganisms, plants, or animals. In this book, we focus on lignocellulosic biomass, because they represent the most abundant of biomass resources. They are mainly composed of cellulose, hemicellulose, and lignin. To differentiate the research of petroleum refinery, a new biorefinery process has been proposed according to biomass-based research activities. Current knowledge of lignocelluose-based biomass and the biorefinery process have been introduced in the first chapter in this book, which presents the basic and whole ideas to convert the biomass into valuable chemicals and materials.

Since the concept of green chemistry was proposed, significant accomplishments have been achieved according to the widely recognized "twelve principles," and recent advances have been introduced in the second chapter in this book. This gives a more in-depth understanding of green chemistry and potential green technologies; those that could be used for biomass processing and conversion. With a better understanding of challenges during biomass processing and conversion, the introduction and exploration of suitable green-chemistry technologies is important to meet the tailored-processing and conversion of biomass. The contributors from different specific research areas provide us with the latest progress and insight in the biomass processing and conversion using green-chemistry technologies. For example, the introduction of green solvents (e.g., ionic liquids, supercritical CO_2, water); sustainable energy sources (e.g., microwave irradiation, sonification); green catalytic technologies; advanced membrane separation technology; etc. We believe that all of these will be strong bases for the foundation and exploration of a cost-competitive and sustainable bioeconomy in the near future.

Traditionally, a focus on the economic assessments of technologies was exercised while social and environmental assessments were often neglected, which is one of the reasons for the ultimate environmental deterioration. The balance of economic assessments, social assessments, and environmental assessments is one of most important issues for any emerging technologies towards a sustainable biorefinery. The last chapter of the book gives us in-depth understanding of environmental assessments of the conversion and use of fuels, chemicals, and materials from biomass.

Research into biomass processing and conversion is a wide-ranging interdisciplinary research field, and the book presents an up-to-date multidisciplinary treatise for the utilization of biomass from a sustainable chemistry point of view. We thank all the people who made valuable contributions and suggestions, from the esteemed contributors to the diligent reviewers, which laid the foundations for a successful project and publication of this book.

DR. HAIBO XIE and DR. NICHOLAS GATHERGOOD

CONTRIBUTORS

Matthew T. Agler, Department of Biological and Environmental Engineering, Cornell University

Thomas E. Amidon, Department of Paper and Bioprocess Engineering, College of Environmental Science and Forestry, State University of New York

Largus T. Angenent, Institute for Environmental Research, RWTH Aachen University

Dimitris S. Argyropoulos, Organic Chemistry of Wood Components Laboratory, Department of Forest Biomaterials, North Carolina State University; Department of Chemistry, Laboratory of Organic Chemistry, University of Helsinki, Finland

Ian Beadham, School of Chemical Sciences, Dublin City University

Weihui Bi, Changchun Institute of Applied Chemistry, Chinese Academy of Sciences, Key Lab Ecomaterials of Chinese Academy of Sciences

Biljana Bujanovic, Department of Paper and Bioprocess Engineering, College of Environmental Science and Forestry, State University of New York

Kerstin Bluhm, Institute for Environmental Research, RWTH Aachen University

Leming Cheng, Department of Biosystems Engineering and Soil Science, The University of Tennessee

Jiangjiang Duan, Department of Materials Science and Engineering, College of Materials, Xiamen University

Qirong Fu, Organic Chemistry of Wood Components Laboratory, Department of Forest Biomaterials, North Carolina State University

Nicholas Gathergood, School of Chemical Sciences, Dublin City University, National Institute for Cellular Biotechnology, Dublin City University, Solar Energy Conversion Strategic Research Cluster, University College Dublin

Mukund Ghavre, School of Chemical Sciences, Dublin City University, Dublin

Mangesh J. Goundalkar, Department of Paper and Bioprocess Engineering, College of Environmental Science and Forestry, State University of New York

David Grewell, Agricultural and Biosystems Engineering, Iowa State University

Sebastian Heger, Institute for Environmental Research, RWTH Aachen University

Henner Hollert, Institute for Environmental Research, RWTH Aachen University

Cuimin Hu, Dalian Institute of Chemical Physics, Chinese Academy of Sciences

Birgit Kamm, Research Institute Bioactive Polymer Systems e. V. and Brandenburg University of Technology

Melissa Montalbo-Lomboy, Agricultural and Biosystems Engineering, Iowa State University

Changzhi Li, State Key Laboratory of Catalysis, Dalian Institute of Chemical Physics, Chinese Academy of Sciences

Shenghai Li, Changchun Institute of Applied Chemistry, Chinese Academy of Sciences, Key Lab Ecomaterials of Chinese Academy of Sciences

Wujun Liu, Dalian Institute of Physical Chemistry, Chinese Academy of Science

Fang Lu, Bioenergy Division, Dalian National Laboratory for Clean Energy, State Key Laboratory of Catalysis, Dalian Institute of Chemical Physics, Chinese Academy of Sciences

Haile Ma, School of Food and Biological Engineering, Jiangsu University

Hong Ma, Bioenergy Division, Dalian National Laboratory for Clean Energy, State Key Laboratory of Catalysis, Dalian Institute of Chemical Physics, Chinese Academy of Sciences

Sibylle Maletz, Institute for Environmental Research, RWTH Aachen University

Ray Marriott, The Biocomposites Centre, Bangor University Gwynedd

Tomohiko Mitani, Research Institute for Sustainable Humanosphere, Kyoto University

Xuejun Pan, Department of Biological Systems Engineering, University of Wisconsin-Madison

Thomas-Benjamin Seiler, Institute for Environmental Research, RWTH Aachen University

Andreas Schäffer, Institute for Environmental Research, RWTH Aachen University

Emily Sin, The Biocomposites Centre, Bangor University Gwynedd; Department of Chemistry, University of York

Aiqin Wang, State Key Laboratory of Catalysis, Dalian Institute of Chemical Physics, Chinese Academy of Sciences

Feng Wang, Bioenergy Division, Dalian National Laboratory for Clean Energy, State Key Laboratory of Catalysis, Dalian Institute of Chemical Physics, Chinese Academy of Sciences

Takashi Watanabe, Research Institute for Sustainable Humanosphere, Kyoto University

Haibo Xie, Bioenergy Division, Dalian National Laboratory for Clean Energy, Dalian Institute of Physical Chemistry, Chinese Academy of Sciences

Xiaopeng Xiong, Department of Materials Science and Engineering, College of Materials, Xiamen University

Jie Xu, Bioenergy Division, Dalian National Laboratory for Clean Energy, State Key Laboratory of Catalysis, Dalian Institute of Chemical Physics, Chinese Academy of Sciences

X. Philip Ye, Department of Biosystems Engineering and Soil Science, The University of Tennessee

Weiqiang Yu, Bioenergy Division, Dalian National Laboratory for Clean Energy, State Key Laboratory of Catalysis, Dalian Institute of Chemical Physics, Chinese Academy of Sciences

Zongbao K. Zhao, Bioenergy Division, Dalian National Laboratory for Clean Energy, Dalian Institute of Physical Chemistry, Chinese Academy of Sciences

Tao Zhang, State Key Laboratory of Catalysis, Dalian Institute of Chemical Physics, Chinese Academy of Sciences

Suobo Zhang, Changchun Institute of Applied Chemistry, Chinese Academy of Sciences, Key Lab Ecomaterials of Chinese Academy of Sciences

Mingyuan Zheng, State Key Laboratory of Catalysis, Dalian Institute of Chemical Physics, Chinese Academy of Sciences

ABOUT THE EDITORS

Dr. Haibo Xie is currently an associate professor at Dalian National Laboratory for Clean Energy and the Dalian Institute of Chemical Physics (DICP), Chinese Academy of Sciences (CAS). He received his PhD from Changchun Institute of Applied Chemistry, CAS in 2006 and BSc from Xiangtan University in 2001. He worked as a postdoctoral researcher at the Department of Forest Biomaterials, North Carolina State University (2006–2007) and an IRCSET-Embark Initiative research fellow at the National Institute for Cellular Biotechnology Research Center and the School of Chemical Sciences, Dublin City University (2008–2010). In 2009, he obtained a Career Start Program Fellowship from Dublin City University. He joined the faculty of DICP, Chinese Academy of Sciences under the One Hundred Talents Program of DICP from March 2010. His main research interests focus on the use of green solvents and green chemistry technologies in the processing and conversion of biomass into biofuels, value-added chemicals and sustainable materials.

Dr. Nicholas Gathergood is a lecturer at the School of Chemical Sciences at Dublin City University (DCU). He received his PhD in 1999 from the University of Southampton, under the guidance of Prof. R. Whitby. Postdoctoral research with Prof. K. A. Jørgensen, Centre for Catalysis, Aarhus University, Denmark and Prof. P. J. Scammells, Victorian College of Pharmacy, Monash University, Australia, followed. Since 2004, Dr Gathergood has established a large research group (15+) at DCU and supervised 19 PhD students.

Positions of responsibility have included Chairman of the Society of Chemical Industry (SCI)—All Ireland group and Irish representative of the EUCHeMS Division of Organic Chemistry. He initiated the SCI sponsored Green Chemistry in Ireland conference series and works closely with the EPA in Ireland. Many postdoctoral fellows he has supported have begun their own academic careers in the United Kingdom, France, and China. Dr Gathergood is especially proud of the 100% success rate for his PhD students finding employment.

His research interests focus on using green chemistry as a tool to realize safer and more sustainable organic chemistry, medicinal chemistry (including drug discovery), and ultimately to develop environmentally friendly pharmaceuticals.

CHAPTER 1

Introduction of Biomass and Biorefineries*

BIRGIT KAMM

The development of biorefineries represents the key to access the integrated production of food, feed, chemicals, materials, goods, fuels, and energy in the future. Biorefineries combine the required technologies for biogenic raw materials from agriculture and forestry with those of intermediate and final products. The specific focus of this chapter is the combination of green agriculture with physical and biotechnological processes for the production of proteins as well as the platform chemicals lactic acid and lysine. The mass and energy flows (steam and electricity) of the biorefining of green biomass into these platform chemicals, proteins, and feed as well as biogas from residues are given. The economic and ecologic aspects for the cultivation of green biomass and the production of platform chemicals are described.

1.1 INTRODUCTION

One hundred and fifty years after the beginning of coal-based chemistry and 50 years after the beginning of petroleum-based chemistry, industrial chemistry is now entering a new era. An essential part of the sustainable future will be based on the appropriate and innovative use of our biologically based feedstocks. It will be particularly necessary to have a substantial conversion industry in addition to research and development investigating the efficiency of producing raw materials and product lines, as well as sustainability.

Whereas the most notable successes in research and development in the field of biorefinery system research have been in Europe and Germany, the first significant

* Dedicated to Michael Kamm, Founder of Biorefinery.de GmbH.

The Role of Green Chemistry in Biomass Processing and Conversion, First Edition.
Edited by Haibo Xie and Nicholas Gathergood.

industrial developments were promoted in the United States of America by the President and Congress [1–5]. In the United States, it is expected that by 2020 at least 25% (compared to 1995) of organic carbon-based industrial feedstock chemicals and 10% of liquid fuels will be obtained from a biobased product industry [6]. This would mean that more than 90% of the consumption of organic chemicals and up to 50% of liquid fuel requirements in the United States would be supplied by biobased products [7]. The US Biomass Technical Advisory Committee (BTAC)—in which leading representatives of industrial companies such as Dow Chemical, E.I. du Pont de Nemours, Cargill, Dow LLC, and Genecor International Inc., as well as corn growers' associations and the Natural Resources Defence Council are involved, and which acts as an advisor to the US government—has made a detailed step-by-step plan of the targets for 2030 with regard to bioenergy, biofuels, and bioproducts [8–10].

Research and development are necessary to

(1) increase the scientific understanding of biomass resources and improve the tailoring of those resources;
(2) improve sustainable systems to develop, harvest, and process biomass resources;
(3) improve the efficiency and performance in conversion and distribution processes and technologies for a multitude of product developments from biobased products; and
(4) create the regulatory and market environment necessary for the increased development and use of biobased products.

BTAC has established specific research and development objectives for feedstock production research. Target crops should include oil- and cellulose-producing crops that can provide optimal energy content and usable plant components. Currently, however, there is a lack of understanding of plant biochemistry as well as inadequate genomic and metabolic information on many potential crops. In particular, research to produce enhanced enzymes and chemical catalysts could advance biotechnological capabilities.

In Europe, there are existing regulations regarding the substitution of nonrenewable resources by biomass in the field of using biofuels for transportation as well as the "Renewable energy law" [11, 12]. According to the EC Directive "On the promotion of the use of biofuels," the following products are considered as "biofuels": (a) "bioethanol," (b) "biodiesel," (c) "biogas," (d) "biomethanol," (e) "bio-dimethylether," (f) "bio-ETBE (ethyl-*tert*-butylether)" based on bioethanol, (g) "bio-MTBE (methyl-*tert*-butylether)" based on biomethanol, (h) "synthetic biofuels," (i) "biohydrogen," and (j) pure vegetable oil.

Member states of the EU have been asked to define national guidelines for the minimum usage quantities of biofuels and other renewable fuels (with a reference value of 2% by 2005 and 5.75% by 2010, calculated on the basis of the energy content of all petrol and diesel fuels for transport purposes). Currently, there are no guidelines for biobased products in the EU or in Germany. However, after passing directives for bioenergy and biofuels, such activities are on the political agenda.

Recently, the German Government has announced the biomass action plan for substantial use of renewable resources, and the German Chemical Societies have published the position paper "Raw material change," including nonfood biomass as raw material for the chemical industry [13, 14]. The European Technology Platform for Sustainable Chemistry has created the EU Lead Market initiative [15]. The directive for biofuels already includes ethanol, methanol, dimethylether, hydrogen, and biomass pyrolysis, which are fundamental product lines of the future biobased chemical industry. A recent paper looking at future developments, published by the Industrial Biotechnology section of the European Technology platform for Sustainable Chemistry, foresaw up to 30% of raw materials for the chemical industry coming from renewable sources by 2025 [16]. The ETPSC has created the EU Lead Market initiative [15].

The European Commission and the US Department of Energy have come to an agreement for cooperation in this field [17]. Based on the European biomass action plan of 2006, both strategic EU-projects (1) BIOPOL, European Biorefineries: Concepts, Status and Policy Implications and (2) Biorefinery Euroview: Current situation and potential of the biorefinery concept in the EU: strategic framework and guidelines for its development, began preparation for the 7th EU framework [18–20].

In order to minimize food–feed–fuel conflicts and to use biomass most efficiently, it is necessary to develop strategies and ideas for how to use biomass fractions, in particular, green biomass and agricultural residues such as straw, more efficiently. Such an overall utilization approach is described in Section 1.2. In future developments, food- and feed-processing residues should therefore also become part of biorefinery strategies, since either specific waste fractions may be too small for a cost-efficient specific valorization (capitalize on nature's resources) treatment *in situ* or the diverse technologies necessary are not available. Fiber-containing food-processing residues may then be pretreated and processed with other cellulosic material from other sources in order to produce ethanol or other platform chemicals. Food-processing residues have, however, a particular feature one has to be aware of. Due to their high water content and endogenous enzymatic activity, food-processing residues have a comparatively low biological stability and are prone to uncontrolled degradation and spoilage including rapid autoxidation. To avoid extra costs for transportation and conservation, the use of food-processing residues should also become part of a regional biomass utilization network [21].

1.2 BIOREFINERY TECHNOLOGIES AND BIOREFINERY SYSTEMS

1.2.1 Background

Biobased products are prepared for economically viable use by a suitable combination of different methods and processes (physical, chemical, biological, and thermal). To this end, base biorefinery technologies need to be developed. For this reason, it is inevitable that there must be profound interdisciplinary cooperation among the individual disciplines involved in research and development. Therefore, it is appropriate to use the term "biorefinery design," which implies that well-founded scientific and technological principles are combined with technologies, products, and product lines inside biorefineries that are close to practice. The basic conversions of each biorefinery can be summarized as follows.

In the first step, the precursor-containing biomass is separated by physical methods. The main products (M1–M*n*) and by-products (B1–B*n*) will subsequently be subjected to further processing by microbiological or chemical methods. The subsequent products (F1–F*n*) obtained from the main products and by-products can be further converted or used in a conventional refinery. Four complex biorefinery systems are currently under testing at the research and development stage:

(1) Lignocellulosic feedstock biorefinery using naturally dry raw materials such as cellulose-containing biomass and wastes.
(2) Whole-crop biorefinery using raw material such as cereals or maize (whole plants).
(3) Green biorefineries using naturally wet biomasses such as green grass, alfalfa, clover, or immature cereal [22, 23].
(4) The two-platforms biorefinery concept, which includes the sugar platform and the syngas platform [24].

1.2.2 Lignocellulosic Feedstock Biorefinery

Among the potential large-scale industrial biorefineries, the lignocellulosic feedstock (LCF) biorefinery will most probably be the most successful. First, there is optimum availability of raw materials (straw, reed, grass, wood, paper waste, etc.), and second, the conversion products are well-placed on the traditional petrochemical as well as on the future biobased product market. An important factor in the utilization of biomass as a chemical raw material is its cost. Currently, the cost for corn stover or straw is US $50/metric ton, and for corn US $80/metric ton [25].

Lignocellulose materials consist of three primary chemical fractions or precursors: (1) hemicellulose/polyoses—a sugar polymer predominantly having pentoses; (2) cellulose—a glucose polymer; and (3) lignin—a polymer of phenols (Fig. 1.1). The lignocellulosic biorefinery system has a distinct ability to create genealogical trees. The main advantages of this method are that the natural structures and structure elements are preserved, the raw materials are cheap, and many product varieties are possible (Fig. 1.2). Nevertheless, there is still a requirement for development and

$$\text{Lignocellulose} + H_2O \rightarrow \text{Lignin} + \text{Cellulose} + \text{Hemicellulose}$$

$$\text{Hemicellulose} + H_2O \rightarrow \text{Xylose}$$

$$\text{Xylose } (C_5H_{10}O_5) + \text{Acid Catalyst} \rightarrow \text{Furfural } (C_5H_4O_2) + 3H_2O$$

$$\text{Cellulose } (C_6H_{10}O_5)_n + n\,H_2O \rightarrow n \text{ Glucose } (C_6H_{12}O_6)$$

FIGURE 1.1 A possible general equation of conversion at the lignocellulosic feedstock (LCF) biorefinery [26].

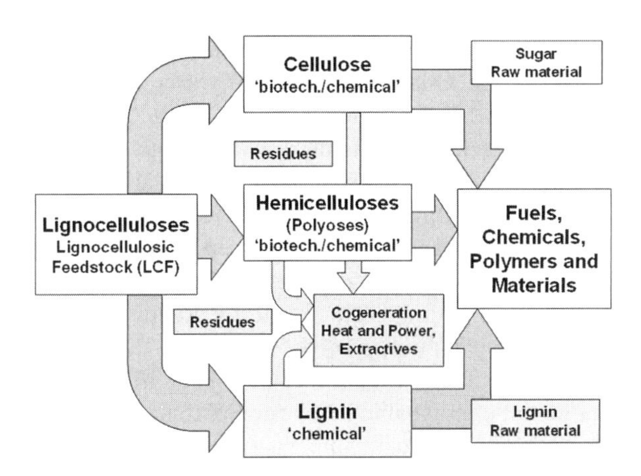

FIGURE 1.2 Lignocellulosic feedstock biorefinery [26].

optimization of these technologies, for example, in the field of separating cellulose, hemicellulose, and lignin, as well as in the use of lignin in the chemical industry.

Furfural and hydroxymethylfurfural, in particular, are interesting products. Furfural is the starting material for the production of Nylon 6,6 and Nylon 6 [27]. The original process for the production of Nylon 6,6 was based on furfural. The last of these production plants in the United States was closed in 1961 for economic reasons (the artificially low price of petroleum). Nevertheless, the market for Nylon 6 is still very large.

However, some aspects of the LCF system, such as the utilization of lignin as a fuel, adhesive, or binder, remain unsatisfactory because the lignin scaffold contains considerable amounts of monoaromatic hydrocarbons which, if isolated in an economically efficient way, could add significant value to the primary process. It should be noted that there are no obvious natural enzymes to split the naturally formed lignin into basic monomers as easily as polymeric carbohydrates or proteins, which are also naturally formed [28].

An attractive accompanying process to the biomass-nylon process is the previously mentioned hydrolysis of cellulose to glucose and the production of ethanol. Certain yeasts produce a disproportionate amount of the glucose molecule while generating glucose out of ethanol. This process effectively shifts the entire reduction ability into the ethanol and makes the latter obtain a 90% yield (w/w; with regard to the formula turnover). Based on recent technologies, a plant was designed for the production of the main products furfural and ethanol from LC-feedstock in West Central Missouri. Optimal profitability can be reached with a daily consumption of about 4360 ton feedstock. Annually, the plant produces 47.5 million gallons ethanol and 323,000 ton furfural [29].

Ethanol may be used as a fuel additive. Ethanol is also a connecting product for a petrochemical refinery, and can be converted into ethylene by chemical methods.

As is well-known from the use of petrochemically produced ethylene, nowadays ethanol is the raw material for a whole series of large-scale technical chemical syntheses for the production of important commodities, such as polyethylene or polyvinylacetate. Other petrochemically produced substances, such as hydrogen, methane, propanol, acetone, butanol, butandiol, itaconic acid, and succinic acid, can similarly be manufactured by substantial microbial conversion of glucose [30–32]. DuPont has entered into a 6-year alliance with Diversa to produce sugar from husks, straw, and stovers in a biorefinery, and to develop processes to coproduce bioethanol and value-added chemicals such as 1,3-propandiol. Through metabolic engineering, the microorganism *Escherichia coli* K12 produces 1,3-propandiol in a simple glucose fermentation process developed by DuPont and Genencor. In a pilot plant operated by Tate and Lyle, the 1,3-propandiol yield reaches $135\,g\,L^{-1}$ at a rate of $4\,g\,L^{-1}\,h^{-1}$ [33]. 1,3-Propandiol is used for the production of polytrimethylene-terephthalate (PTT), a new polymer used in the production of high-quality fibers with the brand name Sorona [33]. Production was predicted to reach $500\,kt\,year^{-1}$ in 2010.

1.2.3 Whole-Crop Biorefinery

Raw materials for whole-crop biorefineries are cereals such as rye, wheat, triticale, and maize (Fig. 1.3). The first step is their mechanical separation into grain and straw, where the portion of grain is approximately 1 and the portion of straw is 1.1–1.3 (straw is a mixture of chaff, stems, nodes, ears, and leaves). The straw represents an LCF and may be processed further in an LCF biorefinery system. Initial separation into cellulose, hemicellulose, and lignin is possible, with their further conversion within separate product lines, as described above for LCF biorefineries. Furthermore, straw is a raw material for the production of syngas via pyrolysis technologies. Syngas is the base material for the synthesis of fuels and methanol (Figs. 1.3 and 1.4).

The corn may either be converted into starch or used directly after grinding into meal. Further processing can take one of the four routes: (1) breaking up, (2) plasticization,

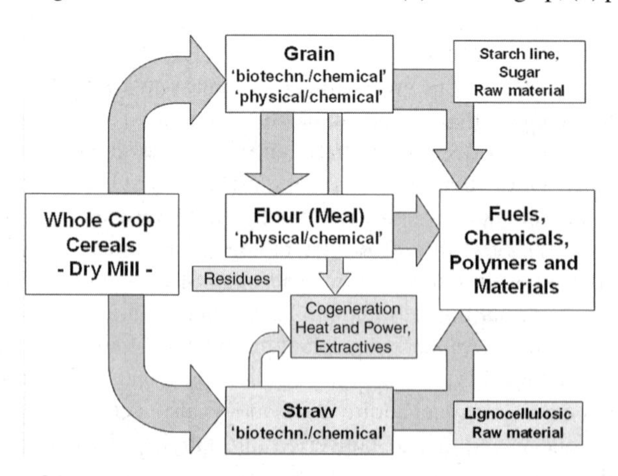

FIGURE 1.3 Whole-crop biorefinery—based on dry milling [26].

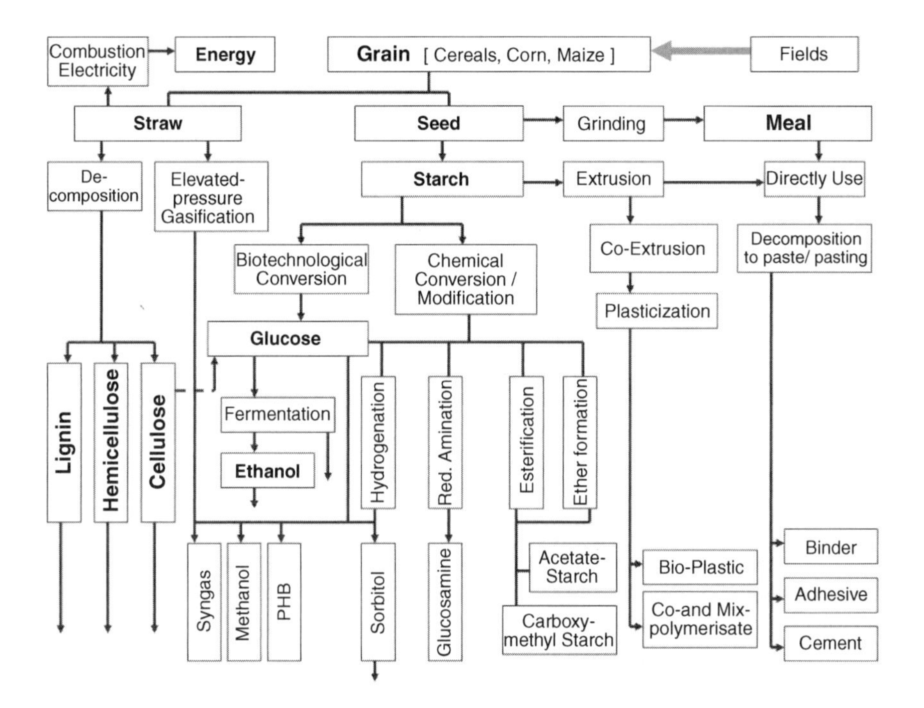

FIGURE 1.4 Products from the whole-crop biorefinery [22, 23].

(3) chemical modification, or (4) biotechnological conversion via glucose. The meal can be treated and finished by extrusion into binder, adhesives, or filler. Starch can be finished via plasticization (co- and mix-polymerization, compounding with other polymers), chemical modification (etherification into carboxy-methyl starch; esterification and re-esterification into fatty acid esters via acetic starch; splitting reductive amination into ethylene diamine), and hydrogenative splitting into sorbitol, ethylene glycol, propylene glycol, and glycerine [34–36]. In addition, starch can be converted by a biotechnological method into poly-3-hydroxybutyric acid in combination with the production of sugar and ethanol [37, 38]. Biopol, the copolymer poly-3-hydroxybutyrate/3-hydroxyvalerate, developed by ICI is produced from wheat carbohydrates by fermentation using *Alcaligenes eutropius* [39].

An alternative to the traditional dry fractionation of mature cereals into sole grains and straw has been developed by Kockums Construction Ltd (Sweden), now called Scandinavian Farming Ltd. In this whole-crop harvest system, whole immature cereal plants are harvested and all the harvested biomass is conserved or dried for long-term storage. When convenient, it can be processed and fractionated into kernels, straw chips of internodes, and straw meal, including leaves, ears, chaff, and nodes (see also Section 1.2.4).

Fractions are suitable as raw materials for the starch polymer industry, the feed industry, the cellulose industry and particle-board producers, as gluten for the chemical industry, and as a solid fuel. This kind of dry fractionation of the whole crop to

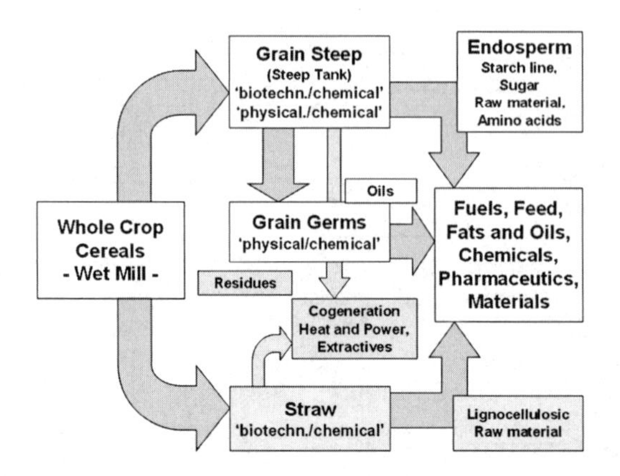

FIGURE 1.5 Whole-crop biorefinery, wet-milling [26].

optimize the utilization of all botanical components of the biomass has been described in Rexen (1986) and Coombs and Hall (1997) [40, 41]. An example of such a biorefinery and its profitability is described in Audsley and Sells (1997) [42]. The whole-crop wet-mill-based biorefinery expands the product lines into grain processing. The grain is swelled and the grain germs are pressed, generating highly valuable oils.

The advantages of the whole-crop biorefinery based on wet milling are that the natural structures and structure elements such as starch, cellulose, oil, and amino acids (proteins) are retained to a great extent, and well-known base technologies and processing lines can still be used. The disadvantages are the high raw material costs and costly source technologies required for industrial utilization. On the other hand, many of the products generate high prices, for example, in pharmacy and cosmetics (Figs. 1.5 and 1.6).

The wet milling of corn yields corn oil, corn fiber, and corn starch. The starch products obtained from the US corn wet-milling industry are fuel alcohol (31%), high-fructose corn syrup (36%), starch (16%), and dextrose (17%). Corn wet milling also generates other products (e.g., gluten meal, gluten feed, oil) [43]. An overview of the product range is shown in Figure 1.6.

1.2.4 Green Biorefinery

Often, it is the economics of bioprocesses that are the main problem because the price of bulk products is affected greatly by raw material costs [44]. The advantages of green biorefineries are a high biomass profit per hectare and a good coupling with agricultural production, combined with low prices for raw materials. On the one hand, simple base technologies can be used, with good biotechnical and chemical potential for further conversions (Fig. 1.7). On the other hand, either fast primary processing or the use of preservation methods such as silage or drying is necessary for both the raw materials and the primary products. However, each preservation method changes the content of the materials.

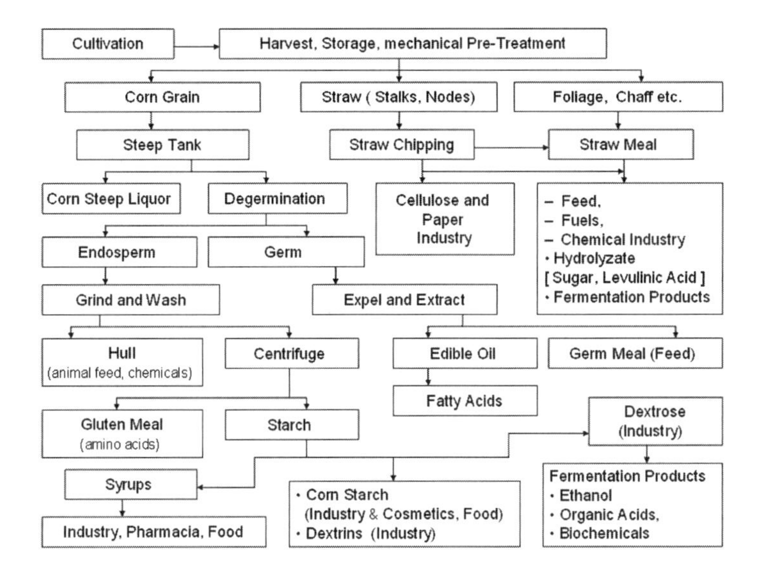

FIGURE 1.6 Products from the whole crop wet mill based biorefinery [26].

Green biorefineries are also multiproduct systems and operate with regard to their refinery cuts, fractions, and products in accordance with the physiology of the corresponding plant material; in other words, maintaining and utilizing the diversity of syntheses achieved by nature. Green biomass consists of, for example, grass from the cultivation of permanent grassland, closed fields, nature preserves, or green crops such as lucerne (alfalfa), clover, and immature cereals from extensive land cultivation.

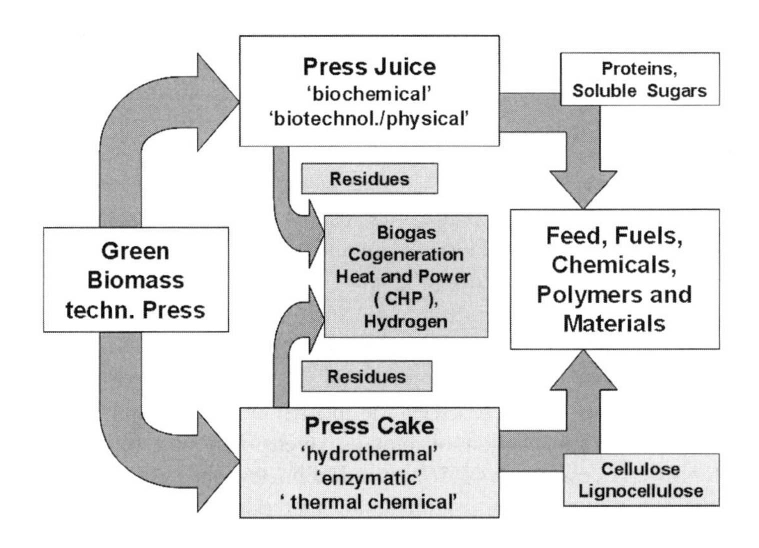

FIGURE 1.7 A green biorefinery system [26].

Today, green crops are used primarily as forage and a source of leafy vegetables. In a process called wet-fractionation of green biomass, green crop fractionation can be used for the simultaneous manufacture of both food and nonfood items [45]. Thus, green crops represent a natural chemical factory and food plant.

Scientists in several countries in Europe and elsewhere have developed green crop fractionation; indeed, green crop fractionation is now studied in about 80 countries [45–48]. Several hundred temperate and tropical plant species have been investigated for green-crop fractionation [48–50]. However, more than 300,000 higher plant species remain to be investigated (for reviews, see Refs. [1, 46, 47,51–54]).

By fractionation of green plants, green biorefineries can process from a few tonnes of green crops per hour (farm-scale process) to more than $100\,t\,h^{-1}$ (industrial-scale commercial process). Wet-fractionation technology is used as the first step (primary refinery) to carefully isolate the contained substances in their natural form. Thus, the green crop goods (or humid organic waste goods) are separated into a fiber-rich press cake (PC) and a nutrient-rich green juice (GJ).

Besides cellulose and starch, PC contains valuable dyes and pigments, crude drugs, and other organics. The GJ contains proteins, free amino acids, organic acids, dyes, enzymes, hormones, other organic substances, and minerals. In particular, the application of biotechnological methods is ideally suited for conversions because the plant water can simultaneously be used for further treatments. When water is added, the lignin–cellulose composite bonds are not as strong as they are in dry lignocellulose feedstock materials. Starting from GJ, the main focus is directed to producing products such as lactic acid and corresponding derivatives, amino acids, ethanol, and proteins. The PC can be used for the production of green feed pellets and as a raw material for the production of chemicals such as levulinic acid, as well as for conversion to syngas and hydrocarbons (synthetic biofuels). The residues left when substantial conversions are processed are suitable for the production of biogas combined with the generation of heat and electricity (Fig. 1.8). Reviews of green biorefinery concepts, contents, and goals have been published [13, 26, 55].

1.2.5 The Two-Platforms Biorefinery Concept

The "two-platform concept" means that first biomass consists on average of 75% carbohydrates, which can be standardized over an intermediate sugar platform as a basis for further conversions, and second that the biomass is converted thermochemically into synthesis gas and further products.

- The "sugar platform" is based on biochemical conversion processes and focuses on the fermentation of sugars extracted from biomass feedstocks.
- The "syngas platform" is based on thermochemical conversion processes and focuses on the gasification of biomass feedstocks and by-products from conversion processes.[24, 46, 56]. In addition to gasification, other thermal and thermochemical biomass conversion methods have also been described: hydrothermolysis, pyrolysis, thermolysis, and burning. The application used depends on the water content of the biomass [57].

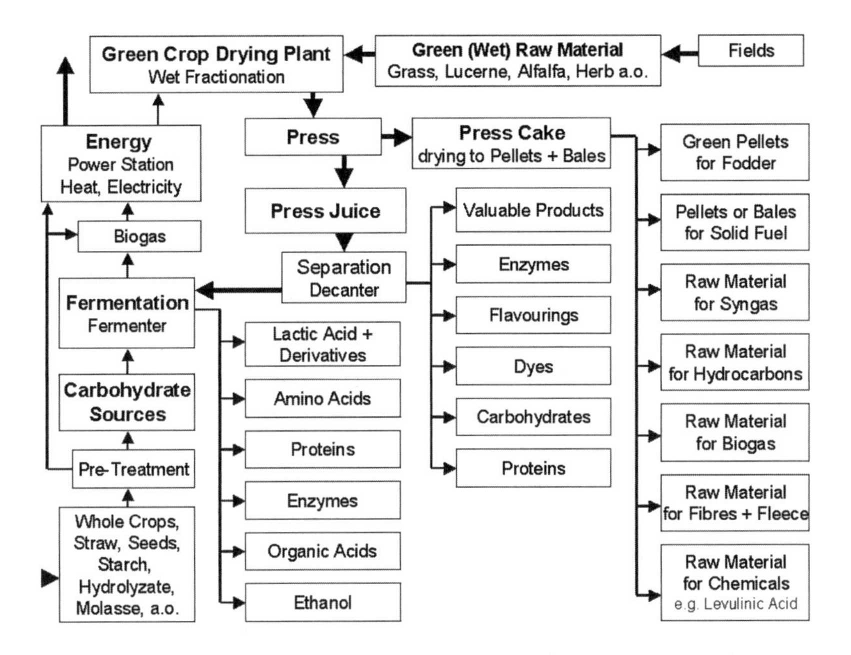

FIGURE 1.8 Products from a green biorefinery system, combined with a green crop drying plant [22, 23].

Gasification and all the thermochemical methods concentrate on the utilization of the precursor carbohydrates as well as their inherent carbon and hydrogen content. The proteins, lignin, oils and lipids, amino acids and general ingredients, as well as the N- and S-compounds occurring in all biomass, are not taken into account in this case (Fig. 1.9).

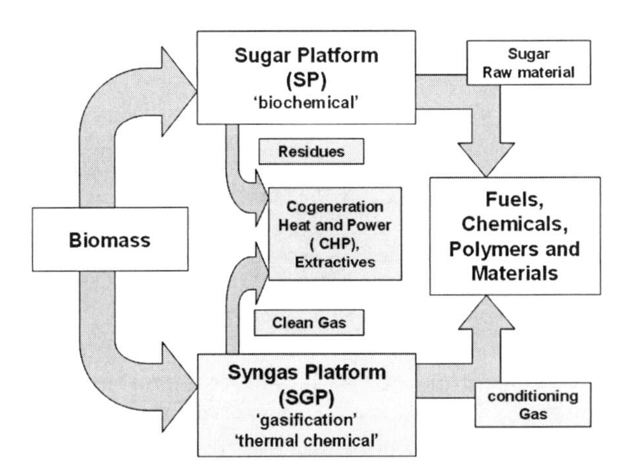

FIGURE 1.9 Sugar platform and Syngas platform [26, 58].

1.3 PLATFORM CHEMICALS

1.3.1 Background

A team from the Pacific Northwest National Laboratory (PNNL) and the National Renewable Energy Laboratory (NREL) submitted a list of 12 potential biobased chemicals [24]. The key areas of the investigation were biomass precursors, platforms, building blocks, secondary chemicals, intermediates, products, and uses (Fig. 1.10).

The final selection of 12 building blocks began with a list of more than 300 candidates. A shorter list of 30 potential candidates was selected using an iterative review process based on the petrochemical model of building blocks, chemical data, known market data, properties, performance of the potential candidates, and the prior industry experience of the team at PNNL and NREL. This list of 30 was ultimately reduced to 12 by examining the potential markets for the building blocks and their derivatives, and the technical complexity of the synthesis pathways.

The selected building-block chemicals can be produced from sugar via biological and chemical conversions. The building blocks can subsequently be converted to a number of high-value biobased chemicals or materials. Building block chemicals, as considered for this analysis, are molecules with multiple functional groups that possess the potential to be transformed into new families of useful molecules. The 12 sugar-based building blocks (Fig. 1.10) are 1,4-diacids (succinic, fumaric, and malic); 2,5-furan dicarboxylic acid; 3-hydroxy propionic acid; aspartic acid; glucaric acid; glutamic acid; itaconic acid; levulinic acid; 3-hydroxybutyrolactone; glycerol; sorbitol; and xylitol/arabinitol [24].

A second-tier group of building blocks was also identified as viable candidates. This group included gluconic acid; lactic acid; malonic acid; propionic acid;

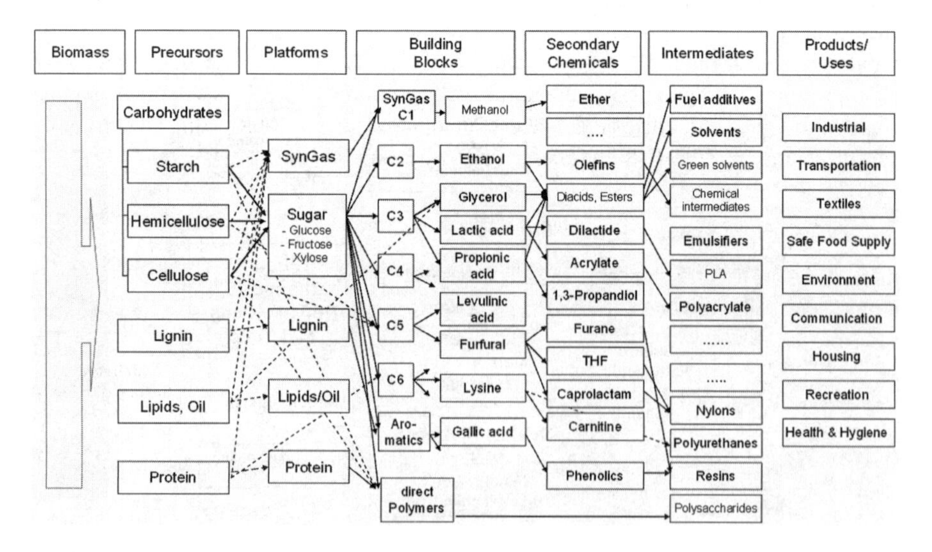

FIGURE 1.10 Model of a biobased product flowchart for biomass feedstock [26].

thetriacids, citric and aconitic acids; xylonic acid; acetoin; furfural; levuglucosan; lysine; serine; and threonine. Recommendations for moving forward include examining top-value products from biomass components such as aromatics, polysaccharides, and oils; evaluating technical challenges related to chemical and biological conversions in more detail; and increasing the number of potential pathways to these candidates. No further products obtained from syngas were selected. For the purposes of this study, hydrogen and methanol are the best short-term prospects for biobased commodity chemical production because obtaining simple alcohols, aldehydes, mixed alcohols, and Fischer–Tropsch liquids from biomass is not economically viable and requires additional development [24].

1.3.2 The Role of Biotechnology in Production of Platform Chemicals

The application of biotechnological methods will be of great importance, and will involve the development of biorefineries for the production of base chemicals, intermediate chemicals, and polymers [59, 60]. The integration of biotechnological methods must be managed intelligently with respect to the physical and chemical conversions of the biomass. Therefore biotechnology cannot continue to be restricted to glucose from sugar plants and starch from starch-producing plants (Fig. 1.11).

One of the main goals is the economical processing of biomass containing lignocellulose and the provision of glucose in the family-tree system. Glucose is a key chemical for microbial processes. The preparation of a large number of family-tree-capable base chemicals is described in the following sections. Among the variety of possible product family trees that can be developed from glucose accessible microbial and chemical sequence products are the C-1 chemicals methane, carbon dioxide, and methanol; C-2 chemicals ethanol, acetic acid, acetaldehyde, and ethylene; C-3 chemicals lactic acid, propandiol, propylene, propylene oxide, acetone, acrylic acid; C-4 chemicals diethylether, acetic acid anhydride, malic acid, vinyl acetate, *n*-butanol, crotone aldehyde, butadiene, and 2,3-butandiol; C-5 chemicals itaconic acid, 2,3-pentane dione, and ethyl lactate; C-6 chemicals sorbic acid, parasorbic acid, citric acid, aconitic acid, isoascorbinic acid, kojic acid, maltol, and dilactide; and the C-8-chemical 2-ethyl hexanol (Fig. 1.12).

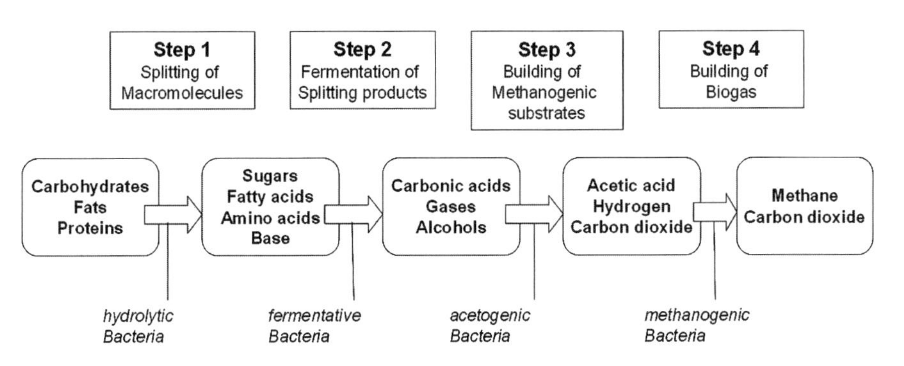

FIGURE 1.11 Simplified presentation of a microbial biomass-breakdown regime [22].

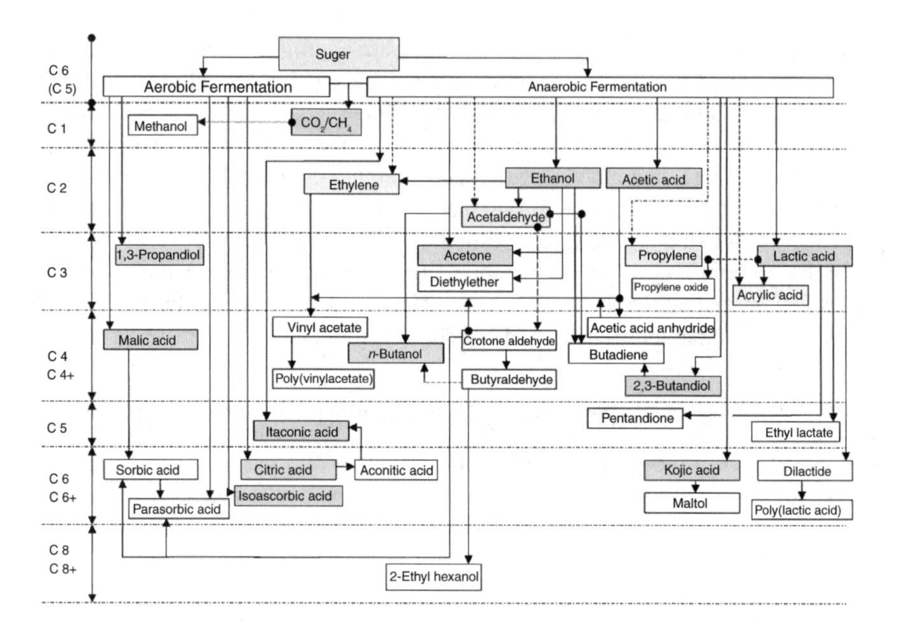

FIGURE 1.12 Biotechnological sugar-based product family tree.

Currently, guidelines are being developed for the fermentation section of a biorefinery. An answer needs to be found to the question of how to produce an efficient technological design for the production of bulk chemicals. The basic technological operations for the manufacture of lactic acid and ethanol are very similar. The selection of biotechnology-based products from biorefineries should be done in a way that they can be produced from the substrates glucose or pentoses. Furthermore, the fermentation products should be extracellular. Fermentors should have a batch, feed batch, or continuous stirred-tank reactor (CSTR) design. Preliminary product recovery may require steps such as filtration, distillation, or extraction. Final product recovery and purification steps may possibly be product-unique. In addition, biochemical and chemical-processing steps should be efficiently connected. Unresolved questions for the fermentation facility include the following: (1) whether or not the entire fermentation facility can/should be able to change from one product to another; (2) can multiple products be run in parallel, with shared use of common unit operations; (3) how should scheduling of unit operations be managed; and (4) how can in-plant inventories be minimized, while accommodating any changeovers required between different products for the same piece of equipment [61].

1.3.3 Green Biomass Fractionation and Energy Aspects

Today, green crops are used primarily as forage and as a source of leafy vegetables. In a process called wet-fractionation of green biomass, green crop fractionation can be used for simultaneous manufacture of both food and nonfood items [45].

The power and heat energy requirements of a forage fractionation of a protein concentrate production system are within practical limits for large farms and dehydrating plants [62]. Mechanical squeezing of the fresh crop results in energy savings of 1.577 MJ ton^{-1} crop input, equal to 52% of the total energy input (compared to energetic drying of green biomass) [63]. Three simplified systems of wet green crop fractionation, which are characterized by the direct use of nutrient-rich green juice or deproteinized juice as feeding supplements for pigs or liquid fertilizer, have been described [64]. Wet green crop fractionation involves an energy saving of 538 MJ ton^{-1} fresh crop, equal to 17.7% of the total energy input of crop drying [63]. Compared with conventional fractionation technology, membrane filtration results in an energy saving of 370 MJ ton^{-1} crop input, which corresponds to 14.8% of the total energy input [64].

Via fractionation of green plants, green biorefineries are able to process amounts in the range of a few tons of green crops per hour (farm scale process) to more than 100 t h^{-1} (industrial-scale commercial process). Careful wet-fractionation technology is used as a first step (primary refinery) to isolate the ingredients in their natural form. Thus, the green crops (or wet organic wastes) are separated into a fiber-rich press cake and a nutrient-rich GJ. Beside cellulose and starch, the PC contains valuable dyes and pigments, crude drugs, and other organics. The GJ contains proteins, free amino acids, organic acids, dyes, enzymes, hormones, further organic substances, and minerals. The application of biotechnological methods is particularly appropriate for conversion processes since the plant water can be used simultaneously for further treatments. In addition, the pulping of lignin–cellulose composites is easier compared to LCF materials. Starting from GJ, the main focus is directed to products such as lactic acid and corresponding derivatives, amino acids, ethanol, and proteins.

The PC can be used for production of green feed pellets; as raw material for production of chemicals, such as levulinic acid; and for conversion to syngas and hydrocarbons (synthetic biofuels). The residues of substantial conversion are applicable to the production of biogas combined with the generation of heat and electricity. Special attention is given to the mass and energy flows of the biorefining of green biomass.

1.3.4 Mass and Energy Flows for Green Biorefining

Green biorefining is described as an example of a type of agricultural factory in greenland-rich areas. Key figures are determined for mass and energy flow, feedstock, and product quantities (Fig. 1.13). Product quantities vary depending on the market and the demand for quality products. Mass flows (Scenario 1, Scenario 2) can be constructed from our own experimental results combined with market demand in the feed, cosmetic, and biotechnology industries. The technical and energy considerations of the fractionation processes of a green biorefinery, and production of the platform chemicals lactic acid and lysine are shown in Figure 1.13.

Using a mechanical press, about 20,000 t press juice [dry matter (DM): 5%] can be manufactured from 40,000 t biomass. First, the juice is the raw material for further

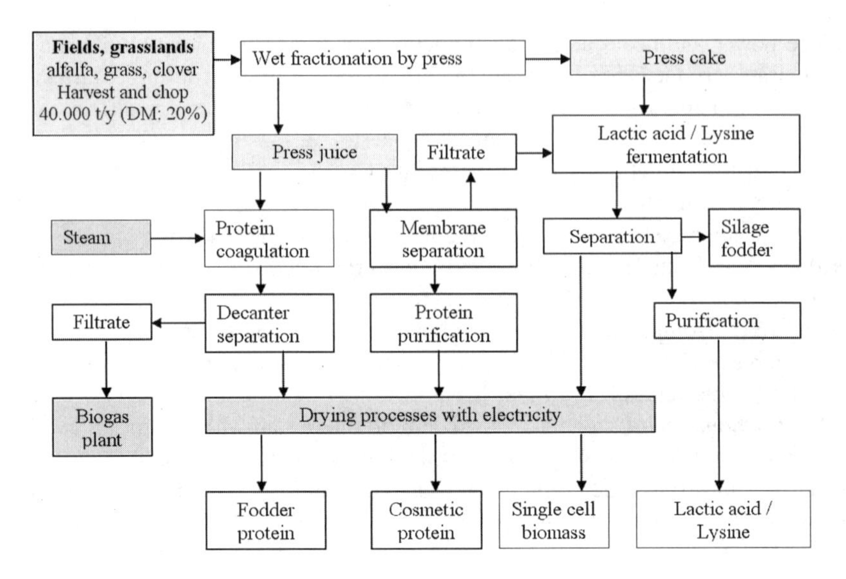

FIGURE 1.13 Selected and simplified processes of a green biorefinery [65].

products; and second, the green cut biomass contains much less moisture. Through fractionation of GJ proteins by different separation and drying processes, high-quality fodder proteins and proteins for the cosmetic industry can be produced [62, 66, 67]. The fodder proteins would be a complete substitute for soy proteins. They even have a nutritional physiological advantage due to their particular amino acid patterns [68]. Utilization of the easily fermentable sugar in the biomass and the available water offers an excellent biotechnological–chemical potential and makes possible the use of basic technologies such as the production of lactic acid or lysine.

In the next step (fermentation), the carbohydrates of the juice and one part of the PC can be used (after hydrolysis) for the production of lactic acid (Scenario 1 [69].) or lysine (Scenario 2 [70]). Thus, single-cell biomass, which can be applied after appropriate drying as a fodder protein, is produced.

The fermentation base in lactic acid fermentation is sodium hydroxide. By means of ultrafiltration, reverse osmosis, [71]. bipolar electrodialysis, and distillation, lactic acid (90%) is recovered from sodium lactate fermentation broth [72–74]. Lysine hydrochloride is the product of lysine fermentation [67]. After separation of the single-cell biomass by ultrafiltration and a membrane separation of water followed by a drying process, lysine hydrochloride (50%) is recovered [70, 71]. The broth that is left after separation from lactic acid or lysine, respectively, and single-cell biomass can be supplied to a biogas plant. Input and output data including required energy were estimated for the production of lysine hydrochloride, lactic acid, proteins for fodder and cosmetics and the utilization of the residue (PC) as silage fodder from 40,000 t green cut biomass (Table 1.1).

By drying, the PC could be manufactured into fodder-pellets. However, this drying is energetically very expensive. From an energy point of view it is far better to

TABLE 1.1 Combined Production of Lactic Acid, Lysine, Cosmetic-Protein, Single-Cell Biomass, Fodder, and Biogas with Energetic Input

Green Biorefinery	Scenario 1	Lactic Acid	
Input		Quantity	Unit
Green biomass (Lucerne, Clover, Grass)	DM: 20%	40,000	t
Steam		2,268	GJ
Electricity		1,300,000	kWh
Output			
Silage fodder	DM: 40%	13,000	t
Fodder-protein 80%	DM: 90%	400	t
Cosmetic-protein 90%	DM: 90%	29.6	t
Lactic acid 90%	DM: 90%	660	t
Residue to biogas plant TS: 2%		17,690	t
Single-cell biomass (as fodder-protein 60%)	DM: 90%	33	t
Green Biorefinery	**Scenario 2**	**Lysine**	
Input		Quantity	Unit
Cut Green Biomass (Lucerne, Clover, Grass)	DM: 20%	40,000	t
Steam		2,268	GJ
Electricity		492,000	kWh
Output			
Silage fodder	DM: 40%	13,000	t
Fodder-protein 80%	DM: 90%	400	t
Cosmetic-protein 90%	DM: 90%	29.6	t
Lysine–HCl, 50%	DM: 90%	620	t
Residue to biogas plant DM: 2%		17,770	t
Single-cell-Biomass (as fodder-protein 60%)	DM: 90%	31	t

Source: Ref. [65].
Note: DM, dry matter.

suggest that the PC be used as silage-feed. From an ecological and economical viewpoint, at this stage it has to be concluded that coupling of green biorefineries with green crop drying industry is necessary.

1.3.5 Assessment of Green Crop Fractionation Processes

Green biorefineries use different kinds of energy (steam and electricity) for the treatment of PC and press juice (intermediate products) to produce valuable end products. It is also possible to use the PC together with the press juice as a source of carbohydrate for the fermentation. For the separate processes mass balances were set up and thus the consumption of energy can be calculated by means of power consumption of the facilities (plants and machinery).

A linear programming model used to optimize the profitability and determine an optimized planning process for biorefineries is described in Annetts and Audsley

(2003) [75]. The raw materials are wheat (straw and grain) and rape, and therefore this would be a model for a whole-crop biorefinery and hardly applicable to a green biorefinery. At a capacity of 40,000 tons per year [t $annum^{-1}$ (a)] fresh biomass (lucerne, wild-mixed-grass), and a operation time of 200 working days per year, an average of 200 t are converted per day. Under these conditions, the screw extrusion press used has an energy consumption of 135,000 kWh per year (kWh a^{-1}). It generates 100 t day^{-1} PC with DM ~35% and 100 t day^{-1} press juice with DM ~5%. Around 10 t of the 100 t press juice are fed to membrane-separation for a cosmetic-protein extraction. For separation of feed protein, 90 t press juice are put into a steam-coagulation. The required heat quantity as steam is 2268 GJ a^{-1}. The freshly pressed juice is preheated up to 45°C in a heat exchanger within a counter-current process. Via steaming, a temperature rise of the freshly pressed juice of up to 30°C is reached. Steam coagulation occurs at a temperature of 75°C. The following calculations are carried out according to Bruhn et al. (1978) [62]. For the separation of feeding proteins the following energy input is required: 1,500 kWh a^{-1} for skimming; 15,000 kWh a^{-1} for dehydration to ~50% DM; and 32,000 kWh a^{-1} for drying up to DM = 90%. Separation of cosmetic-proteins via ultrafiltration needs an energy input of 9700 kWh a^{-1}. For subsequent solvent extraction, a further energy input of about 507 kWh a^{-1} is generated via stirring [76].

For the separation via centrifugation 101 kWh a^{-1} are required and 2360 kWh a^{-1} for the subsequent spray-drying to DM = 90% [66, 77]. If the press-juice contains 2% proteins, 400 t feed proteins as protein concentrate and 29.6 t cosmetic proteins can be produced per year. Correspondingly increased quantities can be produced if the press juice contains a higher proportion of proteins. After protein-separation, 100 t fermentation-broth (~96.6 m^3 at a density of about 1035 kg L^{-1}) are available per working day. The energy input required for stirring during fermentation amounts to 150,000 kWh a^{-1} [70].

For lactic-acid fermentation, NaOH is added as a base, resulting in sodium lactate. The purification of lactic acid occurs with the following steps and corresponding energy yields: ultrafiltration 97,000 kWh a^{-1}, reverse osmosis 171,000 kWh a^{-1}, and bipolar electrodialysis 660,000 kWh a^{-1} [71, 72]. Bipolar electrodialysis is particularly energy-intensive. Subsequently, the lactic acid solution (45%) is concentrated up to a 90% lactic acid via vacuum distillation. The energy consumption for this single-stage distillation will amount 26,400 kWh a^{-1} [74]. The energy consumption for 660 t of 90% lactic-acid amounts 1104 MWh a^{-1} using this procedure.

If lysine fermentation is chosen instead of lactic acid, ultrafiltration and reverse osmosis are required for purification with the following corresponding energy yields: ultrafiltration (97,000 kWh a^{-1}) and reverse osmosis (171,000 kWh a^{-1}) [71]. Afterwards the lysine hydrochloride is dried to DM of 90% with an energy requirement of 49,000 kWh a^{-1}) [75]. The energy consumption of 620 t lysine hydrochloride using this method results in 296,000 kWh a^{-1}.

In a biorefining plant processing 40,000 t green biomass for the combined production of 660 t lactic acid, 29.6 t cosmetic-protein, 33 t single-cell biomass, 400 t fodder-protein, 13,000 t silage fodder, and 17,690 t liquid residues for biogas production, the following energy input is required: 2,268 GJ heat, and 1.3 million kWh electricity. The combined production of 620 t lysine, 29.6 t cosmetic protein, 31 t single-cell biomass, 400 t fodder protein, 13,000 t silage fodder, and 17,700 t liquid residues to produce biogas requires the following energy input: 2,268 GJ heat, and 0.492 million MWh electricity.

These results clearly demonstrate the quantity of products a green biorefinery can provide with the help of biotechnology, and the corresponding required energy input. The economic benefits of biorefining green biomass are the high yields of biomass per hectare and year, and synergetic effects via combination with established production processes in the agriculture and feed industries. Therefore, in the mid-term, it is reasonable to combine the economic potential of green agriculture and green-crop-drying-plants.

These data concerning quantity, quality, and required process energy form the basis of further economic considerations in connection with calculation of breakeven points when planning and establishing a green biorefinery. In future, energy inputs will be reduced further due to optimization of the corresponding biorefinery technology. The combination of biotechnological and chemical conversion processes will be a very important aspect in decreasing process energy input. Thus, the biotechnological production of aminium lactates, such as piperazinium dilactates as starting material for high-purity lactic acid and polylactic acid could be a new approach [69].

1.4 GREEN BIOREFINERY: ECONOMIC AND ECOLOGIC ASPECTS

Plant biomass is the only foreseeable sustainable source of organic fuels, chemicals, and materials. A variety of forms of biomass, notably many LCFs, are potentially available on a large scale and are cost-competitive with low-cost petroleum, whether considered on a mass or energy basis, in terms of price defined on a purchase or net basis for both current and projected mature technologies, or on a transfer basis for mature technology [78]. Green plant biomass in combination with LCF represents the dominant source of feedstocks for biotechnological processes for the production of chemicals and materials [24, 70, 79–81]. The development of integrated technologies for the conversion of biomass is essential for the economic and ecological production of products. The biomass industry, or bioindustry, at present produces basic chemicals such as ethanol (15 million t a^{-1}); amino acids (1.5 million t a^{-1}), of which L-lysine amounts to 500,000 million t a^{-1}; and lactic acid (200,000 million t a^{-1}) [82]. The target of a biorefinery is to establish a combination of a biomass–feedstock mix with a process and product mix [24, 80]. A life cycle assessment (LCA) is available for the production of polylactic acid (capacity 140,000 t a^{-1}) [83]. For total assessment of the utilization of biomass, one has to consider that cultivation of the plant has to fulfill certain economic and ecological criteria. Agriculture both creates pressure on the environment and plays an important role in maintaining many cultural landscapes and seminatural habitats [84]. Green crops, in particular, provide especially high yields.

Additionally, grassland can be cultivated in a sustainable way [85, 86]. European grassland experiments have shown that species-rich grassland cultivation provides not only ecological but also economic advantages. With greater plant diversity, grassland is more productive and the soil is protected against nitrate leaching. Of the 71 species examined so far, 29 had a significant influence on productivity. *Trifolium pratense* has an especially important function regarding productivity. On sites where this species occurs, more than 50% of the total biomass has been produced by this species. Legumes such as clover and herbs also play an important role, as do fast-growing grasses [87]. An initial assessment of the concept of a green biorefinery has been carried out by Schidler and colleagues for the Austrian system approach [88, 89]. Furthermore, an Austrian-wide concept for the use of biomass and cultivable land for renewable resources has yet to be developed in Austria, which also holds true for Europe [90]. The size of such plants depends on the rural structures of the different regions. Concepts with more decentralized units would have a size of about 35,000 t a^{-1} and central plants could have sizes of about 300,000–600,000 t a^{-1} [90, 91].

1.5 OUTLOOK: PRODUCTION OF L-LYSINE-L-LACTATE FROM GREEN JUICES

The aminium lactate L-lysine-L-lactate was produced in fractionated juices from a green biorefinery. To investigate the effect of protein separation onto the lactic-acid fermentation, nontreated and deproteinized alfalfa press juice was compared to the MRS medium [92]. At a glucose concentration of 50 g L^{-1}, the production rates indicated that the separation of proteins from the press juice had no significant influence on the lactic-acid formation. Production rates were at the same level as the fermentation with the MRS medium. Experiments with alfalfa press juice reached higher final lactic-acid concentrations due to further carbohydrates in the press juice that could additionally be metabolized by strand DSMZ 2649 [93]. In further research, the complete carbohydrate composition of the alfalfa press juice and its single conversion to lactic acid is investigated. After increasing the glucose concentration up to 100 g L^{-1}, a significant nutrient limitation was observed during the fermentation with deproteinized press juice. The lactic-acid production rate dropped about 33% and the molar yield was 6% lower than in the fermentation with the semisynthetic medium, MRS. L-lysine-L-lactate could not be produced in the theoretical composition, because of the growing buffer capacity of the biomass with increasing substrate concentration. The pH that provides an equimolar composition of the aminium lactate has to be determined in further experiments. The results presented here show that the fermentative production of L-lysine-L-lactate can be integrated into the green biorefinery system, where deproteinized press juice accrues as a product. The usage of deproteinized press juice as a fermentation medium is technically and economically reasonable because of the stabilizing effect on the press juice and the surplus values from the gained proteins [94]. The N-supplementation that is necessary at high substrate concentrations could be realized by using biomass hydrolysates from previous fermentations. In future experiments, D-(+)-glucose will be substituted by hydrolysates from alfalfa press cakes to obtain a complete fermentation medium from a green biorefinery without any additional carbon source [93].

1.6 GENERAL CONCLUSION

There are various requirements for entering the industrial biorefinery technologies and the production of platform chemicals and materials. On the one hand, the production of substances on the basis of biogenic raw material in the already-existing production facilities of cellulose, starch, sugar, oil, and proteins has to be enlarged, on the other hand, the introduction and establishment of biorefinery demonstration plants is required. Conversion processes have to be developed in the biorefinery regime, that is, in defined product lines and product trees (platform chemicals → intermediate products → secondary products). The organic-technical chemistry has the task to position itself inside of the concept of "biobased products and biorefinery systems," among others things focusing itself on the linking of biological and chemical syntheses and technologies, especially integrating the sectors of reaction engineering, process intensification, and heterogenic catalysis.

Besides promoting the necessary research, development, and industrial implementation, a broader establishment of the specializing field "Chemistry of renewable raw materials/Biorefinery systems" in the education and in academic teaching needs to be achieved.

REFERENCES

1. B. Kamm, M. Kamm and K. Soyez, (Eds.), *Die Grüne Bioraffinerie/The Green Biorefinery. Technologiekonzept. Proceedings of the 1st International Symposium Green Biorefinery/Grüne Bioraffinerie*, October 1997, Neuruppin, Berlin, **1998**.
2. B. Kamm, M. Kamm, K. Richter, B. Linke, I. Starke, M. Narodoslawsky, K. D. Schwenke, S. Kromus, G. Filler, M. Kuhnt, B. Lange, U. Lubahn, A. Segert and S. Zierke, *Grüne BioRaffinerie Brandenburg – Beiträge zur Produkt – und Technologieentwicklung sowie Bewertung. Brandenburgische Umwelt-Berichte*. **2000**, pp. 260–269.
3. M. Narodoslawsky, (Ed.), *Green biorefinery*, in *Proceedings of 2nd international Symposium Green Biorefinery*, 13–14 October 1999, Feldbach, Austria. Proceedings, SUSTAIN, Verein zur Koordination von Forschung über Nachhaltigkeit, Graz TU, Austria, **1999**.
4. US President, **1999**. Developing and promoting biobased products and bioenergy. Executive Order 13101/13134, William J. Clinton, The White House, 12 August 1999, http://www.newuse.org/EG/EG–20/20BioText.html
5. US Congress, **2000**, *Biomass research and development*, Act of 2000, June.
6. BRDI, **2006**, Vision for bioenergy and biobased products in the United States. Biomass Research and Development Initiative. http://www1.eere.energy.gov/biomass/pdfs/final_2006_vision.pdf
7. National Research Council, *Biobased Industrial Products: Priorities for Research and Commercialization*, National Academic Press, Washington DC, **2000**.
8. BTAC, *Roadmap for biomass technologies in the United States*, Biomass Technical Advisory Committee, Washington DC. **2002**. http://www.bioproductsbioenergy.gov/pdfs/FinalBiomassRoadmap.pdf
9. BTAC, *Vision for bioenergy and biobased products in the United States*, Biomass Technical Advisory Committee, Washington DC, **2002**. www.bioproductsbioenergy.gov/pdfs/Bio Vision_03_Web.pdf

10. BTAC, *Roadmap for bioenergy and biobased products in the United States*, October 2006. Biomass R&D Technical Advisory Board, **2007**. http://www1.eere.energy.gov/biomass/pdfs/obp_roadmapv2_web.pdf
11. European Parliament and Council, Directive 2003/30/EC on the promotion of the use of biofuels or other renewable fuels for transport; Official Journal of the European Union L123/42, 17.05.2003, Brussels.
12. Gesetz für den, Vorrang erneuerbarer Energien, Erneuerbare Energiegesetz, EEG/EnWGuaWGÄndG., 29 March 2000, BGBI, 305.
13. Bundesministerium für Ernährung, Landwirtschaft und Verbraucherschutz, **2009**. Aktionsplan der Bundesregierung zur stofflichen Nutzung nachwachsender Rohstoffe. BT–Drucksache 16/14061 vom 03.09.2009.
14. Gesellschaft Deutscher Chemiker **2010**, Dechema, DGMK, VCI, Positionspapier Rohstoffbasis im Wandel, Frankfurt, January 2010, http://www.vci.de/default_cmd_shd_docnr_126682_lastDokNr~-1.htm
15. D. Wittmeyer, EU lead market initiative in the frame of European technology platform for sustainable chemistry. Deutscher Bioraffineriekongress, 8 July 2009, Industrieclub Potsdam, http://www.biorefinica.de, http://ec.europa.eu/enterprise/policies/innovation/policy/lead-market-initiative/, **2011**.
16. European Technology Platform for Sustainable Chemistry, Industrial Biotechnology **2005**, Section, http://www.suschem.org
17. US DOE, **2005**, 1st International Biorefinery Workshop, July 20 and 21, US Department of Energy, Washington D.C.; http://www.biorefineryworkshop.com
18. Biomasse Action plan, **2005**, http://www.euractiv.com/en/energy/biomass–action-plan/article–155362
19. EU-Projekt BIOPOL, **2007**, Specific Support Action, Priority Scientific Support to Policies, http://www.biorefinery.nl/biopol
20. EU – Projekt Biorefinery – Euroview, **2007**, Specific Support Action, Priority Scientific Support to Policies, http://www.biorefinery–euroview.eu
21. B. Mahro and M. Timm, *Potential of biowaste from the food industry as a biomass resource. Eng. Life Sci.* **2007**, 7(5), 457–468.
22. B. Kamm and M. Kamm, *Principles of biorefineries. Appl. Microbiol. Biotechnol.* **2004**, 64, 137–145.
23. B. Kamm and M. Kamm, *Biorefinery systems. Chem. Biochem. Eng. Q.* **2004**, 18(1), 1–6.
24. T. Werpy and G. Petersen, (Eds.), *Top value chemicals under the refinery concept: a phase II study*, US Department of Energy, Office of Scientific and Technical Information. No.: DOE/GO-102004-1992, **2004**. http://www.osti.gov/bridge
25. D. G. Tiffany, Economic comparison of ethanol production from corn stover and grain. AURI Energy Users Conference, 13 March 2007, Redwood Falls, MN.
26. B. Kamm, P. R. Gruber and M. Kamm, Biorefineries – industrial processes and products, *Ullmann's Encyclopedia of Industrial Chemistry*, 7th ed., Wiley-VCH, **2011**.
27. B. Kamm, M. Kamm and T. Hirth, Product family trees: Lignocellulosic-based chemical products, in *Biorefineries – Biobased Industrial Processes and Products. Status Quo and Future Directions*, Vol. 2(Eds. B. Kamm, M. Kamm, P. Gruber,), Wiley-VCH, Weinheim, **2010**, pp. 97–149.

28. M. Ringpfeil, Biobased industrial products and biorefinery systems – Industrielle Zukunft des 21. Jahrhunderts, **2001**, http://www.biopract.de
29. D. L.van Dyne, **1999**, Estimating the economic feasibility of converting ligno-cellulosic feedstocks to ethanol and higher value chemicals under the refinery concept: a phase II study, OR22072-58. University of Missouri.
30. J. G. Zeikus, M. K. Jain and P. Elankovan, *Biotechnology of succinic acid production and markets for derived industrial products. Appl. Microbiol. Biotechnol.* **1999**, 51, 545–552.
31. K. D. Vorlop, Th. Wilke and U. Preuße, *Biorefineries – Industrial Processes and Products, Vol. 2: Biocatalytic and Catalytic Routes for the Production of Bulk and Fine Chemicals from Renewable Resources*, (Eds. B. Kamm, M. Kamm, P. Gruber), Wiley-VCH, Weinheim, **2010**, pp. 385–406.
32. T. Werpy, J. Freye and J. Holladay, *Biorefineries – Industrial Processes and Products, Vol. 2: Succinic Acid – a Model Building Block for Chemical Production from Renewable Resources*, (Eds. B. Kamm, M. Kamm, P. Gruber), Wiley, Weinheim, **2010**, pp. 367–379.
33. DuPont, US Patent Application, 5686276, **1997**, http://www.dupont.com/sorona/home.html
34. D. J. Morris and I. Ahmed, *The Carbohydrate Economy, Making Chemicals and Industrial Materials from Plant Matter*, Institute of Local Self Reliance, Washington DC, **1992**.
35. J. J. Bozel, *Encyclopedia of Plant and Crop Science: Alternative Feedstocks for Bioprocessing* (Ed. R. M. Goodman), Dekker, New York, **2006**, doi: 10.1081/E-EPCS-120010437.
36. C. Webb, A. A. Koutinas and R. Wang, *Developing a sustainable bioprocessing strategy based on a generic feedstock. Adv. Biochem. Eng. Biotechnol.* **2004**, 87, 195–268.
37. R. V. Nonato, P. E. Mantellato and C. E. V. Rossel, *Integrated production of biodegradable plastic, sugar and ethanol. Appl. Microbiol. Biotechnol.* **2001**, 57, 1–5.
38. C. E. V. Rossel, P. E. Mantellato, A. M. Agnelli, J. Nascimento, *Biorefineries – Industrial Processes and Products, Vol. 1: Sugar-Based Biorefinery – Technology for an Integrated Production of Poly(3-hydroxybutyrate) Sugar and Ethanol* (Eds. B. Kamm, M. Kamm, P. Gruber), Wiley-VCH, Weinheim, **2006**, pp. 209–226.
39. A. Fiechter, *Plastics from Bacteria and for Bacteria: Poly(β-hydroxyalkonoates) as Natural, Biocompatible, and Biodegradable Polyesters*, Springer, New York, **1990**, pp. 77–93.
40. F. Rexen, Documentation of Svebio Phytochemistry Group (in Danish): New industrial application possibilities for straw. [Fytokemi I Norden, Stockholm, Sweden, 1986-03-06], **1986**, 12.
41. J. Coombs and K. Hall, *Cereals – Novel Uses and Processes: The Potential of Cereals as Industrial Raw Materials: Legal, Technical, Commercial Considerations*, Plenum, New York, **1997**, pp. 1–12.
42. E. Audsley and J. E. Sells;*Cereals – Novel Uses and Processes – Determining the Profitability of a Whole Crop Biorefinery* (Eds. G. M. Campbell, C. Webb, S. L. McKee), Plenum, New York, **1997**, pp. 191–294.
43. A. J. Hacking, *Economic Aspects of Biotechnology: The American Wet Milling Industry*, Cambridge University Press, New York, **1986**, pp. 214–221.
44. T. Willke and K. D. Vorlop, *Industrial bioconversion of renewable resources as an alternative to conventional chemistry. Appl. Microbiol. Biotechnol.* **2004**, 66(2), 131–142.

45. R. Carlsson, *Handbook of Plant and Crop Physiology: Sustainable Primary Production. Green Crop Fractionation: Effects of Species, Growth Conditions, and Physiological Development* (Ed. M. Pessarakli), Dekker, NY, **1994**, pp. 941–693.
46. N. W. Pirie, *Leaf Protein – Its Agronomy, Preparation, Quality, and Use*, Blackwell, Oxford, **1971**.
47. N. W. Pirie, *Leaf Protein and its By-products in Human and Animal Nutrition*, Cambridge University Press, UK, **1987**.
48. R. Carlsson, *The green biorefinery*, in *Proceedings if 1st International Green Biorefinery Conference, Neuruppin, Germany, Status quo of the utilization of green biomass*, **1997**, GO'T, Berlin, ISBN 3-929672-06-5.
49. R. Carlsson, *Leaf Protein Concentrates: Leaf Protein Concentrate from Plant Sources in Temperate Climates* (Eds. L. Telek, H. D. Graham), AVI, Westport, **1983**, pp. 52–80.
50. L. Telek and H. D. Grafham (Eds.), *Leaf Protein Concentrates*, AVI, Westport, CN, **1983**.
51. R. J. Wilkins (Ed.), *Green Crop Fractionation*, British Grassland Society, Hurley, Maidenhead, UK, **1977**.
52. I. Tasaki (Ed.), *Proceedings of the 2nd International Leaf Protein Research Conference: Recent advantages in leaf protein research*, Nagoya, Japan, **1985**.
53. P. Fantozzi (Ed.), *Proceedings of the 3rd International Leaf Protein Research Conference*, Pisa-Perugia-Viterbo, Italy, **1989**.
54. N. Singh (Ed.), *Green Vegetation Fractionation Technology*, Science, Lebanon, NH, **1996**.
55. B. Kamm, M. Kamm, K. Richter, W. Reimann and A. Siebert, *Formation of aminium lactates in lactic acid fermentation, fermentative production of 1,4-piperazinium-(L,L)-dilactate and its use as starting material for the synthesis of dilactide (part 2). Acta Biotechnol.* **2000**, 20, 289–304.
56. D. H. White and D. Wolf, *Research in Thermochemical Biomass Conversion* (Eds. A. V. Bridgewater, J. L. Kuester), Elsevier, New York, **1988**.
57. C. Okkerse and H.van Bekkum, *From fossil to green. Green Chem.* **1999**, 4, 107–114.
58. NREL**2005**, National Renewable Energy Laboratory, http://www.nrel.gov/biomass/ biorefinery.htm
59. EFIB: The European Forum for Industrial Biotechnology & The Biobased Economy, **2011**, http://www.efibforum.com/efib-agenda-2011.aspx
60. BIO**2011**, Biotechnology Industry Organisation; http://www.bio.org/events/conferences/world-congress-industrial-biotechnology-bioprocessing/2800-0
61. D. L.Van Dyne, M. G. Blase and L. D. Clements, *Perspectives on New Crops and New Uses: A Strategy for Returning Agriculture and Rural America to Long-Term Full Employment Using Biomass Refineries* (Ed. J. Janeck), ASHS, Alexandria, VA, **1999**, pp. 114–123.
62. H. D. Bruhn, R. J. Straub and R. G. Koegel, *Proceedings of the International Grain and Forage Harvesting Conference: A systems approach to the production of plant juice protein concentrate*, American Society of Agricultural Engineers, St. Joseph, MI, **1978**.
63. A. Ricci, F. Favati, L. Massignan, R. Fiorentini and R. Ficcanterri, *Energy evaluation of a conventional wet green crop fractionation process*, in *Proceedings of 3rd International Conference on Leaf Protein Research: Pisa, Perugia, Viterbo* (Italy) 1–7 October, **1989**.

64. F. Favati, Energy evaluation of a wet green crop fractionation process utilizing reverse osmosis. Third International Conference on Leaf Protein Research. 1–7 October, **1989**, Pisa, Perugia, Viterbo, Italy.

65. B. Kamm, P. Schönicke and M. Kamm, *Biorefining of green biomass – technical and energetic considerations. Clean*, **2009**, 37(1), 27–30.

66. G. Bohlmann, *Several reports on White Biotechnology processes*, Stanford Research International, Menlo Park, CA, **2002**.

67. H. B. Reismann, *Economic Analysis of Fermentation*, CRC, Boca Raton, **1988**.

68. K.-D. Schwenke, *Die Grüne Bioraffinerie; Beiträge zur ökologischen Technologie, Vol. 5. – Das Funktionelle Potential von Pflanzenproteinen* (Eds. B. Kamm, M. Kamm, K. Soyez), Gesellschaft für ökologische Technologie und Systemanalyse, Berlin, **1998**, pp. 185–195.

69. B. Kamm, P. Schönicke and M. Kamm, *Biorefining of green biomass – technical and energetic considerations. Clean* **2006**, 37(1), 27–30.

70. M. H. Thomsen, D. Bech and P. Kiel, *Manufacturing of stabilised brown juice for L-lysine production – from university lab scale over pilot scale to industrial production. Chem. Biochem. Eng. Q.* 18(1), **2004**, 37–46.

71. M. Patel, M. Crank, V. Dornburg, B. Hermann, L. Roes, B. Hüsing, L. Overbeek, F. Terragni and E. Recchia, *Medium and long-term opportunities and risks of the biotechnological production of bulk chemicals from renewable resources*, The BREW Project, prepared under the European Commission's GROWTH Programme, Utrecht, **2006**, pp. 120–122.

72. Y. H. Kim and S.-H. Moon, *Lactic acid recovery from fermentation broth using one-stage electrodialysis. J. Chem. Technol. Biotechnol.* **2001**, 76, 169–178 12.

73. R. Datta and S.-P. Tsai, *Fuels and chemicals from biomass: lactic acid production and potential uses: a technology and economics assessment* (Eds. B. C. Saha, J. Woodward), American Chemical Society, Washington DC, **1997**, p. 224.

74. G. Lavis, Evaporation in *Handbook of Separation Techniques for Chemical Engineers*, 3rd ed. (Ed. P. A. Schweitzer), McGraw-Hill, New York, **1996**.

75. J. E. Annetts and E. Audsley, Modelling *the value of a rural biorefinery. Part I: the model description. Part II: analysis and implications. Agric. Syst.* **2003**, 76, 39–76.

76. D. P. Petrides, C. L. Cooney and L. B. Evans, *Chemical Engineering Problems in Biotechnology: An Introduction to Biochemical Process Design* (Ed. M. L. Shuler), American Institute of Chemical Engineers, New York, **1989**.

77. W. H. Bartholomew and H. B. Reismann, Economics of fermentation processes, in *Microbial Technology*, 2nd ed., vol. 2(Eds. H. J. Peppler, D. Perlman), Academic, New York, **1979**.

78. P. A. Fowler, A. R. McLauchlin and L. M. Hall, *The Potential Industrial Uses of Forage Grasses Including Miscanthus*, BioComposites Centre, University of Wales, Bangor, Gwynedd, **2003**, http://www.bc.bangor.ac.uk/_includes/docs/pdf/industrial%20use%20of%20grass.pdf

79. B. Kamm and M. Kamm, *The green biorefinery – principles, technologies and products*, in *Proceedings of 2nd International Symposium Green Biorefinery*, 13–14 October 1999 SUSTAIN, Verein zur Koordination von Forschung über Nachhaltigkeit (Hrsg.) Feldbach, Austria, **1999**, pp. 46–69.

80. B. Kamm and M. Kamm, Biorefineries – multi product processes, in *White Biotechnology (Advances in biochemical engineering/biotechnology*, vol. 105 (Eds. R. Ulber, D. Sell), Springer, Heidelberg, **2007**, pp. 175–204.

81. A. Tullo, *Renewable materials, two pacts may help spur biomass plastics. Chem. Eng. News* 28 March **2005**; http://www.CEN–ONLINE.org
82. *Elements Degussa Science Newsletter*, **2005**, 7, 35.
83. E. T. H. Vink, K. R. Rabago, D. A. Glassner and P. R. Gruber, *Applications of life cycle assessment to NatureWorksTMpolylactide (PLA) production. Polym. Degrad. Stability* **2003**, 80, 403–419.
84. IEEP Contribution to the background study agriculture. The Institute for European Environmental Policy, Brussels, **2004**.
85. S. Kromus, *Die Grüne Bioraffinerie Österreich – Entwicklung eines integrierten Systems zur Nutzung von Grünlandbiomasse*, Dissertation, TU Graz, **2005**.
86. S. Kromus, B. Wachter, W. Koschuh, M. Mandl, C. Krotschek and M. Narodoslawsky, *The green biorefinery Austria – development of an integrated system for green biomass utilization. Chem. Biochem. Eng. Q.* **2004**, 18(1), 13–19.
87. A. Hector, B. Schmid, C. Beierkuhnlein, M. C. Caldeira, M. Diemer, P. G. Dimitrakopoulos, J. A. Finn, H. Freitas, P. S. Giller, J. Good, R. Harris, P. Högberg, K. Huss-Danell, J. Joshi, A. Jumpponen, C. Körner. P. W. Leadley, M. Loreau, A. Minns, C. P. H. Mulder, G. O'Donovan, S. J. Otway, J. S. Pereira, A. Prinz, D. J. Read, M. Scherer-Lorenzen, E.-D. Schulze, A.-S. D. Siamantziouras, E. M. Spehn, A. C. Terry, A. Y. Troumbis, F. I. Woodward, S. Yachi and J. H. Lawton, *Plant diversity and productivity experiments in European grasslands. Science.* **1999**, 286, 1123–1127.
88. S. Schidler, Technikfolgenabschätzung der Grünen Bioraffinerie, Teil I: Endbericht, Institut für Techikfolgen-Abschätzung, Österreichische Akademie der Wissenschaften, **2003**.
89. S. Schidler, H. Adensam, R. Hofmann, S. Kromus and M. Will, Technikfolgenabschätzung der Grünen Bioraffinerie, Teil II: Materialsammlung, Institut für Technikfolgen-Abschätzung, Österreichische Akademie der Wissenschaften, **2003**.
90. M. Narodoslawsky and S. Kromus, *Development of decentral green biorefinery in Austria*, in *Biorefinica 2004 Proceedings and Papers* (Eds. B. Kamm, M. Hempel, M. Kamm), 27–28 October, Osnabrück, biopos, Teltow, **2004**, p. 24.
91. L. Halasz, G. Povoden and M. Narodoslawsky, *Process synthesis for renewable resources*, Presented at *PRES 03*, **2003**, Hamilton, Canada.
92. EMD Chemicals, MRS Agar, **2002**.
93. S. Leiß, J. Venus and B. Kamm, *Fermentative production of L-Lysine–L-lactate with fractionated press juices from the green biorefinery. Chem. Eng. Technol.* **2010**, 33(12), 2102–2105.
94. B. Kamm, C. Hille, P. Schönicke and G. Dautzenberg, *Green biorefinery demonstration in Havelland/Germany. Biofuels Bioprod. Biorefin.* **2010**, 4, 253.

CHAPTER 2

Recent Advances in Green Chemistry

NICHOLAS GATHERGOOD

2.1 INTRODUCTION

Protecting the environment is a huge concern for society. Problems including global warming and ozone depletion highlight the negative affects human activity has on the planet. Not only is the environment affected directly by human endeavors but this activity can also have detrimental affects on human health. Pollution of the environment is the cause of these problems. CO_2 emission contributes to global warming and chlorofluorocarbons (CFCs) continue to cause ozone depletion. With potential bioaccumulation and persistence of toxins and carcinogens in the environment there is a need to reduce the amount of hazardous waste produced [1].

2.2 GREEN CHEMISTRY

Green chemistry is a concept that can be described as chemistry that is environment friendly or "cleaner." One aspect is developing chemical syntheses that reduce or avoid hazardous materials when creating chemical products. This has become more widely appreciated in recent years with journals (e.g., RSC Green Chemistry and Wiley ChemSusChem) now dedicated to publishing advances in this field. Green chemistry should lead to benefits for the environment and health, and also help reduce production cost in the chemical industry. Waste treatment and disposal can be expensive and contribute to the overall cost of manufacture. In 1998, Anastas and Warner compiled the "Twelve Principles of Green Chemistry" [1]. Green chemistry aims to reduce waste, use less hazardous methods to generate the desired chemicals, and minimize the amount of materials and energy.

The Role of Green Chemistry in Biomass Processing and Conversion, First Edition.
Edited by Haibo Xie and Nicholas Gathergood.

2.2.1 The Twelve Principles of Green Chemistry [1]

The 12 principles are listed below:

1. *Prevention*: It is better to prevent waste than to treat or clean up waste after it is formed.
2. *Atom Economy*: Synthetic methods should be designed to maximize the incorporation of all materials used in the process into the final product.
3. *Less Hazardous Chemical Syntheses*: Wherever practicable, synthetic methodologies should be designed to use and generate substances that are of little or no toxicity to human health and the environment.
4. *Designing Safer Chemicals*: Chemical products should be designed to preserve efficacy of function while reducing toxicity.
5. *Safer Solvents and Auxiliaries*: The use of auxiliary substances (e.g., solvents, separation agents, etc.) should be made unnecessary wherever possible and, innocuous when used.
6. *Design for Energy Efficiency*: Energy requirements should be recognized for their environmental and economic impacts and should be minimized. Synthetic methods should be conducted at ambient temperature and pressure.
7. *Use of Renewable Feedstocks*: A raw material or feedstock should be renewable rather than depleting wherever technically and economically practicable.
8. *Reduce Derivatives*: Unnecessary derivatization (blocking groups, protection/deprotection, temporary modification of physical/chemical processes) should be avoided whenever possible.
9. *Catalysis*: Catalytic reagents (as selective as possible) are superior to stoichiometric reagents.
10. *Design for Degradation*: Chemical products should be designed so that at the end of their function they do not persist in the environment and break down into innocuous degradation products.
11. *Real-Time Analysis for Pollution Prevention*: Analytical methodologies need to be further developed to allow for real-time, in-process monitoring and control prior to the formation of hazardous substances.
12. *Inherently Safer Chemistry for Accident Prevention*: Substances and the form of a substance used in a chemical process should be chosen so as to minimize the potential for chemical accidents including releases, explosions, and fires.

The 12 principles are the basic means to use green chemistry in chemical processes. There are many examples of the principles being utilized to give more efficient and unique reactions.

The aim of this chapter is to present recent examples applying green-chemistry principles to synthetic chemistry. Conversion of biomass into value-added compounds, transforming function groups, breaking or making bonds, is a major function of a biorefinery. For a recent in-depth analysis of green chemistry technologies for biorefineries see Clark and Deswate [2].

Herein, significant breakthroughs in cleaner and greener synthetic methodologies are presented. The scope includes examples from biomass conversion, consumer product disposal, pharmaceutical manufacture, and natural product total synthesis. These examples highlight what is possible in synthetic chemistry, when a target of improved efficiency and waste reduction is realized. These examples will inspire researchers working in biorefineries to develop "even cleaner and greener" transformations and methodologies.

2.3 EXAMPLES OF THE TWELVE PRINCIPLES OF GREEN CHEMISTRY

2.3.1 Prevention

The cost of waste disposal and treatment is now a huge issue for society. In general, the more hazardous a material is, the greater will be the cost for safe disposal. By preventing the production of waste this issue is avoided. Thus, using pathways that reduce or eliminate the generation of waste is helpful for reducing costs and avoids the necessity of handling hazardous waste. In addition, potential negative impact on the environment can be eliminated [1].

2.3.1.1 Oxidation by Nitrogen Dioxide Gas Benzylic alcohols, solid fatty alcohols, aromatic aldehydes, heterocyclic thioamides, and diphenylamine can be oxidized using NO_2, in quantitative yield with little waste [3].

Scheme 2.1 shows the oxidation of diphenylamine to tetraphenylhydrazine. Two other methods for the oxidation of diphenylamine to tetraphenylhydrazine using PbO_2 in benzene or CuCl and O_2 in pyridine are described as "waste producing" [3]. Benzene and pyridine are also toxic solvents and their use is best avoided. Diphenylamine undergoes dimerization into the tetraphenylhydrazine salt after hydrogen abstraction by NO_2. The reaction occurs at 0°C so the diphenylamine (m.p. = 50–53°C) [4] is a solid, and Scheme 2.1 is an example of a reaction at the gas–solid interface. Tetraphenylhydrazine was obtained in 95% yield after the gas and the nitric acid were removed *in vacuo*. The *in situ* formed anhydrous nitric acid protects the product from further oxidation due to salt formation [3]. There is no waste from this reaction since the reactant gases can be recovered and used in the Ostwald process [5,6]. The method in Scheme 2.1 is therefore an example of waste prevention as waste produced by the alternative methods may now be averted.

2.3.1.2 The Use of Waste Polyvinyl-Alcohol from Liquid Crystal Display Televisions Polyvinyl-alcohol (PVA) is a component of the polarizing films that are used in liquid crystal display (LCD) televisions [7]. In LCDs, the PVA is doped with iodine. Since PVA/iodine is compatible with the human body, it is possible to

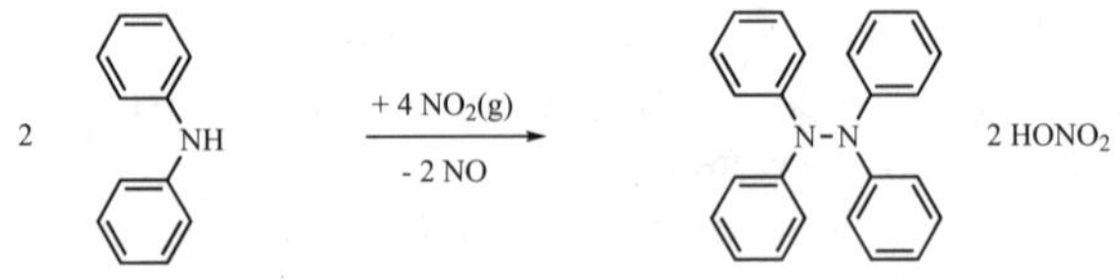

2 NH + 4 NO_2(g) - 2 NO → N-N 2 $HONO_2$

SCHEME 2.1 Oxidation of diphenylamine to tetraphenylhydrazine using NO_2 [3].

TABLE 2.1 Surface Area and Pore Volume of Expanded PVA 4 and LCDPVA

	Expanded PVA 4	LCD PVA
Surface area ($m^2\,g^{-1}$)	143.1	95.0
Total pore volume ($cm^3\,g^{-1}$)	0.47	0.56

Source: Ref. [7].

use this recovered material in a number of applications, including as a support for enzyme immobilization, tissue scaffolding, and in drug-delivery systems [8–10]. All of these applications require large surface areas, and PVA doped with iodine can be expanded to achieve this property. LCD PVA and virgin PVA were compared by Clark et al. [7]. LCD PVA did not require the addition of iodine as the recovered PVA was already iodine doped. PVA 4 was prepared by a four-step process (gelatinization, iodine addition, retrogradation, and precipitation in EtOH). The conditions were optimized and results from the expansion can be seen in Table 2.1 [7].

Both the LCD PVA and expanded PVA 4 achieved a large surface area after expansion. LCD PVA surface area was smaller but the pore volume was larger. There are two advantages of using LCD PVA instead of the virgin PVA for the aforementioned applications. Firstly, waste produced from the production of new PVA is avoided and secondly the reduction and recycling of waste from the large number of discarded LCDs is promoted [7].

2.3.1.3 Synthesis of Sitagliptin (1) The first-generation process reported by Merck research laboratory in 2005 can be seen in Scheme 2.2 [11]. The synthesis requires eight steps starting from *beta*-ketoester **2**, has low atom economy (see Section 2.3.2), and produces large amounts of waste [12].

By developing a one-pot method for the preparation of dehydrositagliptin (**13**), utilizing an alternative asymmetric hydrogenation route, the waste produced was reduced (see Schemes 2.3 and 2.4).

By increasing the pressure of H_2 to 250 psig the chiral catalyst loading was significantly decreased, from 5 mol% to 0.15 mol%, and this became a viable route for the large-scale production of sitagliptin (**1**). The first-generation process gives an overall yield of 52% and the asymmetric hydrogenation route gives an overall yield of 65%. Using this route over the first-generation process (Scheme 2.2) not only has the advantage of giving a higher yield but is also more environment friendly.

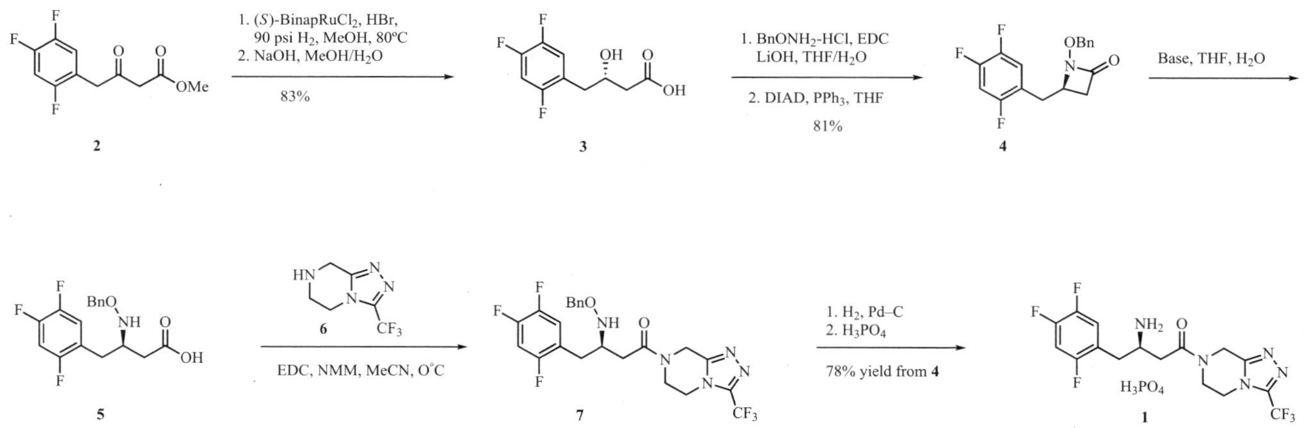

2

1. (*S*)-Binap$RuCl_2$, HBr, 90 psi H_2, MeOH, 80ºC
2. NaOH, MeOH/H_2O

83%

3

1. $BnONH_2$-HCl, EDC, LiOH, THF/H_2O
2. DIAD, PPh_3, THF

81%

4

Base, THF, H_2O

5

6

EDC, NMM, MeCN, O°C

7

1. H_2, Pd–C
2. H_3PO_4

78% yield from **4**

1 (H_3PO_4)

SCHEME 2.2 First-generation process of Sitagliptin (**1**) [12].

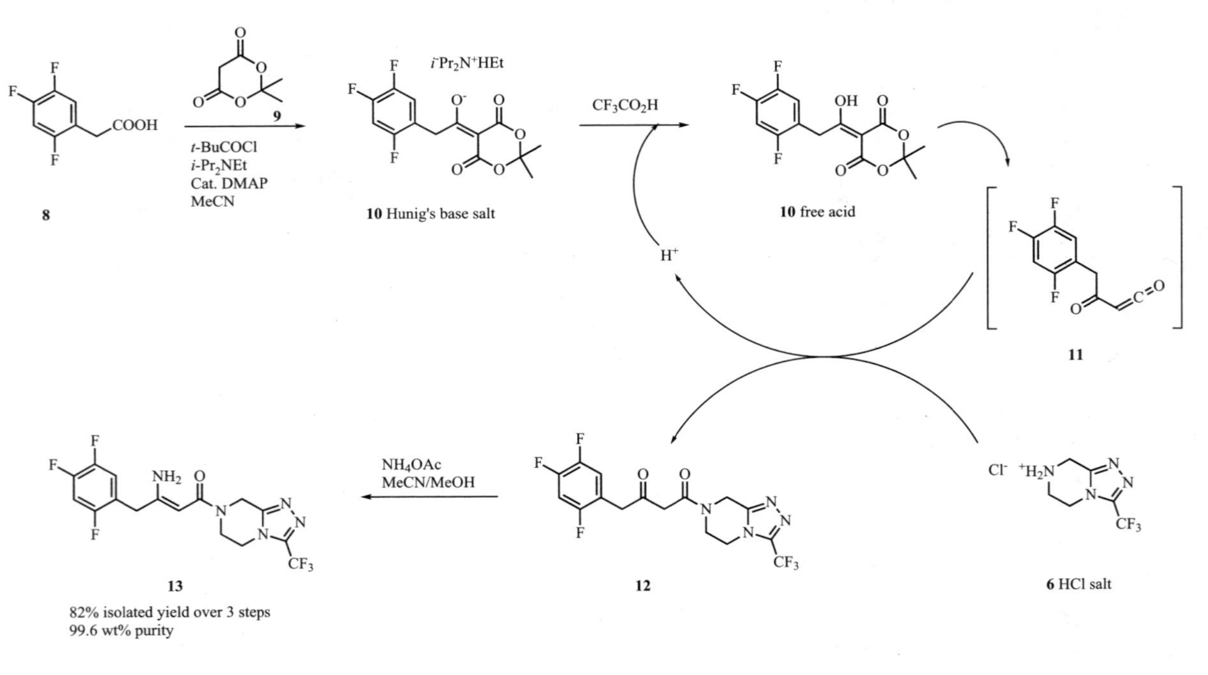

SCHEME 2.3 One pot synthesis of dehydrositagliptin (**13**) [12].

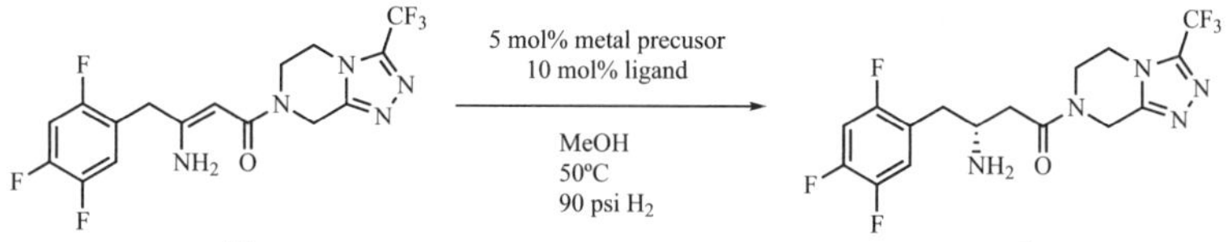

SCHEME 2.4 Synthesis of sitagliptin (**1**) from dehydrositagliptin (**13**) [12].

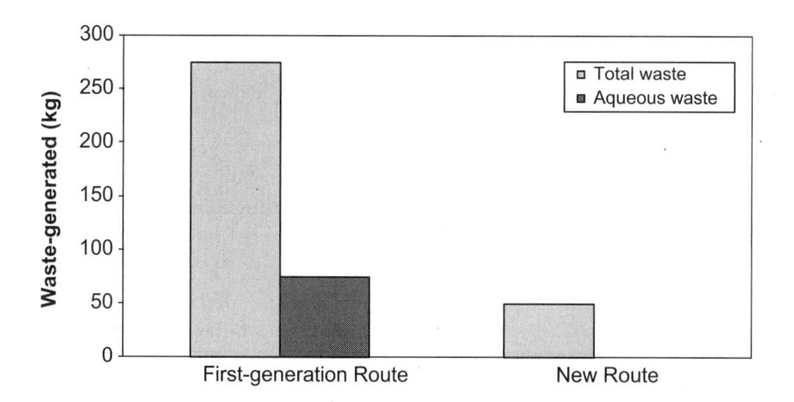

FIGURE 2.1 Waste generated per kg of sitagliptin for the first-generation route and the asymmetric hydrogenation route [12].

The difference in the amount of waste produced by the two routes can be seen in Figure 2.1. There was only 50 kg of waste per 1 kg of sitagliptin (**1**) and no aqueous waste was produced from the new route. A large reduction compared with the first-generation method that produces over 250 kg of waste, including over 50 kg aqueous waste [12].

2.3.2 Atom Economy

The concept of atom economy was first developed by Trost in 1991. A calculation was developed to describe atom economy of reactions which is seen in Equation 2.1 [13]

$$\%\text{Atom economy} = \frac{\text{Formula weight of all atoms utilised}}{\text{Formula weight of all the reactants used}} \times 100 \qquad (2.1)$$

Reactions that have good atom economy are typically addition (Scheme 2.5) and rearrangement reactions, substitution reactions are less economical but elimination reactions usually have bad atom economy. The use of protecting groups should be reduced (see Section 2.3.8) and reagents with high molecular weight (e.g., *N,N′*-Dicyclohexylcarbodiimide) should be avoided.

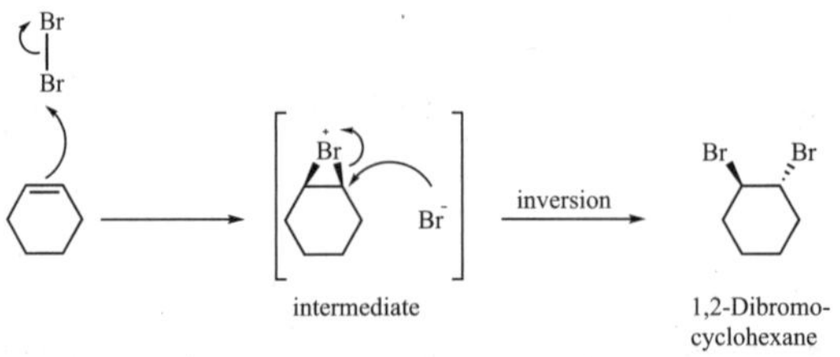

Br
Br
Br
Br
inversion
Br Br
intermediate
1,2-Dibromo-
cyclohexane

SCHEME 2.5 Bromination of cyclohexane with 100% atom economy [32].

2.3.2.1 Ruthenium-Catalyzed Cross-Coupling

Cross coupling of a tertiary propargyl alcohol (**14**) with ω-alkynenitriles (**15a–c**) can be a replacement reaction for an aldol condensation. Aldol condensations of ketones and aldehydes tend to require a large excess of the ketone to prevent side reactions taking place. Most of the reactions described use symmetrical ketones such as acetone. Using unsymmetrical ketones as target nucleophiles in aldol reactions can lead to difficulty in achieving high regioselectivity of the desired product (**16**) [14,15].

Scheme 2.6 shows the $[CpRu(MeCN)_3]PF_6$ catalyzed reaction of the tertiary propargyl alcohol with ω-alkynenitriles to give stereodefined *Z,Z* dienyl ketones (**16**). The *Z,Z* diene isomer cannot be achieved directly by means of conventional aldol condensation. This ruthenium catalyzed cross-coupling reaction is an example of a reaction with excellent atom economy, 100% (see Equation 2.2) [14]. All atoms in the starting materials are present in the product. The *Z,Z* dienyl ketone can also be easily transformed to the *E,E* isomer, by refluxing in tetrahydrofuran (THF) with PhSSPh as a catalyst [14].

2.3.2.2 Manganese Catalyzed Synthesis of Benzene Derivatives

The reaction of 1,3-dicarbonyl compounds with terminal alkynes can produce terphenyl compounds with good atom economy.

Scheme 2.7 shows the reaction pathway; after cycloaddition of the alkyne and dicarbonyl starting material to give a cyclohexadienol (**17**), a water molecule is eliminated to form the *p*-terphenyl derivative (**18**).

After optimizing the reaction conditions a high yield for **18** (87%) was obtained (Scheme 2.8) using 10 mol% $MnBr(CO)_5$ catalyst. Depending on the R groups (e.g., R^1 = Me, *c*-Hex, Ph; R^2 = CO_2Allyl, CO_2Et, COMe; R^3 = Ph) the yields ranged between 0% and 98%. Pentane-2,4-dione reacted with phenylacetylene to give the highest yield (98%) of the *p*-terphenyl aromatic product, in the absence of $MgSO_4$ [16]. The excess (3 equiv.) of the phenylacetylene used for the optimal conditions

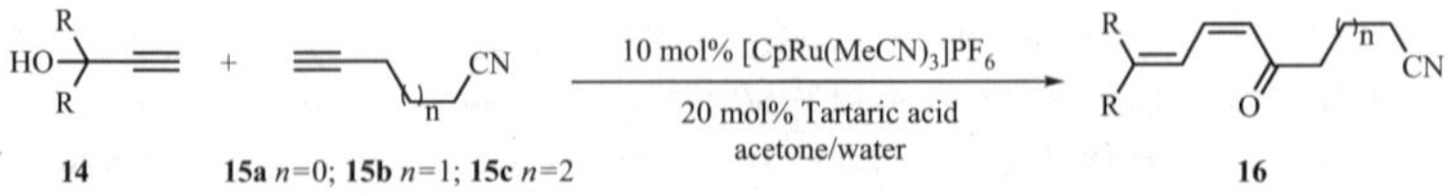

R
HO
R
CN
n
10 mol% $[CpRu(MeCN)_3]PF_6$
20 mol% Tartaric acid
acetone/water
R
R
O
n
CN
14
15a *n*=0; **15b** *n*=1; **15c** *n*=2
16

SCHEME 2.6 Reaction of tertiary propargyl alcohol with ω-alkynenitriles [14].

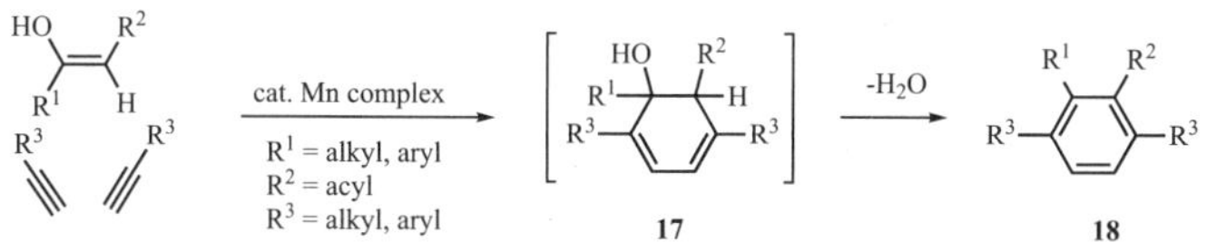

SCHEME 2.7 [2 + 2 + 2] coupling of 1,3-dicarbonyl compounds with terminal alkynes [16].

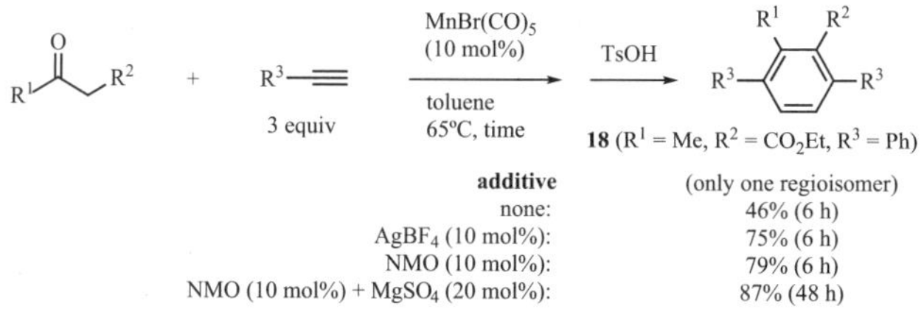

additive	18 (R^1 = Me, R^2 = CO_2Et, R^3 = Ph) (only one regioisomer)
none:	46% (6 h)
$AgBF_4$ (10 mol%):	75% (6 h)
NMO (10 mol%):	79% (6 h)
NMO (10 mol%) + $MgSO_4$ (20 mol%):	87% (48 h)

SCHEME 2.8 Optimum conditions for reaction [16].

reduces the atom economy for this reaction, as does the loss of water; however, the reaction still has good atom economy (72.5%).

2.3.3 Less Hazardous Chemical Syntheses

The generation and use of hazardous reagents should be minimized when possible. Syntheses ideally should be designed to use and generate benign substances [1].

2.3.3.1 Soybean Oil Extraction Currently soybean oil extraction is performed using hexane, which requires a large volume of this solvent. There are health and environmental issues associated with the use of hexane due to its vapor pressure and toxicity to the central nervous system [17]. Elimination of the use of hexane would lead to a less hazardous extraction of soybean oil. A method that is under investigation for this purpose is switchable-hydrophilicity solvents using 1,8-diazabicyclo[5.4.0]undec-7-ene (DBU). Figure 2.2 shows the mechanism of this methodology [17].

From Figure 2.2 we can see that the solvent is potentially recoverable. The main problem with this method is separating the solvent from the water after the CO_2 is removed. Though the hydrophilicity of the solvent does decrease after the CO_2 is removed it is still soluble in water. Using a solvent that is hydrophobic after removing CO_2 from the water is the desired situation shown in Figure 2.2. Work in this area is currently being carried out with amidines and is under investigation. If a suitable amidine solvent can be identified then this method would be a less hazardous extraction than by using large amounts of hexane [17].

2.3.3.2 Arylation of Heteroaromatics Using Carbonate Solvents Direct arylation of a heteroaromatic ring using a carbonate solvent is shown in Scheme 2.9.

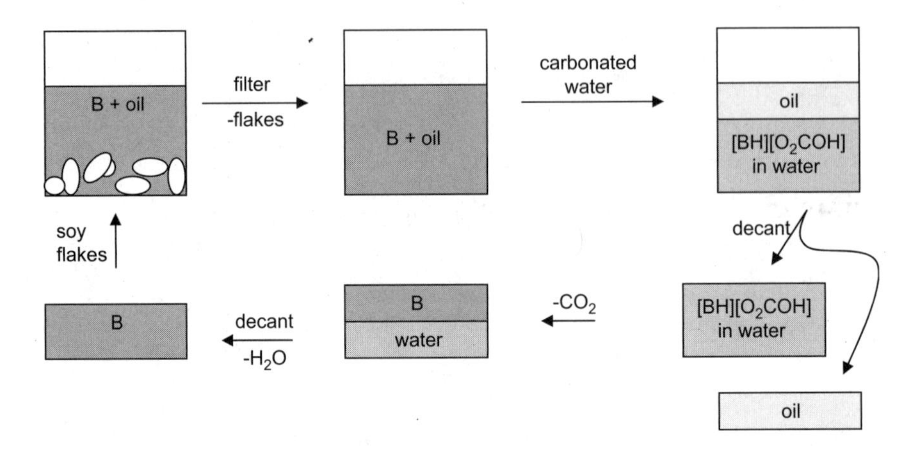

FIGURE 2.2 Soybean extraction using switchable hydrophilicity solvent DBU. B represents the hydrophobic state and [BH] [O_2COH] represents the hydrophilic state [17].

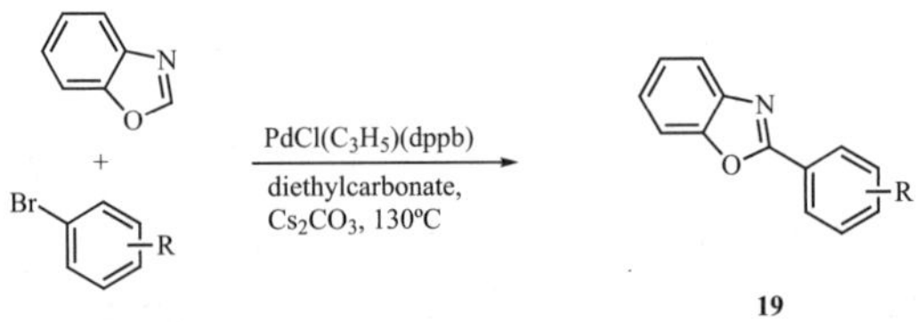

SCHEME 2.9 Coupling of benzoxazole with aryl bromides [18].

Since arylation of a heteroaromatic ring usually involves the use of hazardous solvents (e.g., *N,N*-dimethylformamide (DMF), *N,N*-dimethylacetamide (DMAc), *N*-methyl-2-pyrrolidinone (NMP), or dioxane) this method is a less hazardous chemical synthesis. Carbonate solvents are nontoxic and biodegradable [19,20]. Yields up to 91% for **19** were obtained with the coupling of benzoxazole with 3-bromoquinoline. These yields are higher than when performed with the established toxic solvents [18]. Thus a "greener" synthesis also delivered improved results, a fundamental factor in the uptake of environment friendly synthesis by the chemical industry.

2.3.3.3 Use of Ionic Liquids in the Synthesis of 2-Aryl Benzimidazoles

Scheme 2.10 shows the synthesis of 2-aryl benzimidazoles in the ionic liquid (IL) pentyl methylimidazolium BF_4 [pmim][BF_4]. This process does not use organic solvents in the isolation of the product, except ethanol for the final crystallization. The reaction is performed at room temperature for 4–7 h depending on the aldehyde. High yields are obtained ranging from 82% to 94%. [Pmim][BF_4] can also be recycled two times before losing efficiency. In comparison with current procedures used this is

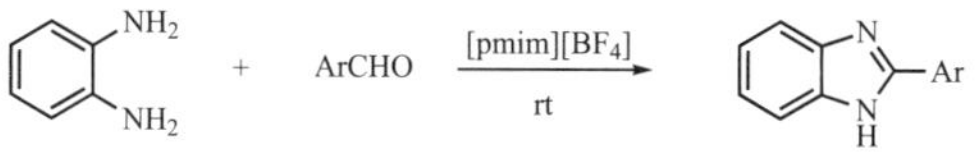

SCHEME 2.10 Synthesis of 2-aryl benzimidazoles [21].

environment friendly and is a less hazardous chemical synthesis. Current procedures use a variety of hazardous organic solvents and reagents including chlorotrimethylsilane/DMF, *p*-toluenesulfonic acid (TsOH)/DMF, and H_2O_2/HCl in MeCN [21].

2.3.3.4 Safer Catalysts The design of a "green" compound, whether having the role of a solvent, [22] reagent, or catalyst [23] should ideally have low toxicity, be readily biodegradable, and not generate toxic persistent metabolites. Of equal importance is the performance of the environmentally benign material. The decision to replace a "toxic" chemical with a "green" replacement is easier if a performance benefit is also attained. The role of a green chemist is to make this decision as easy as possible and avoid the "gray area" where environmental protection comes at a performance cost.

In 2002, Forbes and Davis [24] introduced imidazolium and phosphonium IL's equipped with covalently strong Brønsted acid sulfonic acid functionality (**20**, Fig. 2.3). These materials represented a successful (from both catalytic and operational perspectives) marriage of the excellent solvent properties and non-volatility of IL's with the convenience and activity of traditional solid acids such as (for instance) *p*-TSA. This seminal report spawned considerable interest in the design of similarly devised Bronsted acid based imidazolium-based IL catalysts and catalytic solvents [25]. Two other general strategies have emerged for the design of organic imidazolium-ion-based acidic ILs: the use of protonated imidazolium ion ILs (i.e., **21**, Fig. 2.3) [26] and the installation of an acidic counteranion (**22**, Fig. 2.1) [27–29]. While these materials can—when appropriately designed—exhibit excellent catalytic activity and reduce some of the risks (both practical and environmental) associated with the use of more volatile liquid acids, they remain in effect (particularly in the cases of **20** and **22**) strong Brønsted

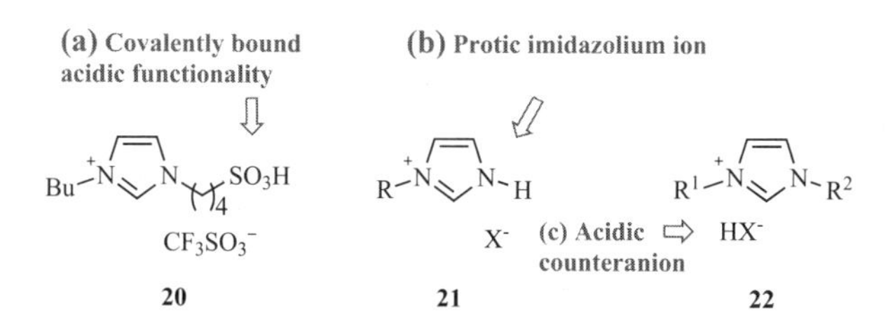

FIGURE 2.3 Imidazolium ion-based acidic ILs: current strategies (a–c) [25–27].

catalyst

MeOH (0.38 M)
rt, 24 h

23 24

catalysts

25

26 X = BF_4
27 X = Br
28 X = PF_6
29 X = NTf_2
30 X = octylsulfate

31 R^1 = H, R^2 = H, X = Br
32 R^1 = H, R^2 = H, X = NTf_2
33 R^1 = Ph, R^2 = H, X = Br
34 R^1 = Ph, R^2 = H, X = NTf_2
35 R^1 = H, R^2 = Me, X = Br
36 R^1 = H, R^2 = Me, X = NTf_2

FIGURE 2.4 Catalysts for acetalization of benzaldehyde (**23**).

acids and their environmental impact and toxicity have not yet, to the best of our knowledge, been unambiguously established.

Figure 2.4/Table 2.2 show the use of ILs as acid catalysts in the acetalization of benzylaldehye (**23**). High yield for acetal product (**24**) was found for catalyst **26**. Antimicrobial toxicity data (Table 2.3 and 2.4) show that catalyst **26**, **27**, and **31** have low toxicity to he bacterial and fungal strains screened. Biodegradation data for these imidazolium salts show they have poor biodegradation [30] (Closed Bottle test) and the development of biodegradable examples is an active research area. The design philosophy of this group is to determine the toxicity and biodegradation of the new catalysts as part of the effort to improve catalyst performance; with the expectation that safer catalysts can be identified [31, 32].

TABLE 2.2 Results Determined by ^{1}HNMR

Entry	Catalyst	Yield (%)
1	**25**	0
2	**26**	85
3	**27**	11
4	**28**	33
5	**29**	51
6	**30**	12
7	**31**	13
8	**32**	23
9	**33**	9
10	**34**	35
11	**35**	11
12	**36**	42

TABLE 2.3 Antifungal Activity of 26, 27, 31 (MIC [mM])

Organisms	Time (h)	**26**	**27**	**31**
Candida albicans	24	>2.0	>2.0	>2.0
ATCC 44859	48	>2.0	>2.0	>2.0
Candida albicans	24	>2.0	>2.0	>2.0
ATCC 90028	48	>2.0	>2.0	>2.0
Candida parapsilosis	24	>2.0	>2.0	>2.0
ATCC 22019	48	>2.0	>2.0	>2.0
Candida krusei	24	>2.0	>2.0	>2.0
ATCC 6258	48	>2.0	>2.0	>2.0
Candida krusei	24	>2.0	>2.0	>2.0
E28	48	>2.0	>2.0	>2.0
Candida tropicalis	24	>2.0	>2.0	>2.0
156	48	>2.0	>2.0	>2.0
Candida glabrata	24	>2.0	>2.0	>2.0
20/I	48	>2.0	>2.0	>2.0
Candida lusitaniae	24	>2.0	>2.0	>2.0
2446/I	48	>2.0	>2.0	>2.0
Trichosporon beigelii	24	>2.0	>2.0	>2.0
1188	48	>2.0	>2.0	>2.0
Aspergillus fumigatus	24	>2.0	>2.0	>2.0
231	48	>2.0	>2.0	>2.0
Absidia corymbifera	24	>2.0	>2.0	>2.0
72	48	>2.0	>2.0	>2.0
Trichophyton mentagrophytes	72	>2.0	>2.0	>2.0
445	120	>2.0	>2.0	>2.0

2.3.4 Designing Safer Chemicals

Eliminating the use of transition-metal-based oxidizing agents such $KMnO_4$, CrO_3, OsO_4, $K_2Cr_2O_7$, or the Swern oxidation methodology [33,34] (dimethyl sulfoxide (DMSO), oxalyl chloride) and replacing them with environment friendly oxygen is one method of designing safer chemicals. Oxygen is a greener oxidant than the toxic transition-metal-based or hazardous oxidizing agents listed above. The development of new oxidation reactions that utilize "green" oxidants to replace the previously established toxic oxidants is one example of designing safer chemicals. Catalysis has an important role to play in these transformations (see also Section 2.3.9).

2.3.4.1 Wacker Oxidation Yields of up to 92%, depending on the R group, were obtained using the method in Scheme 2.11. Increasing the reaction temperature (from 80°C to 100°C) and time (from 12 h to 24 h) lead to an increase in yields. In the Wacker oxidation shown, a terminal alkene is converted to a methyl ketone. Normally, organic solvents DMF and NMP are used in the Wacker oxidation but these have been replaced with the environment friendly ethylene carbonate (EC) making the oxidation process in Scheme 2.11 more environment friendly. Using

TABLE 2.4 Antibacterial activity of 26, 27, 31 (MIC [mM])

Organisms	Time (h)	**26**	**27**	**31**
Staphylococcus. aureus	24	>2.0	>2.0	>2.0
CCM 4516/08	48	>2.0	>2.0	>2.0
Staphylococcus aureus	24	>2.0	>2.0	>2.0
H 5996/08	48	>2.0	>2.0	>2.0
Staphylococcus epidermidis	24	>2.0	>2.0	>2.0
H 6966/08	48	>2.0	>2.0	>2.0
Enterococcus sp.	24	>2.0	>2.0	>2.0
J 14365/08	48	>2.0	>2.0	>2.0
Escherichia coli	24	>2.0	>2.0	>2.0
CCM4517	48	>2.0	>2.0	>2.0
Klebsiella pneumoniae	24	>2.0	>2.0	>2.0
D 11750/08	48	>2.0	>2.0	>2.0
Klebsiella pneumoniae	24	>2.0	>2.0	>2.0
J 14368/08	48	>2.0	>2.0	>2.0
Pseudomonas aeruginosa	24	>2.0	>2.0	>2.0
CCM 1961	48	>2.0	>2.0	>2.0

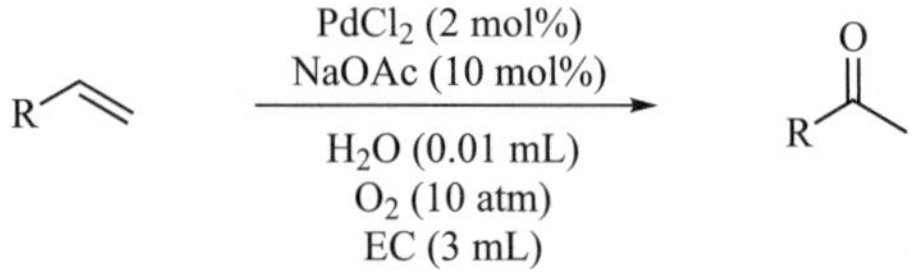

R
$PdCl_2$ (2 mol%)
NaOAc (10 mol%)
H_2O (0.01 mL)
O_2 (10 atm)
EC (3 mL)
O
R

SCHEME 2.11 Wacker oxidation in ethylene carbonate (EC) [35].

oxygen as the oxidant also eliminates the need to use more dangerous oxidizing agents [35].

2.3.4.2 One-Step Oxidation of Primary Alcohols to Terminal Esters

Rossi and coworkers [36] used oxygen as the oxidant to form methyl esters from primary alcohols. Methyl esters are important in the flavoring and fragrance industry [37]. Up to 100% conversion and selectivity of the alcohols to the methyl esters was observed. A reusable SiO_2-supported gold nanoparticle catalyst promoted the one-step oxidation to go to completion. The traditional method of preparing methyl esters from primary alcohols involves two steps. First, the oxidation of the alcohol with oxidizing agents (e.g. $Na_2Cr_2O_7$) to the carboxylic acid intermediate and then the esterification. $Na_2Cr_2O_7$ is toxic (chromium (VI) salt) and environment unfriendly. The dichromate reagent is also required in stoichiometric amounts to perform the oxidation with high yield. The supported gold-nanoparticle-catalyzed oxidation of the alcohol to the ester utilizes oxygen as the oxidizing agent and is a good example of the design for the use of safer chemicals [36].

2.3.5 Safer Solvents and Auxiliaries

Solvents and auxiliaries are used widely in chemical synthesis. Solvents are also used in separation and purification techniques. Potentially large amount of solvent waste may be produced from chemical processes. Using environmentally benign solvents and reducing the amounts used can help tackle this problem [1].

The search for alternative solvents to VOC (i.e., hexane and benzene) and toxic polar aprotic examples (i.e., DMF and DMSO) is a major research area. Supercritical fluids (CO_2 or H_2O), ionic liquids, fluorous solvents, water, switchable solvents, or even solvent-free approaches are under study. Chapters (3–8) give more detail, describing biomass conversion and manipulation in these reaction media. Below are examples of greener synthetic methods – reactions that work effectively in safer solvents compared to conventional procedures.

2.3.5.1 Organic Reactions "On Water" In 2005, Sharpless was the first to describe reactions that occur "on water" [38]. "On water" reactions take place when insoluble reactants are stirred with water as the reaction media. Sustainable rate acceleration is typically observed. Diels–Alder reaction, [39] Claisen rearrangements, [40] cycloaddition reactions, [41] and ene reactions [42] have been shown to go to completion faster and in many cases with higher yields when performed "on water."

Scheme 2.12 shows this reaction occurring "on water" with air or oxygen as the oxidant. Less by-products are produced when the oxidation is performed "on water" compared with the oxidation in neat aldehyde. By performing the oxidation "on water" the use of organic solvents is averted. Water is considered a safer reaction media than most organic solvents, although consideration of the disposal of contaminated aqueous waste is important. Yields for this oxidation are dependent on the aldehyde (e.g., benzaldehyde 83%, cyclohexanal 87%, 2-ethyl hexanal 86%, pentanal 11%). The reaction was performed at room temperature and took 12 h in the presence of air. A reduced reaction time of 2 h was sufficient when oxygen gas was used as an oxidant [43].

2.3.5.2 Arylation of Phenols by Aryl Halides Comparing Schemes 2.13 and 2.14, both depicting diarylether synthesis, Scheme 2.13 has several advantages. Methyl isobutyl ketone (MIBK) is used instead of DMF, and is a more environment friendly solvent [46]. Also only 5 mol% Cu catalyst is required in Scheme 2.13. The use of MIBK instead of DMF in the reaction means a safer solvent is used to produce diarylethers using the Cu-mediated Ullmann condensation shown. MIBK in addition to its low toxicity was also shown to exhibit low leaching of the Cu catalyst compared to DMF [44].

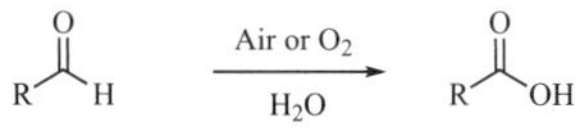

$$\mathrm{RCHO} \xrightarrow[H_2O]{\text{Air or } O_2} \mathrm{RCOOH}$$

SCHEME 2.12 Aldehyde oxidation "on water" [43].

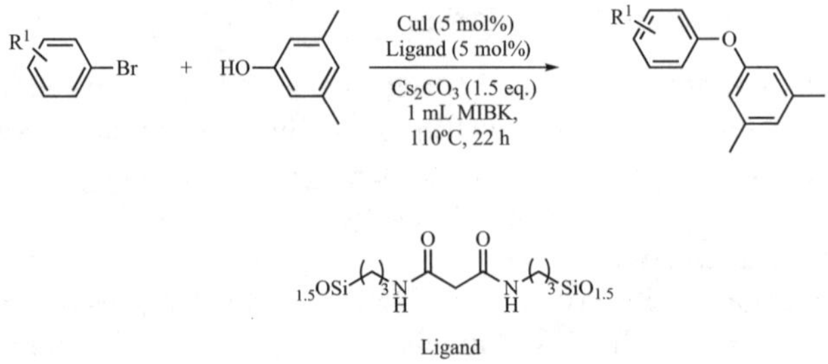

SCHEME 2.13 Coupling of phenols with aryl bromides to form diarylethers [44].

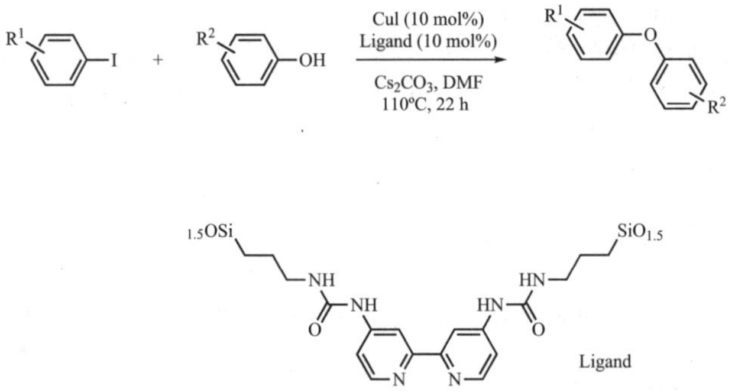

SCHEME 2.14 Coupling of phenols with aryl iodides to form diarylethers [45].

2.3.5.3 Photo-Friedel–Crafts Acylations Photoacylations of quinones with aldehydes are in many cases performed in benzene [47]. Benzene is a toxic and carcinogenic solvent and the search for suitable replacements is a major research effort. In Scheme 2.15 benzene has been replaced with an IL [48]. A comparison between ILs and benzene was carried out, with [bmim][OTf] giving the best results. A yield of 91% was obtained for the photoacylation product **39**, and no photo-reduction by-product **40** was present (c.f. Benzene, 44% yield of **39** and 7% **40**). Not only is the IL a safer solvent but it also achieves better yields. The IL can be reused up to three times and the reaction is performed at room temperature [49].

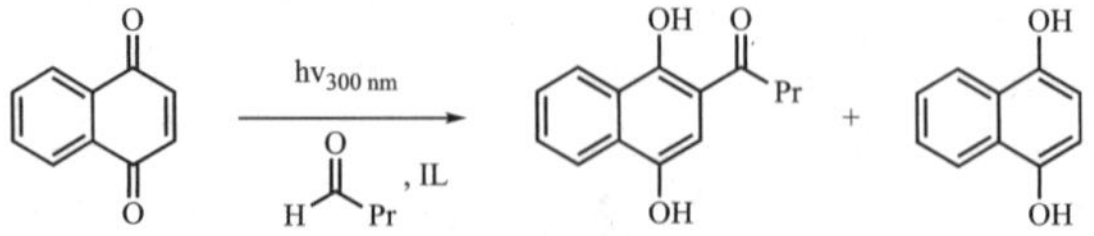

SCHEME 2.15 Photoacylation of 1,4-naphthoquinone (**37**) with butyraldehyde (**38**) [49].

2.3.6 Design for Energy Efficiency

Many chemical reactions require a specific amount of energy before they are thermodynamically favored, known as the activation energy. Addition of catalysts can lower the energy requirements for a reaction, therefore, increasing the energy efficiency.

Energy costs are expected to rise, and chemists are investigating microwave, light, ultrasound, and electricity as other ways to provide the energy required. Of note, some reaction pathways can only proceed under electrochemical or photochemical conditions, whereas ultrasound and microwave generally promote thermal reactions.

The cooling of highly exothermic reactions and the energy requirements for separation/purification processes are two other uses of energy in the chemical industry. In general, the best energy efficiency is achieved when reactions are performed at room temperature and atmospheric pressure, and when separation or purifications processes are minimized [1].

2.3.6.1 Metal-Free Oxidative Cross-Coupling of Unfunctionalized Aromatic Compounds Scheme 2.16 shows the cross-coupling of a thiophene (**41**) with an arene (**43**) to form a biaryl compound (**44**). Thiophene (**41**) is converted to an iodonium salt (**42**) with the hypervalent iodine reagent (PhI(OH)OTs) in hexafluoroisopropanol (HFIP). Trimethylsilyl bromide (TMSBr) and 1-methoxy-naphthalene (**43**) are then added and the biaryl product is formed. High yields can be obtained for these biaryls (with 86% yield obtained for the biaryl **44** Scheme 2.16). The hypervalent (III) iodine reagent is a mild oxidant, and there was no need for a metal catalyst. Cross-coupling reactions are regarded as an environment friendly approach to the synthesis of biaryls. The reaction is performed at room temperature, under mild conditions, that is, the use of trivalent organoiodine oxidants, and only takes 3 h to go to completion [50].

2.3.6.2 Synthesis of Furans and Pyrroles Scheme 2.17 shows the synthesis of pyrroles and furans in good to excellent yields at room temperature in 2 h [51]. Alternative methods, for example, the Paal–Knorr procedure (refluxing acetic acid for long periods of time) are less energy efficient [52]. This intramolecular cyclization synthesis of furans and pyrroles is therefore more energy efficient when compared to the Paal–Knorr procedure.

Hex Hex S **41** —PhI(OH)OTs / HFIP→ Hex Hex PhI S OTs **42** —OMe **43** / TMSBr→ Hex Hex S OMe **44** (86%)

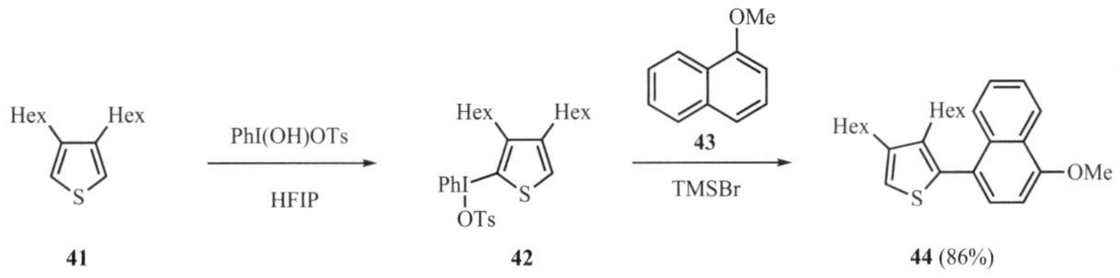

SCHEME 2.16 Cross-coupling reaction of a thiophene with an arene [50].

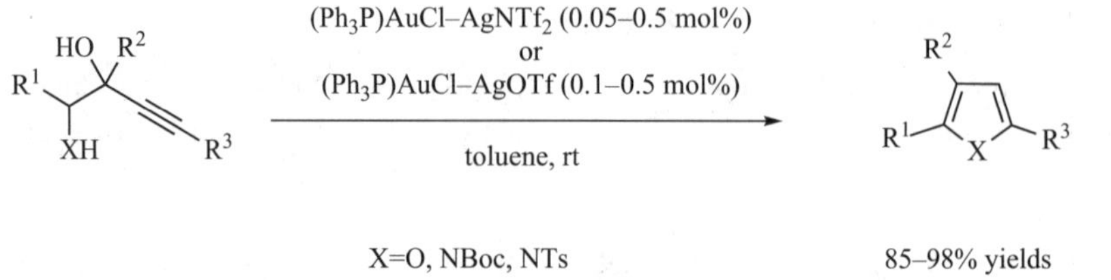

(Ph$_3$P)AuCl–AgNTf$_2$ (0.05–0.5 mol%) or (Ph$_3$P)AuCl–AgOTf (0.1–0.5 mol%), toluene, rt

X=O, NBoc, NTs — 85–98% yields

SCHEME 2.17 Gold-catalyzed cyclizations of alkynediols and aminoalkynols to furans and pyrroles [51].

2.3.7 Use of Renewable Feedstocks

Using renewable feedstocks in reactions will prevent other sources from being depleted, that is, finding alternatives to fossil fuels. The following chapters discuss this principle of Green Chemistry in detail, as only three such examples are give below [1].

2.3.7.1 Direct Conversion of Cellulose to Ethylene Glycol Ethylene glycol (EG) is used in plastics, antifreeze, and polyester fibers [53] EG is produced by the petrochemical industry and 17.8 million tonnes were produced in 2007. Using cellulose instead of petroleum as a source of EG would lead to production from a renewable feedstock. Using nickel-promoted tungsten carbide catalysts, a higher yield of EG was produced from cellulose than previously reported in the literature [54]. Other methods used precious metal catalysts, which are rare and expensive [54]. These methods do not provide a viable alternative to EG production as they all gave low yields. The tungsten carbide was formed on activated carbon (W_2C/AC). These catalysts were prepared using carbothermal hydrogen reduction at elevated temperatures (973–1123 K). Scheme 2.18 represents the conversion of cellulose to EG [55]. The highest yield (61%) was obtained using 2% Ni–30%W_2C/AC-973. Cellulose was completely converted to the polyols after 30 min at 518 K and 6 MPa H_2 [55].

2.3.7.2 Metathesis of Renewable Products Using unsaturated esters and acids from renewable products (e.g., castor oil and oleic acid), amino acid and amino acid ester derivative synthesis have been demonstrated. This is performed by cross metathesis of the ester/acid with acrylonitrile or fumaronitrile followed by hydrogenation of the CN bond and C=C bonds in **45** (Scheme 2.19) [56]. An efficient process, with very low catalyst loadings, can convert renewable feedstocks to form linear amino acid monomers (**46**), fundamental building blocks for polyamides synthesis [56].

2.3.7.3 Formation of Lactate/Lactic Acid A chemical route to prepare lactic acid can be seen in Scheme 2.20. There are no renewable feedstocks used in this method and the starting materials are toxic, volatile, and noxious, including HCN a poisonous gas. Due to these problems, lactic acid is mainly produced by fermentation. In the fermentation process, carbohydrates are transformed into calcium lactate and

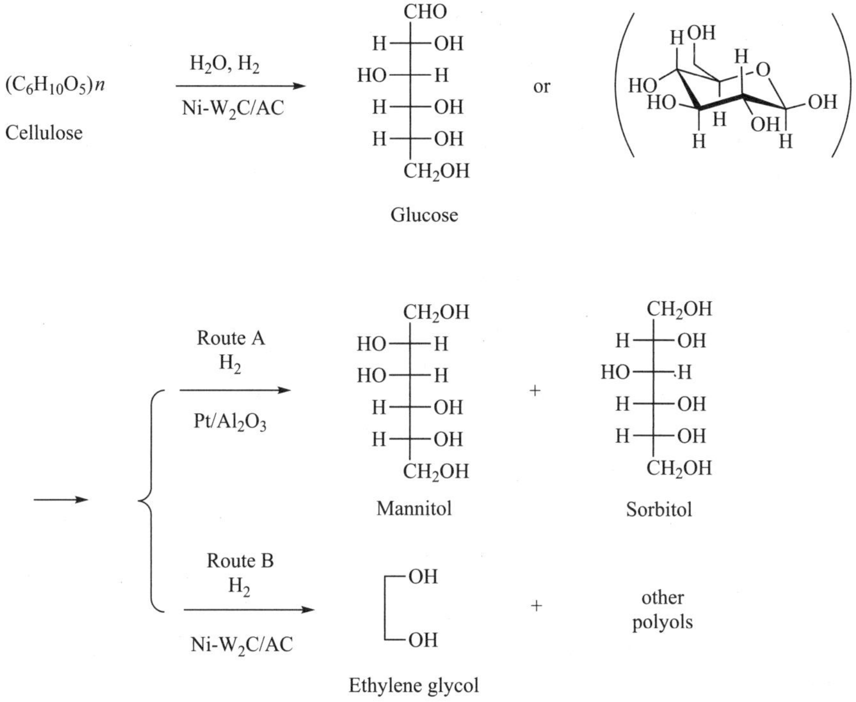

SCHEME 2.18 Catalytic conversion of cellulose to polyols [55].

then converted to lactic acid by reacting the lactate with an acid [57]. However, fermentation has low productivity compared with the chemical routes. A new way to produce lactate is by using biorenewable glycerol from the production of biodiesel. Lactic acid can be used to produce biodegradable polymers [58].

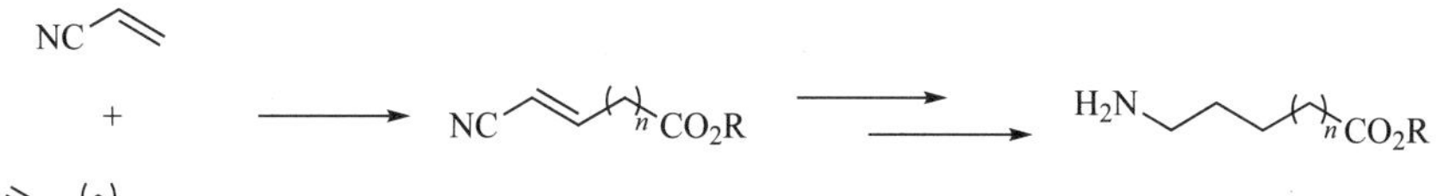

SCHEME 2.19 Cross metathesis then hydrogenation of renewable sources to form amino acids and esters [56].

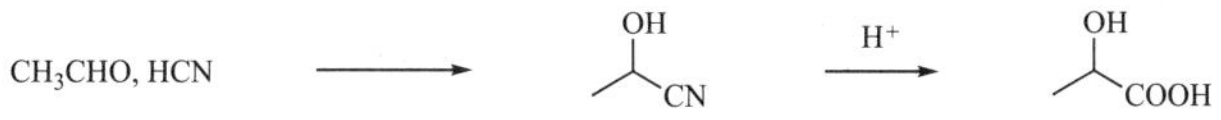

SCHEME 2.20 Industrial chemical route to lactic acid [59].

Glycerol from the biodiesel industry can be reduced to 1,2-propanediol. Using selective oxidation, 1,2-propanediol can then be converted to sodium lactate. Au–Pd catalyst is added to the NaOH/1,2-propanediol solution, this is oxidized by O_2 to form sodium lactate. Sodium acetate and sodium formate are by-products from this process. After determining optimum conditions, a conversion of 94% with a selectivity of 96% for sodium lactate is achieved. This is an acceptable green alternative to the fermentation process [58].

2.3.8 Reduce Derivatives

In chemical synthesis, modifications are made to substances in order to achieve the desired target. These modifications include addition/removal of protecting groups and converting substances into salts. This method of derivatization in a chemical synthesis leads to extra steps to generate the product, therefore, more waste is produced. Thus the use of derivatives in chemical synthesis should be limited or avoided when possible [1] Below is an example of how complex natural products can be prepared without the use of protecting groups and illustrates the ingenuity of chemists to solve challenging problems.

2.3.8.1 Protecting Group Free Synthesis of (−)-Hapalindole, (−)-Fischerindole, (+)-Welwitindolinone, and (+)-Ambiguine (−)-Hapalindole, (−)-fischerindole, (+)-welwitindolinone, and (+)-ambiguine are natural marine products that are obtained from cyanobacteria [60,61]. They have activity as antifungal, antibacterial, antimycotic, and anticancer agents [60–64]. There are a number of ways to synthesize these products, including the use of protecting groups, protecting group-free synthesis, and bio-mimetic synthesis.

Since these are complex molecules, a number of steps are required for their synthesis. (−)-Hapalindole U and (+)-ambiguine H are produced in an eight-step synthesis. Ambiguine H had not been previously produced but hapalindole U was generated as a racemic product by Patterson and coworkers [63]. It took 20 steps with protecting groups being employed. In the protecting group free synthesis of (−)-hapalindole U a single diastereomer was generated. (+)-Welwitindolinone A (**56**) and (−)-fischerindole I (**52**) have been produced with overall yields of 5.7% and 13%. They are prepared in 10 steps (**47–56**) using commercially available materials. Herein the protecting group free synthesis reported in *Nature* by Baran and coworkers (Scheme 2.21) is shown for the production of (−)-fischerindole I (**52**) [65].

Scheme 2.21 also shows the preparation of a single diastereomer of welwitindolinone A [65]. When protecting groups were used for the synthesis of racemic welwitindolinone A an overall yield of 2.5% was achieved in 25 steps, with six protecting groups required. This synthetic route with 25 steps generates more waste than the 10-step method, and more reagents/materials would be necessary. Extra steps required include the protection and then the subsequent deprotection on route to the target the compound. So performing syntheses free of protecting groups reduces the amount of steps required and therefore the amount of derivatives

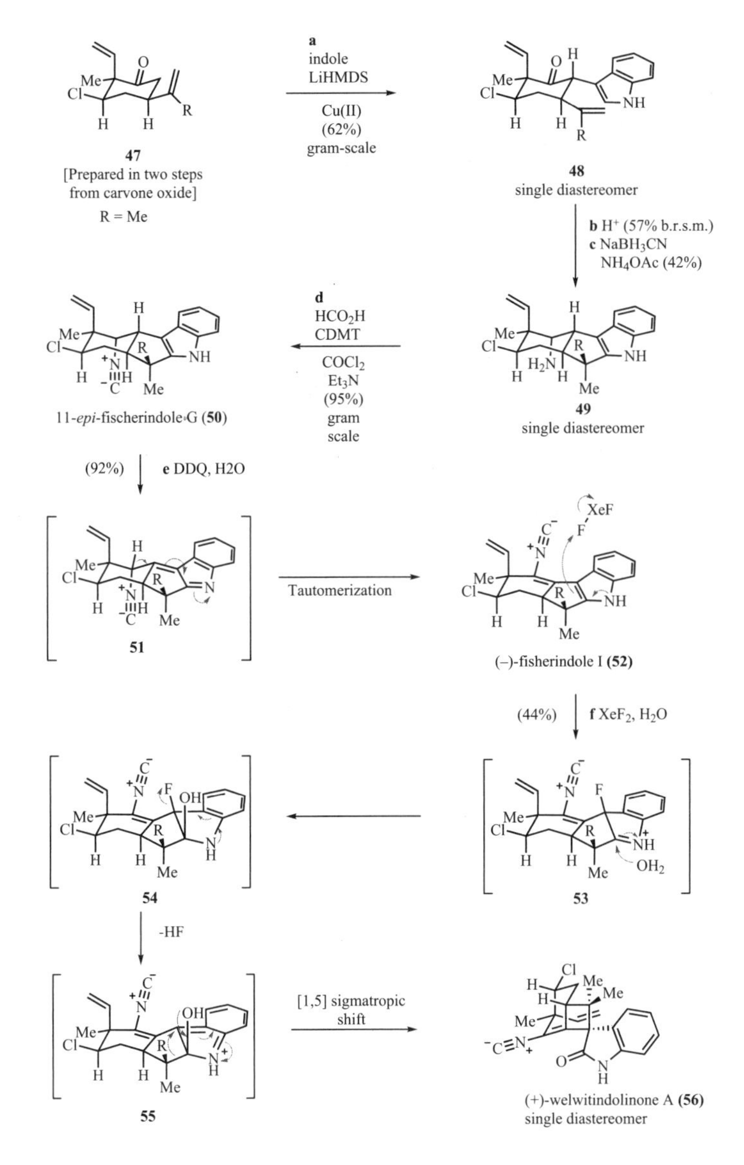

SCHEME 2.21 Total protecting group free synthesis of (+)-welwitindolinone A (**56**) and (−)-fischerindole I (**52**) [65].

produced [65,66]. Protecting group free synthesis is a difficult goal to achieve; however, the green benefits can be considerable when attained.

2.3.8.2 Synthesis of Polyhydroxylated Pyrrolidines Imino-sugars with five-membered rings are known as hydroxypyrrolidines. Scheme 2.22 represents a protecting group free synthesis of hydroxypyrrolidines. Imino-sugars are

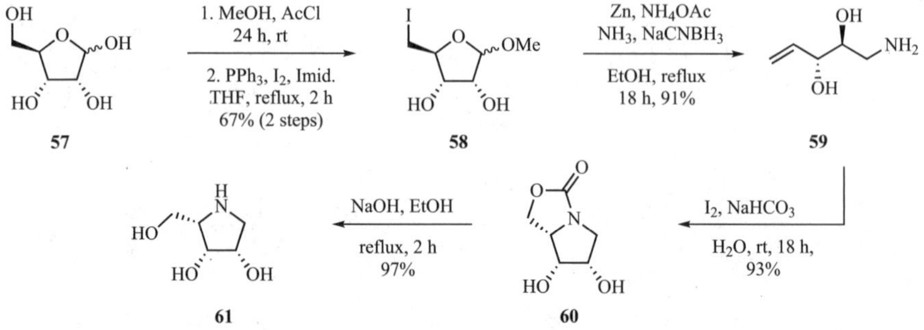

SCHEME 2.22 Synthesis of *cis*-2,3-disubstituted hydroxypyrrolidine (**61**) from D-ribose (**57**) [67].

potential therapeutics in cancer, diabetes, viral infections, and bacterial infections [68]. These hydroxypyrrolidines are produced by a five-step synthesis; (1 and 2) a two-step conversion of a pentose sugar (**57**) to the methyl iodofuranoside (**58**) is performed; (3) Vasella reductive amination of the furan to give an olefinic amine (**59**); (4) carbonylated/halocyclized to give a carbamate (**60**) *via* iodine-promoted annulations; (5) carbamate (**60**) is hydrolyzed to give the *cis*-2,3-disubstituted hydroxypyrrolidine (**61**) [67]. An overall yield of 55%, with no protection groups used, was realized. There are several alternative methods for the production of hydroxypyrrolidines but they employ protecting groups in their syntheses [67].

2.3.9 Catalysis

Using a catalyst is beneficial as it eliminates the need to use stoichiometric reagents in a chemical synthesis. A catalyst performs many turnovers while active. This means less waste is generated while using a catalyst compared with a stoichiometric reagent. Catalysts can also minimize the energy required for a reaction to occur by reducing the activation energy [1]. There are many different types of catalysts, including enzymes, organocatalysts, solid-supported catalysts, and metal-based catalysts.

2.3.9.1 Domino Reaction Catalyzed by Enzymes *Agaricus bisporus*, commonly known as the button mushroom, contains the enzymes laccase and tyrosinase [69]. These enzymes catalyze the domino reaction of phenol with 1,3-dicarbonyls (**62**) shown in Scheme 2.23. Both enzymes are required because catechols (not phenols) are suitable substrates to undergo laccase-catalyzed oxidation to initiate the domino reaction with the 1,3-dicarbonyls (**62**). The tyrosinase oxidizes phenols to catechols, to prepare the substrate for the second enzymatic transformation. As phenols are more widely available than catechols, this expands the scope and synthetic utility of the methodology.

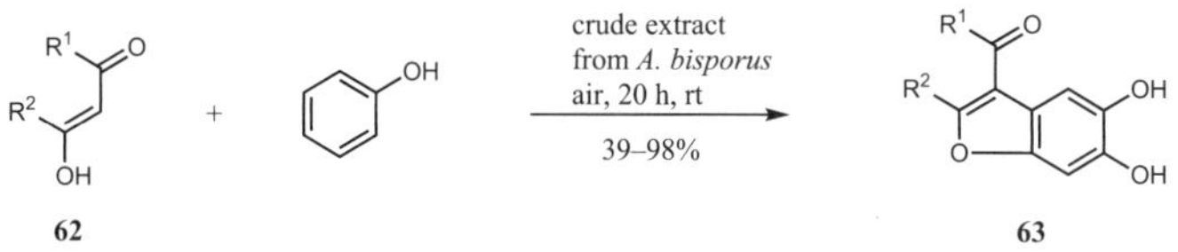

crude extract from *A. bisporus* air, 20 h, rt 39–98%

62 63

SCHEME 2.23 Domino reaction of phenol with 1,3-dicarbonyl (**62**), enzyme catalyzed with *Agaricus bisporus* crude extract [70].

Since the crude extract of *A. bisporus* contains both enzymes, oxidation of the phenol to a catechol can occur and the domino reaction continues on reaction with 1,3-dicarbonyls (**62**) to create *O*-heterocycles (**63**) [70].

This is an environment friendly reaction since the oxidant is oxygen in air and is performed at room temperature using a biocatalyst. High yields for the *O*-heterocycles (**63**), between 39% and 98%, depending on the 1,3-dicarbonyl (e.g., cyclic 1,3-diketones (39–44%), 6-methyl-4-hydroxy-2*H*-pyran-2-one (48%), substituted 4-hydroxy-2*H*-chromen-2-ones (72–98%)) were attained from phenol [70].

2.3.9.2 Organocatalytic Asymmetric Aldol Reactions (*S*)-Proline is an organocatalyst used in asymmetric aldol reactions [71]. The environment friendly process is performed at room temperature and "green" solvents can be employed. Cyclic carbonates, such as propylene carbonate and ethylene carbonate, are nontoxic green solvents [72] that can be used instead of organic solvents (e.g., DMSO or acetonitrile), in the catalyzed aldol reactions. Scheme 2.24 shows the (*S*)-proline catalyzed aldol reaction. Organocatalysts, especially (*S*)-proline have advantages over some metal catalysts. Proline is cheap as it is an available natural product, and is nontoxic. While metal catalysts can be toxic and/or expensive (e.g., some precious metals) [71].

2.3.9.3 Iron Catalyzed Synthesis of 2-Alkyl-Dioxolanes Corn-based ethanol is a commercially available biofuel and 2-alkyl-dioxolanes could replace or supplement ethanol biofuels in the future. These dioxolanes can be generated from renewable resources. Cellulose can be used to produce EG which reacts with syngas (a 1:2 mixture of CO and H_2) to give 2-alkyl-dioxolanes in one step (Scheme 2.25) [73].

The catalyst used for this reaction is magnetic iron nanoparticles, which is prepared from cheap and low-toxicity iron. Determining the toxicity of nanoparticles is an important research area and discussed in detail elsewhere [74]. The selectivity for the

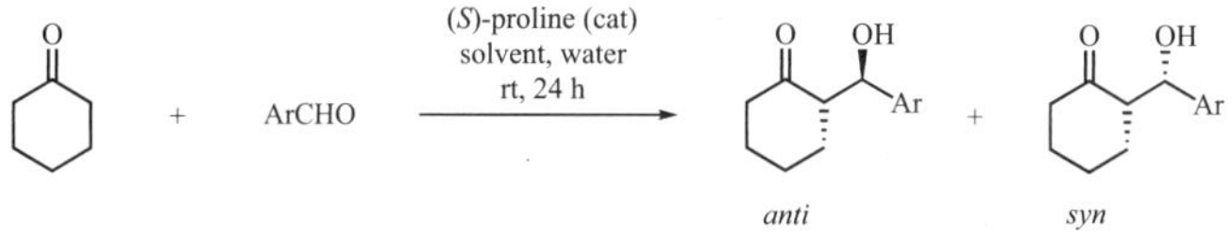

ArCHO (*S*)-proline (cat) solvent, water rt, 24 h

anti *syn*

SCHEME 2.24 (*S*)-Proline catalyzed aldol reaction [71].

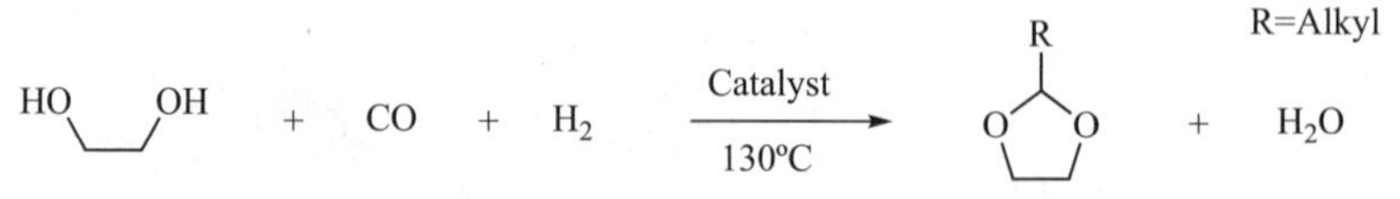

HO‿OH + CO + H_2 —Catalyst, 130°C→ 2-R-1,3-dioxolane + H_2O (R=Alkyl)

SCHEME 2.25 Synthesis of 2-alkyl-dioxolanes from ethylene glycol and syngas [73].

dioxolanes was 68 wt%. The catalyst could also be reused, due to facile removal from the product by a magnet. When using iron nanoparticles as the catalyst, a relatively low temperature of 130°C was sufficient, while 230–270°C was required for the same results using classical iron catalysts. The iron nanoparticles had a higher selectivity for dioxolanes compared with the other catalysts tested [73]. Facile recovery of the magnetic iron nanoparticles reduces the risk of release into the environment.

2.3.9.4 Heck Reaction Catalyzed by Mesoporous Silica Supported Pd

A Pd catalyst is supported by a mesoporous silica (SBA-15), and immobilized on the SBA-15 with the ionic liquid, 1,1,3,3-tetramethylguanidinium lactate (SBA–TMG–Pd catalyst). The catalyst is used in the Heck arylation of olefins with aryl halides under solvent-free conditions. Many different types of Pd supports have been developed; including polymeric, metal oxides, and zeolites. The advantage of using the SBA–TMG–Pd catalyst in the Heck reaction was high conversion at very low catalyst loadings. An isolated yield of 93% was obtained when the Heck reaction of iodobenzene and methyl acrylate (Scheme 2.26) was performed at 140°C for 2 h with 0.001 mol% of Pd, using Et_3N as the base. At 0.01 mol% the reaction time was 65 min with a 91% yield. Pd loading of 0.001 mol% is very low and still gave excellent results. [75].

SBA–TMG–Pd can also be recycled up to six times without losing efficiency. The authors report that the gray color of the catalyst still remained and there was no apparent deactivation. This was due to the immobilization using the 1,1,3,3-tetramethylguanidinium IL. When Pd was supported by SBA-15 without any IL immobilization (SBA–Pd), there was a decrease in the activity of the SBA–Pd catalyst after the first recycle. After the second recycle, the SBA–Pd catalyst turned to a white powder. Figure 2.5 shows that the IL stabilizes the catalyst since both SBA–TMG–Pd and SBA–Pd have the same activity in the first run. This demonstrates that modifying the Pd-supported catalysts with ILs helps improve the recyclability of the catalyst [75].

2.3.9.5 Pd/C Catalyzed Hydrogenation of trans-Cinnamaldehyde While it is straightforward to achieve selective olefin reduction in α,β-unsaturated ketones using simple Pd/C reduction, with α,β-unsaturated aldehydes the situation is more

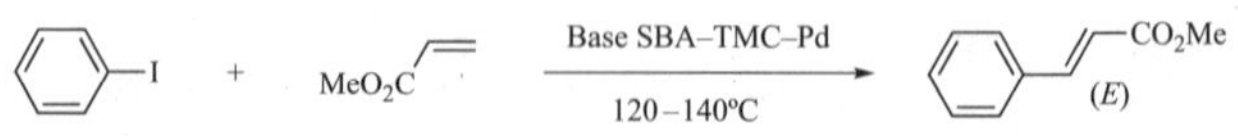

PhI + MeO_2CCH=CH$_2$ —Base SBA–TMC–Pd, 120–140°C→ PhCH=CHCO_2Me (*E*)

SCHEME 2.26 Heck arylation of methyl acrylate with iodobenzene [75].

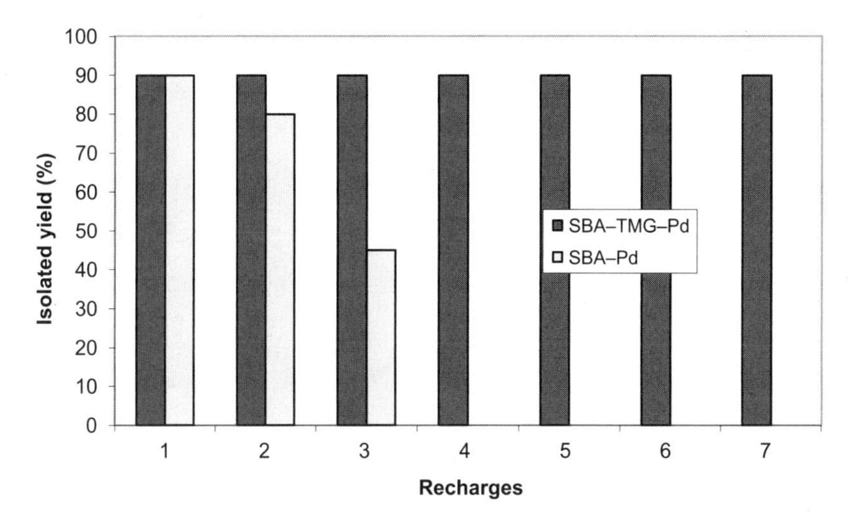

FIGURE 2.5 Recycling of SBA–TMG–Pd and SBA–Pd in the Heck reaction [75].

challenging and product mixtures often arise [76,77]. *Trans*-cinnamaldehyde is widely used as a model substrate because the products of its hydrogenation are extensively used in the fine chemical industry [76,78,79]. The selectivity varies due to the possibility of reduction of either the olefin moiety, to give hydrocinnamaldehyde, or the aldehydic moiety, to give 3-phenylpropanol. (Scheme 2.27) Owing to the thermodynamically favored reduction of the C=C bond over the C=O bond, [76,80] the selectivity towards cinnamyl alcohol is generally poor. This can be frustrating as cinnamyl alcohol is used widely in the flavoring and perfume industries. Hydrocinnamaldehyde is also an important chemical and has uses in the perfumery industry and in the synthesis of anti-HIV compounds [81,82]. In the field of metal-catalyzed hydrogenation ruthenium complexes generally lead to hydrogenation of the carbonyl moiety, [83] whereas rhodium [84–86,87] or ruthenium [88,89] complexes lead to hydrogenation of the olefin moiety but typically high pressures of hydrogen are required.

Palladium on carbon is well known as a universal catalyst for olefin hydrogenation. However, its efficient catalytic activity may lead to poor selectivity. Furthermore, the product ratio of unsaturated alcohol to saturated aldehyde depends on many factors, including the nanostructure of the metal catalyst, [90,91] the degree of polarization of the carbonyl group, [12] and the phase-distribution of reagent, substrate, and catalyst [86,92]. The Pd–H bond is also amphipolar and can transfer hydrogen as either $Pd^{\delta-}$–$H^{\delta+}$ or $Pd^{\delta+}$–$H^{\delta-}$, the latter being favored by catalyst poisons, such as quinoline and the Lindlar catalyst [93]. With so many factors influencing selectivity it is not surprising that some of the solutions to the problem have been elaborate. Tessonier et al. [90] for instance, calcined hydrated palladium (I) nitrate at 350°C (which had previously been added to carbon nanotubes and distilled water) for 2 h to generate Pd nanoparticles. Conversion upto 100% was achieved, resulting in 90% hydrocinnamaldehyde (**65**) and 10% undesired

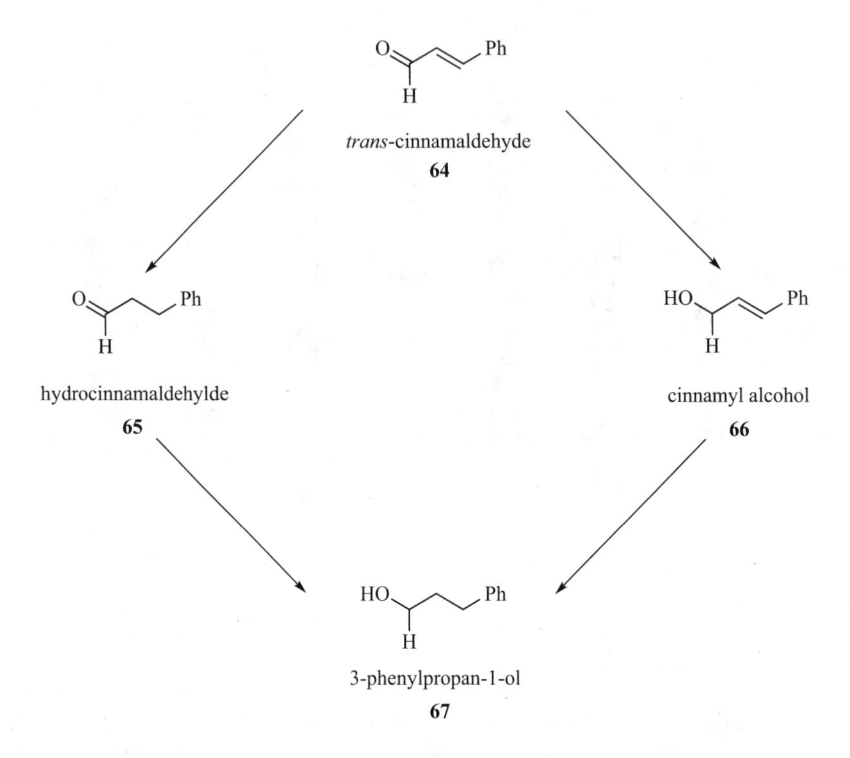

SCHEME 2.27 Reduction pathway for *trans*-cinnamaldehyde [21].

3-phenylpropanol (**67**). This compares very favorably with a commercial Pd–C catalyst that gave a 1:1 ratio of hydrocinnamaldehyde (**65**) to 3-phenyl-1-propanol (**67**) [90]. A more selective process, also involving a palladium catalyst, was developed by Ledoux et al. [91]. After first impregnating a carbon nanofibre support with the palladium salt, palladium(I) nitrate was dried overnight at 110°C and reduced under a flow of hydrogen for 2 h. This support was the major difference from Tessonier's procedure, with the nanofibres, prepared by a gas-phase deposition using ethane and hydrogen over a nickel catalyst (from Al_2O_3, $Ni(NO_3)_2$/glycerol), followed by sonication of the fibers and cleaning with acid at 650°C. In this case, the hydrogenation was 98% selective for the C=C double bond of cinnamaldehyde (**64**), compared with a commercial Pd–C catalyst that gave mixtures of C=C hydrogenation, C=O hydrogenation, and complete reduction to 3-phenyl-1-propanol.

An alternative approach to solve the problem of selectivity in the reduction of *trans*-cinnamaldehyde (**64**) was adopted by Gathergood [22]. By modifying the structure of the solvent (ionic liquid), they investigated the effect on selectivity when using Pd/C as the catalyst. The ionic liquids selected for the study were based on their low antimicrobial toxicity and in some cases were also readily biodegradable [94]. Gathergood and coworkers demonstrated that an imidazolium IL (containing an ester and ether group) displays superior selectivity in olefin hydrogenation compared with conventional organic solvents and alkyl substituted ILs. Furthermore *trans*-cinnamaldehyde (**64**) was selectively reduced to hydrocinnamaldehyde (**65**) with little or no over-reduction of the aldehydic moiety. The reactions were performed with only the use of the IL, substrate, a simple heterogeneous catalyst (Pd/C), and under H_2 at 1 atm. pressure. Successful recycling of the systems was achieved without significant loss of activity. This demonstrates that one preferred approach for the development of green catalytic reactions is to select ambient reactions conditions (1 atm, RT), a simple catalyst, and change the solvent.

2.3.10 Design for Degradation

Chemical products can accumulate in the environment if they do not degrade. These products can be harmful to human health and wildlife. The accumulation of plastic waste is a common example of this problem. Designing for degradation ensures that the products or waste will not accumulate in the environment. The degradation products and metabolites ideally should not be harmful or toxic [1].

2.3.10.1 Biodegradability of Ionic Liquids (ILs) ILs have been described as molten salts that are entirely ionic in nature, comprising both a cationic and anionic species and by definition having a melting point below 100°C [23,95]. ILs are also known as "designer solvents." The physical, chemical, and biological properties of these salts can be "tuned" for a particular function. These reagents have been designed as replacements for conventionally used organic solvents, and much research has been reported on their applications in the area of "green" chemistry [23,95–97].

This has subsequently led ionic liquid researchers to assess the various biological properties, namely toxicity, [98–100] biostability, and biodegradability, [101–105] of these "green" solvents. Investigations by Gathergood and Scammells to develop biodegradable imidazolium ionic liquids led to ionic liquids (**68–70**) being prepared (Fig. 2.6). These ionic liquids pass (>60%) the "CO_2 Headspace" test (ISO 14593) [30] and can be classed as readily biodegradable. Their subsequent work focused on

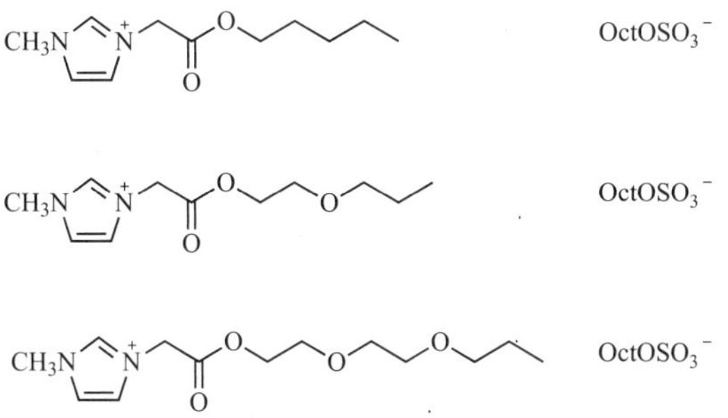

CH_3N … $OctOSO_3^-$ **68**

CH_3N … $OctOSO_3^-$ **69**

CH_3N … $OctOSO_3^-$ **70**

FIGURE 2.6 Scammells (**68**) and Gathergood (**69–70**) combined either short-chain esters, or ether-linked esters with the biodegradable octylsulfate anion.

modifying/replacing the imidazolium ring, changing the counterion and effect of functional groups in the side chain.

Overcoming the inertness of 1-substituted-3-methylimidazolium cation was a goal. It has been reported that 2-methyl substitution on a 1-methylimidazole ring dramatically improves biodegradation from 0% after 31 days for 1-methylimidazole, to 100% after 31 days in the case of 1,2-dimethylimidazole in the mixed microbial community of an activated sludge [106]. However, when Gathergood investigated the biodegradability of 2-methyl substituted imidazolium ionic liquids, the benefit from incorporation of the 2-methyl substituent [94] was negligible.

Bearing the Twelve Principles of Green Chemistry in mind [1]. Scammells prepared pyridine-based ionic liquids (in particular those with 1-methyl carboxy ethylester (89% biodegradable; $OctOSO_3$ salt, 28 days) or 3-carboxylic ester groups (84% biodegradable; $n=3$, $OctOSO_3$ salt, 28 days) (Fig. 2.7)) [107] that can be metabolized by a wide variety of bacteria in the environment, and so have favorable green credentials. Scammells found that for these pyridinium ILs halide and hexafluorophosphate salts exhibited a percentage biodegradation almost as high as the octylsulfates. In more recent work, Scammells showed that even the ditriiflimide salt of pyridinium-3-carboxybutyl esters can be rendered readily biodegradable (68–74% over five experiments) if the pyridinium nitrogen is quaternized with a 2-hydroxyethyl group [108]. However, the bioaccumulation of ditrifliimide anion is a potential problem.

Similarly, Stasiewicz [109] evaluated pyridinium salts containing an *N*-alkoxymethyl group (reminiscent of Bodor's soft antimicrobials) [110] using the closed bottle test OECD 301D [30,105]. By including a 3-hydroxy group on the pyridine ring to improve biodegradability, as well as an innocuous saccharinate counter anion, it was possible to enhance biodegradability to the extent of 72% after 28 days. This represents a substantial improvement over either simple *N*-alkoxymethylpyridinium chloride (48%) or acesulfamate (49%) salts (Fig. 2.8).

Another strategy to reduce toxicity and maximize biodegradability is to make use of cations and anions from nature, which has a minimal impact on the environment. Examples include ionic liquids based on choline, [111] with counteranions such as

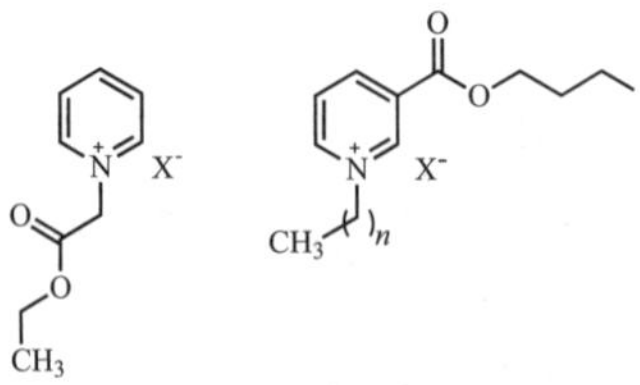

71 X = Br
72 X = PF_6
73 X = $OctOSO_3$

74 $n = 3$, X = $OctOSO_3$
75 $n = 0$, X = $OctOSO_3$
76 $n = 0$, X = PF_6
77 $n = 0$, X = I

FIGURE 2.7 Scammells introduced biodegradable pyridinium ILs with ester substituents.

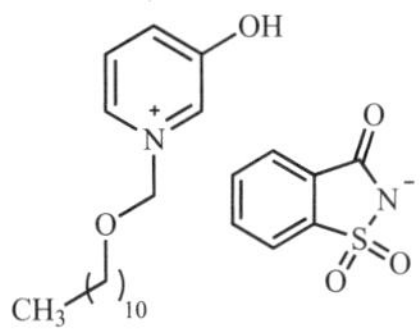

78

FIGURE 2.8 Stasiewicz combined several favorable features to obtain a biodegradable IL.

acetate, succinate, or malate that can be readily incorporated into biochemical pathways. Choline hexanoate is not only completely biodegradable, [112] but is also effective in dissolving powdered cork waste and recovering the major biopolymer from cork, suberin [113].

2.3.11 Real-time Analysis for Pollution Prevention

The use of real-time and in-process analysis can reduce pollution. By monitoring hazardous/toxic by-products from a chemical synthesis it may be possible to change parameters to eliminate their production. The completion of reactions can also be determined by monitoring the reaction. This can eliminate the use of excess reagents in reactions and therefore reduce pollution generated [1]. ReactIR is one example of a real-time analysis method. ReactIR is an *in situ* Fourier Transform Infra-Red Spectroscopy (FTIR) analysis; monitoring the initiation, progression, and completion of reactions [114].

2.3.11.1 Development of a Manufacturing Process for LY518674

ReactIR was used in the development of LY518674 (Fig. 2.9), an agonist peroxisome proliferator-activated receptor alpha (PPARα) [115].

A semicarbazide (**80**) is required to produce LY518674. Originally, a three-step procedure was used to generate the semicarbazide but after ReactIR investigation, a one-pot synthesis was possible. The one-pot process started with addition of aqueous KOCN to a solution of 4-methybenzylhydrazine (**79**). Scheme 2.28 shows the products of this one-pot reaction and some of the impurities produced. Only the

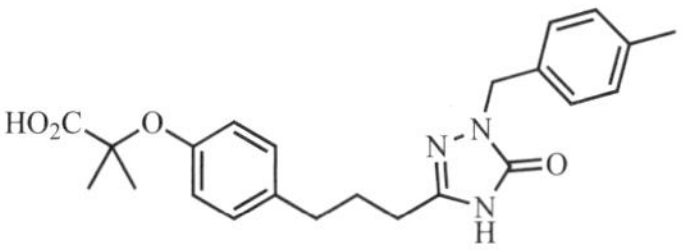

FIGURE 2.9 LY518674 [115].

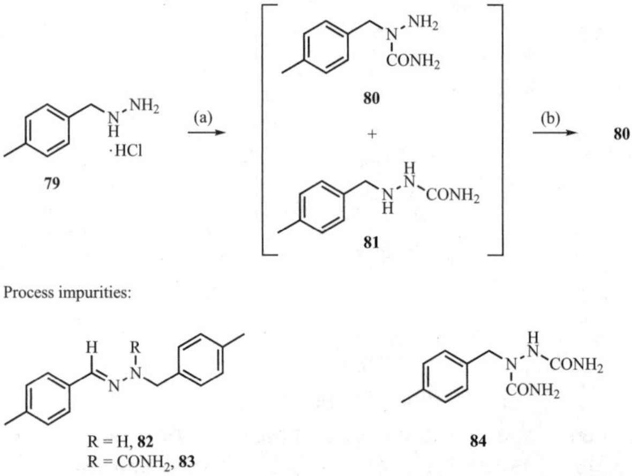

SCHEME 2.28 One-pot synthesis of the semicarbazide **80** [115]. Reagents and conditions for the optimized process: (a) KOCN (0.98 equiv), *n*-BuOH/H_2O (1:10 v/v), 20°C, 8 h, 92% (not isolated, **80/81**, 80:20); (b) HCl (0.3 equiv), 20°C, 3 h then filtered, 59% (**80/81**, 99.9:0.1).

semicarbazide **80** is required to produce LY518674 and the regioisomer **81** is an undesired by-product and must be removed. This is performed by adding aqueous HCl to the semicarbazides. The HCl reacts with **81** to form the HCl salt, which is then separated by filtration. However, a number of impurities are produced from this reaction (**82–84**). Biscarboxamide **84** is an insoluble by-product that is formed when HCl is added to the process when KOCN is still present. ReactIR was used to measure the concentrations of cyanate present during the reaction. From this, a kinetic model was developed to determine reaction completion. The completion time of the reaction could then be determined using temperature, fill time, and stoichiometry. Cyanate levels were therefore controlled to less than 0.5% before the addition of HCl. As a result, levels of **84** were also less than 0.5%. ReactIR was successfully used to minimize the amount of **84** that was generated as an impurity [115].

2.3.12 Inherently Safer Chemistry for Accident Prevention

Accident prevention is very important in chemistry. Accidents in the chemical industry have resulted in the loss of human lives. By using less volatile, less toxic, and less explosive substances the risk can be reduced. The safer the chemical process the less likely an accident will occur [1].

2.3.12.1 One-Pot Synthesis of Nitrones Urea-hydrogen peroxide (UHP) is a safer, solid alternative to hydrogen peroxide. Unlike H_2O_2 it can be stored at room temperature without decomposition. Aqueous H_2O_2 can become explosive upon concentration and is used with great caution. Methyltrioxorhenium (MTO) is a

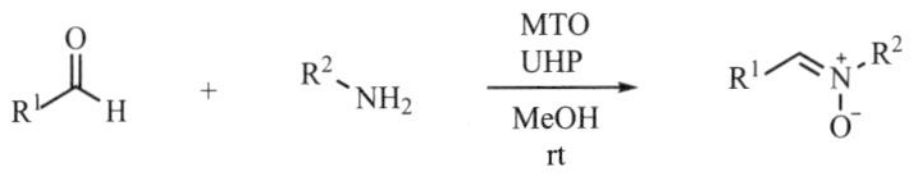

SCHEME 2.29 One-pot synthesis of nitrones using UHP as oxidant [116].

catalyst that is combined with UHP to perform catalyzed oxidation. (e.g., one-pot synthesis of nitrones, Scheme 2.29). Applications of nitrones include spin traps in biological studies, as drugs for age-related illnesses, and as useful intermediates for organic chemistry. Current procedures for the creation of nitrones employ H_2O_2 as the oxidant [117].

2.3.12.2 Synthesis of Bisphenol-A Bisphenol-A (BPA) is important in the production of epoxy resins and polymers [118]. Scheme 2.30 shows the traditional way in which BPA is synthesized. This process is hazardous since cumene hydroperoxide (CHP) is an explosive. The conversion step to CHP and the subsequent concentration of the CHP has previously caused explosions. Changing this process to avoid the hazards caused by the production and concentration of the CHP would lead to safer chemistry [119].

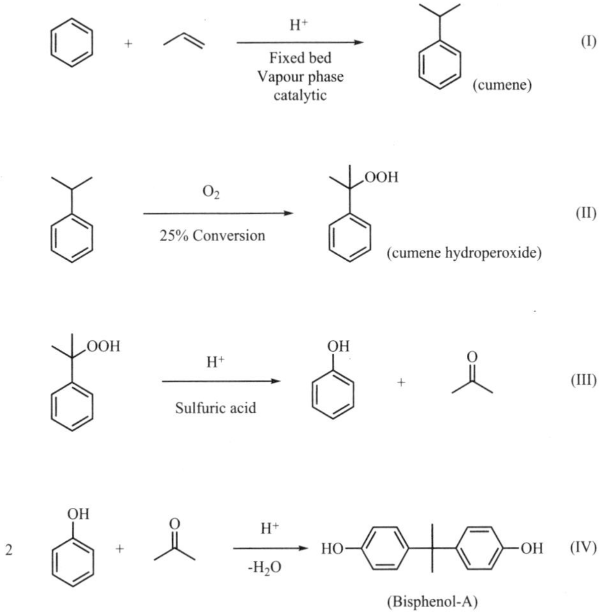

SCHEME 2.30 Hock process for manufacturing of phenol (steps I–III) and manufacturing of BPA from phenol and acetone (IV) [119].

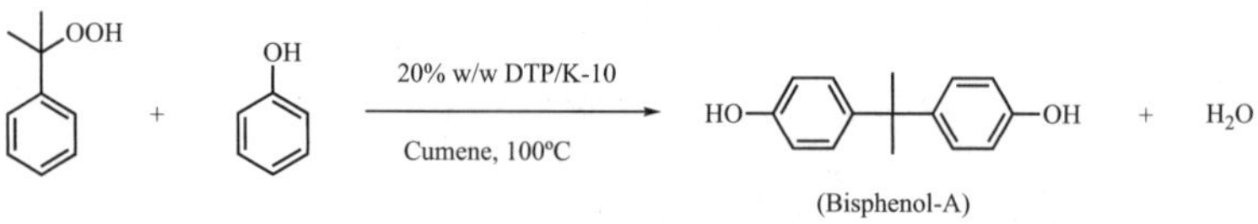

OOH + OH → (20% w/w DTP/K-10; Cumene, 100°C) HO–C₆H₄–C(CH₃)₂–C₆H₄–OH + H_2O

(Bisphenol-A)

SCHEME 2.31 Cascade engineered synthesis of BPA from CHP and phenol [119].

Scheme 2.31 shows a new cascade synthesis of BPA. This eliminates the hazards associated with the handling and concentration of CHP. In the cascade synthesis, CHP no longer needs to be separated from cumene and the reaction is at a lower temperature. By switching to the cascade-engineered synthesis safer chemistry for accident prevention is being employed [119].

2.4 CONCLUSION

The development and growth of green chemistry in industry is essential in improving environmental friendly, safer, and more economical chemical syntheses. From the recent examples presented herein it is generally seen that to make the reaction "greener" a completely novel approach may need to be developed. These examples all show that applying the Twelve Principles of Green Chemistry to chemical synthesis has significant benefits for the chemical industry. Application of waste prevention in the examples discussed demonstrate that using a different process, recycling waste, or reducing the number of steps in a reaction can prevent excess waste from being generated. For atom economy, two examples illustrate that when all the stoichiometric compounds are incorporated into the product the reaction yields high atom economy.

The use of less hazardous chemical syntheses by changing to less toxic solvents in reactions or purification procedures, minimized hazards associated with the reactions. Hazards associated with handling the toxic solvents are also eliminated. The design of safer chemicals showed that by using O_2/air as an oxidant a safer alternative chemical methodology was developed. O_2/air is less toxic than traditional transition-metal-based oxidizing agents. Designing reactions to use O_2/air as oxidant is therefore a major improvement. The safer solvents section looked at the replacement of toxic solvents with environmentally benign alternatives. Also, performing reactions without the use of organic solvents was highlighted that is "on water" reactions. Two examples for energy efficiency were presented when performed at room temperature under mild conditions, a third example showed how using microwave energy reduces overall the amount of energy required for the reaction.

Using renewable feedstocks instead of depleting feedstocks can reduce pollution, be more cost effective, and is sustainable. Cellulose is converted to EG, which is currently produced commercially from petroleum. Petroleum is a finite resource, depleting and increasing in cost, while cellulose is ubiquitous in nature and sustainable.

The reduction of derivatives was shown by eliminating the use of protecting groups. Protecting-group-free synthesis reduced the number of steps required for a reaction to occur. This then leads to a reduction in waste and increases atom economy. The catalysts classes discussed herein include enzymes, organocatalysts, metal nanoparticles, and solid-supported metal catalysts. From these examples it was seen that catalysts reduce energy requirements, eliminate the need for stoichiometric reagents, and can also be recyclable. The design for degradation section highlighted improved biodegradability of ILs by addition of functional groups and choice of benign counter-ions. The use of real time analysis (e.g., ReactIR) assisted in the determination of a mechanistic course of a reaction. ReactIR also stopped the need for in-process monitoring of a contaminant in a reaction due to optimization of the reaction. Finally, safer chemistry for accident prevention by replacing volatile or explosive chemicals with more stable derivatives can reduce the risk of accidents. Changing the process to avoid handling the dangerous chemical also helped reduce the risk. Overall, the application of any of the 12 principles can help reduce waste generated, eliminate use of toxic substances, decrease energy requirements, prevent accidents, and use renewable feedstocks. All the actions listed will help reduce cost of production by reduced waste disposal. Green chemistry is beneficial to the entire chemical industry.

2.5 OUTLOOK

The core principle of green chemistry is the reduction of waste. Conversion of raw materials from renewable feedstocks into value-added compounds, via short and efficient synthetic routes is now emerging.

2.5.1 Ranitidine Synthesis from Renewable 5-(Chloromethyl)furfural

Mascal and coworkers have reported the digestion of sugars, cellulose, or cellulosic biomass with HCl in a biphasic reactor which resulted in the isolation of a mixture of 5-(chloromethyl)furfural (**85**) and levulnic acid (**86**) [120,121]. 5-(Chloromethyl) furfural (**85**) is the major product (Scheme 2.32) (70–90% of the organic material isolated, depending on the feedstock loading), while **86** is less than 10% of the product mixture. A wide range of platform molecules can be efficiently prepared from **85** (Fig. 2.10).

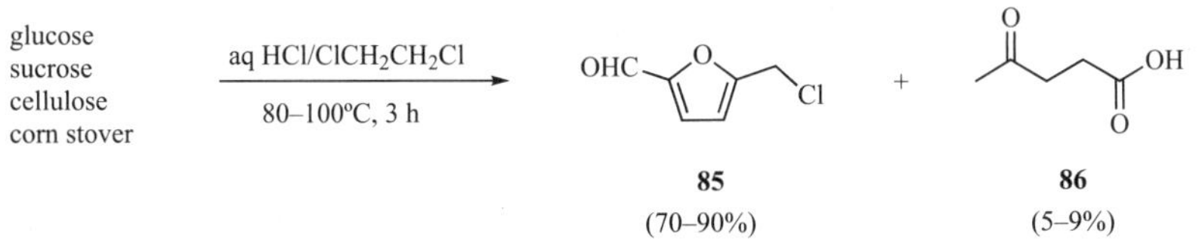

glucose
sucrose
cellulose
corn stover

aq $HCl/ClCH_2CH_2Cl$, 80–100°C, 3 h

OHC, O, Cl — **85** (70–90%)

\+

O, OH, O — **86** (5–9%)

SCHEME 2.32 5-(Chloromethyl)furfural (**85**) synthesis from sustainable resources.

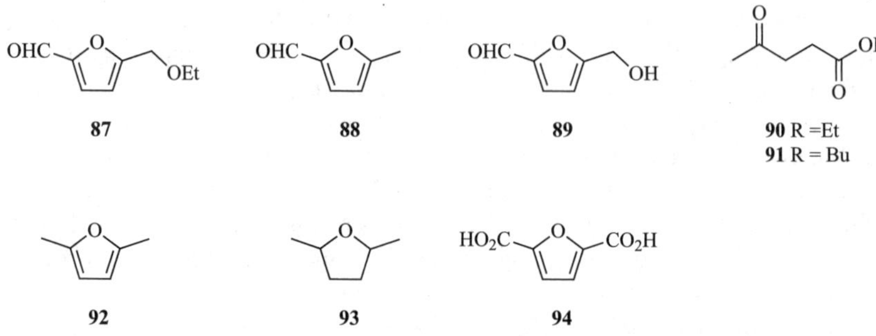

FIGURE 2.10 Platform molecules from **85**.

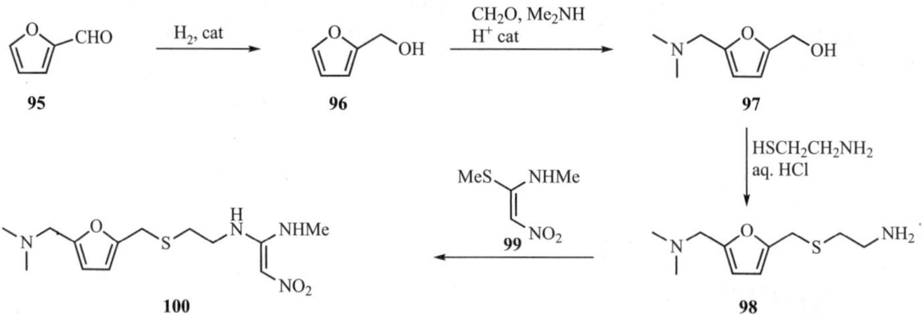

SCHEME 2.33 Mascal's synthesis of Ranitidine (**100**) from platform molecule furfural (**95**).

Mascal selected Ranitidine (**100**), sold under the trade name Zantac, as a target molecule. Zantac is a histamine H2-receptor antagonist, introduced by Glaxo (now GlaxoSmithKline) in 1981, and by 1986 in excess of $1 billion of total sales was achieved. Although as a prescription drug, this has been superseded by other proton pump inhibitors (e.g., omeprazole and esomeprazole), it is now available over-the-counter as a general antacid preparation. Despite great interest in the synthesis of (**100**), the most straightforward approach [122] (up to 2010) was described in the original patent (Scheme 2.33) [123].

Mascal described a more efficient approach starting from **85** (Scheme 2.34). Apart from using a renewable starting material, all the reaction yields are high (average 91%), with 68% overall yield of **100** attained. In addition, no chromatography is required at any stage of the synthesis.

2.5.2 "One-Pot" Organocatalysis

Throughout the past decade, one of the hot topics in organic chemistry, asymmetric organocatalysis, [124,125] has introduced a fresh approach to the design and development of one-pot processes in enantioselective transformations. Jørgensen in a recent review describes the benefits of a "one-pot strategy" versus traditional "stop-and-go" approach to synthesis [125].

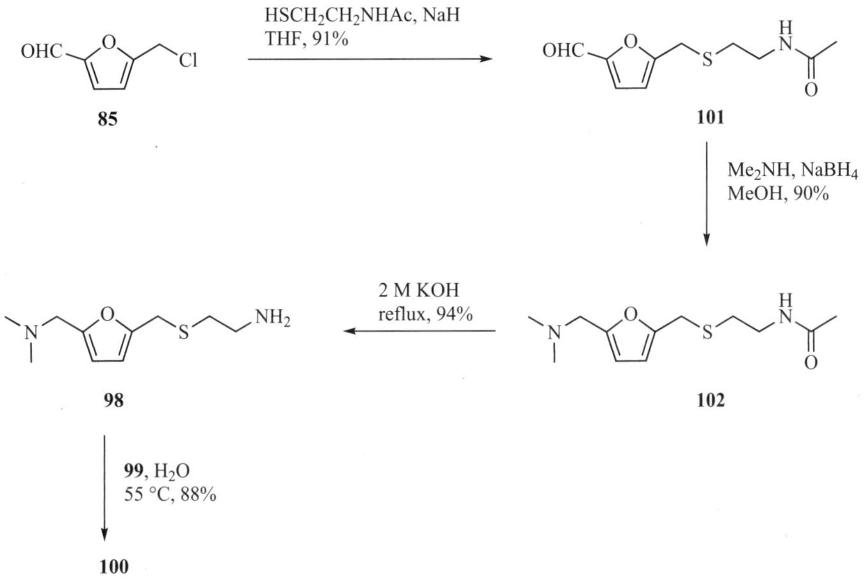

HSCH$_2$CH$_2$NHAc, NaH
THF, 91%

OHC Cl
85

OHC S N H O
101

Me$_2$NH, NaBH$_4$
MeOH, 90%

2 M KOH
reflux, 94%

N S NH$_2$
98

N S N H O
102

99, H$_2$O
55 °C, 88%

100

SCHEME 2.34 Mascal's synthesis of Ranitidine (**100**) from platform molecule 5-(chloromethyl)furfural (**85**).

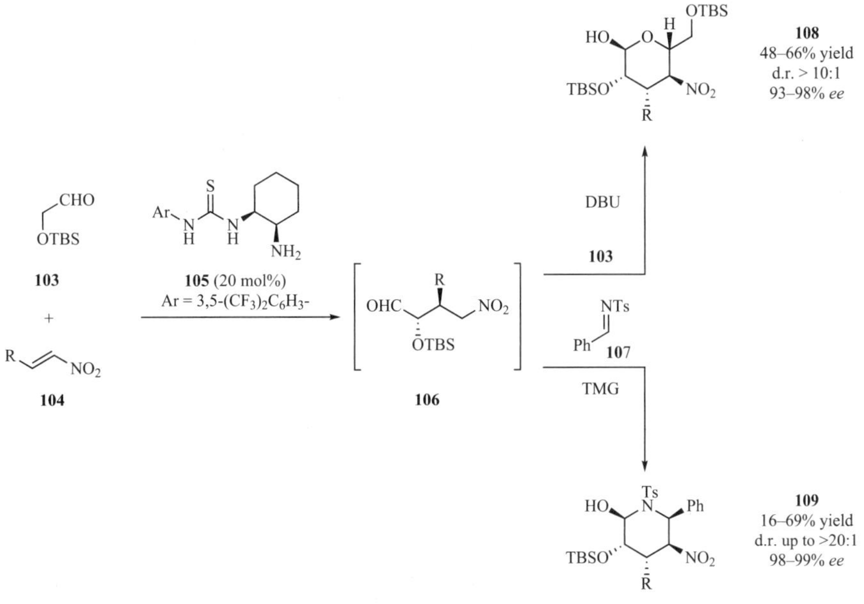

HO O H OTBS
TBSO NO$_2$
R

108
48–66% yield
d.r. > 10:1
93–98% *ee*

DBU

CHO
OTBS
103
+
R NO$_2$
104

Ar N H S N H NH$_2$
105 (20 mol%)
Ar = 3,5-(CF$_3$)$_2$C$_6$H$_3$-

OHC R NO$_2$
OTBS
106

103

NTs
Ph **107**

TMG

HO Ts N Ph
TBSO NO$_2$
R

109
16–69% yield
d.r. up to >20:1
98–99% *ee*

SCHEME 2.35 Synthesis of sugars (**108**) and iminosugars (**109**).

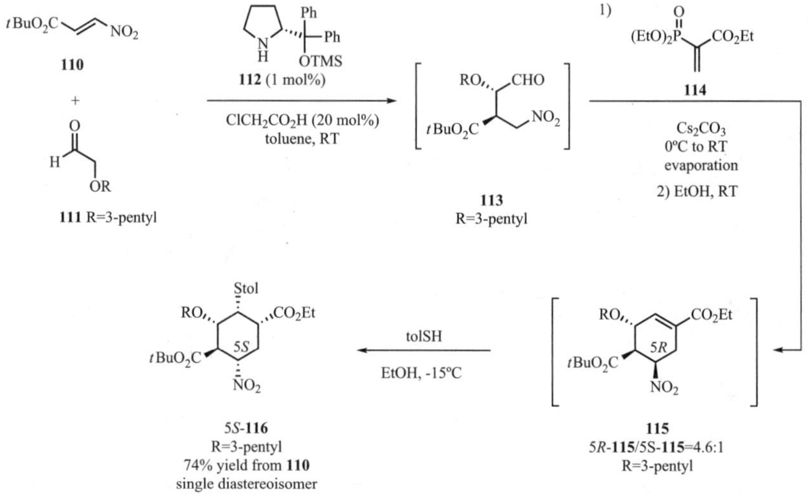

SCHEME 2.36 Hayashi's synthesis of (−)-oseltamivir (**121**). First one-pot cascade reaction, synthesis of (**116**).

2.5.2.1 Sugar and Iminosugar Synthesis Scheme 2.35 shows two methods for preparing sugars and iminosugar derivatives [126]. These procedures involve initial *anti*-selective conjugative addition of aldehyde (**103**) to nitroalkenes (**104**) in the presence of the bifunctional amino-thiourea organocatalyst (**105**). From the common intermediate (**106**), a Henry reaction gives 3,4-deoxy sugar **108**, while an aza-Henry the imino sugar **109** derivative. Of note, the two initial chiral centres direct the diasteroselectivity of the Henry/aza-Henry ring-formation sequence to enable formation of the five contiguous stereocentres with nearly perfect stereochemistry. This methodology showcases how modified sugars can be efficiently prepared from simple building blocks. As natural sugars (and their value-added synthetic derivatives) are important products from the bio-refinery, it is important for chemists to evaluate which manufacturing process is optimal for a specific compound. Which targets are best suited to each method, what suitable starting materials for organocatalysis are renewable, and can organocatalytic methods expand the range of platform molecules from a bio-refinery. The following Schemes 2.36–2.38 describe remarkably efficient synthesis using organocatalysis.

2.5.2.2 (−)-Oseltamivir Synthesis (−)-Oseltamivir phosphate (marketed as Tamiflu) is an antiviral influenza drug and many synthetic routes have been described [127]. Hayashi and coworkers recently reported a straightforward and elegant synthesis of (−)-oseltamivir (**121**) in two one-pot operations [128]. (Schemes 2.36 and 2.37). The total synthesis of (−)-oseltamivir (**121**) was performed in two reaction

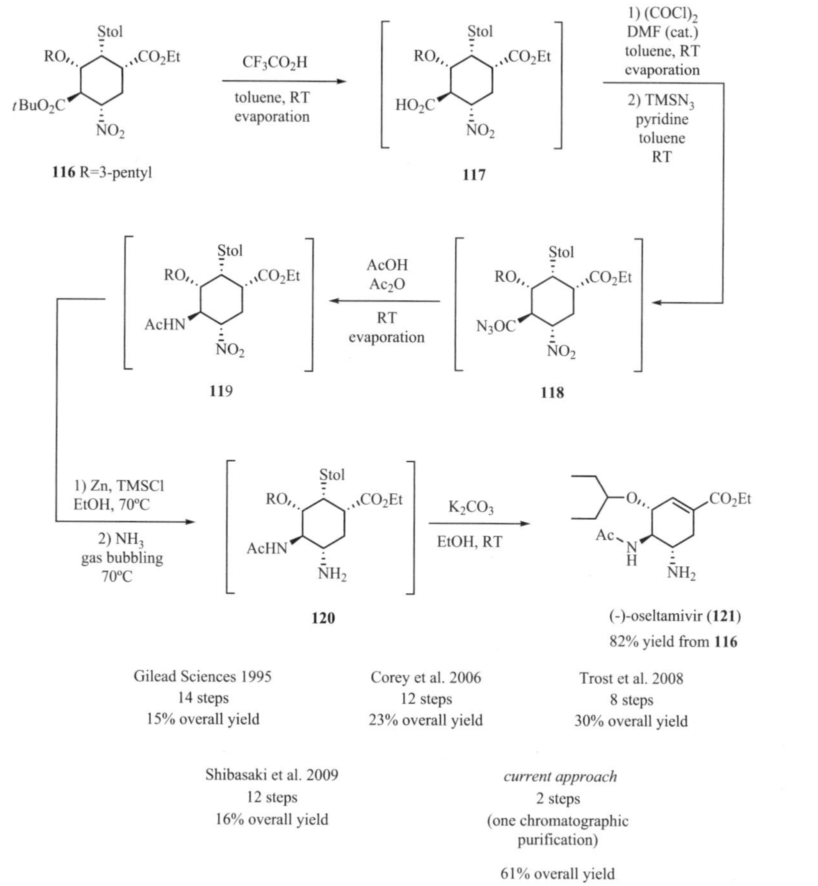

(-)-oseltamivir (**121**)
82% yield from **116**

Gilead Sciences 1995
14 steps
15% overall yield

Corey et al. 2006
12 steps
23% overall yield

Trost et al. 2008
8 steps
30% overall yield

Shibasaki et al. 2009
12 steps
16% overall yield

current approach
2 steps
(one chromatographic purification)
61% overall yield

SCHEME 2.37 Hayashi's synthesis of (−)-oseltamivir (**121**). Second one-pot cascade reaction.

vessels in excellent overall yield (61%). Summaries of selected total synthesis of **121** by classical approaches are also given in Scheme 2.37.

2.5.2.3 ABT-341 Synthesis A similar methodology was adopted by the same group for the synthesis of the structurally related example ABT-341 (**131**) [129]. Remarkably, Hayashi's one-pot cascade reaction (six consecutive reactions and nine manual operations) gave the target ABT-341 (**131**) in 63% overall yield. To attain this result, each step of the synthesis must proceed in nearly quantitative yield [125]. For comparison, Pei and coworkers reported an 11 step synthesis to enantiomerically enriched ABT-341 (**131**) in 11 steps from **123** [130]. For further details about classification, nomenclature, and critical analysis of organocatalytic

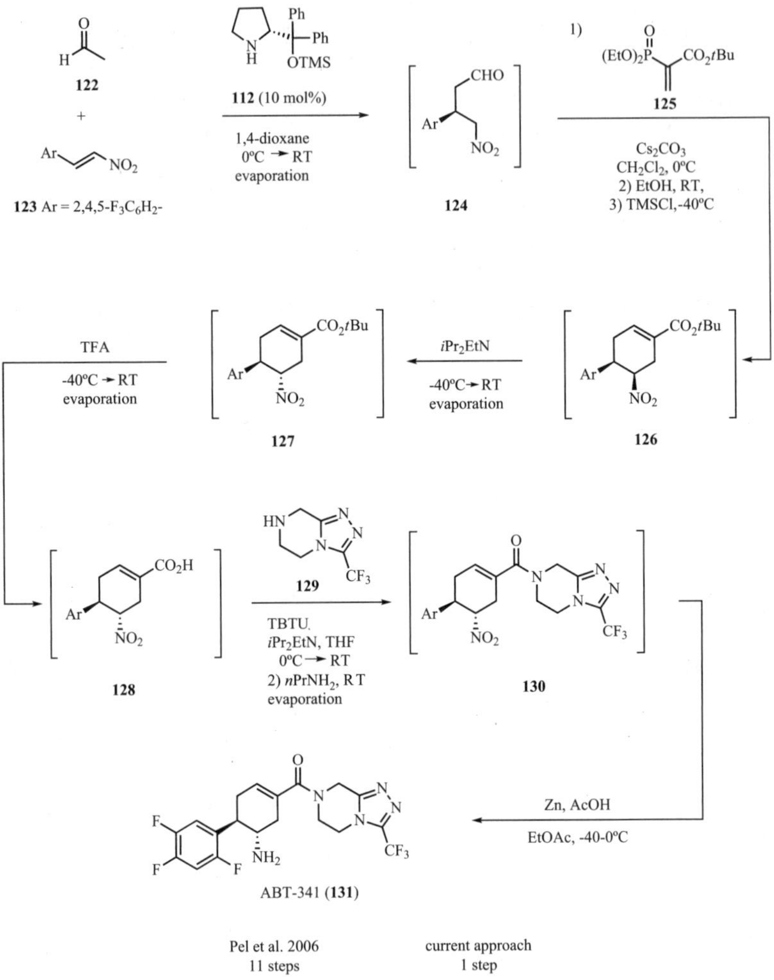

SCHEME 2.38 Hayashi's synthesis of ABT-341 (**131**) by a one-pot cascade reaction.

one-pot reactions see a seminar review by Jørgensen [125]. There is no doubt that one-pot reaction cascades offer chemists the opportunity to convert simple raw materials into complex molecules.

2.5.2.4 Getting the Message Out There Over a decade has passed since the publication of the Twelve Principles of Green Chemistry [1]. These were complemented by the Twelve Principles of Green Engineering in 2003, [131] which highlight the importance of chemical engineering in developing greener and more sustainable chemical processes rather than emphasizing basic chemistry [132]. Experience gained from presenting these formal statements in an educational

Principles of Green Engineering	Principles of Green Chemistry
I – Inherently non-hazardous and safe	**P** – Prevent wastes
M – Minimize material diversity	**R** – Renewable materials
P– Prevention instead of treatment	**O** – Omit derivatization steps
R– Renewable material and energy inputs	**D** – Degradable chemical products
O– Output-led design	**U** – Use safe synthetic methods
V– Very simple	**C** – Catalytic reagents
E – Efficient use of mass, energy, space & time	**T** – Temperature, Pressure ambient
M – Meet the need	**I** – In-Process Monitoring
E – Easy to separate by design	**V** – Very few auxiliary substances
N – Networks for exchange of local mass and energy	**E** – E-factor, maximize feed in product
	L – Low toxicity of chemical products
T – Test the life cycle of the design	**Y** – Yes it's safe
S – Sustainability throughout product life cycle	

FIGURE 2.11 Green chemistry acronyms.

setting, and indeed to the general public, has lead to a more succinct form of these principles; the acronyms IMPROVEMENTS [133] and PRODUCTIVELY [134] (Fig. 2.11). Through effective communication of advances in cleaner and greener chemistry, we can reduce waste, and improve our quality of life.

ABBREVIATIONS

Ac	Acetyl
Bmim	1-butyl-3-methylimidazolium
BINAP	2,2′-bis(diphenylphosphino)-1,1′-binaphthyl
Boc	*t*-Butyloxycarbonyl

b.r.s.m.	Based on recovered starting material
BPA	Bisphenol-A
CDMT	2-Chloro-4,6-dimethoxy-1,3,5-triazine
CFCs	Chlorofluorocarbons
CHP	Cumene hydroperoxide
Cp	Cyclopentadienyl
DBU	1,8-Diazobicyclo[5.4.0]undec-7-ene
DDQ	2,3-Dichloro-5,6-Dicyanobenzoquinone
DIAD	Diisopropyl azodicarboxylate
DMAc	*N,N*-Dimethylacetamide
DMAP	4-Dimethylaminopyridine
DMF	Dimethylformamide
DMSO	Dimethyl sulfoxide
dppb	Diphenylphosphino butane
DTP	dodecatungstophosphoric acid
EC	Ethylene carbonate
EDC	1-Ethyl-3-(3-dimethylaminopropyl)carbodiimide
EG	Ethylene glycol
FTIR	Fourier Transform Infra-Red spectroscopy
HDMS	Hexamethyldisilazane
HFIP	Hexafluoroisopropanol
IL	Ionic liquid
LCD	Liquid crystal display
MIBK	Methyl isobutyl ketone
mp	Melting point
MTO	Methyltrioxorhenium
NMM	*N*-Methylmorpholine
NMP	*N*-Methyl-2-pyrrolidinone
NTf_2^-	Trifluoromethanesulfonimide
$OctOSO_3^-$	Octyl sulfate
pmim	1-Pentyl-3-methylimidazolium
PPARα	Peroxisome proliferator-activated receptor alpha
PVA	Polyvinyl-alcohol
SBA-15	Mesoporous silica
TBS	*tert*-Butyldimethylsilyl
TBTU	*O*-(Benzotriazol-1-yl)-*N,N,N′,N′*-tetramethyluronium tetrafluoroborate
TFA	Trifluoroacetic acid
THF	Tetrahydrofuran
TMGL	1,1,3,3-tetramethylguanidinium lactate
TMS	Trimethylsilyl
Tol	Tolytl
Ts	(*p*-Toluenesulfonyl)
TsOH	*p*-Toluenesulfonic acid
UHP	Urea-hydrogen peroxide.

ACKNOWLEDGMENTS

The author thanks A. O'Driscoll for her assistance in preparing the manuscript. The author (NG) wish to thank Enterprise Ireland (EI), the Irish Research Council for Science, Engineering and Technology (IRCSET), Science Foundation Ireland (SFI) and the Environmental Protection Agency (EPA) in Ireland for funding green chemistry research in Nicholas Gathergood's group.

REFERENCES

1. P. T. Anastas and J. C. Warner, *Green Chemistry: Theory and Practice*, Oxford University Press, New York, **1998**.
2. J. Clark and F. Deswarte, *Introduction to Chemicals from Biomass*, Wiley, Chichester, UK, **2008**.
3. M. R. Naimi-Jamal, H. Hamzeali, J. Mokhtari, J. Boy and G. Kaupp, *Chem. Sus. Chem.* **2009**, 2, 83–88.
4. *CRC, Handbook of Chemistry and Physics*, 93rd ed. W. M. Haynes (Ed.) CRC Press, **2012**.
5. M. Thiemann, E. Scheibler and K. W. Wiegand, *Ullmann's Encyclopedia of Industrial Chemistry*, 5th ed., Vol. A17 (Eds. B. Elvers, S. Hawkins and G. Schulz), Wiley VCH, Weinheim, **1991**, 293–325.
6. S. L. Clark and W. J. Mazzafro, *Kirk–Othmer Encyclopedia of Chemical Technology*, 4th ed., Vol. 17 (Ed. M. Howe-Grant), Wiley, New York, **1996**, 84–96.
7. A. J. Hunt, V. L. Budarin, S. W. Breeden, A. S. Matharu and J. H. Clark, *Green Chem.* **2009**, 11, 1332–1336.
8. J. Drobnik, *Adv. Drug Del. Rev.* **1989**, 3, 229–245.
9. L. Wu, X. Yuan and J. Sheng, *J. Membr. Sci.* **2005**, 250, 167–173.
10. C. R. Nutterman, S. M. Henry and K. S. Anseth, *Biomaterials* **2002**, 23, 3617–3626.
11. K. B. Hansen, J. Balsells, S. Dreher, Y. Hsiao, M. Kubryk, M. Palucki, N. Rivera, D. Steinhuebel, J. D. Armstrong III, D. Askin and E. J. J. Grabowski, *Org. Process Rev. Dev.* **2005**, 9, 634–639.
12. K. B. Hansen, Y. Hsiao, F. Xu, N. Rivera, A. Clausen, M. Kubryk, S. Krska, T. Rosner, B. Simmons, J. Balsells, N. Ikemoto, Y. Sun, F. Spindler, C. Malan, E. J. J. Grabowski and J. D. Armstrong III, *J. Am. Chem. Soc.* **2009**, 131, 8798–8804.
13. B. M. Trost, *Science* **1991**, 254, 1471–1477.
14. B. M. Trost, N. Maulide and M. T. Rudd, *J. Am. Chem. Soc.* **2009**, 131, 420–421.
15. G. Guillena, C. Nájera and D. J. Ramón, *Tetrahedron-Asymmetr.* **2007**, 18, 2249–2293.
16. H. Tsuji, K. Yamagata, T. Fujimoto and E. Nakamura, *J. Am. Chem. Soc.* **2008**, 130, 7792–7793.
17. L. Phan, H. Brown, J. White, A. Hodgson and P. G. Jessop, *Green Chem.* **2009**, 11, 53–59.
18. J. Roger, C. Verrier, R. Le Goff, C. Hoarau and H. Doucet, *ChemSusChem.* **2009**, 2, 951–956.

19. P. Tundo and M. Selva, *Acc. Chem. Res.* **2002**, 35, 706–716.
20. T. Sakakura and K. Kohno, *Chem. Commun.* **2009**, 11, 1312–1330.
21. D. Saha, A. Saha and B. C. Ranu, *Green Chem.* **2009**, 11, 733–737.
22. S. Morrissey, I. Beadham and N. Gathergood, *Green Chem.* **2009**, 11, 466–474.
23. (a) Selected reviews: T. Welton, *Chem. Rev.* **1999**, 99, 2071–2084; (b) P. Wasserscheid and W. Keim, *Angew. Chem. Int. Ed.* **2000**, 39, 3772–3789; (c) R. Sheldon, *Chem. Commun.* **2001**, 23, 2399–2407; (d) J. S. Wilkes, *Green Chem.* **2002**, 4, 73–80; (e) J. H. Davis and P. Fox, *Chem. Commun.* **2003**, 11, 1209–1212; (f) R. A. Sheldon, *Green Chem.* **2005**, 7, 267–278; (g) A. Riisager, R. Fehrmann, M. Haumann and P. Wasserscheid, *Eur. J. Inorg. Chem.* **2006**, 4, 695–706; (h) C. Hardacre, J. D. Holbrey, M. Nieuwenhuyzen and T. G. A. Youngs, *Acc. Chem. Res.* **2007**, 40, 1146–1155; (i) A. A. H. Padua, M. F. Costa Gomes and J. N. A. Canongia Lopes, *Acc. Chem. Res.* **2007**, 40, 1087–1096; (j) X. Han and D. W. Armstrong, *Acc. Chem. Res.* **2007**, 40, 1079–1086; (k) J. Ranke, S. Stolte, R. Störmann, J. Arning and B. Jastorff, *Chem. Rev.* **2007**, 107, 2183–2206; (l) F. van Rantwijk and R. A. Sheldon, *Chem. Rev.* **2007**, 107, 2757–2785; (m) M. Smiglak, A. Metlen and R. D. Rogers, *Acc. Chem. Res.* **2007**, 40, 1182–1192; (n) V. I. Pârvulescu and C. Hardacre, *Chem. Rev.* **2007**, 107, 2615–2665; (o) T. L. Greaves and C. J. Drummond, *Chem. Rev.* **2008**, 108, 206–237; (p) M. A. P. Martins, C. P. Frizzo, D. N. Moreira, N. Zanatta and H. G. Bonacorso, *Chem. Rev.* **2008**, 108, 2015–2050; (q) Y. Gua and G. Li, *Adv. Synth. Catal.* **2009**, 351, 817–847; (r) N. V. Plechkova and K. R. Seddon, *Chem. Soc. Rev.* **2008**, 37, 123–150; (s) H. Xie, T. Hayes and N. Gathergood, Catalysis of reactions by amino acids, *Amino Acids, Peptides and Proteins in Organic Chemistry* (Ed. Andrew B. Hughes), Volume 1 – Origins and Synthesis of Amino Acids, Wiley-VCH, Weinheim, Germany, **2009**.
24. A. C. Cole, J. L. Jensen, I. Ntai, K. L. T. Tran, K. J. Weaver, D. C. Forbes and J. H. Davis, *J. Am. Chem. Soc.* **2002**, 124, 5962–5963.
25. (a) For representative recent examples of strategy A see: D-Q Xu, J. Wu, S-P. Luo, J-X. Zhang, J-Y. Wu, X-H. Du and Z-Y. Xu, *Green Chem.* **2009**, 11, 1239–1246; (b) L Yang, L-W Xu and C-G Xia, *Synthesis* **2009**, 29, 1969–1974; (c) J. Akbari and A. Heydari, *Tetrahedron Lett.* **2009**: 50, 4236–4238; (d) Y. W. Zhao, J. X. Long, F. G. Deng, X. F. Liu, Z. Li, C. G. Xia and J. J. Peng, *Catal. Commun.* **2009**, 10, 732–736.
26. (a) For representative recent examples of Figure 2–3, strategy B see: N. B. Darvatkar, A. R. Deorukhkar, S. V. Bhilare and M. M. Salunkhe, *Synth. Commun.* **2006**, 36, 3043–3051; (b) A. R. Hajipour, L. Khazdooz and A. E. Ruoho, *Catal. Commun.* **2008**, 9, 89–96.
27. (a) For representative recent examples of Figure 2–3, strategy C see: A. Arfan and J. P. Bazureau, *Org. Process Res. Dev.* **2005**, 9, 743–748; (b) D-Q. Xu, W-L. Yang, S-P. Luo, B-T. Wang, J. Wu and Z.-Y. Xu, *Eur. J. Org. Chem.* **2007**, 6, 1007–1012; (c) Z. Duan, Y. Gu and Y. Deng, *Synth. Commun.* **2005**, 35, 1939–1945.
28. For an example of a catalyst which embodies both Figure 2–3, strategies **A** and **C** see: Y. Wang, D. Jiang and L. Dai, *Catal. Commun.* **2008**, 9, 2475–2480.
29. For an example of the use of an ionic liquid with a basic anion in catalysis see: B. C. Ranu and S. Banerjee, *Org. Lett.* **2005**, 7, 3049–3052.
30. (a) OECD 310 (**2006**) Guidelines for Testing of Chemicals. Ready Biodegradability — CO_2 in sealed vessels (Headspace Test and Closed Bottle Test) (b) ISO 14593: Water quality, Evaluation of ultimate aerobic biodegradability of organic compounds in aqueous medium. Method by analysis of inorganic carbon in sealed vessels CO_2, Headspace test, 1999.

31. N. Gathergood, R. Gore, L. Myles and S. Connon, submitted.
32. S. Connon, L. Myles, R. Gore and N. Gathergood, submitted.
33. A. J. Mancuso, S.-L. Huang and D. Swern, *J. Org. Chem.* **1978**, 43, 2480–2482.
34. J. M. Harris, Y. Liu, S. Chai, M. D. Andrews and J. C. Vederas, *J. Org. Chem.* **1998**, 63, 2407–2409.
35. J.-L. Wang, L.-N. He, C.-X. Miao and Y.-N Li, *Green Chem.* **2009**, 11, 1317–1320.
36. R. L. Oliveira, P. K. Kiyohara and L. M. Rossi, *Green Chem.* **2009**, 11, 1366–1370.
37. J. Otera, *Esterification*, Wiley-VCH, Weinheim, Germany, **2003**.
38. S. Narayan, J. Muldoon, M. G. Finn, V. V. Fokin, H. C. Kolb and K. B. Sharpless, *Angew. Chem. Int. Ed.* **2005**, 44, 3275–3279.
39. R. Breslow, *Acc. Chem. Res.* **1991**, 24, 159–164.
40. J. J. Gajewski, *Acc. Chem. Res.* **1997**, 30, 219–225.
41. N. Rieber, J. Alberts, J. A. Lipsky and D. M. Lemal, *J. Am. Chem. Soc.* **1969**, 91, 5668–5669.
42. Y. Leblanc R. Zamboni and M. A. Bernstein, *J. Org. Chem.* **1991**, 56, 1971–1972.
43. N. Shapiro and A. Vigalok, *Angew. Chem. Int. Ed.* **2008**, 47, 2849–2852.
44. S. Benyahya, F. Monnier, M. Wong Chi Man, C. Bied, F. Ouazzani and M. Taillefer, *Green Chem.* **2009**, 11, 1121–1123.
45. S. Benyahya, F. Monnier, M. Taillefer, M. Wong Chi Man, C. Bied and F. Quazzani, *Adv. Synth. Catal.* **2008**, 350, 2205–2208.
46. http://www.emea.europa.eu/pdfs/vet/vich/050299en.pdf (18/02/12).
47. G. A. Kraus and M. Kirihara, *J. Org. Chem.* **1992**, 57, 3256–3257.
48. D. Zhao Y. Liao and Z. Zhang, *Clean* **2007**, 35, 42–48.
49. B. Murphy, P. Goodrich, C. Hardacre and M. Oelgemoller, *Green Chem.* **2009**, 11, 1867–1870.
50. Y. Kita, K. Morimoto, M. Ito, C. Ogawa, A. Goto and T. Dohi, *J. Am. Chem. Soc.* **2009**, 131, 1668–1669.
51. M. Egi, K. Azechi and S. Akai, *Org. Lett.* **2009**, 11, 5002–5005.
52. G. Minetto, L. F. Raveglia and M. Taddei, *Org. Lett.* **2004**, 6, 389–392.
53. H. Yue, Y. Zhao, X. Ma and J. Gong, *Chem. Soc. Rev.* **2012**, 41, 4218-4244.
54. A. Fukuoka and P. L. Dhepe, *Angew. Chem. Int. Ed.* **2006**, 45, 5161–5163.
55. N. Ji, T. Zhang, M. Zheng, A. Wang, H. Wang, X. Wang and J. G. Chen, *Angew. Chem.* **2008**, 120, 8638–8641.
56. R. Malacea, C. Fischmeister, C. Bruneau, J.-L. Dubois, J.-L. Couturier and P. H. Dixneuf *Green Chem.* **2009**, 11, 152–155.
57. R. Datta and M. Henry, *J. Chem. Technol. Biotechnol.* **2006**, 81, 1119–1129.
58. N. Dimitratos, J. A. Lopez-Sanchez, S. Meenakshisundaram, J. M. Anthonykutty, G. Brett, A. F. Carley, S. H. Taylor, D. W. Knight and G. J. Hutchings, *Green Chem.* **2009**, 11, 1209–1216.
59. L. Prati and M. Rossi, *J. Catal.* **1998**, 176, 552–560.
60. K. Stratmann, R. E. Moore, R. Bonjouklian, J. B. Deeter, G. M. L. Patterson, S. Shaffer, C. D. Smith and T. A. Smitka, *J. Am. Chem. Soc.* **1994**, 116, 9935–9942.
61. T. A. Smitka, R. Bonjouklian, L. Doolin, N. D. Jones and J. B. Deeter, *J. Org. Chem.* **1992**, 57, 857–861.

62. R. E. Moore, C. Cheuk and G. M. L. Patterson, *J. Am. Chem. Soc.* **1984**, 106, 6456–6457.
63. R. E. Moore, C. Cheuk, X. Q. G. Yang, G. M. L. Patterson, R. Bonjouklian, T. A. Smitka, J. S. Mynderse, R. S. Foster, N. D. Jones, J. K. Swartzendruber and J. B. Deeter, *J. Org. Chem.* **1987**, 52, 1036–1043.
64. A. Raveh and S. Carmeli, *J. Nat. Prod.* **2007**, 70, 196–201.
65. P. S. Baran, T. J. Maimone and J. M. Richter, *Nature*. **2007**, 446, 404–408.
66. S. E. Reisman, J. M. Ready, A. Hasuoka, C. J. Smith and J. L. Wood, *J. Am. Chem. Soc.* **2006**, 128, 1448–1449.
67. E. M. Dangerfield, M. S. M. Timmer and B. L. Stocker, *Org. Lett.* **2009**, 11, 535–538.
68. N. Asano, R. J. Nash, R. J. Molyneux and G. W. Fleet, *Tetrahedron-Asymmetr*. **2000**, 11, 1645–1680.
69. X. Zhang and W. H. Flurkey, *J. Food Sci.* **1997**, 62, 97–100.
70. H. Leutbecher, S. Hajdok, C. Braunberger, M. Neumann, S. Mika, J. Conrad and U. Beifuss, *Green Chem.* **2009**, 11, 676–679.
71. M. North, F. Pizzato and P. Villuendas, *ChemSusChem*. **2009**, 2, 862–865.
72. B. Schaffner, V. Andrushko, J. Holz, S. P. Verevkin and A. Borner, *ChemSusChem*. **2008**, 1, 934–940.
73. X.-B., Fan, N. Yan, Z.-Y. Tao, D. Evans, C.-X. Xiao and Y. Kou, *ChemSusChem*. **2009**, 2, 941–943.
74. M. Auffan, J. Rose, M. R. Wiesner and J-Y. Bottero, *Environ. Pollut.* **2009**, 157, 1127–1133.
75. X. Ma, Y. Zhou, J. Zhang, A. Zhu, T. Jiang and B. Han, *Green Chem.* **2008**, 10, 59–66.
76. P. Gallezot and D. Richard, *Catal. Rev. Sci. Eng.* **1998**, 40, 81–126.
77. H. Miura, *Shokubai* **2007**, 49, 232.
78. X. Chen, L. Hexing, D. Weilin, W. Jie, R. Yong and Q. Minghua, *Appl. Catal. A-Gen.* **2003**, 253, 359–369.
79. P. Maki-Arvela, J. Haje, T. Salmi and D. Murzin, *Appl. Catal. A-Gen.* **2005**, 292, 1–49.
80. U. Singh and M. Vannice, *Appl. Catal. A-Gen.* **2001**, 213, 1–24.
81. (a) A. Muller, J. Bowers, J. Eubanks, C. Geiger and J. Santobianco, **1999,** WO/1999/008989; (b) F. Bennett, A. Ganguly, V. Girijavallabhan and N. Patel, **1993**, EP0533342;
82. P. Jadhav and H.-W. Man, *Tetrahedron Lett.* **1996**, 37, 1153–1156.
83. P. Kluson and L. Cerveny, *Appl. Catal. A-Gen.* **1995**, 128, 13–31.
84. I. Kostas, *J. Organomet. Chem.* **2001**, 634, 90–98.
85. K. Nuithitikul and M. Winterbottom, *Catal. Today* **2007**, 128, 74–79.
86. K. Nuithitikul and M. Winterbottom, *Chem. Eng. Sci.* **2006**, 61, 5944–5953.
87. K. Nuithitikul and M. Winterbottom, *Chem. Eng. Sci.* **2004**, 59, 5439–5447.
88. J. Hajek, N. Kumar, P. Maki-Arvela T. Salmi and D. Murzin, *J. Mol. Catal. A-Chem.* **2004**, 217, 145–154.
89. J. Qui, H. Zhang, X. Wang, H. Han, C. Liang and C. Li, *React. Kinet. Catal. Lett.* **2006**, 88, 269–275.

90. J. P. Tessonnier, L. Pesant, G. Ehret, M. Ledoux and C. Pham-Huu *Appl. Catal. A Gen.* **2005**, 288, 203–210.
91. C. Pham-Huu, N. Keller, L. Charbonniere, R. Ziessel and M. Ledoux, *Chem. Commun.* **2000**, 19, 1871–1872.
92. M. Bhor, A. Panda, S. Jagtap and B. Bhanage, *Catal. Lett.* **2008**, 124, 157–164.
93. T. Mallat and A. Baiker, *Appl. Catal. A-Gen.* **2000**, 200, 3–22.
94. S. Morrissey, B. Pegot, D. Coleman, M. T. Garcia, D. Ferguson, B. Quilty and N. Gathergood, *Green Chem.* **2009**, 11, 475–483.
95. R. D. Rogers and K. R. Seddon, *Ionic Liquids; Industrial applications to Green Chemistry*, American Chemical Society, Washington, DC, **2002**.
96. P. Wasserscheid and T. Welton (Eds.), *Ionic liquids in synthesis*, Wiley-VCH, Weinheim, **2003**.
97. P. Wasserscheid, R. Van Hal and A. Bösmann, *Green Chem.* **2002**, 4, 400–404.
98. J. Pernak, I. Goc and I. Mirska, *Green Chem.* **2004**, 6, 323–329.
99. M. Rebros, H. Q. Nimal Gunaratne, J. Ferguson, K. R. Seddon and G. Stephens, *Green Chem.* **2009**, 11, 402–408.
100. D. Zhao, Y. Liao and Z. Zhang, *Clean: Soil, Air, Water* **2007**, 35, 42–48.
101. N. Gathergood and P. J. Scammells, *Aust. J. Chem.* **2002**, 55, 557–560.
102. N. Gathergood, M. T. Garcia and P. J. Scammells, *Green Chem.* **2004**, 6, 166–175.
103. N. Gathergood, M. T. Garcia and P. J. Scammells, *Green Chem.* **2006**, 8, 156–160.
104. S. Stolte, S. Abdulkarim, J. Arning, A. Blomeyer-Nienstedt, U. Bottin-Weber, M. Matzke, J. Ranke, B. Jastorff and J. Thoeming, *Green Chem.* **2008**, 10, 214–224.
105. D. Coleman and N. Gathergood, *Chem. Soc. Rev.* **2010**, 39, 600–637.
106. S. Stolte, S. Abdulkarim, J. Arning, A-K. Blomeyer-Nienstedt, U. Bottin-Weber, M. Matzke, J. Ranke, B. Jastorff and J. Thöming, *Green Chem.* **2008**, 10, 214–224.
107. J. R. Harjani, R. D. Singer, M. T. Garcia and P. J. Scammells, *Green Chem.* **2009**, 11, 83–90.
108. L. Ford, J. R. Harjani, F. Atefi, M. T. Garcia, R. D. Singer and P. J. Scammells, *Green Chem.* **2010**, 12, 1783–1789.
109. M. Stasiewicz, E. Mulkiewicz, R. Tomczak-Wandzel, J. Kumirska, E. M. Siedlecka, M. Gołebiowski, J. Gajdus, M. Czerwicka and P. Stepnowski, *Ecotox. Environ. Saf.* **2008**, 71, 157–165.
110. N. Bodor, J. J. Kaminski and S. Selk, *J. Med. Chem.* **1980**, 23, 469–474.
111. J. Pernak, A. Syguda, I. Mirska, A. Pernak, J. Nawrot, A. Pradzyńska, S. T. Griffin and R. D. Rogers, *Chem. Eur. J.* **2007**, 13, 6817–6827.
112. M. Petkovic, J. L. Ferguson, H. Q. N. Gunaratne, R. Ferreira, M. C. Leitão, K. R. Seddon, L. P. N. Rebelo and C. S. Pereira, *Green Chem.* **2010**, 12, 643–649.
113. H. Garcia, R. Ferreira, M. Petkovic, J. L. Ferguson, M. C. Leitão, H. Q. N. Gunaratne, K. R. Seddon, L. P. N. Rebelo and C. S. Pereira, *Green Chem.* **2010**, 12, 367–369.
114. http://ie.mt.com/eur/en/home/products/L1_AutochemProducts/Level_2_ALR_Calorimeters_RC1_LabMax/chemical-syn/synthesis-a/ReactIR-15.html (18/02/12).
115. M. D. Argentine, T. M. Braden, J. Czarnik, E. W. Conder, S. E. Dunlap, J. W. Fennell, M. A. LaPack, R. R. Rothhaar, R. B. Scherer, C. R. Schmid, J. T. Vicenzi, J. G. Wei and J. A. Werner, *Org. Process Res. Dev.* **2009**, 13, 131–143.

116. F. Cardona, M. Bonanni, G. Soldaini and A. Goti, *ChemSusChem.* **2008**, 1, 327–332.

117. H. Mitsui, S.-I. Zenki, T. Shiota and S.-I. Murahashi, *J. Chem. Soc. Chem. Commun.* **1984**, 874–875.

118. AA *Ullmann's Encyclopedia of Industrial Chemistry*, 6th ed., Wiley-VCH Verlag GmbH, Weinheim, Germany, **2002**.

119. G. D. Yadav and S. S. Salgaonkar, *Org. Process Res. Dev.* **2009**, 13, 501–509.

120. M. Mascal and E. B. Nikitin, *Green Chem.* **2010**, 12, 370–373.

121. M. Mascal and E. B. Nikitin, *ChemSusChem.* **2009**, 2, 859–861.

122. M. Mascal and S. Dutta, *Green Chem.* **2011**, 13, 3101–3102.

123. B. J. Price, J. W. Clitherow and J. Bradshaw, US Patent 4128658, **1978**.

124. (a) For reviews on organocatalysis, see, for example: M. J. Gaunt, C. C. C. Johansson, A. McNally and N. T. Vo, *Drug Discov. Today* **2007**, 12, 8–27; (b) *Chem. Rev.* **2007**, 107, 5413–5883 (special issue on organocatalysis); (c) A. Dondoni and A. Massi, *Angew. Chem.* **2008**, 120, 4716–4716; *Angew. Chem. Int. Ed.* **2008**, 47, 4638–4660; (d) D. W. C. MacMillan, *Nature* **2008**, 455, 304–308; (e) S. Bertelsen and K. A. Jørgensen, *Chem. Soc. Rev.* **2009**, 38, 2178–2189; (f) M. Nielsen, D. Worgull, T. Zweifel, B. Gschwend, S. Bertelsen and K. A. Jørgensen, *Chem. Commun.* **2011**, 47, 632–649; (g) A. Berkessel and H. Grcger, *Asymmetric Organocatalysis*, Wiley-VCH, Weinheim, **2004**; (h) *Enantioselective Organocatalysis* (Ed. P. I. Dalko), Wiley-VCH, Weinheim, **2007**; (i) P. Melchiorre, M. Marigo, A. Carlone and G. Bartoli, *Angew. Chem.* **2008**, 120, 6232–6265; *Angew. Chem. Int. Ed.* **2008**, 47, 6138–6171; (j) B. List and J-W. Yang, *Science* **2006**, 313, 1584–1586; (k) M. S. Taylor and E. N. Jacobsen, *Angew. Chem.* **2006**, 118, 1550–1573; *Angew. Chem. Int. Ed.* **2006**, 45, 1520–1543; (l) *Asymmetric Phase Transfer Catalysis* (Ed. K. Maruoka), Wiley-VCH, Weinheim, **2008**; (m) T. Akiyama, *Chem. Rev.* **2007**, 107, 5744–5758.

125. L. Albrecht, H. Jiang and K. A. Jørgensen, *Angew. Chem. Int. Ed.* **2011**, 50, 8492–8509.

126. (a) H. Uehara, R. Imashiro, G. Hernµndez-Torres and C. F. Barba-s III, *Proc. Natl. Acad. Sci. USA* **2010**, 107, 20672–20677; (b) R. Imashiro, H. Uehara and C. F. Barbas III, *Org. Lett.* **2010**, 12, 5250–5253; for related studies, see: (c) T. Urushima, D. Sakamoto, H. Ishikawa and Y. Hayashi, *Org. Lett.* **2010**, 12, 4588–4591; (d) H. Ishikawa, S. Sawano, Y. Yasui, Y. Shibata and Y. Hayashi, *Angew. Chem.* **2011**, 123, 3858–3863;*Angew. Chem. Int. Ed.* **2011**, 50, 3774–3779.

127. (a) M. Shibasaki and M. Kanai, *Eur. J. Org. Chem.* **2008**, 11, 1839–1850; (b) J. Magano, *Chem. Rev.* **2009**, 109, 4398–4438; for selected examples, see: (c) C. U. Kim, W. Lew, M. A. Williams, H. Liu, L. Zhang, S. Swaminathan, N. Bischofberger, M. S. Chen, D. B. Mendel, C. Y. Tai, W. G. Laver and R. C. Stevens, *J. Am. Chem. Soc.* **1997**, 119, 681–690; (d) Y-Y. Yeung, S. Hong and E. J. Corey, *J. Am. Chem. Soc.* **2006**, 128, 6310–6311; (e) B. M. Trost and T. Zhang *Angew. Chem.* **2008**, 120, 3819–3821; *Angew. Chem. Int. Ed.* **2008**, 47, 3759–3761; (f) K. Yamatsugu, L. Yin, S. Kamijo, Y. Kimura, M. Kanai and M. Shibasaki, *Angew. Chem.* **2009**, 121, 1090–1096; *Angew. Chem. Int. Ed.* **2009**, 48, 1070–1076.

128. (a) H. Ishikawa, T. Suzuki, H. Orita, T. Uchimaru and Y. Hayashi, *Chem. Eur. J.* **2010**, 16, 12616–12626; (b) H. Ishikawa, T. Suzuki and Y. Hayashi, *Angew. Chem.* **2009**, 121, 1330–1333; *Angew. Chem. Int. Ed.* **2009**, 48, 1304–1307.

129. H. Ishikawa, M. Honma and Y. Hayashi, *Angew. Chem.* **2011**, 123, 2876–2879; *Angew. Chem. Int. Ed.* **2011**, 50, 2824–2827.

130. Z. Pei, X. Li, T. W. von Geldern, D. J. Madar, K. Longenecker, H. Yong, T. H. Lubben, K. D. Stewart, B. A. Zinker, B. J. Backes, A. S. Judd, M. Mulhern, S. J. Ballaron, M. A. Stashko, A. M. Mika, D. W. A. Beno, G. A. Reinhart, R. M. Fryer, L. C. Preusser, A. J. Kempf- Grote, H. L. Sham and J. M. Trevillyan, *J. Med. Chem.* **2006**, 49, 6439–6442.

131. P. T. Anastas and J. B. Zimmerman, *Environ. Sci. Technol.* **2003**, 37, 94A–101A.

132. N. Asfaw, Y. Chebude, A. Ejigu, B. B. Hurisso, P. Licence, R. L. Smith, S. L. Y. Tang and M. Poliakoff, *Green Chem.* **2011**, 13, 1059–1060.

133. S. L. Y. Tang, R. A. Bourne, M. Poliakoff and R. L. Smith, *Green Chem.* **2008**, 10, 268–269.

134. S. L. Y. Tang, R. L. Smith and M. Poliakoff, *Green Chem.* **2005**, 7, 761–762.

CHAPTER 3

Biorefinery with Ionic Liquids

HAIBO XIE, WUJUN LIU, IAN BEADHAM, and NICHOLAS GATHERGOOD

3.1 INTRODUCTION

The energy needs of our society, including the chemical and material industries, are currently over-dependent on the utilization of fossil resources. With diminishing fossil resources, it is pivotal to explore alternatives to supply chemicals and energy sustainably. Regarding energy and material chemistry, the intrinsic problem is the chemistry of carbon. Representing the major carbon sources on the planet, biomass has an estimated global production of around 1.0×10^{11} tons per year. Recently, biomass has attracted considerable attention as an alternative source for both fuels and chemicals production. It is estimated that by 2025, up to 30% of raw materials for the chemical industry will be produced from renewable sources [1]. Associated with the high demand of biomass utilization, a new definition of "biorefinery" has been proposed in parallel with the conventional petroleum refineries processing, which produce fuels and a wide range of chemicals from oil and gas. A biorefinery is a facility that integrates biomass-conversion processes and equipment to produce fuels, power, and chemicals from biomass [2]. Industrial biorefineries have been identified as the most promising route to the creation of a new domestic biobased industry. For an efficient biorefinery, innovative processing technologies, separation and depolymerization processes, as well as catalytic conversion systems are requisite [3]. Ionic liquids (ILs) [4] based technology is one of the promising emerging biorefinery technologies, which have paved the way towards an environment-friendly and homogenous methodology to use biomass [5]. Great effort has been devoted to using ionic liquids as solvents and/or catalysts for the conversion of biomass into materials, chemicals, and biofuels. This involves sustainable materials preparation, catalytic conversion of ligno(cellulose) into their monomeric constituents and/or valuable platform chemicals and biodiesel productions [6]. The primary focus of this chapter is to provide an overview on the opportunities of the ILs-based

The Role of Green Chemistry in Biomass Processing and Conversion, First Edition.
Edited by Haibo Xie and Nicholas Gathergood.

biorefinery from a chemical point of view, and to highlight the key issues and challenges encountered to a sustainable biorefinery with ILs, and current progress addressing these barriers.

3.2 IONIC LIQUIDS AND THEIR GREENNESS LEADING TO A SUSTAINABLE BIOREFINERY

Over the past two decades, the application of ILs in catalysis, separation and materials has experienced a tremendous growth, which was stimulated by the pursuit of sustainable and greener technologies in chemical industry due to the increasing concerns of environmental protection [7]. Generally, ILs are a group of salts that exist as liquids at relatively low temperatures (<100°C) and they usually consist of an organic cation and an organic or inorganic anion [8]. Their typical structures are depicted in Figure 3.1.

Due to their ionic nature, they have extremely low vapor pressure, which is regarded as one of the most significant advantages compared with traditional organic solvents. For instance, the use of ILs in catalysis can reduce the release of volatile organic compounds. In addition to this inherent "green" feature, ILs are often referred to as "designable solvents," mainly because their physical properties, such as melting point, viscosity, density, solubility, and acidity/coordination properties, can be tuned for different reactions or processes by altering the structure and combination of their cations and anions. It is estimated that 10^{18} different ILs are theoretically possible, which present a great opportunity to design or optimize the most suitable system for specific applications. Although huge amounts of potential ILs have been proposed, not all ILs are capable of dissolving cellulose or biomass. Only a limited number of ionic liquids have been studied for dissolution of cellulose. Some representative structures of ILs are shown in Figure 3.2.

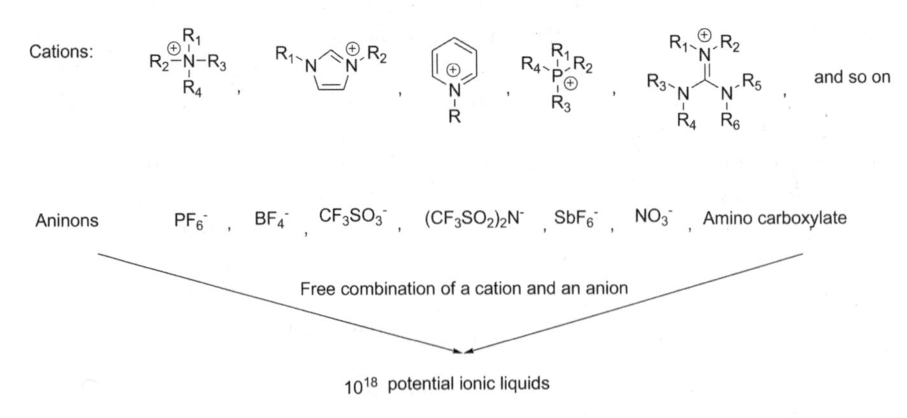

FIGURE 3.1 Free combination of a cation and an anion results in ~10^{18} potential ILs [9].

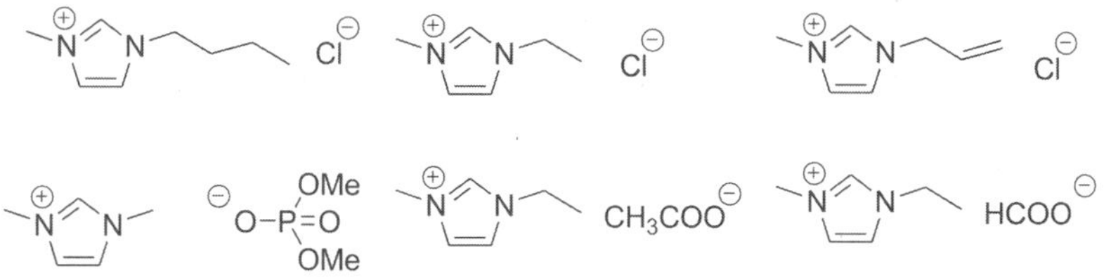

FIGURE 3.2 Representative ILs for biomass dissolution and processing.

In addition to the specific advantages of ILs for catalysis and separation technologies, the extra green benefits or sustainability of ILs for biomass processing and conversion are:

- Good solubility to biomass

The insolubility of biomass in single traditional organic solvent is the entry-point challenge of biomass utilization, and the dissolution of biomass in ILs leads to the transformation of traditional heterogeneous conversion of biomass into homogeneous conversion, which will result in new sciences of biomass utilization.

- The use of biomass

Produced by nature's photosynthesis, biomass is regarded as the most abundant renewable source of carbon and hydrogen. The efficient utilization of biomass is therefore regarded as a major step towards a foundation of sustainable chemical industry.

Recent comprehensive reviews have addressed the basic science, properties, and potential applications of ILs [7, 10]. Considering the main issues of conversion of biomass into valuable chemical feedstocks and sustainable materials, the introduction of ILs into biorefinery will offer various opportunities, greenness and advantages for the foundation of sustainable biorefinery [11].

3.3 IONIC LIQUIDS FOR BIOMASS PROCESSING AND CONVERSION

3.3.1 Mechanism of Dissolving Biopolymers by Ionic Liquids

Despite a broad spectrum of ILs which have been reported capable of dissolving biomass, the underlying mechanism of the dissolving process remains poorly understood. A few theoretical and experimental studies were applied to investigate this topic, such as NMR analysis, molecular dynamic studies and XRD analysis. The first study through ^{13}C and $^{35/37}Cl$ NMR relaxation measurements implied that the solvation of cellulose by the IL, 1-*n*-butyl-3-methylimidazolium chloride [Bmim]Cl involved hydrogen-bonding between the carbohydrate hydroxyl proton and the IL's chloride ion in a 1:1 stoichiometry, and the interaction was further supported by multinuclear NMR-spectroscopy experiments [11b]. The role of cation of ILs has been elucidated by NMR study and molecular dynamic simulations study. Although the analysis of the ^{13}C relaxation rate of the imidazolium carbons showed there is no strong correlation with sugar concentration regardless of the structure of the cation or the cellulose, indicating that their role is not essential to the solvation process [12]. The study of molecular dynamics simulations, however, showed that weak hydrogen bonding interactions with ILs cation occurred through the acidic hydrogen at the C(2) position on the imidazolium ring [13]. The study of molecular dynamics simulations with an all-atom-force-field in 1-ethyl-3-methylimidazolium acetate ([Emim][OAc]) [12a, 14] showed that the interaction energy between the polysaccharide chain and the ILs was stronger (three times) than water or methanol. With the interaction between the ILs and carbohydrates, the preferred β-(1,4)-glycosidic linkage conformation of the cellulose was altered when dissolved after the formation of solution compared with that found in crystalline cellulose dispersed in water. The conformation changes and conversion of cellulose fibers from the a black cellulose I to cellulose II crystal phase are not reversible during the regeneration process, which is one of the key factors in the observed increase in enzymatic hydrolysis rates after ILs pretreatment [14–15]. With the disruption of the natural interactions among the cellulose chains via the formation of interactions between ILs and hydroxyl groups on cellulose, the cellulose chains were released and a cellulose–ILs solution is formed (Fig. 3.3). Recently publications of nonimidazolium-based ILs demonstrated good solubility to cellulose, which argued against the key role of the structure of cations in ILs to a certain extent [16]. Coarse-grain analysis was used to acquire both the structures and strengths of solvent–glucan interactions in ILs (1-butyl-3-methylimidazolium chloride, [Bmim]Cl) and water, and the authors simulated cellulose deconstruction by peeling off an 11-residue glucan chain from a cellulose microfibril and computed the free-energy profile in water and in ILs. For this deconstruction process, the calculated free-energy cost/reduction in water/[Bmim]Cl is ~2 kcal mol^{-1} per glucose residue, respectively [17]. The results demonstrated that solvent–glucan interactions are dependent on the deconstruction state of cellulose. Water couples to the hydroxyl and side-chain groups of glucose residues more strongly in the peeled-off state but lacks driving forces to interact with sugar rings and linker oxygens. Conversely, [Bmim]Cl demonstrates versatility in targeting glucose

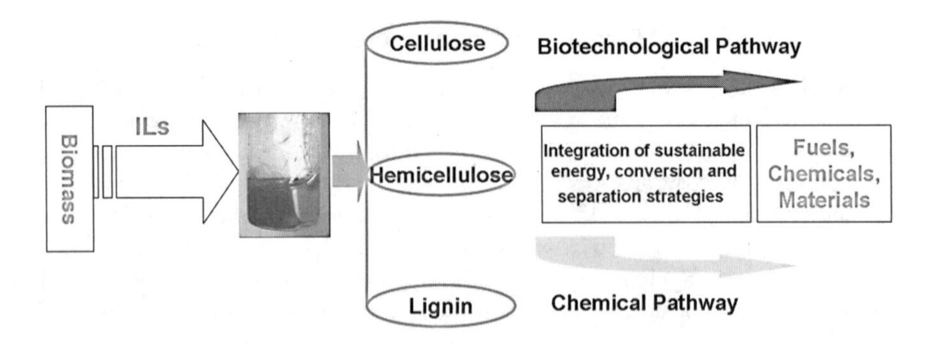

FIGURE 3.3 The concept of ILs-based biorefinery.

residues in cellulose. Anions strongly interact with hydroxyl groups, and the coupling of cations to side chains and linker oxygens is stronger in the peeled-off state. Therefore, with the aim to increase the pretreatment efficiency of ILs, other than enhancing anion–hydroxyl group coupling, configuring cations to target side chains and linker oxygen atoms is a useful design strategy. In fact, the significant fundamental molecular knowledge of the role of ILs, should enable the design and rapid computational screening of a wide range of ILs for biomass pretreatment and facilitate the development of new efficient and economic IL-based biorefinery technologies [18].

3.3.2 The Concept of Ionic Liquids-Based Biorefinery

With the aim to address the conversion of biorenewable resources into sustainable materials, fuels and commodity compounds, a biorefinery concept has been proposed by the US national renewable energy laboratory, which is defined as a facility that integrates biomass-conversion processes and equipment to produce fuels, power, and chemicals from biomass. The biorefinery concept is analogous to today's petroleum refineries, which produce fuels and a wide range of chemical products from petroleum. It is anticipated that industrial biorefineries would be the most promising route to the creation of a new domestic biobased and sustainable industry [2]. Considering the benefits of using biomass as a feedstock, investigations at this research frontier have elicited great attention. Due to the good solubility of cellulose and lignocellulose in some ILs, the chemistry of lignocellulose, and conversion of (ligno)cellulose and their components into sustainable materials and chemicals have been investigated extensively [19]. As such, this is one of the most active research fields for both ILs and biomass utilization.

Ten years research and development activities in this area has cumulated in a primary scheme of ILs-based biorefinery, [6d, 11c, 20] and provides a new homogeneous pathway to convert biomass into valuable, sustainable materials and chemicals with the integration of sustainable energy, conversion and separation strategies (Fig. 3.3).

3.3.3 Wood Chemistry in Ionic Liquids

The efficient utilization of biomass as feedstock for the production of materials and chemicals requires an in-depth understanding of plant cell wall natural structures and their constituents, which can guide the development of mild pretreatment technologies and robust catalysts for the separation of the constituents and further catalytic conversion. From a biological point of view, an in-depth understanding of biomass structure can provide new avenues to rationally design bio-energy crops with improved processing properties by either reducing the amounts of lignin present or providing a lignin that is easier to degrade. However, traditional lignin separation and analysis process (e.g., Klason Method) only provides destroyed structural information of lignin after hydrolysis of carbohydrates under strong acidic conditions due to the insolubility of lignocellulose in conventional organic solvents, and the elucidation of natural structural information of lignocellulosic materials are one of the biggest challenges of wood chemistry. With the good solubility of lignocellulosic materials and lignin in ILs, detection of the structure of lignocellulose and lignin is possible with modern analytic technologies under homogenous conditions, such as NMR [19a].

In 2007, Pu Y. et al. investigated the solubility of lignin in different ILs, [21] and a map of ^{1}H-NMR and ^{13}C-NMR signals of lignin was obtained using ILs as the solvent [22]. It was found that the ^{13}C-NMR signals shift up-field by δ 0.1–1.9 ppm in comparison to spectra acquired using dimethyl sulfoxide (DMSO-d_6) as the solvent. A mixed solvent consisting of perdeuterated pyridinium ILs and DMSO-d_6 was investigated for the direct dissolution and NMR analysis of plant cell walls. For example, a mixture of 1:2 [Hpyr-d_5]Cl/DMSO-d_6 was able to dissolve Poplar up to 8 wt% at 80°C in 6 h. The solution can be used for *in situ* ^{1}H-NMR and ^{13}C-NMR analysis of lignocellulosic materials, and a full structural map of signals from their different composites. For example, the signals at δ 61.5, 74.1, 75.8, 76.9, 80.1, and 103.0 ppm were in part assigned to cellulose. The lignin methoxyl group corresponding to the signals at δ 57 ppm and δ 58–88 ppm could be assigned, in part, to C_β in β-O-4, C_γ/C_α in β-O-4, β-5, and β-β. The signal at δ 106 ppm was assigned to C2/6 resonance of syringyl-like lignin structures, and between 110 and 120 ppm to C2, C5, and C6 resonance of guaiacyl-like lignin structures. The properties and easy preparation of perdeuterated pyridium molten salt [Hpyr-d_5]Cl offer significant benefits over imidazolium molten salts for NMR analysis of plant cell walls, furthermore, the use of non-ball-milled samples in this study can provide a more efficient and accurate characterization of lignin in the plant cell walls compared with the results from traditional methods [23]. Although lignin can provide a renewable source of phenolic polymers, a high lignin content has proved to be a major obstacle not only in the processing of plant biomass to biofuels, but also in other processes such as chemical pulping and forage digestibility. Therefore, precise analytic techniques for efficient lignin-content assessment of a large number of samples are in high demand [19a]. Further study from Ragauskas's group reported a linear extrapolation method for the measurement of lignin content by the addition of a specific amount of isolated switchgrass lignin to the biomass solution, and the integration ratio changes could be measured in the quantitative ^{1}H-NMR spectra with nondeuterated DMSO as the internal standard. The results showed comparable lignin contents as the traditional Klason lignin contents. They demonstrated that this direct dissolution and NMR analysis of biomass provided a new venue for rapid assessing of the lignin contents in large numbers of "new" plants in biofuel research [23]. Recently, this system was extended to evaluate the efficiency of pretreatment technologies by semiquantitative ^{13}C–^{1}H heteronuclear single quantum correlation (HSQC) spectroscopy of untreated, steam, dilute acid, and lime-pretreated poplar biomass samples [24].

^{31}P-NMR analysis has an important role during the study of hydroxyl functionality distribution in lignocellulosic materials. ILs are demonstrated as good solvents for ^{31}P-NMR analysis, and *in situ* quantitative analysis. ^{31}P-NMR analysis of spruce dissolved in ILs showed that there is 13.3 mmol g^{-1} hydroxyl groups [25]. This is close to the theoretically calculated value of 15.7 mmol g^{-1} based on traditional methods [26]. Analysis of different degrees of pulverization provided semiempirical data to chart the solubility of Norway spruce in IL [amim]Cl, and further refinement afforded an optimized method of analysis of the lignin phenolic functionalities, without prior isolation of the lignin from the wood fiber [27].

3.3.4 Sustainable Materials from Biomass in Ionic Liquids

Polymeric materials play an important role in our life, while biomass resources contain libraries of renewable polymeric materials, for example, cellulose, silk fiber, wool, chitin, etc. Among them, cellulose is the most common organic polymer, representing about 1.5×10^{12} tons of the total annual biomass production, and is considered as an almost inexhaustible source of raw material for the increasing demand for environment-friendly and biocompatible products, readily produced by photosynthesis [2]. For millennia, cellulose has been used in the form of wood and plant fibers as an energy source, in building materials, and in clothing [28]. With the aim to increase the processing properties and tailoring function to specific applications, chemical modification of cellulose is necessary [29]. The research is likely to be extended to lignocellulosic materials, which is much more available than cellulose and also avoids ethical problems associated with using a food source as a resource for the chemical industry. With the dissolution of cellulose, [30] lignocellulose (and other biopolymers), [31] the development of new biopolymer-derived materials has been studied extensively, and the literature publications can be classified as follows:

(1) Regenerated cellulose fiber.
(2) Small molecular modification "fine-tuning" of cellulose into cellulose and lignocellulosic derivatives (e.g., acetates, carboxymethylates, benzoylates, urethanes, methacrylates, carbonates, sulfates, sulfonates, phthalates, tritylates, furorates, maleated esters, or ester resins).
(3) Polymers grafted cellulose synthesized by polymerization (e.g., UV radiation, atom-transfer radical polymerization) [22, 32].
(4) Cellulose composites through blending on the basis of formation of interwoven networks between the cellulose chains and synthetic polymer strands during the regeneration process (e.g., cellulose/branched polyamidoamine, cellulose) [10a, 33].

A recent review paper by Pinkert et al. has provided a comprehensive overview in this field [34]. Therefore, herein, we only highlight selected research and recent studies. Cellulose, as the most environment-friendly polymeric material for clothing, can be regenerated in different forms, such as membrane materials, fibers, and gel, depending on the choice of antisolvents and dry processes. The properties of these regenerated materials (e.g., mechanical strength, crystallinity, and surface area) are dependent on the regenerated conditions, such as dissolution time, precipitation rate, initial cellulose concentration, type of regenerating solvents, and drying methods. It was reported that the ligno cellulose fiber, [35] chitin, [36] silk fibroin fiber, [31b]

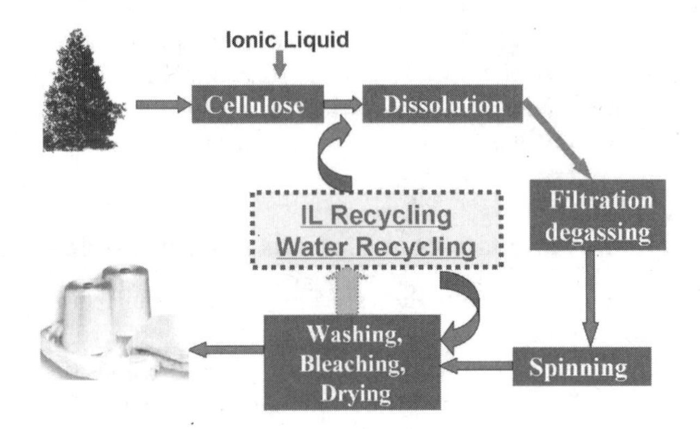

FIGURE 3.4 Preparation of regenerated cellulose fiber from ILs.[1]

cellulose/carbon nanotube, [37] and cellulose/wool keratin fiber/membranes [31c] can be prepared by traditional wet spinning process (Fig. 3.4) and electrospinning, which have emerged as a leading technology for preparation of regenerated cellulose fiber.

The essence of chemical modification of cellulose is the reaction of hydroxyl groups in cellulose with reagents. The full dissolution of cellulose in ILs leads to a homogeneous solution of cellulose, which enables the greatest accessibility of hydroxyl groups in cellulose, thus an increased modification efficiency is anticipated.

A series of derivatives of cellulose or other biopolymers can be obtained, such as acetates, [38] carboxymethylates, [16] benzoylates, [39] methacrylates, [40] sulfates, [30] phthalates, [4] tritylates, [41] furorates, [42] maleated esters, [43] or ester resins [25, 44]. These materials have been widely synthesized and studied in traditional solvents, because they are extensively used in coatings, membranes, binders, fillers, composites, explosives, optical films or separation process, medical applications, and the food industry. Figure 3.5 provides a selected overview of cellulose modification in ILs [34]. It was found that the degree of substitution of targeted derivatives and regioselectivity can be influenced by the choice of modifying reagents, the reaction time, and the mole ratios of the reagents. It is worthy to mention that the ILs can catalyze the acylation and carbanilation of cellulose using anhydrides and isocyanates as modifying reagents. For effective modification in traditional solvents, a basic catalyst is required, such as pyridine. Apart from the well-known functionalization routes of cellulose in traditional solvents, new reactions can be accomplished in ILs. For example, water-soluble cellulose sulfates can be prepared with different sulfating agents in ILs, which presented a first direct sulfation route of biopolymer under completely homogeneous reaction conditions [45].

The chemical modification of cellulose in ILs can be extended to lignocellulosic material. In 2007, Xie et al. reported the full dissolution of lignocellulosic materials

[1] The authors thank Prof. Jun Zhan from Institute of Chemistry of Chinese Academy of Sciences for providing the pictures.

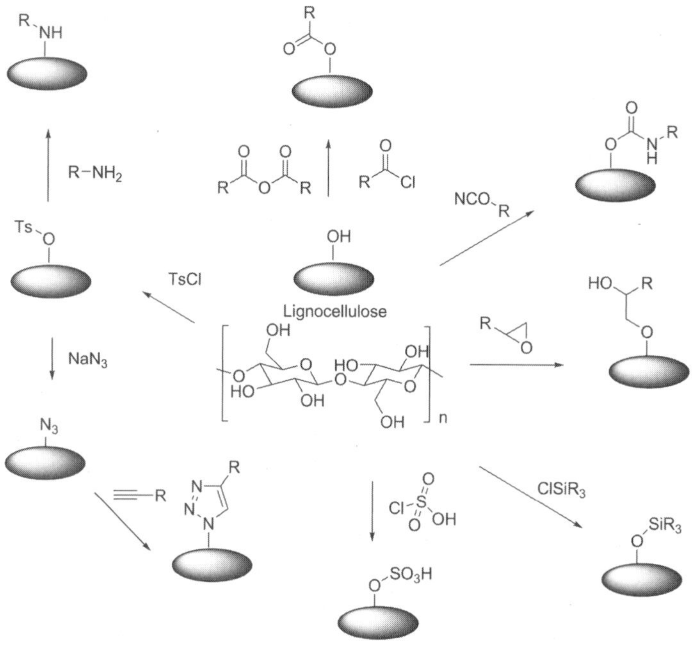

FIGURE 3.5 Selected examples of modification of cellulose and lignocellulose in ILs [34].

[31a] and firstly studied a thorough modification of lignocellulosic materials in ILs [26b]. The results showed that over 95% degree of substitution was obtained, and almost all the hydroxyl groups in wood can be modified. The conversion of polar hydroxyl groups into hydrophobic groups results in a high compatibility in the thermo composites between the lignocellulosic derivatives with polyethylene, polypropylene, and polystyrene [44]. In order to increase the compatibility of cellulose derivatives with man-made polymers and process properties, an interesting approach is the introduction of a related polymer chain grafted onto the cellulose structure or chemical modification and converting the cellulose into thermoplastic polymer materials. Among them, grafting biodegradable aliphatic polyesters onto cellulose has received extensive attention, due to the possibility of creating a new kind of material combining the advantages of these two groups of polymers simultaneously by grafting biodegradable polymers onto the underivatized cellulose backbone. In our opinion, this is a research area with considerable potential. Dong et al. synthesized cellulose–*graft*–polylactide (cellulose-*g*-PLLA) copolymers by ring-opening graft polymerization of LA onto cellulose with tin(II) 2-ethylhexanoate catalyst in an ionic liquid 1-allyl-3-methylimidazolium chloride ([amim]Cl) [46]. The results showed that the amount of grafted PLLA in synthesized copolymers was relatively low, and the graft copolymers could not be melted and were only soluble in DMSO and water. In 2010, Zhang et al. [47] reported that using 4-dimethylaminopyridine (DMAP) as an organic catalyst, cellulose-*g*-PLLA copolymers with a

wide range of compositions can be prepared by the homogeneous graft reaction of cellulose with LA in [Amim]Cl. It was found that the grafting content of PLLA in copolymers increased with the increase of molar ratio of LA to cellulose in feed. When the molar ratio of LA to cellulose was 10, the molar substitution of PLLA (MSPLLA) in the graft product was up to 12.28. The resultant cellulose-*g*-PLLA copolymers exhibited single composition-dependent glass-transition temperature (T_g) in their corresponding DSC thermograms, which are thermoplastic and amorphous polymeric materials.

3.3.5 Value-Added Chemicals from Biomass in Ionic Liquids

Petroleum-based chemicals have taken the dominant position in our daily life since the development of modern chemical industry. However, with the depletion of fossil resources along with the increasing price of petroleum-based fuels, chemicals, and solvents, the chemical industry requires access to low-cost feedstocks/raw materials. Clearly, alternative feedstocks/raw materials must be identified to support our fuel and chemical production. Due to the wide availability of renewable biomass, they have been proposed as a viable feedstock for the fuel and chemical industries [48]. In the traditional petrochemical industry, crude oil is fractionated and refined to produce various grades of liquid fuel, and hydrocarbon feedstocks are refined to produce intermediates and specialty chemicals. The concept of biorefinery is similar in terms of the utilization of biomass to produce heat, sustainable fuels, and value-added chemicals. For a successful conversion of biomass into fuels and chemicals, some important issues need to be addressed: (1) to depolymerize biopolymers, (2) to reduce the oxygen content of the parent feedstock or depolymerized intermediates, and (3) to create C–C bonds between biomass-derived intermediates to increase the molecular weights of the end products [48]. With the full dissolution of biomass in ILs, great efforts have been devoted to the production of valuable feedstocks from biomass via hydrolysis, dehydration, hydrogenolysis, etc., by taking advantage of this new homogenous platform. Many fundamental starting materials and fuels have been synthesized, including monosugars, 5-hydroxymethyl furfural, levulinic acid, and long-chain alkyl glycosides (Fig. 3.6).

3.3.5.1 Monosugars Production from Cellulose and Lignocellulose

Hydrolysis of cellulose to fermentable sugars is an essential step in any practical cellulosic biofuel production and chemicals production. Two methods including acid hydrolysis and enzymatic hydrolysis are currently known for cellulosic biomass hydrolysis. To overcome lignocellulose recalcitrance, pretreatment is required. The goal of pretreatment is to alter the physical features and chemical composition of the lignocellulose to make it more digestible by acids and enzymes [49]. Presently, three types of research have been carried out based on the ILs platform to produce fermentable sugars. First is enzymatic hydrolysis of cellulosic materials regenerated from ILs by addition of an antisolvent; second is *in situ* enzymatic hydrolysis of cellulose in biocompatible ILs; and the third pathway is to use acid catalysts to hydrolyze the dissolved carbohydrates in ILs [19a].

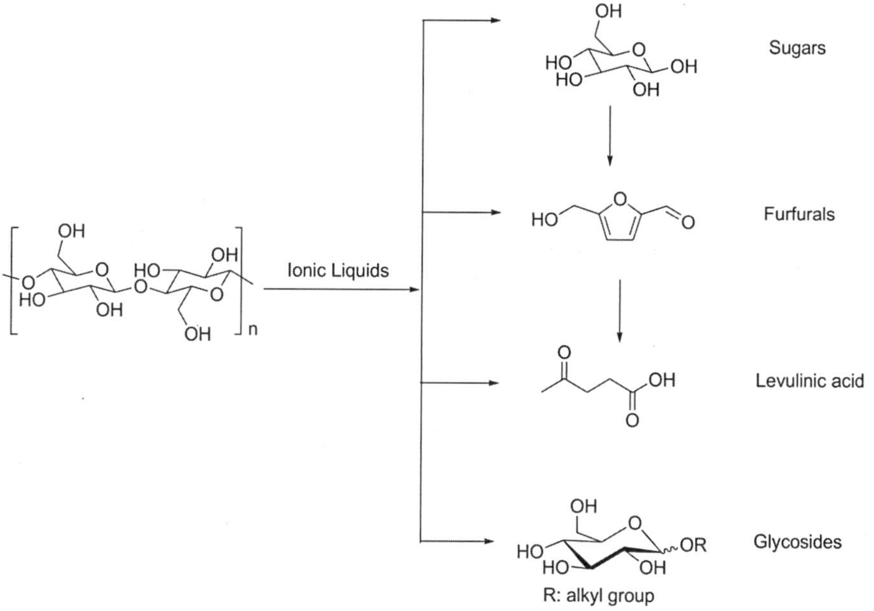

FIGURE 3.6 Catalytic conversion of (ligno)cellulose into value-added chemicals in ILs.

Enzymatic Hydrolysis of (Ligno)Cellulose Regenerated from Ionic Liquids By simple regeneration of cellulose or lignocellulose on addition of anti-solvents, such as water and ethanol, it was found that the crystalline structure of cellulose was destroyed. This should lead to a greater accessibility of the hydrolytic enzymes to more effectively penetrate and hydrolyze the (ligno)cellulose glycoside bonds. Liu et al. first investigated the enzymatic hydrolysis of cellulose samples that had been pretreated with ILs [50]. The hydrolysis rate of wheat straw and steam-exploded wheat straw treated with $[C_4mim]Cl$ were improved significantly. An amorphous and porous morphology and conversion of cellulose fibers from the cellulose I to the cellulose II crystal phase of regenerated materials has been shown by a Wide-angle X-ray diffraction (WXRD) study and morphology study by scanning electron microscope (SEM). Take the spruce thermomechanical pulp (TMP) samples regenerated from 1-methyl-3-allylimidazolium ([Amim]Cl) solution by precipitation which upon water addition gave a glucose yield of 60% by the enzymatic hydrolysis, while only 12% was obtained for the untreated sample [31a]. Although the biomass can be regenerated by adding antisolvents, the anti-solvent must be removed for the ILs to be recovered and recycled. In 2011, Blanch et al. described the use of aqueous kosmotropic salt (such as phosphate, carbonate, or sulfate) solutions to form a three-phase system that precipitates the biomass, forming IL-rich and salt-rich phases. 95.0% of the ILs in K_3PO_4-containing systems can be recovered. This process reduces the amount of water to be evaporated from recycled ILs, permitting efficient recycling of the ILs [51]. Compared with the cellulose obtained from biomass pretreated with ILs and precipitated with water, a more rapid and higher yield of conversion of cellulose to glucose was obtained by the use of the three-phase system. The results showed that around 85% of cellulose pretreated with [Emim][AcO] at 70°C for 18 h, and precipitated with 40 wt% K_3PO_4 can be hydrolyzed to glucose in 8 h. On increasing the pretreating temperature from 70 to 140°C in 1 h, a result of 95% of cellulose converted into glucose in 24 h was obtained. They demonstrated that this pretreatment method not only removes the physical barrier that results from lignin occlusion of the hemicellulose, but also reduces the opportunity for unproductive binding of cellulolytic enzymes to the lignin that occurs with other pretreatment approaches. This makes the substrate more accessible, enhancing the enzymatic hydrolysis step. Recent study showed that with the addition of acetic acid acting both as co-solvent and a catalyst, polyoxometalates ($[PV_2Mo_{10}O_{40}]^{5-}$) can facilitate the delignication of lignocellulose and result in an enhanced enzymatic hydrolysis of regenerated biomass [52].

Apart from the dissolution-pretreatment process using the ionic liquids, ionic liquids aqueous solution can also be used for an extraction-pretreatment process. A mixture of ionic liquid [Emim][OAc] and water was demonstrated to be effective for the pretreatment of lignocellulosic biomass, evidenced by the removal of lignin and a reduction in cellulose crystallinity. An 81% fermentable sugar yield was obtained, which is higher than that (67%) for pure ILs pretreatment under the same conditions. Aqueous ILs pretreatment has the advantages of reduced quantities and easier recycling of ILs, and reduced viscosity [53].

Enzymatic Hydrolysis of Cellulose in Biocompatible Ionic Liquids ILs have also been investigated as suitable solvents for biocatalysis [54]. It is reasonable to propose direct enzymatic hydrolysis of cellulose using ILs as the reaction media, which will avoid the requirement of an anti-solvent to regenerate the biomass. To realize this idea, we need to balance enzymatic biocompatibility of the IL and solubility to biomass though modifying the cation and anion of the IL. However, conventional ILs capable of dissolving cellulose are usually composed of anions such as chloride, dicyanamide, formate, and acetate [10a]. While these anions on one hand can facilitate the dissolving of cellulose through formation of strong hydrogen bonds, on the other hand, they also tend to denature the enzyme. Therefore, the design and synthesis of ILs which are capable of dissolving cellulose, and are enzyme-compatible (especially compatible with cellulase), have obtained much attention aiming to pursue an *in situ* enzymatic hydrolysis of cellulose [55].

In 2008, Kamiya et al. dissolved cellulose samples in the ionic liquid 1-ethyl-3-methylimidazolium diethyl phosphate [Emim][DEP] first, [56] then mixed the ILs–cellulose solution with different volumes of citrate buffer (10 mM, pH 5.0), followed by addition of cellulase directly at 40°C. It was found that the volumetric ratio of ILs affects the activity of cellulase significantly. A satisfactory reducing sugar yield over 70% was obtained after 24 h when the ILs to water ratio was set to 1:4. In comparison, when the ionic liquid [Emim][OAc] was applied under these conditions, cellulase activity was approximately half the value (70% conversion of cellulose) obtained for [Emim][DEP], suggesting that the anion structure played a key role regarding enzyme compatibility of ionic liquids. However, recent publication from

Wang et al. demonstrated that [Emim][OAc] was compatible with the cellulase mixture. A mixture of cellulases and *beta*-glucosidase (Celluclast1.5L, from *Trichoderma reesei*, and Novozyme188, from *Aspergillus niger*, respectively) retained 77% and 65% of its original activity after being preincubated in 15% and 20% (w/v) IL solutions, respectively, at 50°C for 3 h. The cellulases mixture also retained high activity in 15% [Emim][OAc] to hydrolyze Avicel, a model substrate for cellulose analysis, with conversion efficiency of approximately 91%. Although [Bmim]Cl has good solubility to cellulose, it is toxic to cellulase [56]. Recent publications [57] demonstrated that a new approach to improve the cellulase stability against [Bmim]Cl, based on coating of immobilized enzyme particles with hydrophobic ILs can be successful. Hydrophobic ILs clearly enhanced the enzyme thermal stability. For example, butyltrimethyl-ammonium *bis*(trifluoromethylsulfonyl)imide ([N(1114)][NTf_2]) enhances half-life time of the immobilized enzyme at 50°C up to four times, while [Bmim]Cl behaved as a powerful enzyme-deactivating agent. Thus, the stability of cellulase in hydrophobic IL/[Bmim]Cl mixtures was greatly improved compared to [Bmim]Cl alone. A stabilized cellulase derivative obtained by coating immobilized enzyme particles with [N(1114)][NTf_2] has also successfully been used for the saccharification of dissolved cellulose in [Bmim]Cl (i.e., up to 50% hydrolysis in 24 h) at 50°C and 1.5 w/v water content.

Besides the effect of ILs's structure, viscosity of the ILs and enzyme thermo stability can impact the hydrolysis remarkably. In 2010, Bose et al. [5] investigated the activity and stability of commercial cellulase samples in eight ILs by optical and calorimetric techniques. Cellulose hydrolysis was observed only in 1-hydrogen-3-methylimidazolium chloride and *tris*-(2-hydroxyethyl) methyl ammonium methylsulfate (HEMA). Further investigation into the relationship of enzymatic activity and thermal stability as a function of viscosity and reaction temperature demonstrated that this difference in the rate of hydrolysis is largely attributed to both the viscosity of the ILs and enzyme stability. The inherent high viscosity of the ILs slows the diffusion of the substrate to the enzyme, and thus a lower activity was observed. Reversibility of the enzyme unfolding process was studied by the gradual cooling of the denatured enzyme from temperature greater than $T_{1/2}$. It was found that cellulase unfolding was irreversible in buffer (pH = 4.8) and accompanied by precipitation of the enzyme. In HEMA, folding and unfolding processes were reversible even at up to 120°C, which indicates that excellent cellulase thermostability can be achieved in this IL. However, in 1-hydrogen-3-methylimidazolium chloride, the unfolding process was completely irreversible.

Acidic Hydrolysis of (Ligno)Cellulose in Ionic Liquids Acid-catalyzed hydrolysis of cellulose is one important route to produce monosugars, processes of which are usually performed under heterogeneous and harsh conditions. Traditional acid hydrolysis of (ligno)cellulose is inefficient and cost-intensive due to important restrictions on the kinetics of the hydrolysis caused by the supramolecular and crystalline structure of cellulose [58]. So far, both solid acids and mineral acids have been applied to hydrolyze cellulose in ILs (see Table 3.1) [19a].

In 2007, Zhao et al. firstly reported the hydrolysis behavior of cellulose in ILs in the presence of mineral acids under homogenous conditions [59]. It was found that

TABLE 3.1 **Catalytic Hydrolysis of (Ligno)Cellulose into Monosugars in ILs**

Raw Materials	Acid	Ionic Liquids	Regeneration Solvent	TRS Yield (%)	Glucose Yield (%)	References
Avicel	H_2SO_4	[Bmim]Cl	Water	73%	32%	[71]
α-Cellulose	H_2SO_4	[Bmim]Cl	Water	63%	39%	[71]
Spruce	H_2SO_4	[Bmim]Cl	Water	71%	28%	[71]
Sigmacell	H_2SO_4	[Bmim]Cl	Water	66%	28%	[71]
Corn stalk	HCl	[Bmim]Cl	Water	66%	—	[59b]
Rice straw	HCl	[Bmim]Cl	Water	74%	—	[59b]
Pine wood	HCl	[Bmim]Cl	Water	81%	—	[59b]
Bagasse	HCl	[Bmim]Cl	Water	66%	—	[59b]
Eucalyptus grandis	HCl	[Amim]Cl	Water, methanol, or ethanol	95%[a]		[60b]
Southern pine	HCl	[Amim]Cl	Water, methanol, or ethanol	67%[a]		[60b]
Norway spruce	HCl	[Amim]Cl	Water, methanol, or ethanol	82%[a]		[60b]
Thermomechanical pulp	HCl	[Amim]Cl	Water, methanol, or ethanol	82%[a]		[60b]
Cellulose	HCl	[Emim]Cl	Water	—	89%	[69]
Corn stover	HCl	[Emim]Cl	Water	70–80%	—	[69]
Miscanthus grass	CH_3SO_3H	[Emim]Cl	Water	—	68%	[26b]
Cellobiose	$H_3PW_{12}O_{40}$	—	Water[b]	—	51%	[67]
Cellulose	$Sn_{0.75}PW_{12}O_{40}$	—	Water[b]	23%	—	[67]
Lignocellulose	$H_3PW_{12}O_{40}$	—	Water[b]	32%	—	[67]
Cellulose	Nafion® NR50	[Bmim]Cl	Water	35%	—	[68]
α-Cellulose	HY zeolite	[Bmim]Cl	Water	46.9%	34.9%	[57]
Avicel cellulose	HY zeolite	[Bmim]Cl	Water	47.5%	36.9%	[57]
Spruce cellulose	HY zeolite	[Bmim]Cl	Water	44.4%	34.5%	[57]
Sigma Cell cellulose	HY zeolite	[Bmim]Cl	Water	42.4%	32.5%	[57]

β-Cellulose	HY zeolite	[Bmim]Cl	Water	—	12.5%	[57]
Cellulose	$Si_{33}C_{66}$-673-SO_3H	—	Water	—	50% glucose	[41b]
Microcrystalline cellulose	$FeCl_2$	$[(CH_2)_4SO_3Hmim][HSO_4]$	Diethyl ether and water	8.35	—	[41a]
Western red cedar	O_2	[Emim][Cl]	DMSO	N.C.[d]	N.C.[c]	[72]
Cellulose	Proton acid	[Bmim]Cl–water	Water	97%	N.C.[c]	[73]
Loblolly pine						
Wood	TFA	[Bmim]Cl	Water	79%	N.C.[c]	[60a]
Corn stover	Boronic acids	[Emim][OAc]	Hot water	N.C.[d]	>97% glucose	[70]

Source: Ref. [19a].

[a]Carbohydrates were hydrolyzed at 1.4–1.5 mol of HCl g^{-1} wood acid concentration.

[b]The reaction was carried out in aqueous solution.

[c]N.C., not characterized.

catalytic amounts of acid were sufficient to drive the hydrolysis reaction. For example, when the acid/cellulose mass ratio was set to 0.46, yields of total reducing sugar (TRS) and glucose were 64% and 36%, respectively, after 42 min at 100°C. A preliminary kinetic study indicated that the cellulose hydrolysis in [B_4mim]Cl catalyzed by H_2SO_4 followed a consecutive first-order reaction sequence, where k_1 for TRS formation and k_2 for TRS degradation were 0.073 min^{-1} and 0.007 min^{-1}, respectively. Further studies demonstrated that the catalytic hydrolysis system can be extended to different (ligno)cellulosic materials [59–60]. For example, the acidic pretreatment of woody biomass species (*Eucalyptus grandis*, Southern pine, and Norway spruce) in [Amim]Cl resulted in the near-complete hydrolysis of cellulose, hemicellulose, and a significant amount of lignin [59b, 60b]. The hydrolytic kinetics of cellobiose in the 1-ethyl-3-methylimidazolium chloride ([E_2mim]Cl) as a model compound shows the rates of the two competitive reactions, [26b] polysaccharide hydrolysis and sugar decomposition, varied with acid strength. In acids with an aqueous p*Ka* below approximately zero, the hydrolysis reaction is significantly faster than the degradation of glucose, thus allowing hydrolysis to be performed with a high selectivity in glucose, which was consistent with the results obtained in Li's work [59b]. The basicity of the acetal *O*-site, the hydroxyl *O*-sites, and the p*Ka* of acidic catalyst also play important roles during the hydrolysis process, as the proposed hydrolysis of cellulose usually consists of three consecutive processes, (1) protonation of the glycosidic oxygen; (2) formation of a cyclic carbocation; (3) nucleophilic attack of water on the cyclic carbocation species by a water molecule (Fig. 3.7). Although the second step is rate-limiting, the protonation of the glycosidic oxygen is difficult and can be slow with electron-deficient acetals during the hydrolysis process. Furthermore, it is believed that the full dissolution of cellulose in ILs makes the glycosidic oxygen more accessible for external acidic catalysts, resulting in a fast hydrolysis rate.

The acidic strength plays an important role during the hydrolysis process of cellulose, although the acidity in ILs is a concept still in its infancy, and the relative acidity order of several organic and inorganic acids in ILs was found identical to that in water, [62] recent publication demonstrated that the water content had substantial effects on glucose and cellobiose yields. It can be speculated that the water in ILs can

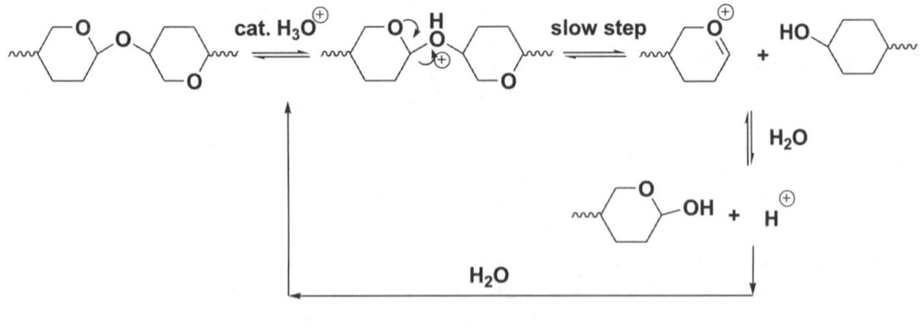

FIGURE 3.7 Proposed mechanism for hydrolysis of cellulose [61].

play a role as H^+ support and facilitate the mobility of H^+, thus a better hydrolytic behavior was achieved. Other acids, such as trifluoroacetic acid, Brönsted acidic ionic liquids have also been investigated and satisfactory results were obtained in the hydrolysis of cellulose in ionic liquids [63]. Among all these significant contributions into the production of monosugars from biomass with the ILs platform, the cellulose could be hydrolyzed in the absence of acidic catalysts, with yields up to 97% for the reducing sugars when the temperature was increased to 140°C at 1 atm. A combined study of experimental methods and *ab initio* calculations demonstrated that the K_w value of water in the IL–water mixture was up to three orders of magnitude higher than that of the pure water under ambient conditions. They hypothesized this increased K_w value was responsible for the remarkable performance in the absence of acid catalysts and that the increased $[H^+]$ was attributed to the enhanced water auto ionization by ILs under mild conditions [64].

It is well recognized that under homogeneous hydrolytic conditions, the recycling of the acidic catalysts is one of the main drawbacks, and the introduction of heterogeneous catalyst will result in a much lower-cost process for the catalyst separation and recycle. Following the homogeneous hydrolysis of cellulose in ILs, different solid acid catalysts have also been investigated for the hydrolysis of cellulose. In 2008, Rinaldi et al. [65] reported that a solid acid (Amberlyst 15 DRY) catalyzed hydrolysis of cellulose and (ligno)cellulose in ILs, in which depolymerized cellulose was precipitated and recovered by addition of water. The degrees of polymerization were estimated by gel-permeation chromatography and the size of recovered cellulose fibers became smaller successively over time. The depolymerization of cellulose proceeded progressively. For example, cellooligomers consisted of approximately ten anhydroglucose units (AGU) after 5 h. It was also interesting to observe that there was an induction period for the production of glucose. Further titration study of the ILs separated from a suspension of Amberlyst 15DRY in [Bmim]Cl suggested that proton was progressively released into the bulk liquid within an hour upon through an ion-exchange process involving $[Bmim]^+$ of the ionic liquid and H^+ species of the solid acid.

The design of solid catalysts that are suitable for both heterogeneous and homogeneous conversion is one of the greatest challenges for biomass utilization [66]. It was found that the H^+ species and reaction media are highly related to their catalytic activity towards the hydrolysis of cellulose. For example, Shimizu K. et al. developed $H_3PW_{12}O_{40}$ and $Sn_{0.75}PW_{12}O_{40}$ for the hydrolysis of lignocellulose, which presented higher TRS yield than conventional H_2SO_4 in water [67]. Other solid acids, such as Nafion® NR50, sulfonated silica/carbon nanocomposites, have also been studied for the hydrolysis of cellulose in ILs. It was found that the crystalline cellulose partially loosened and transformed to cellulose II from cellulose I, then to glucose assisted by Nafion® NR50. Afterwards, a catalyst was recycled and the residual (hemi)cellulose solid, which could be hydrolyzed into monosugars by enzymes, was separated by adding anti-solvents [68]. Due to the presence of strong, accessible Brønsted-acid sites and the hybrid surface structure of sulfonated silica/carbon nanocomposites, it was found that a 42.5% glucose yield was achieved after three recycles of this catalyst in ILs [41b].

Solid acids-catalyzed hydrolysis of cellulose in ILs was greatly promoted by microwave heating. The results showed that H-form zeolites with a lower Si/Al molar ratio and a larger surface area exhibited better performance than that of the sulfated ion-exchanging resin NKC-9. The introduction of microwave irradiation at an appropriate power significantly reduced the reaction time and increased the yields of reducing sugars. A typical hydrolysis reaction with Avicel cellulose could produce glucose with $\sim$37% yield within 8 min [57].

Monosugars are intermediates linking the sustainable biomass and clean energies, such as bioethanol and microbial biodiesel. In 2010, Binder J. B. et al. first investigated the fermentation potential of sugars produced from cellulose in ILs after separation of ionic liquids by ion-exclusion chromatography. The results showed that adding water gradually to a chloride ionic liquid-containing catalytic HCl led to a nearly 90% yield of glucose from cellulose and 70–80% yield of sugars from untreated corn stover. Ion-exclusion chromatography allowed the recovery of ILs and delivered sugar feedstocks that support the vigorous growth of ethanologenic microbes. This simple chemical process presents a full pathway from biomass to bioenergy based on the ionic liquids platform, although the development of much more economic technologies for the recovery and separation of ILs and sugars is still in high demand [69].

Recent work has demonstrated that the recovery of sugars from ILs could be fulfilled by extraction based on the chemical affinity of sugars to boronates such as phenyl boronic acid and naphthalene-2-boronic acid [70]. Up to 90% of mono- and disaccharides could be extracted by boronate complexes from aqueous ILs solutions, pure ILs systems, or hydrolysates of corn stove containing ILs.

3.3.5.2 5-Hydroxylmethyl Furfural and Furfural Derivatives from Carbohydrates 5-Hydroxymethylfurfural (HMF) is an important intermediate in the conversion of hexose sugars into a spectrum of valuable chemicals, such as 2,5-furandicarboxylic acid, 2,5-furancarboxaldehyde, 2,5-dihydroxymethyl furan, and levulinic acid [6d, 74]. In the past few years, the preparation of HMF through the dehydration of biomass-based sugars has received much attention, especially in ILs, due to the good solubility of carbohydrates in ILs [6f]. The dehydration of fructose into HMF has been investigated for many years due to its high efficiency and selectivity catalyzed by mineral acids and solid acids. ILs can be used as solvents and/or catalysts for the preparation of HMF from fructose and inulin (see Table 3.2). However, the challenge of the efficient dehydration of glucose and cellulose into HMF still remains. Herein, we focus on the studies of HMF production from glucose and cellulose in ILs.

In 2007, Zhang et al. [92] reported that chromium(II) chloride could catalyze the dehydration of glucose into HMF in 1-ethyl-3-methylimidazolium chloride ([Emim] Cl) with a good yield of 68%. It was proposed that glucose was isomerized through a $CrCl_2$-associated enediol intermediate to fructose followed by dehydration reaction. Less than 2% yield was achieved when a strong ligand, 2,2′-bipyridine or glyceraldehydes was added. The authors concluded that the metal catalyst interacted only with the hemiacetal portion of glucopyranose, and not with the polyalcohol portion

TABLE 3.2 HMF from Fructose in ILs and Organic Solvents

Ionic Liquids	Co-Solvent	Catalyst	Condition	By-Product	HMF Yield (Selectivity)	References
[Bmim][BF_4]	DMSO	Amberlyst-15	60°C, 24 h	N.C.[a]	80%	[75]
[ASBI][TfO]	DMSO	[ASBI][TfO]	160°C, 5 min	N.C.[a]	92%	[76]
[ASCBI][TfO]		[ASCBI][TfO]				
[Cho]Cl	AcOEt	Citric acid	80°C, 1 h.	N.C.[a]	90%	[77]
[Bmim]Cl	—	Amberlyst 15	80°C, 10 min	N.C.[a]	83.3%	[78]
[Bmim]Cl	Acetone	Amberlyst 15	25°C	N.C.[a]	78–82% (85–89%)	[79]
[Bmim]Cl	$CHCl_3$	Hydrochloric acid	25°C, 24 h	N.C.[a]	72%	[80]
[Bmim]Cl	—	Hydrochloric acid	80°C, 8 min	N.C.[a]	97%	[81]
[Bmim]Cl	—	HNO_3	80°C, 5 min	N.C.[a]	93%	[81]
[Bmim]Cl	—	[Sbmim][HSO_4]	80°C, 26 min	N.C.[a]	91%	[81]
[NMM][CH_3SO_3]	DMF–LiBr	[NMM][CH_3SO_3]	90°C, 2 h	N.C.[a]	74.8%	[82]
[NMP][CH_3SO_3]	DMSO	[NMP][CH_3SO_3]	90°C, 2 h	N.C.[a]	72.3% (87.2%)	[20b]
[NMP][HSO_4]	DMSO	[NMP][HSO_4]	90°C, 2 h	N.C.[a]	69.4% (70.4%)	[20b]
IL-HSO_4	DMSO	Si-IL-HSO_4	130°C, 30 min	N.C.[a]	63% (99.9%)	[83]
[Bmim]Cl	—	H_2SO_4	120°C, 240 min	Humin	93%	[3a]
[Bmim]Cl	—	NHC–Cr	100°C, 6 h	N.C.[a]	96%	[84]
[Hmim]Cl	—	[HMIM]Cl	90°C, 45 min	N.C.[a]	92%	[6a]
[Emim][HSO_4]	—	[Emim][HSO_4]	100°C 30 min	N.C.[a]	88%	[85]
[Bmim]Cl	—	WCl_6	50°C, 240 min	N.C.[a]	72%	[86]
[Emim][BF_4]	—	$SnCl_4 \cdot 5H_2O$	100°C, 180 min	N.C.[a]	62%	[87]
[Emim]Cl	—	$PtCl_2$	80°C, 180 min	N.C.[a]	83%	[8a]
[Emim]Cl	—	$CrCl_2$	80°C, 180 min	N.C.[a]	65%	[8a]
[Emim]Cl	—	$CrCl_3$	80°C, 180 min	N.C.[a]	69%	[8a]
—	H_2O	[Bmim][CH_3SO_3]	80°C, 120 min	N.C.[a]	~75%	[6b]
—	H_2O	[Bmim][CH_3SO_3] + CH_3SO_3H	80°C, 120 min	N.C.[a]	~85%	[6b]
—	DMA	H_2SO_4/NaBr	100°C, 120 min	N.C.[a]	93%	[8a]
—	DMSO	HCOOH	150°C, 6 h	N.C.[a]	99%	[88]
—	H_2O	$B(OH)_3$/NaCl	150°C, 45 min	N.C.[a]	60% (92%)	[89]
[Bmim]Cl	—	$IrCl_3$	150°C, 30 min	N.C.[a]	86.1% (100%)	[90]
[Bmim]Cl	—	$GeCl_4$	100°C, 5 min	N.C.[a]	92.1%	[91]

[a]N.C., not characterized.

of the sugar. Zhang also proposed that the $CrCl_3^-$ anion played an important role in proton transfer, facilitating mutarotation of glucose to fructose. Since this original disclosure, great effort has been devoted into developing more efficient and environment-friendly processes for the conversion of glucose (and cellulose) into HMF and elucidating the mechanism in ILs [6d, 6f].

The integrated hydrolysis of cellulose and dehydration of obtained glucose into HMF will present significant advantages for the development of chemicals from biomass. In 2009, Zhang and Zhao's group simultaneously reported this tandem conversion of cellulose and even lignocellulosic materials into HMF, [6c, 93] furthermore, it was found that the conversion can be significantly enhanced by microwave irradiation [94]. For example, a 62% yield could be achieved when using Avicel cellulose as feedstock in the presence of $CrCl_3{\cdot}6H_2O$ in [Bmim]Cl under microwave irradiation for only 2 min [6c]. Interestingly, when these components were heated at 100°C over an oil-bath for 240 min, HMF yield was only 17%, but total reducing sugar yield was 45%. These results suggested that microwave irradiation was critical to high yields of HMF, and $CrCl_3{\cdot}6H_2O$ was capable of catalyzing hydrolysis of cellulose. Over 90% yield of HMF was obtained when glucose was treated under these conditions. This system can be extended to the conversion of other common lignocellulosic materials (e.g., corn Stover, Pine wood) into HMF and furfural products with satisfactory yields ranging from 45 to 52% and 23 to 31%, respectively [94].

In Zhang's work, they selected a pair of metal chlorides ($CuCl_2/CrCl_2$, $\chi_{CuCl_2} = 0.17$) as the catalyst for direct conversion of cellulose into HMF in [Emim]Cl under mild conditions. It is believed that the metal chlorides can accelerate the depolymerization of cellulose in ionic liquids, thus facilitating the conversion [93]. HMF yields were around 55% when the reaction was held between temperatures of 80 and 120°C for 8 h. Solid acid catalysts with moderate acidic property, such as Zeolite, can be used to promote cellulose hydrolysis and to slow down the decomposition of HMF during this tandem conversion associated with chromium catalyst towards converting biomass to HMF in ILs [12b].

Although a lot of effort has been devoted to screening lower toxic and more efficient catalysts than chromium (II)-based catalysts, successful examples are rare. Therefore, the elucidation of the mechanism of $CrCl_2$ catalysis in ILs is necessary for the development of more efficient catalytic system. In 1970s, the mechanism of the conversion of glucose into HMF in acidic water was investigated by isotopic labeling experiments [95]. A 1,2-hydride shift was a major contributor in the process of formation of fructose from glucose, followed by dehydration of fructose into HMF. In 2010, Pidko and coworkers [96] reported the molecular-level details of the unique reactivity of chromium(II) chloride towards selective glucose dehydration in ILs through a combination of kinetic experiments, *in situ* X-ray absorption spectroscopy, and density functional theory calculations, which was inspired by the enzyme-catalyzed process. They proposed that the chloride anions of the ILs play the role of basic mediators through formation of a hydrogen-bonding network with the hydroxyl groups of the carbohydrate, thus facilitating the proton transfer; the rate-controlling H-shift reaction of the open form of the carbohydrate is facilitated by the transient self-organization of the Lewis acidic Cr^{2+} centers into a binuclear complex. This research provided evidence of a cooperative nature of the Cr complexes and the presence of moderately basic sites in the ILs. In addition to these models, two variations for the mechanism were proposed recently [97]. First, that the enolization pathway goes through an enediol(ate) intermediate that is protonated at C-1 to yield a ketose assisted by Cr^{2+}, which is then dehydrated into HMF via a furanosyl oxocarbenium ion. Second, the hydride shift pathway that goes through a chromium-assisted 1,2-hydride shift to form the ketose from the aldose in a single step. Their further study using isotopic labeled glucose demonstrated the ketose formation occurring via 1,2-hydride shift, similar to that found in the catalytic mechanism of xylose isomerase. Zhao et al. considered that in both of these two pathways, the cooperative action of Cr^{2+} and halide anion contributed to the formation of the enediol intermediate and the 1,2-hydride shift to form ketose from aldose [6c]. Collectively, a putative mechanism for the chromium-catalyzed conversion of glucose into HMF was presented in Figure 3.8. It is worthy of mention that although $CrCl_2$ has been very successful in terms of HMF production from glucose and cellulose, $CrCl_3$ or $CrBr_3$ gave similar, even better results for the same chemistry [6c, 94]. Zhao et al. believe that the good performance of $CrCl_2$ might be due to the oxidation of $CrCl_2$ into $CrCl_3$ by residual oxygen in the reaction system [98].

Apart from the chromium catalyst for the dehydration of glucose into HMF, other catalytic systems have been investigated and the results are summarized in Table 3.3. Especially, a series of heteropoly acids, such as 12-tungstophosphoric acid (12-TPA ($H_3PW_{12}O_{40}$)), 12-molybdophosphoric acid (12-MPA ($H_3PMo_{12}O_{40}$)), 12-tungstosilicic acid (12-TSA ($H_3SiW_{12}O_{40}$)), and 12-molybdosilicic acid (12-MSA ($H_3SiMo_{12}O_{40}$)) were also found to dehydrate carbohydrates efficiently in ILs. Nearly 100% yield of HMF could be achieved using 12-MPA in a solution of [Emim]Cl and acetonitrile. The addition of acetonitrile to [Emim]Cl can suppress the formation of humins from glucose. The authors proposed that the high HMF selectivity achievable with heteropolyacid catalysts is ascribed to the stabilization of 1,2-enediol and other intermediates involved in the dehydration of glucose and the avoidance of forming the 2,3-enediol intermediate leading to furylhydroxymethyl ketone (FHMK). Furthermore, carbon-supported metals, for example, Pd/C, were effective in promoting the hydrogenation of HMF dissolved in [Emim]Cl into 2,5-dimethyl furan, which presented a tandem conversion of biomass into biofuels in ILs. Another example of tandem conversion of biomass into biofuels is a single-step conversion of agar into biofuel candidates, such as 5-ethoxymethyl-2-furfural (EMF), and levulinic acid ethyl ester (LAEE) in the presence of mixed catalytic system including Dowex, [Emim]Cl, and $CrCl_2$ under mild conditions. The GC–MS results revealed that the crude product contained only EMF and LAEE (EMF/LAEE = 5:2), in 30 wt% yield [99]. Considering the inconvenient separation of products of biomass conversion of ionic liquids, the tandem conversion of biomass into valuable chemicals and biofuels will be a right direction in the future.

On the basis of an in-depth understanding of the conversion of cellulose into HMF through hydrolysis, enolization, and dehydration process, the corporation of pairs of metal catalysts, especially associated with $CrCl_2$ was studied. The effect of the

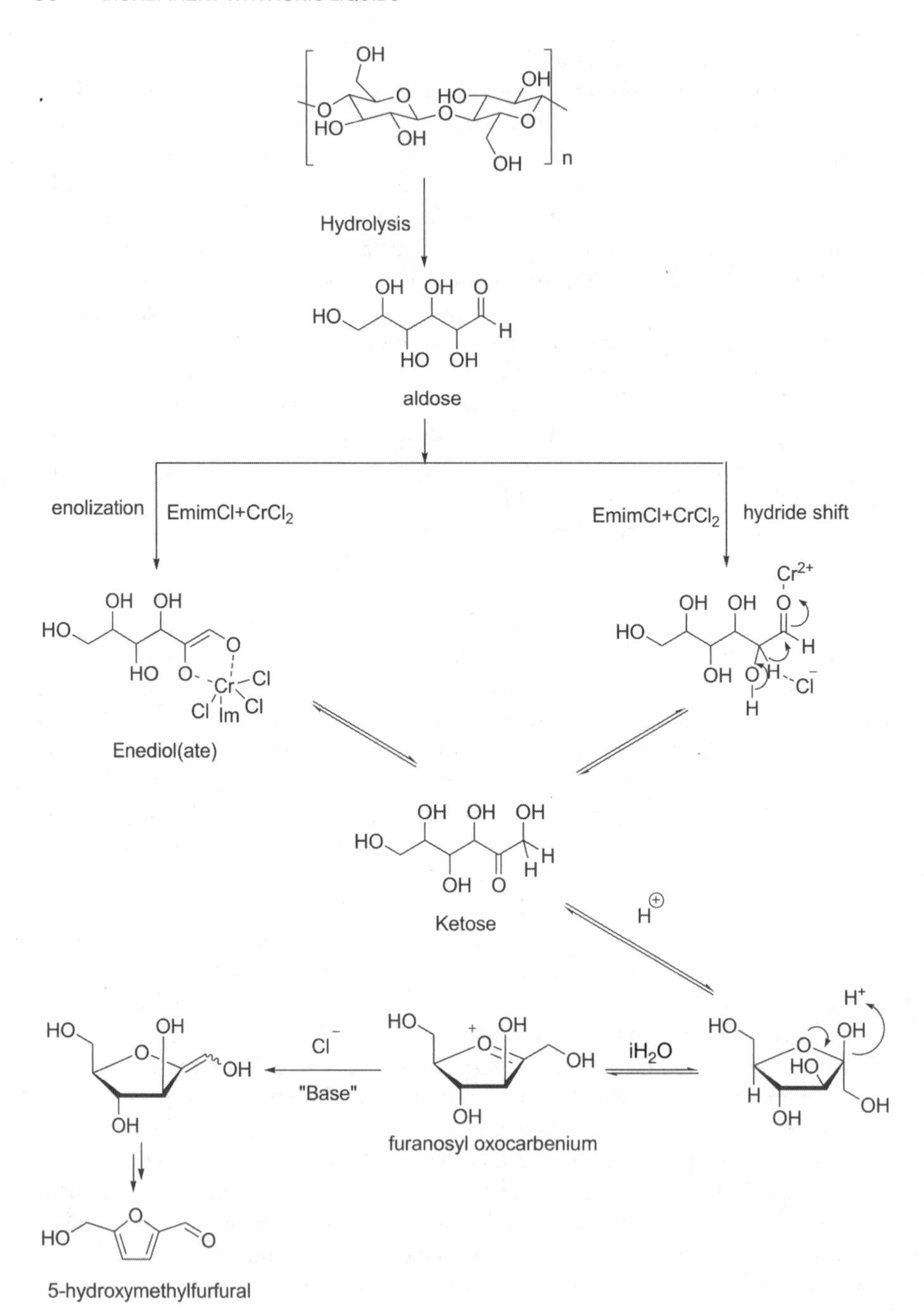

FIGURE 3.8 Proposed mechanism of $CrCl_2$ promotes the dehydration of glucose in ILs.

TABLE 3.3 HMF from Glucose, Cellulose, and Other Carbohydrates in ILs

Substrate[a]	Catalyst	Ionic Liquids	Co-Solvent	Condition	By-Product	HMF Yield (Selectivity)	References
Glucose	$CrCl_2$	[Emim]Cl	–	100°C, 180 min	–	68%	[101]
Glucose	$CrCl_3$	[Emim]Cl	–	100°C, 180 min	–	45%	[101]
Glucose	NHC/$CrCl_2$	[Bmim]Cl	–	100°C, 6 h	–	81%	[102]
Glucose	$CrCl_2$/LiBr	–	DMA	100°C, 240 min	–	79%	[97a]
Glucose	$SnCl_4{\cdot}5H_2O$	[Emim][BF_4]	–	100°C, 180 min	–	60%	[103]
Glucose	$CrCl_3$	[Bmim]Cl	–	MI (400 W)[b], 1 min	–	91%	[6c]
Avicel	$CrCl_3$	[Bmim]Cl	–	MI (400 W)[b], 240 min	16% TRS	61%	[6c]
Spruce	$CrCl_3$	[Bmim]Cl	–	MI (400 W)[b]	26% TRS	53%	[6c]
Sigmacell	$CrCl_3$	[Bmim]Cl	–	MI (400 W)[b]	23% TRS	55%	[6c]
α-Cellulose	$CrCl_3$	[Bmim]Cl	–	MI (400 W)[b]	20% TRS	62%	[6c]
Glucose	$CrCl_3{\cdot}6H_2O$	[Bmim]Cl	–	100°C, 240 min	–	91%	[104]
Glucose	H_2SO_4	[Bmim]Cl	–	120°C, 120 min	–	11.9	[17]
Xylan	$CrCl_2$, HCl	[Emim]Cl	–	140°C, 2 h	–	25%	[105]
Glucose	CF_3COOH	[Bmim]Cl	–	100°C, 180 min	–	75% (16% Humins)	[99]
Glucose	12-MSA	[Bmim]Cl	–	100°C, 180 min	–	91% (9% Humins)	[99]
Tapioca starch	HCl/$CrCl_2$	[Omim]Cl	–	120°C, 90 min	–	73%	[106]
Sucrose	HCl/$CrCl_2$	[Moim]Cl	–	120°C, 30 min	–	82%	[107]
Agarose	Dowex 50WX8/$CrCl_2$	[Emim]Cl	DMSO	90°C, 10 h	EMF and LAEE 30%	25%	[108]
D-xylose	NaCl/HCl	–	H_2O	200°C	–	79.2% (87.7%)	[109]
Glucose	$CuCl_2$/$CrCl_2$	[Bmim]Cl	–	80°C	–	81%	[100]
Glucose	$CrCl_2$	[Emim]Cl	–	100°C, 3 h	–	62%	[96]
Glucose	$CrCl_3{\cdot}6H_2O$	[Bmim]Cl	–	140°C MI[b], 30 s	–	71%	[110]
Sucrose	$CrCl_3{\cdot}6H_2O$	[Bmim]Cl	–	100°C MI[b], 5 min	–	76%	[110]
Cellobiose	$CrCl_3{\cdot}6H_2O$	[Bmim]Cl	–	140°C MI[b], 5 min	–	55%	[110]
Cellulose	$CrCl_3{\cdot}6H_2O$	[Bmim]Cl	–	150°C MI[b], 10 min	–	54%	[110]
Inulin	Amberlyst 15	[Bmim]Cl	–	80°C, 65 min		82%	[111]
Glucose	$Yb(OTf)_3$	[Bmim]Cl	–	140°C, 6 h	–	24% (37%)	[112]
Cellulose	$CrCl_3{\cdot}6H_2O$	[Bmim]Cl	–	160°C, 5 min	–	55%	[113]

(*continued*)

TABLE 3.3 (*Continued*)

Substrate[a]	Catalyst	Ionic Liquids	Co-Solvent	Condition	By-Product	HMF Yield (Selectivity)	References
Cellulose	$CrCl_3 \cdot 6H_2O$	[Bmim]Cl	–	MI[b] (400 W), 2.5 min	–	62%	[94]
Corn stalk	$CrCl_3 \cdot 6H_2O$	[Bmim]Cl	–	MI[b] (400 W), 3 min	Furfural 23%	45%	[94]
Rice straw	$CrCl_3 \cdot 6H_2O$	[Bmim]Cl	–	MI[b] (400 W), 3 min	Furfural 25%	47%	[94]
Pine wood	$CrCl_3 \cdot 6H_2O$	[Bmim]Cl	–	MI[b] (400 W), 3 min	31%	52%	[94]
Glucose	$B(OH)_3$/NaCl	–	H_2O	150°C, 5 h	–	14% (34%)	[89]
Cellulose	$[C_4SO_3Hmim]HSO_4$	[Bmim]Cl	–	80–120°C,	–	13%	[114]
Raw acorn	$CrBr_3/CrF_3$	[Omim]Cl	Ethyl acetate	80°C, 2 h	–	58.7%	[115]
Glucose	boric acid	[Emim]Cl	–	120°C, 3 h.	–	42% (43%)	[116]
Sucrose	boric acid	[Emim]Cl	–	120°C, 24 h.	–	66%	[116]
Cellulose	$CuCl_2/PdCl_2$	[Emim]Cl	–	120°C, 0.5 h.	–	8%	[117]
Cellulose	Ipr-$CrCl_2$	[Bmim]Cl	–	120°C, 12 h	–	47.5%	[118]
Microcrystalline cellulose	$CoSO_4$/MIBK	[4-SBMI][HSO_4]	–	150°C, 300 min	Furfural 7%; Levulinic acid 8%; sugar 4%	24% (84%)	[119]
Glucose	$Cr(NO_3)_3$	[Bmim]Cl	–	100°C, 3 h	–	37.2% (81.9)	[90]
Sucrose	$IrCl_3$	[Bmim]Cl	–	100°C, 3 h	–	37.4%	[90]
Chicory roots	HCl	[Omim]Cl	Ethyl acetate	100°C, 3 h	–	50.9%	[120]
Glucose	5 Å molecular Sieves/$GeCl_4$	[Bmim]Cl	H_2O	100°C, 30 min	humins	48.4%	[91]
Cellobiose	$GeCl_4$	[Bmim]Cl	H_2O	100°C, 30 min	–	41%	[91]
Sucrose	$GeCl_4$	[Bmim]Cl	H_2O	100°C, 30 min	–	55.4%	[91]
Cellulose	$GeCl_4$	[Bmim]Cl	H_2O	100°C, 30 min	–	35%	[91]
Cellulose	$Cr[(DS)H_2PW_{12}O_{40}]_3$	–	H_2O	150°C, 2 h	–	52.7% (77.1%)	[121]
Corn stover	$Cr[(DS)H_2PW_{12}O_{40}]_3$	–	H_2O	150°C, 2 h	–	30.8%	[121]
Husk of *Xanthoceras sorbifolia Bunge*	$Cr[(DS)H_2PW_{12}O_{40}]_3$	–	H_2O	150°C, 2 h	–	35.5%	[121]

[a]N.C., not characterized.

[b]MI, microwave irradiation. TRS, total reducing sugars.

association of $CuCl_2$ into $CrCl_2$ was first identified by Zhang et al. [21]. One recent publication by Kim et al. [9] demonstrated that the combination of $CrCl_2$ and $RuCl_3$ is outstanding among the combination of different metal salts (e.g., $MnCl_2$, $FeCl_2$, $CoCl_2$, $NiCl_2$, $ZnCl_2$) with $CrCl_2$, which is an effective catalyst for the conversion of cellulose into HMF with nearly 60% yield. Molecular-level insight into the structural and coordination properties of chromium(II) chlorides and copper(II) and their complexes with glucose in the ionic liquids has been investigated through a combined density functional theory and *in situ* X-ray absorption spectroscopy study [100]. It is believed that both of these metal chlorides are able to promote mutarotation of glucose that involves the glucopyranose ring opening. However, only Cr^{2+} can catalyze further isomerization to fructose that is required for the selective HMF production. There is no special chemical bonding between the metal centers and cationic part of the ionic liquids. The promotion effect of $CuCl_2$ on the glucopyranose ring-opening, mutarotation, and potential low-selective dehydration paths are explained by the high mobility of basic Cl^- ligands in the dominant $CuCl_4{}^{2-}$ species and substantially predicted stability of undercoordinated $CuCl_3{}^-$ species. XAS measurements indicate the reduction of bivalent copper species in the presence of glucose at 80°C. High glucose conversion by $CuCl_2$ is most likely associated with the oxidation of the sugar substrate and reduction of the active species to Cu^+ [100].

In order to increase the recycle ability of chromium catalyst, stabilization and immobilization of the $CrCl_2$ active phase in a thin layer of an ionic liquid covalently grafted to mesoporous silica and hydroxyapatite had also been reported, with moderate HMF yields and recyclability achieved [15].

3.3.5.3 Catalytic Conversion of Lignin or Lignin Model Compounds into Valuable Compounds Lignin is the second largest component of biopolymers in biomass, however, the catalytic conversion of lignin into aromatic compounds in ILs has only recently started. Presently, the study usually starts with lignin model compounds with the aim to exploit ILs as media for depolymerization of lignin. Lignin contains several aryl–alkyl ether linkages, among which the *beta-O-*4 linkage is dominant. In 2009, Joseph B [122] et al. investigated reactions of lignin model compounds in 1-ethyl-3-methyl-imidazolium triflate [Emim][OTf] using Bronsted acid catalysts at moderate temperatures (below 200°C). They obtained up to 11.6% molar yield of the dealkylation product 2-methoxyphenol from the model compound 2-methoxy-4-(2-propenyl)phenol and cleaved 2-phenylethyl phenyl ether, a model for lignin ethers. However, the acid catalyst failed in the dealkylation of the saturated-chain model compound 4-ethyl-2-methoxyphenol and did not produce monomeric products from organosolv lignin. A series of organic bases of various basicity and structure were used to investigate the cleavage of the *beta*-O-4 bond in a lignin model compound guaiacylglycerol-*beta*-guaiacyl ether using 1-butyl-2,3-dimethylimidazolium chloride as a solvent. Their results showed that among all the tested N-bases, 1,5,7-triazabicyclo[4.4.0]dec-5-ene was the most active, leading to more than 40% O-4 ether bond cleavage, and suggest the higher activity is probably associated with the accessibility of the N-atoms [85].

ILs based on the 1-methylimidazolium cation with chloride, bromide, hydrogen-sulfate, and tetrafluoroborate counterions along with 1-butyl-3-methylimidazolium hydrogen-sulfate were employed to degrade two lignin model compounds, guaiacylglycerol-*beta*-guaiacyl ether and veratrylglycerol-*beta*-guaiacyl ether. The acidity of each ionic liquid was approximated using 3-nitroaniline as an indicator to measure the Hammett acidity ($H(0)$). Although all of the tested ILs were strongly acidic ($H(0)$ between 1.48 and 2.08), the relative acidity did not correlate with the ability of the ILs to catalyze *beta*-O-4 ether bond hydrolysis. The reactivity of the model compounds in the ILs is dependent not only on the acidity, but also on the nature of the ions and their interaction with the model compounds [123].

Guaiacylglycerol-guaiacyl ether (GG), which contains a predominant inter-unit linkage of lignin, could be converted into a corresponding glycerol type enol–ether (EE), 3-(4-hydroxy-3-methoxyphenyl)-2-(2-methoxyphenoxy)-2-propenol, by the heat treatment in ionic liquids. EE was formed as a primary reaction product in all ILs used in this research at 120°C, although the decomposition rate and secondary decomposition products of GG varied with the ILs used. NMR data suggested that dehydration reaction of GG progressed stereospecifically and the [*Z*] isomer was formed.

The catalytic conversion of lignin in ILs is seldom reported possibly due to their complex structure and products distribution, however, structural modification by oxidation catalyzed by metal salts in ILs have been investigated recently [1a, 65]. For example, when Alcell and soda lignin were dissolved in the ionic liquid 1-ethyl-3-methylimidazolium diethylphosphate [Emim][DEP] and subsequently oxidized using several transition-metal catalysts and molecular oxygen under mild conditions. $CoCl_2$ in [Emim] [DEP] proved particularly effective for the oxidation. The catalyst rapidly oxidized benzyl and other alcohol functionalities in lignin, but left phenols, and 5-5′, *beta*-O-4 and phenylcoumaran linkages intact, as determined by analysis of various lignin model compounds and ATR–IR spectroscopy. The catalyst system oxidized the alcohol functionality contained in cinnamyl alcohol to form cinnamaldehyde or cinnamic acid, and/or disrupted the double bond to form benzoic acid or an epoxide. The benzyl functionality in veratryl alcohol, a simple nonphenolic lignin model compound, was selectively oxidized to form veratraldehyde at a maximum turnover frequency of 1440 h^{-1}, compared to 10–15 h^{-1} reported for earlier systems. Phenolic functional groups contained in guaiacol, syringol, and vanillyl alcohol remained intact, although the benzyl alcohol group in the latter was oxidized to form vanillin [65]. The system represents a potential method in a biorefinery scheme to increase the oxygen functionality in lignin prior to depolymerization or additional functionalization of already depolymerized lignin.

3.3.5.4 Other Platform Molecules ILs are ideal solvents for both biomass dissolution and catalytic conversion [124]. It is expected that various valuable chemicals can be achieved with different catalysts in ILs. Besides the furfural derivatives, the preparation of other valuable chemicals such as levulunic acid and esters, sugar alcohols, and long-chain alkyl glycosides has been investigated.

Sugar alcohols, which are a family of important bioproducts, are usually produced from the hydrogenation of monosaccharides. After the dissolution of carbohydrates in ILs, two kinds of catalysts have been tried for the reductive depolymerization of cellulose for the production of sugar alcohols. In 2010, Ignatyev et al. [125] reported that a combination of the heterogeneous catalysts Rh/C and Pt/C in the presence of Lewis acid $BF_3{\cdot}Et_2O$ at 20°C under 2 MPa hydrogen gas is efficient for the hydrogenation of 1,1-diethoxycyclohexane to ethoxycyclohexane, while ineffective for the hydrogenation of cellubiose. However, a 43% yield of sorbitol was achieved if the heterogeneous catalysts were replaced by a homogeneous catalyst precursor ($HRuCl(CO)(PPh_3)_3$). Accordingly, a 51% of sorbital yield was obtained when using cellulose as the feedstock under optimized conditions.

Zhu and coworkers [126] reported ILs with boronic acid functionality capable of reversibly binding cellulose and stabilizing transition-metal nanoparticles. The reversible binding of boronic acids with the multiple hydroxyl groups on cellulose could break up the crystal packing of cellulose, thus improving solubility and catalytic activity. Only 15% conversion was achieved for cellulose hydrogenation to hexitols using a Ru nanocluster catalyst in [Bmim]Cl in the absence of the designed ionic liquid. However, the yields range from 76% to 93% in the presence of the designed ionic liquids. This ionic liquid is efficient for the hydrolysis of cellulose in ionic liquids, with a 95% glucose yield achieved after 5 h at 80°C. Furthermore, the catalytic system can be easily reused without loss of activity even after five runs.

Long-chain alkyl glycosides are nonionic compounds with excellent surfactant properties, low toxicity, and good biodegradability. They have a couple of applications in cosmetics and detergents, food emulsifiers, and pharmaceutical dispersing agents. One-pot catalytic conversion of cellulose into them was reported in ILs under mild conditions catalyzed by acidic resins [127]. It was found that the catalytic system is applicable to various alcohols, such as butanol, hexanol, and octanol, with total yields of surfactants ranging from 7.2% to 91%. The amount of water in the reaction medium played an important role. It was found that the removal of water from the system by reducing the reaction pressure could facilitate the glycosidation significantly.

Levulinic acid and levulinic-acid esters have been particularly recognized as important bioderived platform chemicals that may provide a starting point for the production of chemicals and fuels. Saravanamurugan et al. [128] reported the catalytic transformation of sugars fructose, glucose, and sucrose to ethyl levulinate with different sulfonic acid-functionalized ILs (SO_3H-ILs) as catalysts, in the presence of ethanol as reactant and solvent. Traditional catalysts, such as acidic resins, Y-type zeolites, Fe-pillared montmorillonite catalysts usually suffer from disadvantages including low yields, low selectivity, and low thermal stability of catalysts. The acidic ILs have high thermal stability and catalytic activity and good reusability. In all these reactions, ethyl levulinate was found to be the predominant product, and yields of 68, 70, and 74% were obtained with the ionic liquids 1-methyl-3-(4-sulfobutyl)imidazolium hydrogensulfate ([Bmim-SO_3H][HSO_4]), 1-(4-sulfobutyl)pyridinium hydrogensulfate ([BPyr-SO_3H][HSO_4]), and *N,N,N*-triethyl-4-sulfobutan-ammonium hydrogensulfate ([NEt_3B-SO_3H][HSO_4]), respectively, and with full fructose conversion. ILs based on the [NTf_2] anion exhibited slightly higher yield of ethyl levulinate (77%) compared to other ILs, thus suggesting that the reaction progression is correlated with the acid strength of the ILs [128]. Solid acids were also efficient for this conversion [129].

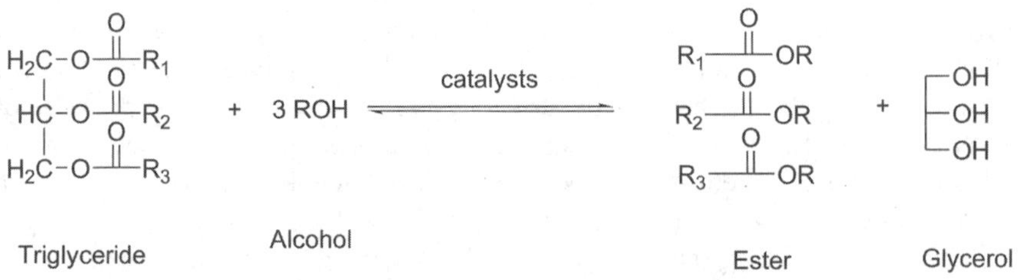

$$\begin{matrix} H_2C-O-C(=O)-R_1 \\ HC-O-C(=O)-R_2 \\ H_2C-O-C(=O)-R_3 \end{matrix} + 3\,ROH \xrightleftharpoons{\text{catalysts}} \begin{matrix} R_1-C(=O)-OR \\ R_2-C(=O)-OR \\ R_3-C(=O)-OR \end{matrix} + \begin{matrix} -OH \\ -OH \\ -OH \end{matrix}$$

Triglyceride Alcohol Ester Glycerol

FIGURE 3.9 Acid-, base-, and lipases-catalyzed transesterfication of triglycerides to biodiesel.

3.3.6 Production of Biodiesel with Ionic Liquids

Biodiesel, chemically defined as monoalkyl esters of long-chain fatty acids, are derived from renewable feedstocks such as vegetable oils and animal fats. Recently, the production of biodiesel from lipid from oleaginous yeast and algae has obtained much attention. The biodiesel was produced by both chemical and biological conversions of oil or fat with a monohydric alcohol in the presence of an acid or base catalysts, and lipases (Fig. 3.9) [130]. Downstream processing costs and environmental problems associated with biodiesel production and by-products recovery have stimulated the search for alternative production methods and alternative substrates.

ILs are recognized as green solvents due to their special properties compared with traditional organic solvents, such as tunability, nondetectable vapor point, and performance benefits over molecular solvents. Their properties can be designed to suit a particular need, for example, it is easy to make either the cation, anion, or both acidic [131] or basic for special synthesis. The principle was widely used for the biodiesel production from lipids [6e]. The published literatures on biodiesel production in ILs can be classified as follows:

(1) Acid and base-catalyzed esterification or transesterification reaction in ILs;
(2) Acidic and basic functionalized ILs catalyzed reaction;
(3) Enzyme-catalyzed esterification or transesterification reaction in ILs.

The simplest way to use ILs for the biodiesel synthesis is to perform the transesterification reaction under multiphase acidic and basic conditions by adding acidic or basic catalysts in ILs directly, [132] such as K_2CO_3, NaOH, hydroxide salts of ammonium cations, sodium methoxide, lithium diisopropylamide, and H_2SO_4. At the end of the reaction, a two-phase system (a glycerol–methanol–ionic liquids–catalyst phase and biodiesel phase) usually forms due to the immiscibility of biodiesel with the ionic liquid used. The catalytic system can be reused after decanting the biodiesel. The miscibility of ionic liquids with biodiesel and glycerol can also be used to extract the glycerol by-product during traditional process for high-purity biodiesel production [133]. However, the chemical stability of ionic liquids needs to be considered for this purpose under the drastic acidic or basic condition. For example, the tetrafluroborate and hexaflurorophosphate-based ionic

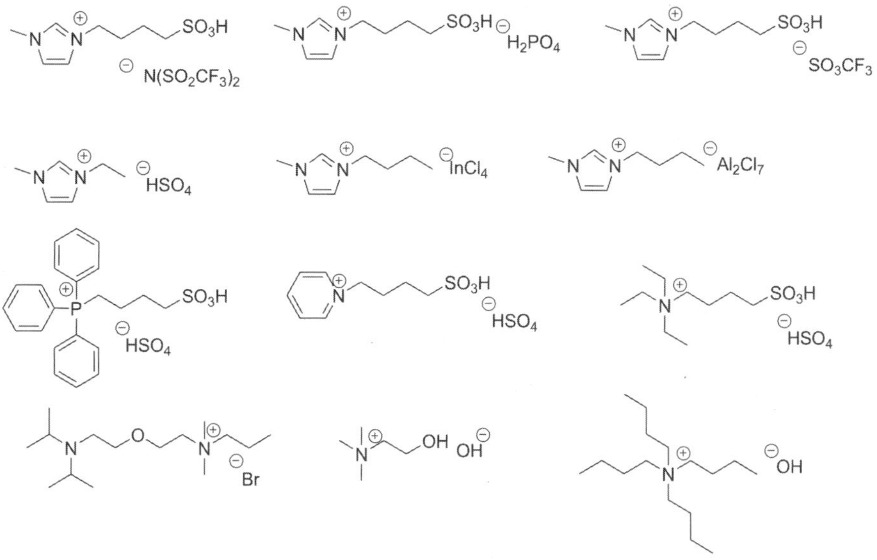

FIGURE 3.10 Typical acidic and basic ILs for biodiesel synthesis.

liquids should be avoided due to their decomposition by the formation of HF during the reaction [54].

Besides the external addition of acidic or basic catalysts, the usage of intrinsically acidic or basic ILs as both catalysts and solvents for the synthesis of biodiesel are also popular. Figure 3.10 has listed ILs commonly used in the literatures [6e, 133b]. The ILs can be synthesized by introduction of acidic functional groups into either the cation or anion, or adding a Lewis-acid catalysts in ILs to form a catalytic active Lewis acid ILs. Regardless of the use of ILs as solvents or catalysts, the process provides an efficient and facile pathway for the biodiesel synthesis, with high yields and purities usually achieved, which have been collected in Table 3.4.

ILs are also good solvents for biocatalytic conversion. Compared with the acids/bases catalytic processes, the enzyme-catalyzed process is safer and less corrosive [134]. The design and application of lipase-compatible ILs for biodiesel synthesis have also been widely investigated [132, 135]. For example, *Candida antarctica* lipase and *Pseudomonas cepacia* lipase were successfully immobilized in different structural ILs for the methanolysis of soyabean oil. It is demonstrated that the usage of ILs can resolve the problems of low stability of enzyme and decreased enzyme activity which are usually the case in traditional organic solvents. Furthermore, the reaction can be performed at room temperature, and the biodiesel is separated by simple decantation resulting in a facile separation and reuse of ILs/enzyme catalytic systems; the system can be reused at least four times without the loss of catalytic activity and selectivity. However, it was also found that most available ILs (especially hydrophobic examples) have poor capability in dissolving lipids, while hydrophilic ILs tend to cause enzyme inactivation. Zhao et al. [136]

TABLE 3.4 Biodiesel Production in ILs

Raw Materials[a]	Catalyst	Ionic Liquids	Conditions	By-Product	Biodiesel Yield (%)	References
Biodiesel/glycerol	–	$EtMe_3NCl$	$EtMe_3NCl$/glycerol (1:1)	Glycerol 100%	N.C.[a]	[133a]
Biodiesel/glycerol	–	$EtNH_3Cl$	$EtNH_3Cl$/glycerol (1:1)	Glycerol 100%	N.C.[a]	[133a]
Soybean oil	Immobilized *Candida antarctica* Lipase-catalyzed	[Emim][TfO]	50°C, 12 h	<20%	80%	[134]
Soybean oil	*Pseudomonas cepacia lipase*	[Bmim][NTf_2]	r.t.	N.C.[a]	96.3%	[135b]
Soybean oil	Acid/base-catalyzed	[Bmim][NTf_2]	H_2SO_4/K_2CO_3 (40 mol%)	N.C.[a]	>98%	[132]
Waste oils	Brønsted acidic IL	[$(CH_2)_4SO_3HPy$][HSO_4]	170°C for 4 h; methanol:oils: catalyst 12:1:0.06 (molar ratios)	N.C.[a]	>93.5%	[137]
Soybean oil	Chloroaluminate IL	[Et_3NH]Cl–$AlCl_3$ ($x(AlCl_3) = 0.7$),	70°C, 9 h	N.C.[a]	98.5%	[138]
Soybean oil	Lipase-producing filamentous fungi immobilized on biomass support particles	[Emim][BF_4] or [Bmim][BF_4]	24 h in biphasic systems	N.C.[a]	60%	[139]
Long-chain fatty acids	Brønsted acidic ionic liquid	[NMP][CH_3SO_3]	70°C, 8 h.	N.C.[a]	93.6–95.3%	[140]
Soybean oil	Fungus whole-cell Biocatalysts	[Bmim][BF_4]	72 h	N.C.[a]	60%	[135a]
Rapeseed oil or free fat acid	Brønsted acidic Ionic liquid	Zwitterion IL	70°C, 7 h	N.C.[a]	98%	[141]
Soybean oil,	Choline Chloride·$x$$ZnCl_2$ ILs	Choline Chloride·$x$$ZnCl_2$ ILs	70°C, 72 h	N.C.[a]	54.52%	[142]

Triolein	Novozym 435	$[C_{18}mim][NTf_2]$	60°C, 6 h	N.C.[a]	96%	[143]
Triolein or waste canola oil	Novozym 435	$[Bmim][PF_6]$	48°C	Triacetylglycerol	72%	[144]
Triolein	Lipase	$[Bmim][PF_6]$	48–55°C.	Triacetylglycerol	80%	[145]
Miglyol® oil 812	Novozym 435	$[Me(OEt)_3–Et_3N][OAc]$	50°C, 96 h	N.C.[b]	98%	[136]
Triolein	Novozym 435	$[C_{16}mim][NTf_2]$	60°C, 24 h	N.C.[b]	99.39%	[146]
Olive	Novozym 435	$[C_{16}mim][NTf_2]$	60°C, 24 h	N.C.[b]	93.38%	[146]
Sunflower	Novozym 435	$[C_{16}mim][NTf_2]$	60°C, 24 h	N.C.[b]	92.78%	[146]
Palm	Novozym 435	$[C_{16}mim][NTf_2]$	60°C, 24 h	N.C.[b]	94.11%	[146]
Cooking waste	Novozym 435	$[C_{16}mim][NTf_2]$	60°C, 24 h	N.C.[b]	96.91%	[146]
Crude palm oil	KOH	$[Bmim][HSO_4]$	1.0% KOH, 50 min, 60°C	N.C.[b]	98.4%	[147]
Oleic acid/EtOH	$[TMEDAPS][HSO_4]$	$[TMEDAPS][HSO_4]$	70°C, 6 h	N.C.[b]	96%	[148]
Stearic acid/EtOH	$[TMEDAPS][HSO_4]$	$[TMEDAPS][HSO_4]$	70°C, 6 h	N.C.[b]	94%	[148]
Myristic acid/EtOH	$[TMEDAPS][HSO_4]$	$[TMEDAPS][HSO_4]$	70°C, 6 h	N.C.[b]	94%	[148]
Palmitic acid/EtOH	$[TMEDAPS][HSO_4]$	$[TMEDAPS][HSO_4]$	70°C, 6 h	N.C.[b]	95%	[148]
Myristic acid/palmitic acid (1:1)/EtOH	$[TMEDAPS][HSO_4]$	$[TMEDAPS][HSO_4]$	70°C, 6 h	N.C.[b]	94%	[148]
Soybean oil/EtOH	$[TMEDAPS][HSO_4]$	$[TMEDAPS][HSO_4]$	70°C, 6 h	N.C.[b]	94%	[148]
Canola	$[HBSSB][HSO_4]$	3,3′-(hexane-1,6-diyl) *bis*(6-sulfo-1-(4-sulfobenzyl)-*1H*-benzimidazolium) hydrogensulfate	60°C, 5 h	N.C.[b]	95.1%	[149]
Palm		$[HBSSB][HSO_4]$	60°C, 5 h	N.C.[b]	84.5	[149]
Soybean	$[HBSSB][HSO_4]$	$[HBSSB][HSO_4]$	60°C, 5 h	N.C.[b]	91.9%	[149]
Sunflower	$[HBSSB][HSO_4]$	$[HBSSB][HSO_4]$	60°C, 5 h	N.C.[b]	93.3%	[149]
Refined corn oil	*Penicillium expansum* lipase	$[Bmim][PF_6]$	40°C, 20 h	N.C.[b]	86%	[150]
Miglyol® oil 812	Novozym 435	Choline acetate	40°C, 3 h, choline acetate/glycerol (1:1.5 molar ratio).	N.C.[b]	97%	[151]

[a]Another substrate was MeOH without special illustration.
[b]N.C., not characterized.

synthesized a new type of ether-functionalized ILs carrying anions of acetate or formate. These ILs are capable of dissolving oils at the reaction temperature (50°C); meanwhile, lipases maintained high catalytic activities in these media even in high concentrations of methanol (up to 50% v/v). High conversions of Miglyol oil were observed in mixtures of ILs and methanol (70/30, v/v) when the reaction was catalyzed by a variety of lipases and different enzyme preparations (free and immobilized), especially with the use of two alkylammonium ILs 2 and 3. The preliminary study on the transesterification of soybean oil in ILs/methanol mixtures further confirms the potential of using oil-dissolving and lipase-stabilizing ILs in the efficient production of biodiesels.

3.4 TOXICITY AND ECOTOXICITY OF IONIC LIQUIDS FOR BIOREFINERY

3.4.1 Introduction

There is great disparity between the toxicity and ecotoxicity studies of ionic liquids and concurrent studies with these solvents in the biorefinery. Many roles that ionic liquids will feature require large volumes of these isoteric materials. Three questions need answers

(1) How toxic are ionic liquids? and,

(2) How toxic are ionic liquids under study for biorefinery applications?

Leading ultimately to,

(3) How do we design and develop low toxicity and biodegradable ionic liquids for the biorefinery?

The previous sections have focused on the scope of applications of ionic liquids and performance of these processes. Herein, we give an overview of the current knowledge of toxicity and ecotoxicity of ionic liquids. The aim is to provide the reader with an understanding and appreciation of the complexity of toxicity and ecotoxicity studies. It is only by careful analysis of existing toxicity data that a reasoned judgment can be made to avoid making poor choices when selecting an ionic liquid for study. While modeling can support this work, there is no doubt that experimental data is essential.

This first Section (3.4.2) encompasses all major classes of ionic liquids. Lessons learnt from the investigations to reduce environmental impact of ionic liquids can be used as a guide for the design of novel ionic liquids for the biorefinery. The second Section (3.4.3) is restricted to toxicity and ecotoxicity data (Section 3.4.4) for ionic liquids investigated for biomass-dissolution properties. Specifically, a subset of 25 ILs was selected to provide an overview of cellulose dissolution studies in a recent review by Rogers [11c]. Comprehensive reviews of biodegradation of ionic liquids were reported by Gathergood [152] in 2010 and Stolte in 2011 [153].

The search for a "non-toxic" ionic liquid and the search for a "biodegradable" ionic liquid are separate yet intrinsically linked problems. As it is not viable to screen an ionic liquid against every organism on the planet, one can only state that an ideal nontoxic compound, in fact has low toxicity to the organisms tested. The compound could have high toxicity to a particular organism, but without data we will never know. Here modeling and extrapolation from known results is the key to make the best judgment call we can to promote safer chemicals. Biodegradation offers a solution to the limitation problems of toxicity screening. If an ionic liquid rapidly biodegrades, with 100% mineralization, then the issue of toxicity is for the most part alleviated. One caviat is that the toxicity of the ionic liquid (and metabolites) must not hinder the pathways leading to biodegradation. The challenge of designing any biodegradable compound, including ionic liquids, is twofold. First, the substance should rapidly "fall apart" or "breakdown" in the environment. Secondly, the substance should be "robust" and "fit for purpose" in the biorefinery. Researchers in the field work on the assumption that these two principles are not mutually exclusive. Before the discussion is expanded to include bioaccumulation, one must highlight the effect of metabolites. A flaw in toxicity screening is the oversimplification of reporting the value (e.g., IC_{50}, LC_{50} etc.) for the compound tested. A scenario can be envisaged where the chemical under test, has low toxicity to the particular organism, yet is converted under the test conditions (by chemical or biochemical pathways) to a toxic compound. The value reported for the toxicity of the (benign?) parent compound is in fact due to a metabolite. Thus the problem of toxicity screening is further exasperated as you can not screen against everything, and if this was possible, you need to identify complex mixtures of metabolites for each and every test. While this paints a very black picture, green chemistry aims to lead to real improvements. A more accurate term would be "greener," "cleaner," or "safer" chemistry. Is it possible to improve the propensity for a chemical to biodegrade? Can a toxic antimicrobial biocide solvent be substituted with a useful more environment-friendly low antimicrobial toxicity example?

Although ILs have repeatedly been described as "green solvents" because of their much lower vapor pressure than conventional reaction media, consideration of end of life factors should be of equivalent importance before reaching such a verdict [154]. To satisfy the requirements of being "green," it is desirable that ILs should neither persist in the environment nor prove toxic should an environmental release occur. Without toxicity and biodegradability studies, there is no way of estimating effects on the environment if, leaching of an IL from a landfill site into groundwater should take place. While computer modeling is an important tool, experimental data is required. Even though ILs may be considered as relatively benign because of the low risk of their atmospheric release, compared with volatile organic compounds (VOCs), their effect on aquatic and terrestrial life is still of key importance and detailed studies into the effects of ILs on aquatic organisms have now been undertaken. Furthermore, in a recent review into the toxic effects of ILs, Zhang [155], asserts that even the hazards represented by volatile emissions from ILs cannot be dismissed because of the high level of uncertainty that exists regarding environmental degradation of ILs, and accompanying gaseous emissions that may occur. It is now widely appreciated that ILs with PF_6 and BF_4 counter ions undergo hydrolysis to give HF (and also POF_3 in the case of PF_6) while an overview of toxicological data on common ILs was published by Ranke et al. [156]. In 2002, [157] Jastorff and coworkers undertook the first in-depth study into the toxicity and hazardous nature of ionic liquids. Since then, with increasing research into the use of ionic liquids for large-scale industrial applications, attention has shifted to the potential for environmental contamination by ILs through accidental spills, or effluent releases. The most commonly-used cations, 1-alkyl-3-methylimidazolium, ammonium, and pyridinium, have now been studied in a variety of systems based on different levels of biological complexity to assess the potential environmental impact of ionic liquids [156]. ILs based on phosphonium cations have also been studied and so far have shown little or no biodegradability [158].

3.4.2 Toxicity Studies

3.4.2.1 Toxicity Testing Using Bacteria The antimicrobial activities of a variety of methylimidazolium ILs containing an alkoxymethyl side chain (Fig. 3.11) ranging from 3 to 16 atoms in length with anions [Cl], [BF_4], and [PF_6] were measured by Pernak et al. [159]. Cocci (*Micrococcus luteus, Staphylococcus epidermidis, Staphylococcus aureus, Staphylococcus aureus (MRSA)*, and *Enterococcus hirae*), rods (*Escherichia coli, P. vulgaris, Klebsiella pneumoniae, Pseudomonas aeruginosa*), and fungi (*Candida albicans, Rhodotorula rubra*) were tested. The minimum inhibitory concentration (MIC) and minimum bactericidal concentration (MBC) values were noted and compared with the antimicrobial surfactant benzalkonium chloride (BAC). The authors demonstrated that ILs with side chains of more than six carbons proved active against all of the organisms, and the activity was dependent on side-chain length and to a lesser degree on the nature of the anion. The most active ILs against rods and cocci were those with side-chains C_{10} to C_{16}. C_{12} ionic liquids were the most active of all, with MIC (μM) values of 25 (Cl^-), 21 (BF_4^-), and 18 (PF_6^-) to *S. aureus*, 99 (Cl^-), 95 (BF_4^-), and 37 (PF_6^-) to *E. hirae*, 99 (Cl^-), 170 (BF_4^-), and 73 (PF_6^-) to *E. coli*, 197 (Cl^-), 170 (BF_4^-), and 147 (PF_6^-) to *K. pneumoniae*, and 395 (Cl^-), 340 (BF_4^-), and 587 (PF_6^-) to *P. aeruginosa*. All three salts containing C_{12} in the alkoxy side chain approached the activity of BAC (MIC values obtained for BAC ranged from as low as 7–11 μM

3-alkoxymethyl-1-imidazolium ionic liquids

R: C_3–C_{16}

X:[Cl], [BF_4], [PF_6]

BAC: Benzalkonium chloride,

R: C_8–C_{18}

FIGURE 3.11 Pernak compared 3-alkoxymethyl-1-imidazolium ionic liquids with the reference antimicrobial surfactant, BAC.

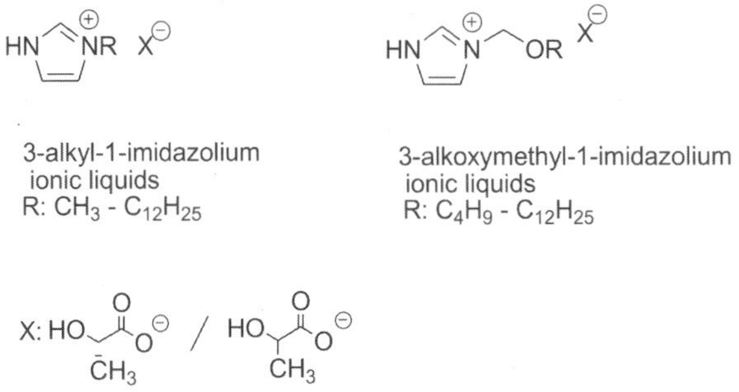

3-alkyl-1-imidazolium ionic liquids
R: CH_3 - $C_{12}H_{25}$

3-alkoxymethyl-1-imidazolium ionic liquids
R: C_4H_9 - $C_{12}H_{25}$

FIGURE 3.12 Lactate ionic liquids.

against *S. aureus*, *E. hirae*, *E. coli*, and *K. pneumoniae* to 54 μM against *P. aeruginosa*).

A further study from Pernak's group demonstrated that the stereochemical configuration of the counteranion can also affect the MIC of an IL. To prove this, in 2004 Pernak et al. [160] investigated the physical properties and biological activities of a series of racemic and L-lactate salts of alkyl and 3-alkoxymethyl-1-imidazolium ILs (Fig. 3.12) with side-chain lengths ranging from C_1 to C_{12}. Five different rod-shaped bacteria (*E. coli, P. vulgaris, K. pneumoniae, P. aeruginosa*, and *Serratia marcescens*), five strains of cocci (*M. luteus, S. epidermis, S. aureus, S. aureus (MRSA), E. hirae*), and two strains of fungi (*C. albicans and R. rubra*) were cultured in the presence of ionic liquids with lactate counter ions. In general, the L-lactates gave lower MIC values than the racemates, with the lowest values recorded for L-lactates with 11 or 12 carbons in the imidazolium side chain. The strongest inhibitor of growth was the C_{11} alkoxy L-lactate IL for which MICs of 11.4 and 45.6 μM were calculated for *S. aureus* and *E. hirae*, respectively. This compares with MIC values for C_{12}-DL-lactate ILs of 88 and 176 μM respectively for *S. aureus* and *E. hirae*, and MIC values for the corresponding C_{12}-L-lactate ILs of 10.9 and 44 μM. The lowest MBC values for cocci were recorded for DL-lactate ILs with 12-carbon ester chains and L-lactates with 11-carbon chains. Compared with the MIC and MBC values of benzalkonium chloride, it was shown that the L-lactates with the longest chains exhibited similar activity. Lactates C_1–C_5 generally proved to be inactive.

The toxicity of imidazolium and pyridinium ILs with varying side-chain lengths was measured using the bioluminescent marine bacterium, *Vibrio fischeri*, for which a decrease in light output signifies an increase in toxicity [161]. The antimicrobial effect of these ILs was also determined using four different bacteria and one yeast (*E. coli*, *S. aureus*, *Bacillus subtilis*, *Pseudomonas fluorescens*, and *Saccharomyces cerevisiae*). Toxicity results showed increasing toxicity with an increase in the alkyl chain length. The hexyl and octyl side chains were shown to be more toxic than commonly used organic solvents, with the butyl side chain being the same as some of the less toxic solvents. The 13 antimicrobial activities for all organisms were shown to increase with increasing alkyl chain length. The octyl containing side chain was shown to be the most effective at inhibiting colony formation, followed by hexyl, with butyl not showing much difference to the buffer. [Omim]Br and [Ompyr]Br were shown to be the most effective antimicrobial ILs tested in the study. All ILs were shown to be most inhibitory to *B. subtilis*. *E. coli*, and *P. fluorescens* were significantly affected by all ILs and *S. aureus* and *S. cerevisiae* were the least affected. Since the same trend of higher toxicity with increasing alkyl chain length can be seen for all organisms, the authors suggest that the toxicity may be related to a cellular structure or process that is common to all the organisms studied.

The rod-shaped bacterium, *V. fischeri* is a commonly used organism to measure toxicity as the acute bioluminescence inhibition assay (DIN EN ISO 11348) is widely used in Europe and for which a wide range of EC values are available for comparison. Using these organisms, increasing alkyl chain length of ILs was found to increase toxicity [162,163,164]. Another luminescent bacterium, *Photobacterium phosphoreum*, was used by Gathergood et al. [165] which again displayed the same trend of toxicity correlating with substituent chain length. While studying a range of aquatic organisms including *V. fischeri*, Stolte et al. [166] investigated the effects of IL head groups, functionalized side chains, and anions of ILs on toxicity. They found that halide anions (chloride and bromide) did not exhibit an intrinsic effect on toxicity; however, the [NTf_2] anion was found to exhibit a certain degree of toxicity. as Apart from confirming that toxicity increased with side-chain length effect, they found that incorporation of hydroxyl or carboxylic acid groups at the terminus of C_7/C_8 side chains of ILs accelerated biodegradation. While investigating the possibility of using imidazolium ILs for the extractive fermentation of lactate, Matsumoto et al. investigated the toxicity of 14 ILs on the lactic-acid-producing bacterium *Lactobacillus rhamnosus* NBRC 3863 [167]. The ILs tested were [Bmim], [hmim], and [omim][PF_6] and batch-test culture experiments were performed to investigate the toxicity. The overall results showed that this bacterium survived in all the ILs tested, but exhibited a low activity. These results were compared with toluene, in which almost no bacteria survived. There was little difference amongst the three side chains, with butyl and hexyl having approximately equal effects and showing slightly more bacterial activity than octyl. To examine the suitability of ILs as a replacement for organic solvents for *in situ* extractive fermentation of lactic acid, Matsumoto [168] further investigated the toxicity of these imidazolium ILs used in their previous lactic-acid-based test on a series of nine bacteria (*L. rhamnosus, Lactobacillus homohiochi* (NRIC 0119 and 1815), *Lactobacillus fructivorans* (NRIC 0224 and 1814), *Lactobacillus delbruekii*, *Lactobacillus pentosaceus*, *Leuconostoc fallax, Bacillus coagulans*). The degree to which bacterial lactic-acid production was inhibited was used as a measure of toxicity. The results demonstrated that all of the bacteria produced lactic acid in the presence of the ILs, with *L. delbruekii* giving the highest concentrations of the acid. However, as was expected, the ability to ferment sugars to lactic acid decreased upon increasing the alkyl chain length of the ionic liquid in the test.

Docherty et al. [169] performed the first Ames test on ILs. The Ames test is a method of elucidating the mutagenicity of a compound. Ten bromide salts of ionic liquids based on common imidazolium, pyridinium, and quaternary ammonium cations were tested using histidine-requiring TA98 and TA100 strains of *Salmonella typhimurium* bacteria. The ability of the IL to cause mutations in these bacteria was measured. After interaction with mutagens, these strains of bacteria revert to histidine-independent strains. The rate of reversion is used as a measure of the mutagenicity of the test sample. The ILs were tested over a broad concentration range (0.01–20 mg per plate); however, none of the ILs could be classified as mutagenic, according to United States Environmental Protection Agency (US EPA) criteria. However, some imidazolium ILs did show potential mutagenicity at much higher doses, and the authors advise that further work is required before any definitive conclusions regarding mutagenic effects can be drawn.

Toxicity, itself can be regarded as an important tunable characteristic of ionic liquids. On one hand, it can have a negative impact on the environment, but it may also be exploited in a number of beneficial applications. Microbial biofilms are ubiquitous in nature and are usually selected as a mode of growth of microorganisms. In 2009, Carson et al. [170] determined the antibiofilm activity of a series of 1-alkyl-3-methylimidazolium chloride ionic liquids [$C_n$mim]Cl ($n = 4, 6, 8, 10, 12, 14, 16$, or 18) by a simple zone of inhibition study against seeded *S. aureus*. In their experiments, the activity of ionic liquids proved to be dependent on the alkyl chain length, in keeping with observations made by numerous studies into ionic liquids with alkyl chain lengths greater than 10 carbon atoms. These ILs exhibited potent, broad-spectrum antimicrobial activity against a panel of pathogenic microorganisms, including clinical isolates of MRSA and other pathogens associated with hospital acquired, or nosocomial, infections. The authors anticipate that the employment of ionic liquid-based antimicrobials may help address the significant patient cost (in terms of morbidity and mortality) and financial burden imposed by the ever-present spectre of hospital acquired infections.

3.4.2.2 Toxicity Testing Using Aquatic Eukaryotes (Daphnia magna)

The water flea, *Daphnia magna* is a popular eukaryotic organism for the investigation of freshwater toxicity. Yet again, studies with *D. magna* confirmed that increasing alkyl chain length in the ILs correlated with higher toxicity [165, 171]. The acute and chronic toxicity of imidazolium-based ILs was measured against these organisms [172]. The ILs used were [Bmim][Br], [Cl], [PF_6], and [BF_4]. These ILs were compared with corresponding sodium salts of the same anion. It was concluded that the cation of the IL played the major role in increasing toxicity. LC_{50} values were an order of magnitude lower for all ILs containing the imidazolium moiety compared to their sodium analogues. Concerning the acute toxicity study, the most toxic species were found to be [Bmim]Br (LC_{50} 8.03 mg L^{-1}), with $NaPF_6$ being the least toxic (LC_{50} 9344.81 mg L^{-1}). All of the imidazolium ILs were found to have a negative impact on the reproduction of *D. magna*. In order to put the relative toxicity of these ILs in perspective, the authors compared their LC_{50} values with those of commonly used laboratory chemicals and found that the ILs demonstrated LC_{50} values comparable with those of phenol (LC_{50} 10–17 mg L^{-1}), tetrachloromethane (LC_{50} 35 mg L^{-1}), and trichloromethane (LC_{50} 29 mg L^{-1}), but lower than those of benzene (LC_{50} 356–620 mg L^{-1}), methanol (LC_{50} 3289 mg L^{-1}), and acetonitrile (LC_{50} 3600 mg L^{-1}), indicating considerable higher toxicity amongst the

ILs. However, the authors did find that the ILs were less toxic than the considerably more hazardous chemicals, chlorine (LC_{50} 0.12–0.15 mg L^{-1}) and ammonia (LC_{50} 2.90–6.93 mg L^{-1}). They stated that the mechanism of toxicity to *D. magna* is unknown, but based on previous studies [173–175] propose enzyme inhibition, disruption of membrane permeability, or structural DNA damage as possible modes of action. A range of ILs containing imidazolium, pyridinium, and ammonium cations were tested for toxicity against two aquatic organisms (*V. fischeri* and *D. magna*) by Couling in order to create quantitative structure property relationship models, to predict toxicity [163]. Neonates of *D. magna* less than 24 h old were used in a 48 h acute toxicity bioassay to measure toxicity of the ILs and the results were used to build a predictive toxicity model for *D. magna* based on Quantitative Structure–Property Relationship (QSPR) modeling in order to be able to design ILs with limited toxicity. QSPR modeling is based on the idea that for a given compound, chemical structure determines physical properties. The experimental results show that toxicity increases with alkyl chain length of the cation, and that the corresponding salts used to synthesize the IL, that is, NaBr, are less toxic than the ILs. This indicates the negligible effect of the anionic species on toxicity. The results also show that ILs containing imidazolium and pyridinium cations are more toxic than those containing quaternary ammonium species. The predictions based on these experiments again indicated that increasing alkyl chain length exacerbates toxicity. Also predicted was the fact that the toxicity would increase with an increase in the number of nitrogen atoms on the cationic species. The addition of methyl groups to the aromatic ring of the cation is predicted to decrease toxicity and monoatomic anions, (e.g. Br) should be less toxic than anions containing regions of positive charge, that is $[NTf_2]$. The imidazolium cation was predicted to be more toxic than the pyridinium, which in turn was predicted to be more toxic than the ammonium (cation toxicity: imidazolium > pyridinium > ammonium). Nockemann et al. [176] investigated the ecotoxicity of their choline saccharinate and acesulfamate ILs (Fig. 3.13).

Choline chloride is used as an additive in chicken feed, and saccharinate and acesulfamate are salts of artificial sweeteners. The authors suggest that because these ILs are derived from "food grade" ions, they should exhibit minimal toxicity. Indeed these two ILs displayed toxicity to *D. magna* two orders of magnitude lower than common imidazolium and pyridinium ILs.

3.4.2.3 *Toxicity Testing Using Algae* The toxicity of imidazolium ILs was investigated in two taxonomically different algal species, *Oocystis submarina* and

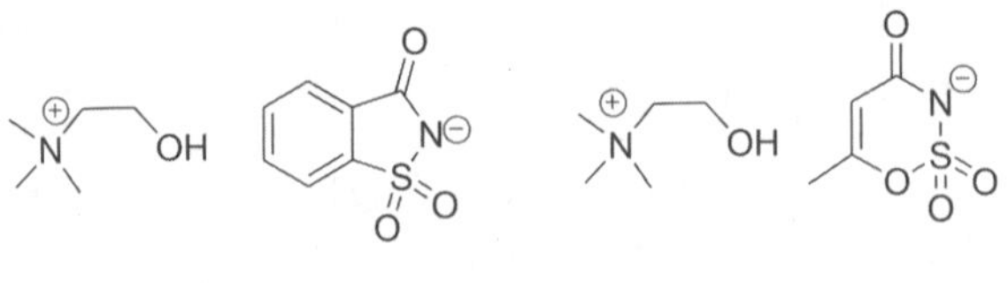

Choline saccharinate   choline acesulfamate

FIGURE 3.13 Choline saccharinate and choline acesulfamate ionic liquids.

the diatom, *Cyclotella meneghiniana* by Latała et al. [177]. The growth inhibition of these algae was measured in relation to ionic liquids, [Emim], [Bmim], [Hmim] and [benzylmim][BF_4]. Diatoms are commonly used when investigating marine toxicity, because they are the algae with highest sensitivity towards organic compounds. Growth inhibition was measured over three concentration values for each IL. With *O. submarina*, the results show the same pattern of cell-growth inhibition for all the ILs tested. The two lowest concentrations showed cell growth to be inhibited initially and then resumed within the timeframe of the experiment, indicating that the alga *Oocystis* can acclimatize to the presence of the IL. The highest concentration however showed a complete inhibition of growth for the duration of the experiment. Unusually, in the case of *O. submarina*, growth inhibition was at its greatest for the shortest alkyl chain length, [Emim][BF_4]. For *C. meneghiniana*, growth inhibition was observed for all concentrations of IL throughout the experiment. The unicellular freshwater alga, *Pseudokirchneriella subcapitata* (formerly known as *Selenastrum capricornutum*), was used as a test organism for measuring the toxicity of ILs [178]. Further study from this group on the toxicity of imidazolium and pyridinium-based ionic liquids towards the algae *Chlorella vulgaris*, *O. submarina* (green algae), and *C. meneghiniana*, and *Skeletonema marinoi* (diatoms) indicated a pronounced alkyl chain toxicity effect. EC_{50} values were linearly very well-correlated with the number of carbon atoms in the IL alkyl chains, but also proved that diatoms are far more sensitive than green algae to ionic liquids. Cell size also played an important role in the intoxication process, with a tenfold difference in cell size resulting in a 100% more sensitive reaction to ILs in both green algae and diatoms. Latała also found no significant difference in toxicity between alkylimidazolium salts and an alkylpy ridinium compound of similar lipophilicity. However, in comparison with the other anions tested in their study, Latała found that the use of BF_4^- and $CF_3SO_3^-$ as counteranions in the IL structure results in a pronounced increase in toxicity. In the case of BF_4^- possible hydrolysis leading to the formation of fluoride may give rise to a further increase in toxicity. In the case of $CF_3SO_3^-$, increased toxicity is likely to be a result of the relatively high lipophilicity of the anion and its strong association with alkylimidazolium cations, which in turn may enhance their cell-wall penetration.

Pham et al. [179] examined the effect of the anion on the toxicity of imidazolium ILs. Their results showed the [SbF_6] anion to be most toxic, while the other anions decreased in toxicity in the order [PF_6] > [BF_4] > [CF_3SO_3] > [$C_8H_{18}OSO_3$] > [Br] ~ [Cl]. Pham [180] also investigated the effect of the side-chain on toxicity using the same organism. Imidazolium bromide salts with octyl, hexyl, butyl, and propyl side chains, [Omim], [Hmim], [Bmim], [Pmim], were found to conform to the established pattern of reduced toxicity with decreasing chain length. Using the same alga as an indicator for toxicity, Pham [181] found the same trend of decreasing toxicity with decreasing alkyl chain length. They also found the pyridinium ILs tested to be more toxic than their imidazolium counterparts. In comparison with organic solvents, most of the ILs they tested were more toxic than methanol, DMF, and 2-propanol. The same trend of decreasing toxicity with decreasing alkyl chain length was found by Kulacki and Lamberti [182] using [Bmim], [Hmim], and [Omim] bromide ILs. They examined the nonmotile *Scenedesmus quadricauda* and the motile freshwater alga

Chlamydomonas reinhardtii in nutrient-rich media and low-nutrient groundwater to uncover whether these parameters affected toxicity. The usual trend was however observed with each of the two algae in both media. *Scenedesmus vacuolatus* was used as a test organism for IL toxicity by Stolte et al. [183] and Matzke et al. [184]. The side-chain effect (in which increased side-chain length/lipophilicity leads to increased toxicity) was demonstrated in both studies.

3.4.2.4 Toxicity Studies Using Plants Jastorff et al. [185] studied the ecotoxicity and genotoxicity of [BF_4] ILs on higher plants using WST-1 cell-viability assays (in which cells' metabolic activity is quantified by their enzymatic conversion of the tetrazolium precursor, WST-1 to a dye, formazan). Two higher plants, lesser duckweed (*Lemna minor*) and garden cress (*Lepidium sativum*), were used to study the toxicity of the ILs [Bmim] and [Omim][BF_4]. Both tests showed [Omim][BF_4] to be more toxic than [Bmim][BF_4]. In the case of *L. minor*, the number of foliaceous fronds produced in comparison to a control plant indicated the toxicity level. [Omim][BF_4] showed an 87% reduction in growth at a concentration of 10 mg kg^{-1}, while the [Bmim] IL showed no reduction even at an IL concentration of 100 mg kg^{-1}. The effect of the ILs on *L. sativum* was monitored by the number of seedlings produced by the plant. At a concentration of 100 mg kg^{-1}, [Omim][BF_4] significantly reduced the number of seedlings produced, while a significant effect was only noted for [Bmim] at a concentration of 1000 mg kg^{-1}. Stolte et al. [183] and Matzke et al. [184] studied the toxic effects of ILs on *L. minor*, wheat (*Triticum aestivum*) and *L. sativum*. The side-chain length effect was observed in both cases. Balczewski et al. [186] synthesized novel CILs and tested them for toxicity using the higher plants, spring barley (*Hordeum vulgare*) and common radish (*Raphanus sativus*). Several of their ILs were toxic and the toxicity was found to be concentration-dependent. In 2009, Studzińska et al. [187] reported that imidazolium ionic liquids are toxic to garden cress *L. sativum* L. However, their toxicity is low compared with other chemical substances. Following the common trend for ionic liquids, the toxicity in aquatic solution was closely linked to IL hydrophobicity, that is the more hydrophobic the ionic liquid, the fewer seeds were capable of germination.

3.4.2.5 Toxicity Testing Using Mammalian Cell Lines The toxicity of imidazolium ILs was tested on the luminescent bacterium, *V. fischeri*, as well as on rat cell lines (C_6 glioma cells) and IPC 81 (human leukemia cells) [164]. These cell lines were chosen to obtain information about the effect of ILs on two physiologically different areas, namely the haematopoietic system and the central nervous system. The effects of increasing chain length of the N_2 (methyl and ethyl) and N_1 (C_3–C_6) position on the imidazolium moiety was investigated, together with the effect of the anionic species on the cell-viability assays. The anions chosen were the popular [PF_6] and [BF_4], which were then compared to [Br], [Cl], and [*p*-OTs] anions. It was shown that toxicity increased with increasing alkyl chain length, both for the N_1 and N_2 positions, for each system tested. The authors propose that this effect is caused by an increase in lipophilicity, resulting in increased cell-membrane permeability. Their results showed an increase in toxicity when the length of the N_1 side chain was increased in comparison with increasing length of the N_2 side chain. Overall, no effect of the different anion on the test systems could be noted therefore rendering the cationic species the determining factor for toxicity. A comparison with the widely used organic solvents, methanol, acetone, acetonitrile, and MTBE, proved that the ILs were more toxic than all of the organic solvents except MTBE, which showed similar toxicity to the ILs studied in the *V. fischeri* assay.

Mammalian blood cells were used by Jastorff to test the genotoxicity of two ILs, [Bmim] and [Dmim][BF_4], using the SCE (sister chromatid exchange) assay [185]. [Bmim][BF_4] was shown to have no genotoxic effects within the given concentration range, while [Dmim][BF_4] showed a dose-dependent trend within half of this concentration range. The group also mentioned their preliminary findings on the anion effects of ILs on toxicity. A significant difference in toxicity between the [BF_4] and [NTf_2] anion was shown. Also, significant differences in toxicity between various headgroups of the ILs, namely between [bmpyr] and [Bmim], and [bmpy] and [Bmim] were evident. Finally, this group predicted possible metabolites of the [Bmim] and [omim] cations. The most probable metabolites were synthesized and their toxicity was measured, using the WST-1 cell-viability assay, to compare the toxicity of the most probable metabolites with the parent IL. Stolte et al. [188] followed this work with a study into the effect that anions of ILs have on cytotoxicity using the WST-1 cell-viability assay, this time using the IPC-81 rat leukemia cell line.

Out of 27 commercially available anions tested, only 10 displayed cytotoxicity. In order to ascertain an effect of the anion combined with typical IL cations, the anions were combined with the widely used [Rmim] cation with side-chain lengths ranging from $R = C_2$ to C_6. The tests revealed that the combination of the anions with these cationic headgroups did indeed increase the toxicity results obtained. Stolte et al. [186] again used the rat cell line IPC-81 to assess the cytotoxicity of a range of ILs. They screened 100 ILs, varying the headgroup, side-chain, and anion of the ILs in order to observe the effect of each respective moiety on toxicity. Using the previously derived correlation between lipophilicity of the IL side-chain and cytotoxicity using HPLC, [189] Ranke used evidence from 23 comparative cases to establish that the side-chain is far more influential than the headgroup or anion in terms of toxicity. This supports the hypothesis that toxicity effects from ILs are a direct result of lipophilic interactions with the cell membrane. It was also apparent that the lipophilic [NTf_2] anion had an intrinsic cytotoxic effect. However, when this anion was combined with a polar cationic moiety, the cytotoxicity was not as severe. A significant result obtained from this study was that the presence of functional groups (ether, hydroxyl and nitrile) in the side chain of the ILs tends to lower cytotoxicity.

Epithelial cells are the sites for the first point of contact in an organism with toxic compounds. Three types of human epithelial cells have been examined in terms of cytotoxicity of ILs, namely, the breast cancer cell line MCF7, the carcinoma cell line HeLa, and the Caco-2 cell line. With the objective of identifying alternatives to imidazolium and pyridinium ILs, Salminen et al. [190] studied the toxicities of a range of hydrophobic pyrrolidinium and piperidinium ILs using the MCF7 cell line.

The same increasing toxicity trend with increasing alkyl chain length was observed and the cytotoxicity of these compounds fell in the same range as imidazolium ILs. With few exceptions, the [NTf_2] anion was found to increase cytotoxicity in comparison with bromide. The cytotoxicity of ILs towards the HeLa cell line was investigated by Wang et al. [191] with imidazolium, pyridinium, triethylammonium, and choline IL derivatives. Increasing toxicity with increasing side-chain length was observed, and it was shown that changing the anion displayed a less significant effect than changing the side-chain length. The cytotoxicity of imidazolium ILs towards the Caco-2 cell line was investigated by Garcia et al. [192]. With increasing alkyl chain length an increase in the toxicity was observed and the ILs tested were shown to be more toxic than acetone, methanol, acetic acid, and benzene.

3.4.2.6 Enzyme Inhibition The purified enzyme, acetylcholinesterase, from the electric eel (*Electrophorus electricus*) was used to test the effect of ILs on the inhibition of this enzyme [193]. The authors tested pyridinium, imidazolium, and phosphonium ILs, with varying side-chain lengths, and also varying anions. Also investigated were the effects of inserting an aromatic substituent into the side chain. Regarding the cationic species of the ILs, results show pyridinium to have the highest inhibitory value, followed by imidazolium and finally phosphonium. The inhibitory effects of the ILs which increase with increasing alkyl chain length on the R_2 position of the cation was shown [193]. Regarding aromatic 24 substituents in the side chain, they conclude that the slight increase in inhibition of the enzyme by the aromatic containing side chain ILs is due to the increased lipophilic nature in comparison with the straight-chain moieties tested. The effect of lengthening the side chain at the R_1 position of the cation was investigated [193]. No measurable difference in inhibitory values was observed, suggesting that the steric position of the lipophilic side chain also influences toxicity. The EC_{50} values for imidazolium and pyridinium ILs regarding the effect of the anion showed conflicting results. For imidazolium, [Br] and [Cl] showed the highest inhibitory effect, followed by [BF_4] and then [PF_6]. However, tests indicated that the pyridinium IL [PF_6] salt was gave higher inhibition than the salt of [BF_4]. Skladanowski et al. [194] used AMP deaminase isolated from rat skeletal muscle to investigate the toxicity of imidazolium-based ILs. Together with the ILs were tested known pollutant musks, such as musk xylene and galaxolide (Fig. 3.14).

IC_{50} values were used as a measure of toxicity of these compounds. Four [Bmim] ILs were tested, each with a different anion, namely [BF_4], [PF_6], [Cl], and [*p*-OTs].

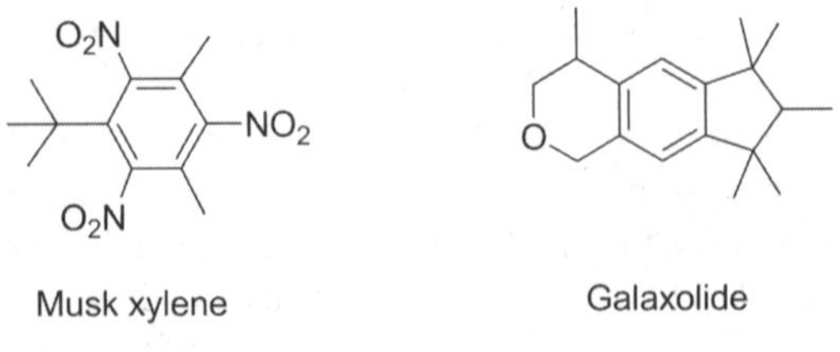

FIGURE 3.14 Synthetic musks.

Perhaps surprisingly, all the ILs displayed IC_{50} values higher than those of the synthetic musks (used in perfumery), indicating lower toxicity. The ILs were however found to be toxic at higher concentrations, with [Bmim][BF_4] and [Bmim][PF_6] ($IC_{50} = 5\,\mu M$), compared with the other two ILs tested ($IC_{50} = 10\,\mu M$).

3.4.2.7 Toxicity Studies Using Higher Organisms The anatomically transparent free-living soil roundworm, *Caenorhabditis elegans* was used to investigate the toxic effects of imidazolium chloride ILs with chain lengths of C_4, C_8, and C_{14} [195]. It was found that the IL containing the longest alkyl chain was the most toxic, with toxicity decreasing thus from C_8 to C_4. The C_4 IL caused no adverse effects to the worms, while the C_{14} IL was lethal to them at all three concentrations of IL. Bernot et al. [196] studied the effects of ILs containing imidazolium and pyridinium cations with varying chain length with anions [Br] and [PF_6], on the freshwater pulmonate snail, *Physa acuta*. The acute toxicity, as well as the behavioral (locomotive and feeding) effects of the ILs on the snail were investigated. Results for the acute toxicity study show [Ompyr]Br to be the most toxic IL investigated, with [TBA]Br being the least toxic. LC_{50} values in general then decreased with increasing alkyl chain length. The range of LC_{50} values obtained for the ILs studied are shown to be in the same range as values for organic solvents such as ammonia and phenol. Imidazolium and pyridinium-containing cations could not be distinguished according to toxicity level, as [Hmim]Br was shown to be more toxic than [Hmpyr]Br, while [Omim]Br was less toxic than [Ompyr]Br. The grazing rate of the snails decreased with increasing concentration of IL and also upon increasing chain length. The results for the movement study showed the movement of the snails to decrease initially with low concentrations of ILs, but then increase above a certain threshold where IL concentration increased. The combination of the results for locomotor and feeding behavior shows an escape response, implying that the organisms search for refuge from the IL when a certain concentration threshold is reached. The toxicity of ILs to zebrafish (*Danio rerio*) was investigated [197]. Ammonium, imidazolium, pyridinium, and pyrrolidinium-based ILs were used. The ammonium-based ILs were shown to be by far the most toxic, with LC_{50} values lower than those of commonly used organic solvents, that is, MeOH, DCM, ACN, and TEA. The fish exposed to these ILs showed erratic behavior compared to the control fish, and numerous abnormalities were also noted in the fish upon histopathological inspection. All other ILs exhibited LC_{50} values greater than $100\,mg\,L^{-1}$; therefore, were not investigated further, and deemed by the authors to be "non-highly lethal towards zebra fish." In 2009, Costello et al. [198] reported how ILs affect mortality and feeding of the zebra mussel (*Dreissena polymorpha*). Six pyridinium and imidazolium-based ILs were studied with 96 h acute bioassays. It was found that the ILs tested caused acute mortality over a wide range of concentrations ($LC_{50} = 21.4$ to $1290\,mg\,L^{-1}$), and ILs with longer alkyl chains were more toxic, and similar toxicities were observed for pyridinium- and imidazolium-based ILs. When the toxicity of the IL [Bmim]Br was compared across various organisms, zebra mussels exhibit one of the highest LC_{50} values ($1290\,mg\,L^{-1}$) and would be among the most resistant aquatic organisms in the event of an ionic liquid spill in a coastal area. It was also found that short-term exposure to any IL tested reduced zebra mussel feeding. For ionic liquids with butyl- and hexyl-chain, feeding was significantly reduced at the acute LC_{50}, whereas the octyl-chain IL reduced feeding at the acute LC_{50}. On the basis of reduced survival and feeding by zebra mussels in the presence of ILs, the authors demonstrated that the potential release of ionic liquids into the ecosystems would have substantial effects on other trophic levels and lead to changes to affect ecosystems. Representative freshwater organisms (*P. subcapitata*, *D. magna*, and *D. rerio*) also have been selected as model organisms to investigate the static acute toxicities of 18 ionic liquids by Pretti et al. Their results showed that long-chain ammonium salts showed higher toxicity to algae, cladocerans, and fish, whereas very low toxicities characterized sulfonium- and morpholinium-based ILs. In imidazolium-based ILs, the introduction of a more electronegative atom, such as chlorine or oxygen, into the longer alkyl chain was found to reduce the acute toxicity for both algae and cladocerans. Amphibians are often regarded as the main vertebrate group at risk for exposure to contaminants in aquatic systems, because their larvae usually live in water. Toxic effects of 1-methyl-3-octylimidazolium bromide ([Omim]Br) on the early embryonic development of the frog *Rana nigromaculata* were evaluated by Li et al. [199]. It was found that the 96 h median lethal concentration values of frog embryos in different developmental stages (early cleavage, early gastrula, or neural plate) were 85.1, 43.4, and 42.4 mg L^{-1}, respectively. In the case of embryos exposed to [Omim]Br, the duration of embryo dechorination was prolonged in both early cleavage and neural plate, but not the early gastrula. Embryos in the neural-plate developmental stage exposed to [Omim]Br were found to have the highest mortality rate.

The previous sections highlighted the wide range of toxicity assays performed on ionic liquids. This data is valuable as the chemist can determine if an ionic liquid has "unacceptably high" toxicity for a role as a bulk chemical (i.e., solvent). Ionic liquids with high antimicrobial toxicities (MIC in the low microM) would be of particular concern. While one can argue that addition cleaning of waste steams to remove ionic liquids could be put in place before entering the treatment plant, this is likely to be costly.

3.4.3 Toxicity of ILs Used in Biorefinery (Rogers Subset)

Table 3.5 shows the lack of toxicity data available for ionic liquids studied for biomass dissolution. By correlating the 25 ILs from Rogers paper [11c] with toxicity of ionic liquid reviews published by Pham [200] and Luis [201] in 2010, only 11 ILs had any toxicity data. Over half, 14, did not have any toxicity data reported in the reviews. Pham selected a wide range of parameters to assess the toxicity and ecotoxicity, "An ecotoxicology battery." This included toxicity of ILs to different levels of biological complexity including enzyme, bacteria, algae, rat cell line, human cell lines, duckweed, and invertebrate. A clear picture emerges from this simple correlation of the Pham and Rogers papers, despite the research efforts to determine toxicity of ionic liquids, scarce data is available for the ionic liquids applied in a biorefinery context.

TABLE 3.5 Toxicity of ILs to Different Levels of Biological Complexities, Including Enzyme, Bacteria, Algae, Rat Cell Line, Human Cell Lines, Duckweed and Invertebrate

ILs	Acetylcholin Esterase	*Vibrio fischeri*	*Escherichia coli*	*Pseudo kirchneriella subcapitata*	*Scenedesmus vacuolatus*	IPC-81	HeLa	*Lemna minor*	*Daphnia magna*
[Emim]Cl	2.06	4.55, 4.33 ± 0.11	nd	nd	2.78 ± 0.06	nd	nd	nd	nd
[Pmim]Cl	2.27	nd	nd	nd	nd	>4.30	nd	nd	nd
[Bmim]Cl	1.91 ± 0.04	3.71 ± 0.14	nd	2.34 ± 0.01	2.26 ± 0.08	3.55	nd	2.82	1.93
		2.95							1.93 ± 0.06
		3.34 ± 0.13							
		3.27 ± 0.09							
[Bmim]Br	1.90 ± 0.02	4.01 ± 0.05	nd	3.46 ± 0.062	nd	3.43	3.44 ± 0.11	nd	1.57
		3.07 ± 0.03							1.56 ± 0.07
		3.35							1.85 ± 0.06
		3.27 ± 0.09							
[Pnmim]Cl	1.96	nd	nd	nd	nd	>3.00	nd	nd	nd
[Hmim]Cl	1.92	1.94	nd	−1.92 ± 0.01	0.08	2.85	nd	nd	nd
		2.32 ± 0.16							
		2.91 ± 0.09							
[Hpmim]Cl	2.07	nd	nd	nd	nd	2.53	nd	nd	nd
[Omim]Cl	1.60	1.19 ± 0.11	nd	−1.46	−2.67 ± 0.37	2.01	nd	nd	nd
		1.01 ± 0.06							
Pyr4-3MeCl	1.15	nd	nd	nd	nd	nd	nd	nd	nd
[Bmim][OAc]	3.32	nd	nd	nd	nd	nd	nd	nd	nd
[Bmim][HCOO]	3.19	nd	nd	nd	nd	nd	nd	nd	nd

Note: $Log_{10}EC_{50}$ (mM) except $log_{10}LC_{50}$ (mM) in case of *D. magna*. Adapted from Pham [200] and Luis [201]. For specific references see within nd = not determined.

3.4.4 Biodegradation of ILs Used in Biorefinery

If the same list of 25 ILs [11c] is cross-referenced with biodegradation data from reviews by Gathergood [152] and Stolte [153] a similar dearth of experimental data is found. Twenty ILs have no reported biodegradation data, 4 are classed not readily biodegradable, and 1 [Omim]Cl is inherently biodegradable. It is our opinion that the real problem here, is not that the available data records that four out of five are "not readily biodegradable," but that research to establish whether ionic liquids for the biorefinery are biodegradable is marginalized (Table 3.6).

One must always remember that while a "pass" result in a single biodegradation test can signify a compound is biodegradable (within the confines of the property being measured, e.g., primary vs. readily vs. ultimate biodegradation), a "fail" does not mean that the ionic liquid, or indeed any compound, will not biodegrade if released into the environment. A single "fail" signifies a more comprehensive suite of biodegradation tests is first required, and on the body of this work a judgment call made on whether the ionic liquid is likely or not to biodegrade. A reasonable viewpoint is that ionic liquids which pass biodegradation tests should be promoted as candidates for the biorefinery over candidates which fail a range of biodegradation tests.

The third level of toxicity and ecotoxicity studies is bioaccumulation. Although an ionic liquid passes a biodegradation test, do metabolites (toxic or non-toxic?) persist? A comment on bioaccumulation and how it relates to the biorefinery is given in Chapter 2—Green Chemistry introduction. Leaders in the field have acknowledged the importance of the search for safer and greener ionic liquids. Rogers stated "ILs must become available at lower cost and should be of low toxicity and be biodegradable. If the ILs and any necessary processing solvents were themselves prepared from renewable feedstocks in a low energy, sustainable manner, the sustainability of the entire technology platform would be improved. Thus, the search for even more environmentally-friendly ILs should be one of the drivers for research in the area, and fully biodegradable and non-toxic ILs should be a major goal"[11c].

TABLE 3.6 Biodegradation Data for 5 ILs with Reported Biomass Dissolution Studies

ILs	Biodegradation [%]	Classification	References
[Emim]Cl	0[a]	Not readily biodegradable	[202]
[Bmim]Cl	0[a], 0[b]	Not readily biodegradable	[202,203] [a, b]
[Hmim]Cl	11[a]	Not readily biodegradable	[202]
[Omim]Cl	100[a]	Inherently biodegradable	[202]
[Bmim]Br	0[c], <5[d], 1[e], <5[e]	Not readily biodegradable	[204–207] [c, d, e]

Source: Ref. [11c].

Note: No biodegradation data available/reported for other 20 ILs in reviews by Gathergood and Coleman in 2010[152] and Stolte et al. in 2011. [153]

[a] Modified OECD 301D Test.

[b] OECD 301F.

[c] "Die Away" Test OECD 301A.

[d] CO_2 Headspace Test (ISO 14593).

[e] Closed Bottle Test (OECD 301D).

3.4.5 Conclusion for Toxicity and Biodegradation of Ionic Liquids

Toxicity and antimicrobial studies have been performed on a range of species, from bacteria and fungi to higher organisms such as the freshwater snail, frogs, and terrestrial plants. These studies clearly demonstrate that the potential negative environmental impact of ionic liquids and the ionic liquids' structures play a major role in determining toxicity to many organisms, with the cation being more culpable in toxicity than the anion, with longer alkyl chains being more toxic than short chains. The introduction of polar groups into the side chain can reduce the toxicity of ionic liquids. Although significant progress in understanding the toxicity of ILs has been achieved, the exact mechanism of toxicity is still unknown. Cationic surfactants are well-known industrial chemicals, and they usually have bulky, positively charged head-groups with alkyl chains of more than 10 carbons. The structures of ionic liquids are similar to these surfactants and so it is reasonable to hypothesize that they share the same mechanism of toxicity. Purportedly, the toxicity of cationic surfactants is due to integration of their lipophilic alkyl chains into cellular membranes, with concomitant membrane disruption [208, 209]. Ionic liquids with longer alkyl chains are more lipophilic, [210] and it has been suggested that such ILs can integrate more readily into cellular membranes, causing toxicity [211]. There has also been evidence that anions can play a role in toxicity, but in general their effect is insignificant compared with that of any side-chain present on the cation [179]. The best possible scenario for the researcher in ionic liquids is to be able to identify and readily synthesize structures that can be designated "non-toxic" and "readily biodegradable." There are already some examples of ILs which come close to this ideal, based on biological cations such as choline and betaine, and anions that are commonly used in the food industry such as tartrate, citrate, malate, and saccharinate.

However, the Brønsted basicity or acidity of such species severely restricts their applications, as does their extreme hydrophilicity. Until there is a better alternative, imidazolium ILs and their pyridinium analogues (which are in many cases no more biodegradable) will still be relied upon, at least into the near future. In such cases, the 1-octyl-3-methylimidazolium cation, represents a vast improvement over 1-butyl-3-methylimidazolium, albeit at the expense of the toxicity associated with an eight-carbon aliphatic chain. Gathergood, and also Scammells have gone some way towards addressing the question of toxicity, fielding ILs with ester linkages, low toxicity, and biodegradabilities of 66% (Gathergood, imidazolium, closed bottle test, OECD 301D [152, 205]) and 82% (Scammells, pyridinium, CO_2 headspace test, ISO 14593 [152, 205]). In the case of these less toxic ionic liquids, the biodegradability of the heterocyclic core still remains a challenge, as does stability to strong bases, acids, nucleophiles and electrophiles.

The toxicological effects of imidazolium and pyridinium ILs have now been studied on a wide range of simple aquatic organisms, as well as some higher plants and animals. However, in spite of the enormous number of studies now published, data on human exposure to ILs remain scarce. Until human toxicity data can be given with confidence, it will remain difficult for industry to risk exposing workers to ionic liquids [212]. The importance of toxicity data to industry is even greater now that the

27 states of the European Union, as well as three nonmember states are requiring compliance with "REACh" (Registration, Evaluation, Authorisation and restriction of Chemicals). This set of regulations, introduced in 2006, requires all chemicals for which more than 1 tonne is sold per year in Europe to be toxicologically tested. Compliance with REACh is proving very costly, with an enormous "back-catalogue" of existing compounds [213] (including ionic liquids) requiring testing, let alone new ionic liquids that are yet to be introduced to the market. These difficulties may be circumvented if industrial solvents or catalysts are chosen from structures already within the REACh database, such as AMMOENGs™ 100-102 and 110-112.

With this ever growing database of toxicity and biodegradation data for ionic liquids, it is upto researchers in all areas of environmental science, organic synthesis, and sustainability to work together for the common good.

3.5 CONCLUSIONS AND PROSPECTS

Biorenewable resources have been recognized as potential sources for the production of green fuels, sustainable materials, and chemicals. Considering the structural difference of biomass and petroleum-based materials and chemicals, key issues, such as derivation through the reaction of hydroxyl groups on carbohydrates, hydrolysis, dehydration, hydrogenations, need to be addressed with the aim to produce sustainable materials and chemicals from biomass. The full dissolution of biomass in ionic liquids has provided a significant platform for all of these purposes, and although significant progress has been obtained, for example, generated cellulose, and cellulose composite cellulose fiber for textile industry, high-efficient production of HMF, levulinic acid, levulinic acid esters from carbohydrates, green production of biodiesel from lipids. The research into biomass by ILs has just started, especially on wood chemistry, pretreatment technology, catalytic conversion of both carbohydrates and lignin. This chapter provides a general overview of the wide variety of sustainable materials and chemicals from biomass by using ILs as solvents and/or catalysts, which offers significant promise for the foundation of sustainable biorefinery process based on ILs. It is quite evident from the foregoing discussion that ILs platform have enormous potentials to provide innovative pathways for the utilization of biomass by virtue of their unique properties, especially in combination with other sustainable energy, conversion, and separation technologies. However, for their successful implementation as commercially viable technologies, there are still a number of challenges ahead on their potential industrial applications, including,

(1) The design and economic preparation of cheaper, low toxicity, enzyme-compatible ILs capable of dissolving cellulose, on the basis of in-depth understanding of dissolution mechanism of cellulose in ILs.
(2) Design and development of new catalytic systems for more efficient transformations, such as depolymerization, dehydration, hydrogenolysis, or alcoholysis, of (ligno)cellulose and related materials.
(3) New chemistry of (ligno) cellulose or other biomass.

(4) Tandem conversion of carbohydrates into valuable chemicals and biofuels.
(5) Integration of sustainable energy methodologies, advanced catalytic technologies, and separation technologies into the ILs platforms. Integration of ionic liquids based biorefinery into existing refinery process.
(6) Developing new strategies for ILs recycling and product recovery.
(7) Catalytic conversion of lignin into value-added chemicals.
(8) Toxicity, ecotoxicity, and biodegradation evaluation.

There is still no "perfect" green ionic liquid and the search for improvements that combine chemical robustness with favorable biodegradability and toxicology remains an active field of research.

The continuing evolution of green ionic liquids has underscored the requirement to limit the number of steps in the synthesis of an IL, which must be practical to make on a large scale. In addition, the ionic liquid cannot truly be described as "green' if it is prepared by a multistep synthesis in which each operation requires halogenated reagents, solvents, or chromatography. In such cases, the overall environmental impact of a molecule must encompass the sum of its synthetic steps, as well as its industrial life cycle, which hang over the final molecule like a kind of "chemical karma." The importance of a full life-cycle analysis/assessment (LCA) [214] is now increasingly being recognized in academic circles, as well as in industry where cumulative energy demand (CED) and costing must be considered alongside direct environmental effects [215].

Even if an ionic liquid has demonstrated nontoxicity towards a particular organism, continued testing may still uncover toxic effects against other species at a sufficiently high test concentration. The concept of "non-toxicity" is not really meaningful, and it is more realistic to target the synthesis of ionic liquids that have a known, but low toxicity to particular organisms and can biodegrade rapidly, should an accidental release into the environment take place.

It is our opinion that the search for environment-friendly ionic liquids for use in biorefineries is challenging. However, through the combined efforts from researchers in many disciplines we believe a major breakthrough is imminent.

3.6 RELATED IONIC LIQUIDS: FULL NAME AND ABBREVIATION

[Amim]Cl	1-allyl-3-methylimidazolium chloride
[ASBI][OTf]	3-allyl-1-(4-sulfobutyl)imidazolium trifluoromethanesulfonate
[ASCBI][OTf]	3-allyl-1-(4-sulfurylchloride butyl)imidazolium trifluoromethane sulfonate
[Bmim]Cl	1-butyl-3-methylimidazolium chloride
[Bmim][BF_4]	1-butyl-3-methylimidazolium tetrafluoroborate
[Bmim][NTf_2]	1-*n*-butyl-3-methylimidazolium bis(trifluoromethylsulfonyl)imide
[Bmim][PF_6]	1-butyl-3-methylimidazolium hexafluorophosphate

$[Bmim\text{-}SO_3H][HSO_4]$	1-methyl-3-(4-sulfobutyl)imidazolium hydrogensulfate
$[BPyr\text{-}SO_3H][HSO_4]$	1-(4-sulfobutyl)pyridinium hydrogensulfate
$[(CH_2)_4SO_3Hmim][HSO_4]$	1-(4-sulfonic acid)butyl-3-methylimidazolium hydrogen sulfate
$[C_{18}mim][NTf_2]$	1-octadecyl-3-methylimidazolium *bis*(trifluoromethylsulfonyl)imide
$[C_{16}mim][NTf_2]$	1-hexadecyl-3-methylimidazolium *bis*(trifluoromethylsulfonyl)imide
$[(CH_2)_4SO_3HPy]HSO_4$	1-(4-sulfonic acid)butyl-pyridinium hydrogen sulfate
[Emim]Cl	1-ethyl-3-methylimidazolium chloride
[Emim][OAc]	1-ethyl-3-methylimidazolium acetate
$[Emim][HSO_4]$	1-ethyl-3-methylimidazolium hydrogensulfate
$[Emim][BF_4]$	1-ethyl-3-methylimidazolium tetrafluoroborate
[Emim]Cl	1-ethyl-3-methylimidazolium chloride
[Emim][OAc]	1-ethyl-3-methylimidazolium acetate
[Emim][TfO]	1-ethyl-3-methylimidazolium trifluoromethanesulfonate
$[Emim][BF_4]$	1-ethyl-3-methylimidazolium tetrafluoroborate
$[HBSSB][HSO_4]$	3,3′-(hexane-1,6-diyl)*bis*(6-sulfo-1-(4-sulfobenzyl)-*1H*-benzimidazolium) hydrogensulfate
HEMA	*tris*-(2-hydroxyethyl) methyl ammonium methylsulfate
[Hmim]Cl	1-hexyl-3-methylimidazolium chloride
[Hpmim]Cl	1-heptyl-3-methylimidazolium chloride
$[HNMP][CH_3SO_3]$	*N*-methyl-2-pyrrolidonium methyl sulfonate
[Moim]Cl	1-methyl-3-octylimidazolium chloride
$[N(1114)][NTf_2]$	butyltrimethyl-ammonium *bis*(trifluoromethylsulfonyl)imide
$[NMM][CH_3SO_3]$	*N*-methylmorpholinium methylsulfonate
$[NMP][CH_3SO_3]$	*N*-methyl-2-pyrrolidonium methyl sulfonate
$[NMP][HSO_4]$	*N*-methyl-2-pyrrolidonium hydrogen sulfate
$[NEt_3B\text{-}SO_3H][HSO_4]$	*N,N,N*-triethyl-4-sulfobutan-ammonium hydrogensulfate
[Omim]Cl	1-octyl-3-methylimidazolium chloride
[Pmim]Cl	1-propyl-3-methylimidazolium chloride
[Pnmim]Cl	1-pentyl-3-methylimidazolium chloride
[Rmim]Cl	1-alkyl-3-methylimidazolium chloride
$[Sbmi][HSO_4]$	1-(4-sulfonic acid) butyl-3-methylimidazolium hydrogen sulfate
$[TMEDAPS][HSO_4]$	*N,N,N′,N′*-tetramethyl-*N,N′*-dipropanesulfonic acid-1,6-hexanediammonium hydrogensulfate
$[Dmim][BF_4]$	1-decyl-3-methylimidazolium tetrafluoroborate
$[Benzylmim][BF_4]$	1-benzyl-3-methylimidazolium tetrafluoroborate

ACKNOWLEDGMENTS

The authors (HBX, WJL) wish to thank the grants support from the 100 Talent Programm of Dalian Institute of Chemical Physicals; the National Natural Science Foundation of China (NSFC 21002101, 31270637).

The authors (NG, IB) wish to thank Enterprise Ireland (EI), the Irish Research Council for Science, Engineering and Technology (IRCSET), Science Foundation Ireland (SFI) and the Environmental Protection Agency (EPA) in Ireland for funding green chemistry research in Nicholas Gathergood's group.

REFERENCES

1. (a) J. Zakzeski, P. C. A. Bruijnincx and B. M. Weckhuysen, *Green Chem.* **2011**, 13, 671–680; (b) A. Corma, S. Iborra and A. Velty, *Chem. Rev.* **2007**, 107, 2411–2502.
2. A. J. Ragauskas, C. K. Williams, B. H. Davison, G. Britovsek, J. Cairney, C. A. Eckert, W. J. Frederick, J. P. Hallett, D. J. Leak, C. L. Liotta, J. R. Mielenz, R. Murphy, R. Templer and T. Tschaplinski, *Science* **2006**, 311, 484–489.
3. (a) Y. Nagamatsu and M. Funaoka, *Green Chem.* **2003**, 5, 595–601; (b) M. Stocker, *Angew. Chem.-Int. Ed.* **2008**, 47, 9200–9211.
4. C. F. Liu, R.-C. Sun, A.-P. Zhang, M.-H. Qin, J.-L. Ren and X.-A. Wang, *J. Agric. Food Chem.* **2007**, 55, 2399–2406.
5. A. Pinkert, K. N. Marsh, S. Pang and M. P. Staiger, *Chem. Rev.* **2009**, 109, 6712–6728.
6. (a) C. Yang and P. Liu, *Ind. Eng. Chem. Res.* **2009**, 48, 9498–9503; (b) Y. Roman-Leshkov, J. N. Chheda and J. A. Dumesic, *Science* **2006**, 312, 1933–1937; (c) C. Z. Li, Z. H. Zhang and Z. B. K. Zhao, *Tetrahedron Lett.* **2009**, 50, 5403–5405; (d) T. Stahlberg, W. J. Fu, J. M. Woodley and A. Riisager, *ChemSusChem* **2011**, 4, 451–458; (e) M. J. Earle, N. V. Plechkova and K. R. Seddon, *Pure Appl. Chem.* **2009**, 81, 2045–2057; (f) M. E. Zakrzewska, E. Bogel-Lukasik and R. Bogel-Lukasik, *Chem. Rev.* **2011**, 111, 397–417.
7. J. P. Hallett and T. Welton, *Chem. Rev.* **2011**, 111, 3508–3576.
8. (a) J. Y. G. Chan and Y. G. Zhang, *ChemSusChem* **2009**, 2, 731–734; (b) K. Yamaguchi, T. Sakurada, Y. Ogasawara and N. Mizuno, *Chem. Lett.* **2011**, 40, 542–543.
9. B. Kim, J. Jeong, D. Lee, S. Kim, H. J. Yoon, Y. S. Lee and J. K. Cho, *Green Chem.* **2011**, 13, 1503–1506.
10. (a) M. E. Zakrzewska, E. Bogel-Lukasik and R. Bogel-Lukasik, *Energy Fuels* **2010**, 24, 737–745; (b) X. H. Qi, H. X. Guo and L. Y. Li, *Ind. Eng. Chem. Res.* **2011**, 50, 7985–7989.
11. (a) J. Carlos Serrano-Ruiz and J. A. Dumesic, *Energy Environ. Sci.* **2011**, 4, 83–99; (b) V. Degirmenci, E. A. Pidko, P. Magusin and E. J. M. Hensen, *ChemCatChem* **2011**, 3, 969–972; (c) N. Sun, H. Rodriguez, M. Rahman and R. D. Rogers, *Chem. Commun.* **2011**, 47, 1405–1421; (d) J. H. Clark, F. E. I. Deswarte and T. J. Farmer, *Biofuels, Bioprod. Biorefin.* **2009**, 3, 72–90.
12. (a) F. R. Tao, H. L. Song, J. Yang and L. J. Chou, *Carbohydr. Polym.* **2011**, 85, 363–368; (b) M. Tan, L. Zhao and Y. Zhang, *Biomass Bioenergy* **2011**, 35, 1367–1370.

13. J. Guan, Q. Cao, X. Guo and X. Mu, *Comput. Theor. Chem.* **2011**, 963, 453–462.
14. H. Liu, K. L. Sale, B. M. Holmes, B. A. Simmons and S. Singh, *J. Phys. Chem. B* **2010**, 114, 4293–4301.
15. Z. Zhang and Z. Zhao, *Biores. Technol.* **2011**, 102, 3970–3972.
16. F. Yang, L. Z. Li, Q. Li, W. G. Tan, W. Liu and M. Xian, *Carbohydr. Polym.* **2010**, 81, 311–316.
17. H. M. Cho, A. S. Gross and J.-W. Chu, *J. Am. Chem. Soc.* **2011**, 133 (35), 14033–14041.
18. A. Pinkert, K. N. Marsh and S. S. Pang, *Ind. Eng. Chem. Res.* **2010**, 49, 11121–11130.
19. (a) H. B. Xie, W. J. Liu and Z. B. K. Zhao, **2012**, *Biomass Conversion*, Book ISBN: 978-3-642-28417-5, (Eds. C. Baskar, S. Baskar and R. S. Dhillon), Chapter 3, 123–144, Springer-Verlag, Berlin, Heidelberg; (b) H. Tadesse and R. Luque, *Energy Environ. Sci.* **2011**, 4, 3913–3929.
20. (a) A. Stark, *Energy Environ. Sci.* **2011**, 4, 19–32; (b) Y. Zheng, X. Xuan, A. Xu, M. Guo and J. Wang, *Prog. Chem.* **2009**, 21, 1807–1812.
21. Y. Pu, N. Jiang and A. J. Ragauskas, *J. Wood Chem. Technol.* **2007**, 27, 23–33.
22. T. Meng, X. Gao, J. Zhang, J. Yuan, Y. Zhang and J. He, *Polymer* **2009**, 50, 447–454.
23. N. Jiang, Y. Q. Pu and A. J. Ragauskas, *ChemSusChem* **2010**, 3, 1285–1289.
24. R. Samuel, M. Foston, N. Jaing, S. Cao, L. Allison, M. Studer, C. Wyman and A. J. Ragauskas, *Fuel* **2011**, 90, 2836–2842.
25. H. Xie, A. King, I. Kilpelainen, M. Granstrom and D. S. Argyropoulos, *Biomacromolecules* **2007**, 8, 3740–3748.
26. (a) H. Xie, I. Kilpelainen, A. King, T. Leskinen, P. Jarvi and D. S. Argyropoulos, in *Cellulose Solvents: For Analysis, Shaping and Chemical Modification*, Vol. 1033 (Ed. T.F. Liebert, T. J. Heinze and K. J. Edgar), **2009**, 343–363; (b) L. Vanoye, M. Fanselow, J. D. Holbrey, M. P. Atkins and K. R. Seddon, *Green Chem.* **2009**, 11, 390–396.
27. A. W. T. King, L. Zoia, I. Filpponen, A. Olszewska, H. B. Xie, I. Kilpelainen and D. S. Argyropoulos, *J. Agric. Food Chem.* **2009**, 57, 8236–8243.
28. D. Klemm, B. Heublein, H. P. Fink and A. Bohn, *Angew. Chem.-Int. Ed.* **2005**, 44, 3358–3393.
29. B. Kim, J. Jeong, D. Lee, S. Kim, H.-J. Yoon, Y.-S. Lee and J. K. Cho, *Green Chem.* **2011**, 13, 1503–1506.
30. R. P. Swatloski, S. K. Spear, J. D. Holbrey and R. D. Rogers, *J. Am. Chem. Soc.* **2002**, 124, 4974–4975.
31. (a) I. Kilpelainen, H. Xie, A. King, M. Granstrom, S. Heikkinen and D. S. Argyropoulos, *J. Agric. Food Chem.* **2007**, 55, 9142–9148; (b) D. M. Phillips, L. F. Drummy, D. G. Conrady, D. M. Fox, R. R. Naik, M. O. Stone, P. C. Trulove, H. C. De Long and R. A. Mantz, *J. Am. Chem. Soc.* **2004**, 126, 14350–14351; (c) H. B. Xie, S. H. Li and S. B. Zhang, *Green Chem.* **2005**, 7, 606–608.
32. S. S. Y. Tan, D. R. MacFarlane, J. Upfal, L. A. Edye, W. O. S. Doherty, A. F. Patti, J. M. Pringle and J. L. Scott, *Green Chem.* **2009**, 11, 339–345.
33. (a) T. G. A. Youngs, J. D. Holbrey, M. Deetlefs, M. Nieuwenhuyzen, M. F. C. Gomes and C. Hardacre, *ChemPhysChem* **2006**, 7, 2279–2281; (b) T. Q. Yuan, J. He, F. Xu and R. C. Sun, *Prog. Chem.* **2010**, 22, 472–481; (c) M. Zavrel, D. Bross, M. Funke, J. Buchs and A. C. Spiess, *Biores. Technol.* **2009**, 100, 2580–2587.

34. A. Pinkert, K. N. Marsh, S. S. Pang and M. P. Staiger, *Chem. Rev.* **2009**, 109, 6712–6728.
35. N. Sun, W. Li, B. Stoner, X. Jiang, X. Lu and R. D. Rogers, *Green Chem.* **2011**, 13, 1158–1161.
36. Y. Qin, X. Lu and N. Sun, R. D. Rogers, *Green Chem.* **2010**, 12, 968–971.
37. H. Zhang, Z. G. Wang, Z. N. Zhang, J. Wu, J. Zhang and J. S. He, *Adv. Mater.* **2007**, 19, 698–704.
38. (a) J. Wu, J. Zhang, H. Zhang, J. S. He, Q. Ren and M. Guo, *Biomacromolecules* **2004**, 5, 266–268; (b) R. C. Remsing, I. D. Petrik, Z. W. Liu and G. Moyna, *Phys. Chem. Chem. Phys.* **2010**, 12, 14827–14828.
39. J. Zhang, J. Wu, Y. Cao, S. Sang, J. Zhang and J. He, *Cellulose* **2009**, 16, 299–308.
40. C. X. Lin, H. Y. Zhan, M. H. Liu, S. Y. Fu and J. J. Zhang, *Carbohydr. Polym.* **2009**, 78, 432–438.
41. (a) F. R. Tao, H. L. Song and L. J. Chou, *ChemSusChem* **2010**, 3, 1298–1303; (b) S. Van de Vyver, L. Peng, J. Geboers, H. Schepers, F. de Clippel, C. J. Gommes, B. Goderis, P. A. Jacobs and B. F. Sels, *Green Chem.* **2010**, 12, 1560–1563.
42. S. Köhler and T. Heinze, *Cellulose* **2007**, 14, 489–495.
43. C. Vanderghem, P. Boquel, C. Blecker and M. Paquot, *Appl. Biochem. Biotechnol.* **2010**, 160, 2300–2307.
44. H. Xie, P. Jarvi, M. Karesoja, A. King, I. Kilpelainen and D. S. Argyropoulos, *J. Appl. Polym. Sci.* **2009**, 111, 2468–2476.
45. (a) N. Bhatt, P. K. Gupta and S. Naithani, *J. Appl. Polym. Sci.* **2008**, 108, 2895–2901; (b) M. Gericke, T. Liebert and T. Heinze, *Macromol. Biosci.* **2009**, 9, 343–353.
46. H. Dong, Q. Xu, Y. Li, S. Mo, S. Cai and L. Liu, *Coll. Surf., B.* **2008**, 66, 26–33.
47. C. Yan, J. Zhang, Y. Lv, J. Yu, J. Wu, J. Zhang and J. He, *Biomacromolecules* **2009**, 10, 2013–2018.
48. D. M. Alonso, J. Q. Bond and J. A. Dumesic, *Green Chem.* **2010**, 12, 1493–1513.
49. A. T. W. M. Hendriks and G. Zeeman, *Biores. Technol.* **2009**, 100, 10–18.
50. L. Y. Liu and H. Z. Chen, *Chin. Sci. Bull.* **2006**, 51, 2432–2436.
51. K. Shill, S. Padmanabhan, Q. Xin, J. M. Prausnitz, D. S. Clark and H. W. Blanch, *Biotechnol. Bioeng.* **2011**, 108, 511–520.
52. (a) J. van Spronsen, M. A. T. Cardoso, G.-J. Witkamp, W. de Jong and M. C. Kroon, *Chem. Eng. Processing Process Intensification* **2011**, 50, 196–199; (b) N. Sun, X. Jiang, M. L. Maxim, A. Metlen and R. D. Rogers, *ChemSusChem* **2011**, 4, 65–73.
53. (a) D. Fu and G. Mazza, *Biores. Technol.* **2011**, 102, 8003–8010; (b) D. Fu and G. Mazza, *Biores. Technol.* **2011**, 102, 7008–7011.
54. H. Zhao, *J. Chem. Technol. Biotechnol.* **2010**, 85, 891–907.
55. H. Zhao, G. A. Baker, Z. Y. Song, O. Olubajo, T. Crittle and D. Peters, *Green Chem.* **2008**, 10, 696–705.
56. N. Kamiya, Y. Matsushita, M. Hanaki, K. Nakashima, M. Narita, M. Goto and H. Takahashi, *Biotechnol. Lett.* **2008**, 30, 1037–1040.
57. Z. H. Zhang and Z. B. K. Zhao, *Carbohydr. Res.* **2009**, 344, 2069–2072.
58. M. FitzPatrick, P. Champagne, M. F. Cunningham and R. A. Whitney, *Bioresour. Technol.* **2010**, 101, 8915–8922.

59. (a) C. Z. Li and Z. K. B. Zhao, *Adv. Synth. Catal.* **2007**, 349, 1847–1850; (b) C. Z. Li, Q. Wang and Z. K. Zhao, *Green Chem.* **2008**, 10, 177–182.
60. (a) C. Sievers, M. B. Valenzuela-Olarte, T. Marzialetti, D. Musin, P. K. Agrawal and C. W. Jones, *Ind. Eng. Chem. Res.* **2009**, 48, 1277–1286; (b) B. Li, I. Filpponen and D. S. Argyropoulos, *Ind. Eng. Chem. Res.* **2010**, 49, 3126–3136.
61. R. Rinaldi, N. Meine, J. vom Stein, R. Palkovits and F. Schüth, *ChemSusChem* **2010**, 3, 266–276.
62. D. R. MacFarlane, J. M. Pringle, K. M. Johansson, S. A. Forsyth and M. Forsyth, *Chem. Commun.* 2006, 1905–1917.
63. (a) A. S. Amarasekara and O. S. Owereh, *Ind. Eng. Chem. Res.* **2009**, 48, 10152–10155; (b) A. S. Amarasekara and O. S. Owereh, *Catal. Commun.* **2010**, 11, 1072–1075.
64. R. Rinaldi and F. Schuth, *ChemSusChem* **2009**, 2, 1096–1107.
65. J. Zakzeski, A. L. Jongerius and B. M. Weckhuysen, *Green Chem.* **2010**, 12, 1225–1236.
66. Y. C. Lin and G. W. Huber, *Energy Environ. Sci.* **2009**, 2, 68–80.
67. K. Shimizu, H. Furukawa, N. Kobayashi, Y. Itaya and A. Satsuma, *Green Chem.* **2009**, 11, 1627–1632.
68. S. J. Kim, A. A. Dwiatmoko, J. W. Choi, Y. W. Suh, D. J. Suh and M. Oh, *Bioresour. Technol.* **2010**, 101, 8273–8279.
69. J. B. Binder and R. T. Raines, *Proc. Natl. Acad. Sci. USA* **2010**, 107, 4516–4521.
70. T. C. R. Brennan, S. Datta, H. W. Blanch, B. A. Simmons and B. M. Holmes, *Bioenergy Res.* **2010**, 3, 123–133.
71. C. Z. Li and Z. K. B. Zhao, *Adv. Synth. Catal.* **2007**, 349, 1847–1850.
72. A. Nakamura, H. Miyafuji and S. Saka, *J. Wood Sci.* **2010**, 56, 256–261.
73. Y. T. Zhang, H. B. Du, X. H. Qian and E. Y. X. Chen, *Energy Fuels* **2010**, 24, 2410–2417.
74. (a) O. O. James, S. Maity, L. A. Usman, K. O. Ajanaku, O. O. Ajani, T. O. Siyanbola, S. Sahu and R. Chaubey, *Energy Environ. Sci.* **2010**, 3, 1833–1850; (b) A. A. Rosatella, S. P. Simeonov, R. F. M. Frade and C. A. M. Afonso, *Green Chem.* **2011**, 13, 754–793.
75. C. Lansalot-Matras and C. Moreau, *Catal. Commun.* **2003**, 4, 517–520.
76. Q. X. Bao, K. Qiao, D. Tomida and C. Yokoyama, *Catal. Commun.* **2008**, 9, 1383–1388.
77. S. Q. Hu, Z. F. Zhang, Y. X. Zhou, B. X. Han, H. L. Fan, W. J. Li, J. L. Song and Y. Xie, *Green Chem.* **2008**, 10, 1280–1283.
78. X. H. Qi, M. Watanabe, T. M. Aida and R. L. Smith, *Green Chem.* **2009**, 11, 1327–1331.
79. X. H. Qi, M. Watanabe, T. M. Aida and R. L. Smith, *ChemSusChem* **2009**, 2, 944–946.
80. L. K. Lai and Y. G. Zhang, *ChemSusChem* **2010**, 3, 1257–1259.
81. C. Z. Li, Z. K. Zhao, A. Q. Wang, M. Y. Zheng and T. Zhang, *Carbohydr. Res.* **2010**, 345, 1846–1850.
82. X. Tong, Y. Ma and Y. Li, *Carbohydr. Res.* **2010**, 345, 1698–1701.
83. K. B. Sidhpuria, A. L. Daniel-da-Silva, T. Trindade and J. A. P. Coutinho, *Green Chem.* **2011**, 13, 340–349.
84. A. P. Lemes, M. A. Soto-Oviedo, W. R. Waldman, L. H. Innocentini-Mei and N. Duran, *J. Polym. Environ.* **2010**, 18, 250–259.
85. X. Zhang, A. Glusen and R. Garcia-Valls, *J. Membr. Sci.* **2006**, 276, 301–307.

86. X. Zhang, J. Benavente and R. Garcia-Valls, *J. Power Sources* **2005**, 145, 292–297.
87. A. K. Gupta and I. S. Bhatia, *Phytochemistry* **1982**, 21, 1249–1253.
88. T. Thananatthanachon and T. B. Rauchfuss, *ChemSusChem* **2010**, 3, 1139–1141.
89. T. S. Hansen, J. Mielby and A. Riisager, *Green Chem.* **2011**, 13, 109–114.
90. Z. J. Wei, Y. Li, D. Thushara, Y. X. Liu and Q. L. Ren, *J. Taiwan Inst. Chem. Eng.* **2011**, 42, 363–370.
91. Z. H. Zhang, Q. A. Wang, H. B. Xie, W. J. Liu and Z. B. Zhao, *ChemSusChem* **2011**, 4, 131–138.
92. H. B. Zhao, J. E. Holladay, H. Brown and Z. C. Zhang, *Science* **2007**, 316, 1597–1600.
93. Y. Su, H. M. Brown, X. Huang, X. D. Zhou, J. E. Amonette and Z. C. Zhang, *Appl. Catal. A Gen.* **2009**, 361, 117–122.
94. Z. H. Zhang and Z. B. K. Zhao, *Bioresour. Technol.* **2010**, 101, 1111–1114.
95. G. K. Farber, A. Glasfeld, G. Tiraby, D. Ringe and G. A. Petsko, *Biochemistry* **1989**, 28, 7289–7297.
96. E. A. Pidko, V. Degirmenci, R. A.van Santen and E. J. M. Hensen, *Angew. Chem.-Int. Ed.* **2010**, 49, 2530–2534.
97. (a) J. B. Binder and R. T. Raines, *J. Am. Chem. Soc.* **2009**, 131, 1979–1985; (b) J. B. Binder, A. V. Cefali, J. J. Blank and R. T. Raines, *Energy Environ. Sci.* **2010**, 3, 765–771.
98. H. B. Xie and Z. B. Zhao, in *Ionic Liquids: Applications and Perspectives* (Ed. A. Kokorin), InTech, **2011**.
99. M. Chidambaram and A. T. Bell, *Green Chem.* **2010**, 12, 1253–1262.
100. E. A. Pidko, V. Degirmenci, R. A.van Santen and E. J. M. Hensen, *Inorg. Chem.* **2010**, 49, 10081–10091.
101. H. B. Zhao, J. E. Holladay, H. Brown and Z. C. Zhang, *Science* **2007**, 316, 1597–1600.
102. G. Yong, Y. G. Zhang and J. Y. Ying, *Angew. Chem.-Int. Ed.* **2008**, 47, 9345–9348.
103. S. Q. Hu, Z. F. Zhang, J. L. Song, Y. X. Zhou and B. X. Han, *Green Chem.* **2009**, 11, 1746–1749.
104. S. Lima, P. Neves, M. M. Antunes, M. Pillinger, N. Ignatyev and A. A. Valente, *Appl. Catal. A-Gen.* **2009**, 363, 93–99.
105. J. B. Binder, J. J. Blank, A. V. Cefali and R. T. Raines, *ChemSusChem* **2010**, 3, 1268–1272.
106. J. A. Chun, J. W. Lee, Y. B. Yi, S. S. Hong and C. H. Chung, *Starch-Starke* **2010**, 62, 326–330.
107. J. A. Chun, J. W. Lee, Y. B. Yi, S. S. Hong and C. H. Chung, *Kor. J. Chem. Eng.* **2010**, 27, 930–935.
108. B. Kim, J. Jeong, S. Shin, D. Lee, S. Kim, H. J. Yoon and J. K. Cho, *ChemSusChem* **2010**, 3, 1273–1275.
109. G. Marcotullio and W. De Jong, *Green Chem.* **2010**, 12, 1739–1746.
110. X. H. Qi, M. Watanabe, T. M. Aida and R. L. Smith, *ChemSusChem* **2010**, 3, 1071–1077.
111. X. H. Qi, M. Watanabe, T. M. Aida and R. L. Smith, *Green Chem.* **2010**, 12, 1855–1860.
112. T. Stahlberg, M. G. Sorensen and A. Riisager, *Green Chem.* **2010**, 12, 321–325.
113. S. C. Wu, C. L. Wang, Y. J. Gao, S. C. Zhang, D. Ma and Z. B. Zhao, *Chin. J. Catal.* **2010**, 31, 1157–1161.

114. F. Jiang, Q. J. Zhu, D. Ma, X. M. Liu and X. W. Han, *J. Mol. Catal. A: Chem.* **2011**, 334, 8–12.
115. J. W. Lee, M. G. Ha, Y. B. Yi and C. H. Chung, *Carbohydr. Res.* **2011**, 346, 177–182.
116. T. Stahlberg, S. Rodriguez-Rodriguez, P. Fristrup and A. Riisager, *Chem.-A Eur. J.* **2011**, 17, 1456–1464.
117. Y. Su, H. M. Brown, G. S. Li, X. D. Zhou, J. E. Amonette, J. L. Fulton, D. M. Camaioni and Z. C. Zhang, *Appl. Catal. A-Gen.* **2011**, 391, 436–442.
118. M. X. Tan, L. Zhao and Y. G. Zhang, *Biomass Bioenergy* **2011**, 35, 1367–1370.
119. F. R. Tao, H. L. Song and L. J. Chou, *Carbohydr. Res.* **2011**, 346, 58–63.
120. Y. B. Yi, J. W. Lee, S. S. Hong, Y. H. Choi and C. H. Chung, *J. Ind. Eng. Chem.* **2011**, 17, 6–9.
121. S. Zhao, M. X. Cheng, J. Z. Li, J. A. Tian and X. H. Wang, *Chem. Commun.* **2011**, 47, 2176–2178.
122. J. B. Binder, M. J. Gray, J. F. White, Z. C. Zhang and J. E. Holladay, *Biomass Bioenergy* **2009**, 33, 1122–1130.
123. B. J. Cox, S. Y. Jia, Z. C. Zhang and J. G. Ekerdt, *Polym. Degrad. Stabil.* **2011**, 96, 426–431.
124. H. Olivier-Bourbigou, L. Magna and D. Morvan, *Appl. Catal. A-Gen.* **2010**, 373, 1–56.
125. I. A. Ignatyev, C. V. Doorslaer, P. G. N. Mertens, K. Binnemans and D. E. De Vos, *ChemSusChem* **2010**, 3, 91–96.
126. Y. H. K. Zhu, Z. N. Kong, L. P. Stubbs, H. Lin, S. C. Shen, E. V. Anslyn and J. A. Maguire, *ChemSusChem* **2010**, 3, 67–70.
127. N. Villandier and A. Corma, *Chem. Commun.* **2010**, 46, 4408–4410.
128. S. Saravanamurugan, O. N.Van Buu and A. Riisager, *ChemSusChem* **2011**, 4, 723–726.
129. L. Peng, L. Lin, J. Zhang, J. Shi and S. Liu, *Appl. Catal. A-Gen.* **2011**, 397, 259–265.
130. E. M. Shahid and Y. Jamal, *Renew. Sust. Energy Rev.* **2008**, 12, 2484–2494.
131. A. C. Cole, J. L. Jensen, I. Ntai, K. L. T. Tran, K. J. Weaver, D. C. Forbes and J. H. Davis, *J. Am. Chem. Soc.* **2002**, 124, 5962–5963.
132. A. A. M. Lapis, L. F. de Oliveira, B. A. D. Neto and J. Dupont, *ChemSusChem* **2008**, 1, 759–762.
133. (a) A. P. Abbott, P. M. Cullis, M. J. Gibson, R. C. Harris and E. Raven, *Green Chem.* **2007**, 9, 868–872; (b) W. Crocker, *J. Mater. Chem.* **2007**, 17, T41–T41.
134. S. H. Ha, M. N. Lan, S. H. Lee, S. M. Hwang and Y. M. Koo, *Enzyme Micro. Technol.* **2007**, 41, 480–483.
135. (a) S. Arai, K. Nakashima, T. Tanino, C. Ogino, A. Kondo and H. Fukuda, *Enzyme Microb. Technol.* **2010**, 46, 51–55; (b) M. Gamba, A. A. M. Lapis and J. Dupont, *Adv. Synth. Catal.* **2008**, 350, 160–164.
136. H. Zhao, Z. Y. Song, O. Olubajo and J. V. Cowins, *Appl. Biochem. Biotechnol.* **2010**, 162, 13–23.
137. M. Han, W. Yi, Q. Wu, Y. Liu, Y. Hong and D. Wang, *Bioresour. Technol.* **2009**, 100, 2308–2310.
138. X. Z. Liang, G. Z. Gong, H. H. Wu and J. G. Yang, *Fuel* **2009**, 88, 613–616.
139. K. Nakashima, S. Arai, T. Tanino, C. Ogino, A. Kondo and H. Fukuda, *J. Biosci. Bioeng.* **2009**, 108, S43–S43.

140. L. Zhang, M. Xian, Y. C. He, L. Z. Li, J. M. Yang, S. T. Yu and X. Xu, *Bioresour. Technol.* **2009**, 100, 4368–4373.
141. X. Z. Liang and J. G. Yang, *Green Chem.* **2010**, 12, 201–204.
142. T. Long, Y. F. Deng, S. C. Gan and J. Chen, *Chin. J. Chem. Eng.* **2010**, 18, 322–327.
143. P. Lozano, J. M. Bernal, R. Piamtongkam, D. Fetzer and M. Vaultier, *ChemSusChem* **2010**, 3, 1359–1363.
144. N. I. Ruzich and A. S. Bassi, *Energy Fuels* **2010**, 24, 3214–3222.
145. N. I. Ruzich and A. S. Bassi, *Can. J. Chem. Eng.* **2010**, 88, 277–282.
146. T. De Diego, A. Manjon, P. Lozano, M. Vaultier and J. L. Iborra, *Green Chem.* **2011**, 13, 444–451.
147. Y. A. Elsheikh, Z. Man, M. A. Bustam, S. Yusup and C. D. Wilfred, *Energy Convers. Manage.* **2011**, 52, 804–809.
148. D. Fang, J. M. Yang and C. M. Jiao, *ACS Catal.* **2011**, 1, 42–47.
149. M. Ghiaci, B. Aghabarari, S. Habibollahi and A. Gil, *Bioresour. Technol.* **2011**, 102, 1200–1204.
150. K. P. Zhang, J. Q. Lai, Z. L. Huang and Z. Yang, *Bioresour. Technol.* **2011**, 102, 2767–2772.
151. H. Zhao, G. A. Baker and S. Holmes, *Org. Biomol. Chem.* **2011**, 9, 1908–1916.
152. D. Coleman and N. Gathergood, *Chem. Soc. Rev.* **2010**, 39(2), 600–637.
153. S. Stolte, S. Steudte, A. Igartua and P. Stepnowski, *Curr. Org. Chem.* **2011**, 15, 1946–1973.
154. (a) J. Dupont, R. F. de Souza and P. A. Z. Suarez, *Chem. Rev.* **2002**, 102, 3667–3691; (b) R. D. Rogers and K. R. Seddon, *Science* **2003**, 302, 792–793.
155. Z. Zhang, D. Zhao and Y. Liao, *Clean* **2007**, 35, 42–48.
156. J. Ranke, S. Stolte, R. Stormann, J. Arning and B. Jastorff, *Chem. Rev.* **2007**, 107, 2183–2206.
157. J. Ranke and B. Jastorff, *Environ. Sci. Pollut. Res.* **2000**, 7, 105–114.
158. F. Atefi, M. T. Garcia, R. D. Singer and P. J. Scammells, *Green Chem.* **2009**, 11, 1595–1604.
159. J. Pernak, K. Sobaszkiewicz and I. Mirska, *Green Chem.* **2002**, 5, 52–56.
160. J. Pernak, I. Goc and I. Mirska, *Green Chem.* **2004**, 6, 323–329.
161. K. M. Docherty and C. F. Kulpa, Jr., *Green Chem.* **2005**, 7, 185–189.
162. A. Romero, A. Santos, J. Tojo and A. Rodriguez, *J. Hazard. Mater.* **2008**, 151, 268–273.
163. D. J. Couling, R. J. Bernot, K. M. Docherty, J. K. Dixon and E. J. Maginn, *Green Chem.* **2006**, 8, 82–90.
164. J. Ranke, K. Molter, F. Stock, U. Bottin-Weber, J. Poczobutt, J. Hoffmann, B. Ondruschka, J. Filser and B. Jastorff, *Ecotoxicol. Environ. Saf.* **2004**, 58, 396–404.
165. M. T. Garcia, N. Gathergood and P. J. Scammells, *Green Chem.* **2004**, 7, 9–14.
166. S. Stolte, S. Abdulkarim, J. Arning, A.-K. Blomeyer-Nienstedt, U. Bottin-Weber, M. Matzke, J. Ranke, B. Jastorff and J. Thöming, *Green Chem.* **2008**, 10, 214–224.
167. M. Matsumoto, K. Mochiduki, K. Fukunishi and K. Kondo, *Sep. Purif. Technol.* **2004**, 40, 97–101.
168. M. Matsumoto, K. Mochiduki and K. Kondo, *J. Biosci. Bioeng.* **2004**, 98, 344–347.
169. K. M. Docherty, S. Z. Hebbeler and Ch. F. Kulpa, Jr., *Green Chem.* **2006**, 8, 560–567.

170. L. Carson, P. K. W. Chau, M. J. Earle, M. A. Gilea, B. F. Gilmore, S. P. Gorman, M. T. McCann and K. R. Seddon, *Green Chem.* **2009**, 11, 492–497.
171. A. S. Wells and V. T. Coombe, *Org. Process Res. Dev.* **2006**, 10, 794–798.
172. R. J. Bernot, M. A. Brueseke, M. A. Evans-White and G. A. Lamberti, *Environ. Toxicol. Chem.* **2005**, 21, 87–92.
173. B. Jastorff, R. Stormann, J. Ranke, K. Molter, F. Stock, B. Oberheitmann, W. Hoffmann, J. Hoffmann, M. Nuchter, B. Ondruschka and J. Filser, *Green Chem.* **2003**, 5, 136–142.
174. M. K. Agrawal, D. Bagchi and S. N. Bagchi, *Hydrobiologia.* **2001**, 464, 37–44.
175. L. R. Shugart, in *Ecotoxicology: A Hierarchical Treatment*, (Eds. M. C. Newman and C. H. Jagoe), Lewis Publishers, Boca Raton, FL, **1996**.
176. P. Nockemann, B. Thijs, K. Driesen, C. Janssen, K.Van Hecke, L.Van Meervelt, S. Kossmann, B. Kirchener and K. Binnemans, *J. Phys. Chem. B* **2007**, 111, 5254–5263.
177. A. Latała, P. Stepnowski, M. Nędzi and W. Mrozik, *Aquat. Toxicol.* **2005**, 73, 91–98.
178. A. Latała, M. Nędzi and P. Stepnowski, *Green Chem.* **2009**, 11, 580–588.
179. C. W. Cho, T. P. T. Pham, Y. C. Jeon and Y. S. Yun, *Green Chem.* **2008**, 10, 67–72.
180. C. W. Cho, Y. C. Jeon, T. P. T. Pham, K. Vijayaraghavan and Y. S. Yun, *Chemosphere* **2007**, 69, 1003–1007.
181. T. P. T. Pham, C.-W. Cho, J. Min and Y.-S. Yun, *J. Biosci. Bioeng.* **2008**, 105, 425–428.
182. G. A. Lamberti and K. J. Kulacki, *Green Chem.* **2008**, 10, 104–110.
183. S. Stolte, J. Arning, U. Bottin-Weber, A. Müller, W.-R. Pitner, U. Welz-Biermann, B. Jastorff and J. Ranke, *Green Chem.* **2007**, 9, 760–767.
184. M. Matzke, S. Stolte, K. Thiele, T. Juffernholz, J. Arning, J. Ranke, U. Welz-Biermann and B. Jastorff, *Green Chem.* **2007**, 9, 1198–1207.
185. B. Jastorff, K. Molter, P. Behrend, U. Bottin-Weber, J. Filser, A. Heimers, B. Ondruschka, J. Ranke, M. Schaefer, H. Schroder, A. Stark, P. Stepnowski, F. Stock, R. Stormann, S. Stolte, U. Welz-Biermann, S. Ziegert and J. Thoming, *Green Chem.* **2005**, 7, 362–372.
186. P. Balczewski, B. Bachowska, T. Bialas, R. Biczak, W. Wieczorek and A. Balinska, *Agric. Food Chem.* **2007**, 55, 1881–1892.
187. S. Studzińska and B. Buszewski, *Anal. Bioanal. Chem.* **2009**, 393, 983–990.
188. S. Stolte, J. Arning, U. B. Weber, M. Matzke, F. Stock, K. Thiele, M. Uerdingen, U. Welz-Biermann, B. Jastorff and J. Ranke, *Green Chem.* **2006**, 8, 621–629.
189. J. Ranke, A. Muller, U. Bottin-Weber, F. Stock, S. Stolte, J. Arning, R. Stormann and B. Jastorff, *Ecotoxicol. Environ. Saf.* **2007**, 67, 430–438.
190. J. Salminen, N. Papaiconomou, R. A. Kumar, J.-M. Lee, J. Kerr, J. Newman and J. M. Prausnitz, *Fluid Phase Equilibria* **2007**, 261, 421–426.
191. X. Wang, A. Ohlin, Q. Lu, Z. Fei, J. Hu and P. Dyson, *Green Chem.* **2007**, 9, 1191–1197.
192. A. García-Lorenzo, E. Tojo, J. Tojo, M. Teijeira, F. J. Rodríguez-Berrocal, M. P. González and V. S. Martínez-Zorzano, *Green Chem.* **2008**, 10, 508–516.
193. F. Stock, J. Hoffmann, J. Ranke, R. Störmann, B. Ondruschka and B. Jastorff, *Green Chem.* **2004**, 6, 286–290.
194. A. C. Składanowski, P. Stepnowski, K. Kleszczynski and B. Dmochowska, *Environ. Toxicol. Pharmacol.* **2005**, 19, 291–296.

195. R. P. Swatloski, J. D. Holbrey, S. B. Memon, G. A. Caldwell, K. A. Caldwell and R. D. Rogers, *Chem. Commun.* **2004**, 668–669.
196. R. J. Bernot, E. E. Kennedy and G. A. Lamberti, *Environ. Toxicol. Chem.* **2005**, 24, 1759–1765.
197. C. Pretti, C. Chiappe, D. Pieraccini, M. Gregori, F. Abramo, G. Monni and L. Intorre, *Green Chem.* **2005**, 8, 238–240.
198. D. M. Costello, L. M. Brown and G. A. Lamberti, *Green Chem.* **2009**, 11, 548–553.
199. X. Y. Li, J. Zhou, M. Yu, J. J. Wang and Y. C. Pei, *Ecotoxicol. Environ. Saf.* **2009**, 72, 552–556.
200. T. P. Pham, C. W. Cho and Y. S. Yun, *Water Res.* **2010**, 44(2), 352–372.
201. P. Luis, A. Garea and A. Irabien, *J. Mol. Liq.* **2010**, 152, 28–33.
202. S. Stolte, S. Abdulkarim, J. Arning, A. Blomeyer-Nienstedt, U. Bottin-Weber, M. Matzke, B. Jastorff and J. Thöming, *Green Chem.* **2008**, 10(2), 214–242.
203. A. S. Wells and V. T. Coombe, *Org. Process Res. Dev.* **2006**, 10(4), 794–798.
204. K. M. Docherty, J. K. Dixon and C. F. Kulpa, *Biodegradation* **2007**, 18, 481–493.
205. N. Gathergood, P. J. Scammells and M. T. Garcia, *Green Chem.* **2006**, 8(2), 156–160.
206. N. Gathergood, M. T. Garcia and P. J. Scammells, *Green Chem.* **2004**, 6(2), 166–175.
207. M. T. Garcia, N. Gathergood and P. J. Scammells, *Green Chem.* **2005**, 7(1), 9–14.
208. W. Roberts and J. Costello, *QSAR Comb. Sci.* **2003**, 22, 220–225.
209. M. J. Rosen, F. Li, S. W. Morrall and D. J. Versteeg, *Environ. Sci. Technol.* **2001**, 35, 954–959.
210. R. Sheldon, *Chem. Commun.* **2001**, 23, 2399–2407.
211. D. M. Costello, L. M. Brown and G. A. Lamberti, *Green Chem.* **2009**, 11, 548–553.
212. A. Wells, *Green Solvents—Progress in Science and Application*, Lake Constance, Friedrichshafen/Germany, 28 September–01 October **2008**.
213. T. Hartung and C. Rovida, *Nature* **2009**, 460, 1080–1081.
214. (a) P. Wasserscheid and T. Welton (Eds.), *Ionic Liquids in Synthesis*, Wiley-VCH, Weinheim, **2002**; (b) T. Welton, *Green Chem.* **2008**, 10, 483–483.
215. D. Kralisch, A. Stark, S. Körsten, G. Kreisel and B. Ondruschka, *Green Chem.* **2005**, 7, 301–309.

CHAPTER 4

Biorefinery with Water

X. PHILIP YE, LEMING CHENG, HAILE MA, BILJANA BUJANOVIC,
MANGESH J. GOUNDALKAR, and THOMAS E. AMIDON

4.1 INTRODUCTION

There are superficial similarities between the biomass refinery and the petroleum refinery but the feedstock and product complexities are significantly different. The nonvolatile nature of biomass, as opposed to petroleum, suggests that liquid-phase conversion may be better suited for the new paradigm of biorefinery. Water is arguably the most environmentally benign and food-safe solvent that can be used in chemical synthesis [1]. Particularly in recent decades, chemists have begun investigating the possibility of using water as a solvent for green organic reactions because of the advantages of water with respect to cost, safety, and environmental concerns [2]. However, the range of water-soluble substrates in biomass is very limited, making ambient water an unsuitable medium for many chemical reactions. Water in the sub- or supercritical state exhibits unique and tunable properties similar to those of organic solvents and is capable of depolymerizing lignocellulosic biomass for further refining.

"Supercritical water" is defined as a state in which both the temperature and pressure of water are higher than the critical point (374 °C, 22.1 MPa) [3]. "Subcritical water" refers to liquid water under pressure at temperatures between the usual boiling point (100 °C) and the critical temperature (374 °C) [4]; it is also known as "superheated water" or "hot compressed water." The terms sub- or supercritical water are specifically used in this chapter to clearly differentiate these two water states because of the dramatic change in water properties around the critical point. However, the acronym SCW collectively refers to both sub- and supercritical water when a study covers both states. In some of the literature, "hydrothermal" has been used for the SCW conditions.

The Role of Green Chemistry in Biomass Processing and Conversion, First Edition.
Edited by Haibo Xie and Nicholas Gathergood.

4.2 RATIONALE FOR BIOREFINERY WITH WATER

Compared to petroleum, biomass resources from lignocellulosics are bulky and difficult to handle, transport, and store. Up to now, biomass has not been utilized as efficiently as it could be. One of the major obstacles to the efficient utilization of biomass in traditional thermochemical processes, such as pyrolysis, gasification, or combustion conversion routes, is that the water content in most biomass feedstock is much higher than that of fossil resources. Wet biomass must be dried in a process that demands significant energy to volatilize the water as steam. For example in a corn–ethanol energy consumption analysis, Johnson (2006) [5] showed that only about 1/3 of the total energy consumed in producing ethanol is associated with the production of the corn (including planting, harvest, irrigation, fertilizer, and transport); 2/3 of the energy needed is associated with the manufacturing plant that converts corn to ethanol. Approximately 50% of the total energy requirements are consumed by two steps alone: distillation and drying [6]. These two inefficient steps are primarily the result of energy requirements for volatilizing water. Furthermore, the low density and low heating value of biomass are drawbacks that pose great challenges for large-scale biomass transportation, storage, gasification, and combustion processes.

Water in the sub- or supercritical state exhibits exciting physicochemical properties, rendering a suitable and sustainable solvent and reaction medium for biorefinery. Selected properties of SCW in comparison with water steam in a range used in biomass processing are presented in Figure 4.1 (data from ref. [7]).

4.2.1 Energy Efficiency of Processing Biomass in SCW

Because the reaction medium is water, wet biomass can be directly used as feedstock and the contained water can be removed in its liquid phase by a nonevaporative dewatering process at higher pressures [8], in which free water

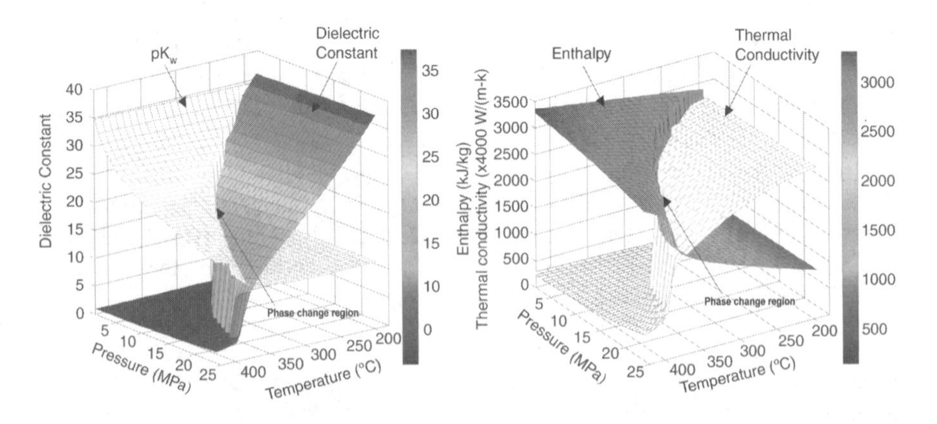

FIGURE 4.1 Selected water properties in sub- and super-critical states in comparison with water steam [7]. (Reproduced by permission of Elsevier.)

can be removed in liquid state without vaporization and then products with low moisture content can be obtained after solid–liquid separation. In conventional processing, water is typically separated thermally, by vaporization in pyrolysis, by distillation in biochemical processing, or by similar means in other processes going through phase change from liquid to gas. These separation steps lead to large energy losses that consume much of the energy content of biomass [6]. As shown in Figure 4.1, when water is removed in its liquid state, about 40% enthalpy can be saved when the process is operated at 300 °C and 25 MPa, compared with steam at 100 °C but ambient pressure. Most of the latent heat can be saved when water is removed in its liquid state.

The thermal conductivity of SCW is significantly higher than that of ambient pressure steam, as shown in Figure 4.1. As water changes from liquid to gaseous state, its thermal conductivity significantly decreases. Thus, it is feasible to recover most of the energy from the SCW by preheating the incoming water in a heat exchanger, thereby cooling the process water. It is possible to recover most of the energy by a heat-recovery unit, leaving only a small part of energy to be added by the heater. Therefore, once the plant is warmed up to operating temperature, the heating requirements could be greatly reduced. Ideally, the product can be separated by controlling the properties of water (such as forming an organic phase that is not dissolved in water) or separated in solid phase. Then, only a small amount of residual water needs to be evaporated in order to dry the product. This higher thermal conductivity also facilitates organic reactions in SCW.

4.2.2 Unique and Tunable Properties of Water at SCW Conditions

With a change in temperature and pressure, water exhibit changes in its properties that are much greater than in the case of other solvents or reaction media, especially around the saturation lines and critical point. Thus, water in either the sub- or supercritical state possesses unique solvent and transport properties different from those of water in the liquid or steam state under ambient pressures, having liquid-like densities and gas-like diffusivities [3]. The viscosity and surface tension of SCW decrease and diffusivity increases with increasing temperature at high pressures [9]. For chemical reactions involving polar solutes, the most important property of the solvent is its ion product. Typical variations of the dielectric constant of water and self-ionization (pK_w) can be found in Figure 4.1. Water at 13.7 MPa holds its ionic properties until its boiling point (335 °C); whereas at 34.5 MPa, water retains its ionic properties to over 400 °C. Compared to the familiar pK_w of 14 at 25 °C and ambient pressure, the pK_w of water at 250 °C and 5 MPa is 11.2, meaning that the concentration of hydronium ion (H_3O^+) is higher at this subcritical condition, and hence the pH is lower (although the level of hydroxide (OH^-) is increased by the same amount so the water is still neutral). The dielectric constant (relative permittivity) decreases sharply as the temperature rises, significantly affecting the behavior of water at high temperatures (Fig. 4.1). These properties make SCW a tunable solvent and reaction medium for the conversion of biomass, which requires high temperature to activate reactions upon, especially for lignocellulosics. Organic molecules often show a dramatic increase in solubility in water as temperature rises, partly because of the polarity changes, and also because the solubility of less soluble materials tends to increase with temperature as they have a high enthalpy of solution. [3,6] SCW exhibits a reduction in dielectric constant and density relative to ambient water; its ability to dissolve both nonpolar organic molecules and inorganic salts is comparable to that of the popular organic solvent acetone. Further, the solvent properties can change from nonpolar behavior as, for example, in organic solvents to highly ionic characteristics as, for example, in melt salts. This opens up many promising opportunities for separation processes and chemical reactions [10].

4.2.3 Suitable Medium for Biomass Extraction, Pretreatment, Fractionation, and Conversion

Biomass, especially the most abundant lignocellulosic biomass, is heterogeneous, complex, and thermodynamically stable, requiring high energy to induce reactions upon. Whereas most hemicelluloses and extractives in lignocellulosics can be extracted at mild SCW conditions, cellulose requires a temperature of 320 °C and a pressure of 25 MPa to become amorphous in water [11]. Deguchi et al. [11] have utilized polarized light microscopy to observe the loss of crystallinity in cellulose fibers using techniques similar to those conventionally used to monitor starch gelatinization, namely a loss of birefringence that corresponds to a loss of crystallinity. Scanning at 11 to 14 °C min^{-1} and 25 MPa, Deguchi's team observed a loss of birefringence at around 320 °C, indicating that the cellulose crystallinity disappeared at this condition, which is in the range of subcritical water. The significant effect that subcritical water has on modifying the crystallinity of cellulose will greatly facilitate the hydrolysis of cellulose into monomer sugars and their derivatives, suggesting a good approach for cellulose conversion using SCW technologies.

Even homogeneous or heterogeneous catalysis is possible in SCW. This has been demonstrated with the cobalt-catalyzed cyclotrimerization of acetylenes to form benzene derivatives or hydroformylation to produce aldehydes from olefins; there, only the addition of CO is necessary, the H_2 required being formed by the equilibrium of the water–gas-shift reaction [10]. After a homogeneous reaction in sub- or supercritical state, the reaction mixture can be separated at subcritical conditions. Moreover, depending on the reaction conditions, ionic or radical reaction pathways can be favored or suppressed, allowing for controlled selectivity.

4.3 WATER PRETREATMENT OF LIGNOCELLULOSICS FOR PRODUCING BIOFUELS/BIOCHEMICALS/BIOMATERIALS

Mature technologies based on lignocellulosics (LCs) are focused on the production of cellulose, while lignin, hemicelluloses, and extractives are mostly considered as energy sources or waste. For example, kraft delignification results in black liquor, which is burnt in a recovery boiler during the regeneration of pulping chemicals. Due to a high heating value of 23.26–25.58 kJ g^{-1} [12], lignin represents the main source of energy in this process. Meister (2002) [13], however, describes the incineration technology as having "little thermodynamic or economic sense. With two-fifths of the plants' absorbed energy being used to make the one quarter of its dry mass that is lignin, lignin represents a large investment of biochemical effort by the plant. Reducing this macromolecule to CO_2 destroys much of that investment." Consistent with this concept is a growing demand for the development of alternative technologies, which would use all LC constituents to produce materials, chemicals, and fuels while replacing increasingly inadequate fossil fuels. [1,14] In accordance with the "biomass refining" philosophy, these new technologies may be envisioned to follow sequential treatments to create separate streams of relatively clean LC components [15]. After purification, carbohydrate-rich streams may undergo chemical or biological conversion to produce a variety of products [14b,16]. Streams rich in lignin and extractives have a high potential for the production of high-value low-volume products to enhance the profitability of this type of biorefinery. [17] Different products including thermoplastics, adhesives, and chemicals with a wide range of applications have been proposed. [18]

Extensive research has been performed in the evaluation of pretreatments, which would be an appropriate first step in sequential refining of LCs. Pretreatment is most commonly intended to remove hemicelluloses and lignin, reduce cellulose crystallinity, and increase porosity to facilitate processability of the cellulose-enriched residue. [14,16b,19] A very important goal in any pretreatment is to minimize the formation of inhibitors for subsequent hydrolysis and fermentation steps. The inhibitors include natural constituents of hemicelluloses (e.g., acetic acid), products of carbohydrate degradation (e.g., furfural and hydroxymethyl furfural), and aromatic compounds (e.g., lignin-degradation products and phenolic extractives) [14a,16b,17,19b,19c,20]. Different methods have been suggested including physical, chemical, and biological pretreatments [14a].

Pretreatment based on hydrothermal treatment, that is the use of water and heat only (SCW), has been recommended frequently to create a cleaner, more environmentally benign process with a low corrosion risk to equipment [14b,16b,21]. A water prehydrolysis stage has also been studied and used earlier in some commercial pulping operations prior to kraft pulping [22]. Accordingly, the effect of high-temperature water on lignocellulosic/wood constituents has intrigued numerous research groups since the 1930s [23].

In the context of sequential fractionation of LCs for the production of energy, materials, and chemicals, water-based pretreatment methods have been investigated in numerous studies and designated as: aqueous pretreatment [24], hydrothermal processing [23f,23g], autohydrolysis [19c,23i,25], hot-compressed water treatment [23h,26], liquid hot-water treatment [19b,20,27], mild autohydrolysis [15], hot-water treatment [16b,28], or pressurized hot-water extraction (PHWE) [23j,23n,29]. These studies have been conducted under a range of conditions including variations in temperature, pressure, and time of the treatment, and water-to-solid ratio (ranging from 2:1 to 20:1 has been used) [23f]. A wide range of raw-material particle size has also been studied. It has been demonstrated that extraction of hemicelluloses from wood chips results in lower yields than from ground wood meal [23j]. Different reactor types used in these studies may be described as percolator [26a,27] or flow through [23h,23n], batch [15,23k,23l], and accelerated extractors [23j,25a,30]. Even though studied under a broad range of conditions, most of the recommended treatments may be categorized as subcritical water treatment ($150 < T < 370\,°C$, $0.4 < P < 22\,MPa$) performed at temperatures lower than 230 °C.

Water pretreatment under subcritical conditions and at temperatures <230 °C is consistent with "the biomass refining" concept because it is a hemicellulose-dissolving/degrading step resulting in carbohydrate-rich extracts, which may be purified and used in polymer form [16] or fermented after hydrolysis [31]. Hydrolyzed hot-water extracts containing primarily xylose have been fermented to ethanol using different xylose-fermenting yeast strains including *Pichia stipitis*. [31,32] For the production of *n*-butanol, saccharolytic strains such as *Clostridium acetobutylicum* and *Clostridium beijerinckii* have been used, as they are able to ferment a variety of sugars and produce neutral solvents consisting of acetone, butanol, and ethanol [33]. Fermentation has also been performed by a variety of microorganisms to produce biopolymers, lactic acid for polylactic acid (PLA), and poly-β-hydroxyalkanoates (PHAs) [16b,34].

An important advantage of water treatments performed under subcritical conditions and at temperatures <230 °C compared to those performed at higher temperatures is a solid stream, for example, extracted wood, which contains relatively undamaged cellulose [23f]. After delignification, the solid stream enriched in cellulose may be hydrolyzed to fermentable glucose or used in polymer form for the production of pulp, cellulose derivatives, or nanocellulose [14b,21,22,28,35]. In pulp production, an increase in kraft-delignification rate has been observed specifically for wood extracted at temperatures lower than 170 °C [16b,22d,22f,35]. However, it has been found that paper produced from this pulp is of lower strength than paper from conventional pulp [22d,35]. To avoid this adverse effect, paper reinforcement may be accomplished by using dry strength agents in wet-end of the paper-making process. In the context of biorefinery, paper may be reinforced in the paper surface treatment with PLA, which is another designed product of the biorefinery. Recent results have shown a superior strength-to-weight ratio of PLA-treated paper made from kraft pulp of extracted sugar maple compared to paper made from conventional kraft pulp of sugar maple. These results demonstrate that the integration of the biorefinery lines in pulp and paper mills may be an attractive option to increase economic returns [35]. Delignification studies have been performed by other methods, including organosolv methods using ethanol and acetone, designed products of the biorefinery [16b,36]. Again, these studies have shown that during water treatment at temperatures lower than 170 °C hardwoods acquire physicochemical properties, which make them easier to delignify. In the case of acetone, the favorable effects include a remarkable stability of cellulose [22c,23f,36]. Extracted wood depleted of hydrophilic hemicelluloses of relatively low calorific value can also be used for the production of pellets, which exhibit higher moisture resistance and higher calorific value than pellets produced from unextracted wood [37].

Medium-density fiberboards from extracted hardwoods have also been confirmed to be of greater dimensional stability and moisture resistance than corresponding products from unextracted wood [38].

Similar to the H-factor (developed for kraft pulping) [39], the *P*-factor aimed to encompass the effects of temperature and time of hydrothermal treatment in a single number has been introduced. The *P*-factor is calculated by taking into account the activation energy typical for the cleavage of glycosidic bonds in hardwood xylan ($E_a = 125.6\,kJ\,mol^{-1}$); $P\text{-factor} = \int \exp(40.48 - (15{,}106/T))dt$, T (K) and t (hours). An increase in the *P*-factor correlates well with the amount of residual xylan in wood at various temperatures [40]. The *P*-factor should also include the effects of heating up and cooling-down periods. Most of the results obtained in investigations of hydrothermal treatment of wood, however, have been reported in correlation with an earlier empirically developed severity factor $R_0 = t * \exp[(T - 100)/14.75]$ (t in min, T in °C), expressed as $\log R_0$ [22a]. The time–temperature effect of wood treatment with water at elevated temperatures has been also reported by the H-factor approach [21,22d,22e].

Hydrothermal treatment of LCs may be described as autocatalyzed hydrolysis or autohydrolysis (AH) [15,16b,21,22c,22f,23i,25,30]. AH refers to the process of hydrolysis catalyzed by the hydronium ions (H^+) originating from water and acidic compounds generated from the treated lignocellulosic material. AH consists of a number of reactions, including deacetylation of acetylated hemicelluloses and hydrolysis of hemicelluloses. Proposed mechanisms of the acidic deacetylation and cleavage of glycosidic bond are presented elsewhere [41]. The H^+ ions resulted from water ionization act as catalysts for these reactions initially. At high temperatures, the H^+ ions are generated at higher concentrations than in ambient liquid water providing an effective medium for acid hydrolysis [3]. Subsequently, acetic acid ($pK_a = 4.76$) formed during deacetylation promotes further random cleavage of hemicellulose/ xylan chains resulting in a mixture of oligosaccharides and monosaccharides. Even though xylan chains are most likely to retain 4-O-MeGlcA (methylglucuronic acid) residues, if accessible to water and/or dissolved, the uronic acid residues would dissociate, increase the H^+ concentration, and enhance hydrolysis [23f,23j,42] (D-glucuronic acid, $pK_a \sim 2.8$) [43]. Pectins containing unesterified and easily hydrolysable methyl esterified galacturonic-acid units are expected to exhibit a similar effect (D-galacturonic acid $pK_a = 3.5$) [44,45]. However, the effects of pectins on autohydrolysis are not well established [23j,23n]. At the end of water treatment of LCs, the pH of hydrolyzate has been reported to range from 3.2 to 3.8 under subcritical conditions and at temperatures $<230\,°C$ [16b,22a,22b,22f,23c,23h,23j,23n,25a].

An increased acidity of the medium at high temperatures may cause degradation of generated monosaccharides, pentoses and hexoses to furfural and hydroxymethyl furfural, respectively [41a,41b]. Moreover, the degradation of hexoses generates formic acid ($pK_a = 3.75$), which increases the concentration of the H^+ ions in the medium. Accordingly, the extent of degradation of monosaccharides is an important factor that affects AH, leading to a chain of undesirable polymerization reactions (self-polymerization of furfural and hydroxymethyl furfural and their participation in lignin-condensation reactions (Fig 4.2 route 2A)), and reducing the efficiency of the following operations on the extract and extracted material [23c,23d,23f].

Hydrothermal treatment of wood has been investigated for the effect on lignin by studying water extracts, lignin isolated from wood, and extracted wood [23a–d,23f,23i,23m,24,46]. The amount of lignin extracted from wood during autohydrolysis depends on the temperature and time of treatment and is species-dependent, but it may be as high as 25% of the amount of lignin originally present in

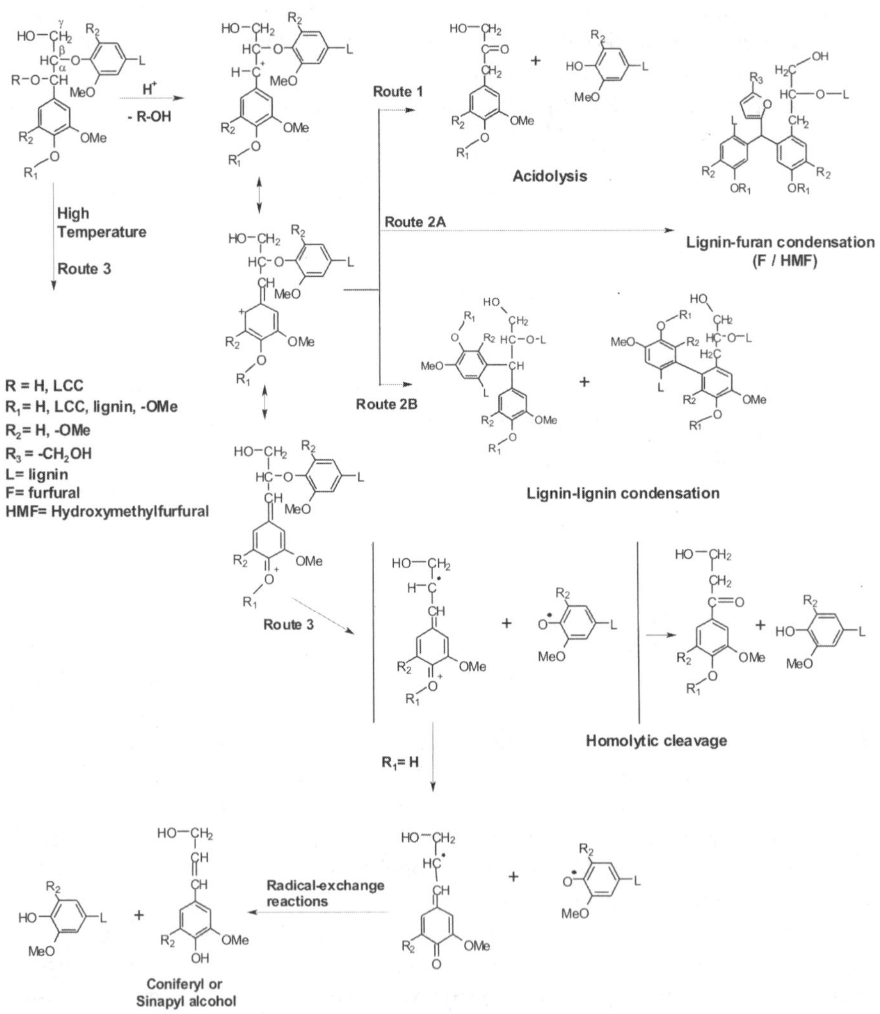

FIGURE 4.2 Scheme of biorefinery based on HWE and potential products F-furfural; HMF–hydroxymethylfurfural; PHAs-polyhydroxyalkanoates; PLA-polylactic acid; RWP-reconstituted wood products; CHP-combined heat and power. (Adapted from Ref. [23i, 49, and 50].)

wood at temperatures <230 °C [23a,23f]. Extracted lignin and lignin-degradation products appear in wood extracts in two fractions: an insoluble and a soluble fraction [23m,46]. In addition, MWL (milled wood lignin is considered to be representative of protolignin in wood and is widely used for structural studies of lignin) [41b,47] was subjected to conditions prevailing during AH [23e]. It has been proposed that during autohydrolysis lignin participates in two types of competing reactions resulting in lignin cleavage (depolymerization) and condensation (repolymerization) (Fig. 4.2) [23i,49, 50]. Depolymerization of lignin is initiated by the proton-induced elimination of water/alcohol from the benzylic position (C_{α}) in lignin/LCC and the formation of resonance-stabilized carbonium (benzylium) ions. The cleavage of the β-O-4 bond is suggested to follow an acidolytic [23i,48] and/or homolytic type of cleavage [23i,23m,49]. Consistent with the cleavage of the β-O-4 bonds, MWL containing fewer components of high-molecular weight than MWL of the original *Eucalyptus globulus* wood has been isolated from the corresponding water-extracted wood (water treatment at 170 °C) [23k]. Condensation reactions involve the reaction between electrophiles, that is, benzylium ions formed during the cleavage of benzyl ether bonds and the lignin nucleophiles formed in acidic medium, that is, aromatic rings with a high electron density at C2 and C6 to form carbon–carbon bonds [23d,50] (Fig. 4.2, route 2B). Electrophilic furfural and hydroxymethyl furfural containing aldehyde group and carbon–carbon double bonds in the furan ring can also participate in lignin-condensation reactions (Fig. 4.2 route 2A) [23d,50].

The stability of lignin–carbohydrate bonds (LC bonds) (several types of LC bonds have been proposed to exist in the native plant cell wall and among them the benzyl ether and ester types have been considered the most probable type of linkage) [51] during autohydrolysis is also of interest. Model experiments have suggested that even though LC bonds are chemically labile in acid, the degradation rate varies depending on the type of structure [52]. Consequently, cleavage of the LC bonds is expected to occur during hydrothermal treatment of wood [23i], but the extent of this process is unknown. However, the documented more efficient delignification of extracted wood is a strong indication of this type of effect [16b,22d,22f,35,36].

4.4 WATER EXTRACTION OF VALUE-ADDED CHEMICALS

Although extractives are detrimental in the pulping process, they have the potential to provide added value to a biorefinery. Extractives are classified based on their chemical composition, structural backbone, and chemical moieties present on the structure. Aromatic extractives present in wood are mainly phenolics appearing in different forms [53]. The nonaromatic extractives are usually comprised of terpenes/terpenoids, fatty acids and waxes. Knots of softwoods and hardwoods may contain exceptionally large amounts of polyphenols, mostly lignans and pinosylvins in softwoods, and flavonoids in hardwoods [54]. For example, aspen knots were shown to be rich in flavonoids [55] whereas willow was shown to be rich in lignans [56]. Of all extractives, flavonoids and lignans are probably the most investigated categories mainly because of their ability to scavenge free radicals and act as potential antioxidizing agents. However, most of the extraction methods employ neat organic solvents or a mixture of organic solvents and water. To reduce the use of organic solvents, PHWE has been proposed for extracting flavonoids from wood. PHWE has already been used to extract bioactive compounds from medicinal plant materials amongst other potential applications [29b,57]. It has been demonstrated that PHWE is more efficient in extraction of flavonoids such as naringenin and dihydrokaempferol from knotwood of aspen than other extraction techniques such as Soxhlet and ultrasonic extraction, and refluxing in organic solvents [29a].

Flavonoids and lignans have been studied for their antioxidant, antiatherogenic, and anticancer activity and other potential medicinal uses. For example, naringenin, a flavonoid found in high concentration in citrus/grapefruits has been studied for its effect on the uptake of glucose by skeletal muscles [58]. Most of the literature considers flavonoids from food, since that is directly related to consumption by humans. However, with knotwood being shown to be rich in flavonoids, attempts to isolate them might compliment the biorefinery concept by adding another stream of value-added chemicals. These value-added chemicals can also be used as platform chemicals to obtain other useful derivatives. For example, hydroxymatairesinol (HMR) obtained from coniferous trees has been shown to metabolize to a known mammalian lignan, enterolactone, which possesses antioxidant and antitumor properties [59]. HMR has also been used as a starting material to synthesize *R*(–) imperanene [60]. Imperanene is a rare, naturally occurring phenolic extracted from rhizomes of the plant *Imperata cylindrical*; it is a biologically active compound showing platelet-aggregation inhibition. Lariciresinol, which is a precursor for conversion to enterolactone by intestinal bacteria in humans and a potential anticancer agent were synthesized from HMR, along with isolariciresinol, a potential anti-inflammatory and antioxidation agent [61]. HMR has also been used as a starting material to synthesize matairesinol, enterolactone, and enterodiol [62]. Chemical properties of HMR were further investigated by treating it with oxidizing agents such as DDQ (2,3-dichloro-5,6-dicyanobenzoquinone), which produced oxomatairesinol amongst other oxygenated products [63]. The chemical structure of naringenin and HMR may be found elsewhere [53].

4.4.1 Hot-Water Extraction (HWE) of Hardwoods

Following the principles of sequential fractionation of lignocellulosics, a biorefinery concept based on hardwoods, such as northeastern hardwoods and plantation-grown willow biomass has been under investigation at the State University of New York College of Environmental Science and Forestry (SUNY-ESF) [14b,16b,21]. The first step in the proposed biorefinery is a hydrothermal treatment under subcritical conditions at temperatures $\leq 170\,°C$ called hot-water extraction (HWE). The two streams produced in HWE, hot-water extract and extracted wood, are envisioned for further processing to produce a palette of high-value low-volume and low-value high-volume products. A simplified scheme of a biorefinery based on HWE is shown in Figure 4.3. The HWE process has been performed at low *P*-factors, that is, under relatively mild conditions to minimize dehydration of released monosaccharides to

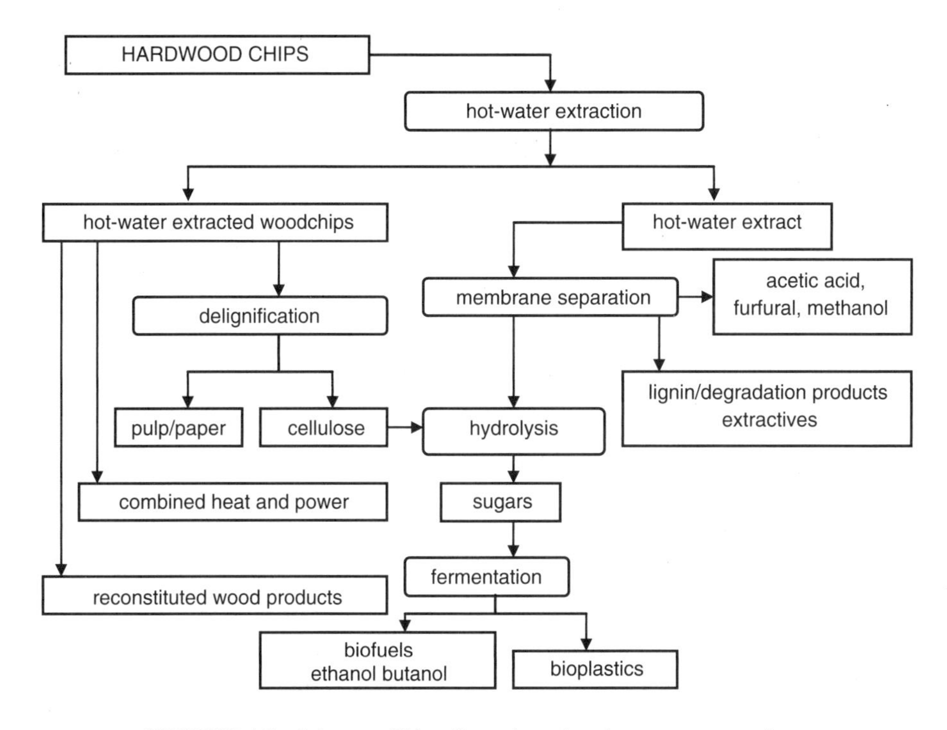

FIGURE 4.3 Scheme of biorefinery based on hot-water extraction.

preserve them and to enhance the results of the following fermentation stage. In fact, it has been demonstrated that only a minor dehydration of xylose to furfural is expected at *P*-factors of up to 600 [25a]. In the HWE of sugar maple performed at 160 °C for 2 h (*P*-factor of ~540; HWE_{540}), the yield of extracted wood was 77% while 75% of xylans and 50% of glucomannans were extracted [16b]. Even though in small amounts, low-molecular weight (LMW compounds) including furfural, hydroxymethyl furfural, formic acid, and methanol were formed. Furfural, hydroxymethyl furfural, and formic acid are most likely the products of monosaccharide degradation [41a,41b]. Methanol may be formed during demethylation of lignin phenylpropanoid units [23c]. In accordance with the design of HWE-based biorefinery, the LMW compounds are removed from the extract by membrane separation (Fig. 4.3.). The LMW compounds are recovered for sale as chemical raw material while their removal increases the yield of the following fermentation phase, because most of these compounds are fermentation inhibitors [16b,64].

4.4.2 Noncarbohydrate-Derived Material Dissolved During HWE

The studies showed that HWE_{540} of sugar maple results in solubilization of ~12% of the total amount of lignin in sugar maple [36]. Dissolved lignin appeared in the hot-water extract in two distinct fractions as previously stated: an insoluble fraction, referred to as Insol fraction, which readily precipitates from the hot-water extract and a soluble fraction, referred to as Sol fraction, which may be extracted from the hot-water extract with organic solvents, such as chloroform or ether [46].

4.4.3 Insol fraction

To separate the material, which readily precipitates after HWE, an ultrafiltration membrane (1000 Da nominal molecular-weight cut-off) was used. The amount of Insol fraction was ~1.8% of the total amount of wood used in HWE or >60% of the total amount of lignin dissolved during HWE_{540} of sugar maple. The 2D NMR, HSQC (Two-Dimensional Nuclear Magnetic Resonance, Heteronuclear Single Quantum Correlation) data recorded on purified Insol fraction after extraction with chloroform to remove LMW compounds showed characteristic lignin peaks indicating a high content of lignin in the Insol fraction [46a]. A certain amount of carbohydrates in this sample was revealed by the presence of characteristic carbohydrate correlations. The β–β and β-5 lignin bonds along with the most prominent β-O-4 bond were detected in this spectrum. In addition, the correlations indicating the presence of benzaldehydes and cinnamyl aldehydes (e.g., vanillin and coniferaldehyde) and aromatic carboxyl acids (e.g., vanillic and ferulic acid), known to be sporadically incorporated into the lignin structure [65] were observed. The lignin content of Insol fraction was 87% (Klason + acid soluble lignin) [46a47]. An increase in the lignin content to ~93% was achieved by mild acid hydrolysis of the Insol fraction (pH = 2, 2 h at 60 °C, N_2) [46a]. These results clearly indicate that the composition of the precipitate formed from hot-water extract may be modified by changing the conditions of separation, that is, the composition of the Insol fraction may be altered in accordance with the requirements of designed use. Based on the carbohydrate analysis (quantitative analysis of sugars by ^{1}H-NMR analysis) [66], xylose was confirmed as the predominant sugar in the purified Insol fraction because more than 75% of remaining carbohydrates were xylose-based [46a]. This result is consistent with the presence of lignin–carbohydrate bonds in the hot-water extract of sugar maple [51]. Nitrobenzene oxidation [47] performed on the purified Insol fraction revealed an S/G ratio of ~2.6 for the extracted lignin [46a], which is more than twice as high as the S/G ratio characteristic for all lignin in sugar maple [67]. This result is consistent with a high susceptibility of S-units to hydrothermal solubilization. [23f] The high content of S-units in lignin dissolved during HWE could also indicate that the readily available lignin from hardwoods is lignin in the fibers because it is predominantly S-lignin [41b]. The content of free phenolic hydroxyl groups (PhOH) evaluated by periodate oxidation [47] was 3.0 mmol g^{-1} lignin. This is a relatively high content of the PhOH groups, which may be associated with low-molecular weight lignin. The molecular weight of the lignin including $M_w = 1.6 \times 10^3$, $M_n = 5.2 \times 10^2$, and polydispersity $M_w/M_n = 3.1$ was measured using size exclusion chromatography [46a,46b]. In general, characterization of the Insol fraction indicated that the lignin dissolved during HWE is similar to organosolv (Alcell) lignin [18a]. This is not surprising since the organosolv process is based on an aqueous ethanol treatment of hardwood species [18a]. During HWE, the PhOH content of the lignin remaining in sugar maple increased from 0.35 mmol g^{-1} lignin to 1.83 mmol g^{-1} lignin. This considerable increase in the content of PhOH groups of lignin remaining in extracted wood is a strong indication of the cleavage of β-O-4 bonds during HWE/autohydrolysis [23i,23k–m, 36].

4.4.4 Sol Fraction

Organic extracts (extraction with chloroform and ether) of permeates obtained during laboratory ultrafiltration of hot-water extracts of sugar maple (the solvent-soluble fraction of hot-water extract, Sol fraction) were analyzed by GC-MS [46a,46b,46d]. The total amount of extracted material was ~1.5% of oven-dry wood used in HWE, or about 6.5% of the total material removed during HWE. The composition of extractives extracted from wood using organic solvents (chloroform, 95% ethanol, or acetone:water (9:1) [41c]) was compared with the composition of the chloroform- and ether-soluble fraction of hot-water extracts. Typical phenolic wood extractives vanillin, guaiacol, and syringol were identified in both samples. However, the results revealed a number of organic LMW compounds, which were present only in the solvent-soluble fraction of the hot-water extracts. Specifically, *p*-hydroxy benzaldehyde, syringaldehyde, *p*-hydroxybenzoic acid, vanillic acid, syringic acid, coniferaldehyde, sinapaldehyde, dihydroconiferyl alcohol, dihydroferulic acid, medioresinol, and syringaresinol were detected only in the solvent-soluble fraction of the hot-water extracts. Identified dissolved aromatic acids would contribute to the autohydrolysis processes during HWE by providing additional sources of the H^+ ions; *p*-hydroxybenzoic acid, $pK_a = 4.3$; ferulic acid, $pK_a = 4.27$; vanillic acid, $pK_a = 4.08$; syringic acid, $pK_a = 3.86$ [68]. In accordance with the earlier results on the aqueous hydrolysis of lignin performed on hemlock wood and Brauns' lignin [23b], these compounds may be attributed to lignin cleavage taking place during HWE. Moreover, evidence for the occurrence of homolytic cleavage of the β-aryl ether bonds was found based on the presence of syringaldehyde, sinapaldehyde, syringic acid, and syringaresinol. These compounds were proposed to be formed via a homolytic cleavage of an initially formed quinone methide (Fig. 4.2 route 3) or to result from oxidation reactions (syringic acid) [23i,49]. Coniferyl and sinapyl alcohol, which are expected products of homolytic cleavage of the phenolic β-aryl ether structure were not identified in the Sol fraction. Their absence may be accounted for by an inherent low stability of coniferyl alcohol (and hence low stability of sinapyl alcohol too) at high temperature and in acid conditions [69]. It is interesting to note that typical products of lignin acidolysis, such as Hibbert ketones [49] were not present in the Sol fraction of sugar maple hot-water extract. These results appeared to validate homolytic cleavage of the β-aryl ether bonds during HWE, whereas no evidence for occurrence of acidolytic cleavage was obtained.

The Sol fraction of sugar maple hot-water extract also contained *p*-hydroquinone, which may be a product of hydrolysis of arbutin, which was identified among sugar-maple extractives [46d]. Nonphenolic low-molecular weight compounds identified in both the Sol fraction and extractives of sugar maple were lactic acid ($pK_a = 3.86$) and hexanoic acid ($pK_a = 4.85$). Other identified carboxylic acids such as 2-hydroxy-acetic acid ($pK_a = 3.83$) may be carbohydrate-degradation products [46d]. These various carboxyl acids released during HWE help in maintaining acid conditions during autohydrolysis.

4.4.5 Potential Use of Noncarbohydrate-Derived Material Dissolved During HWE

Depending on the conditions of HWE, the total amount of Insol and Sol fraction of hot-water extract may amount to more than 3.5% of oven-dry wood used in HWE or 15% of the total material dissolved during HWE_{540}. At the proposed biorefinery capacity of 1000 tons of wood per day [16b], the annual production of these two fractions may equal 10,000 tons. This is a significant amount of valuable material with large potential to increase the economic returns of the HWE of lignocellulosic hardwoods, enhancing the desirability of hardwood biorefineries. With a modest minimum assigned value of $0.30 per pound, this would be over $6 million per year in incremental revenue.

In contrast to kraft lignin, the most readily available source of lignin today, the lignin released during HWE has the significant advantage of being sulfur-free. Namely, due to its considerable sulfur content, the utilization of kraft lignin is limited despite an annual production of over 25×10^6 tons in the United States alone [70]. Most kraft lignin is used for energy production and is burned in the kraft-recovery process. Less than 2% of kraft lignin is currently used for other purposes [18a].

4.4.6 Adhesives

Characterization of the Insol fraction indicated that lignin dissolved during HWE is similar to organosolv (Alcell) lignin, lignin produced during an aqueous ethanol delignification of hardwood species [18a]. Alcell lignin from hardwood species has been readily included into phenol–formaldehyde resins [71] and lignin was cross-linked with formaldehyde (electrophilic) at nucleophilic C2 and C6 positions of the aromatic rings in acid conditions [23d,50]. Similarly, the Insol fraction containing both lignin and sugars/xylose appears to be an excellent candidate to produce resins under acid conditions by formation of cross-linking agents, primarily furfural by dehydration of xylose *in situ*. A consecutive condensation process may be described to follow the mechanism proposed by van der Klashorst (corresponding to Fig 4.2 route 2A at higher acidity) [50]. Alternatively, the added cross-linking agents may be furfural, which is a designed biorefinery product (Fig. 4.3.). The preliminary results in the production and application of the Insol fraction-based adhesive are promising. This type of adhesive would be consistent with an increasing interest in the development of biorenewable green adhesives to substitute for synthetic thermosetting resins.

4.4.7 Lignin-Based Polymer Blends

Lignins of different origin may vary in chemical configuration and physicochemical properties, which may cause pronounced and important differences in mechanical behavior and thermal mobility, affecting the important area of lignin application in the production of plastics [18a]. To improve the processability and mechanical properties of lignin-derived fibers, lignin may be blended with various synthetic or biorenewable polymers, which decreases the glass-transition temperature (Tg) and increases a plastic response to mechanical deformation. Thermoplastics have been made by the blending of different technical lignins, such as kraft and Alcell lignin with different synthetic polymers, among which polyethylene oxide (PEO) showed the most promising results [18a]. Incorporation of small amounts of PEO (5–10%) sufficiently disrupts the lignin complex structure leading to superior mechanical properties, comparable to polystyrene. Blends are prepared by mechanical mixing followed by thermal extrusion. Lignin produced during HWE is expected to have similar thermal properties and processability as Alcell lignin. Being of similar S/G ratio, PhOH content, and MW/polydispersity, these lignins are probably of similar condensation level, which mainly governs lignin thermal mobility [18a,46a,46b]. One important difference between Alcell lignin and lignin produced during HWE is that the Alcell lignin contains an ethoxylated benzylic position, which leads to a reduced T_g and better thermoplasticity of lignin. This effect of lignin ethoxylation is similar to alkylation which results in lignin that, even though quite brittle, can be effectively plasticized by blending with suitable aliphatic polyesters [70].

4.4.8 Production of Oxygenated Aromatic Compounds

Oxidation in alkaline medium is an established process for the depolymerization of lignins to high-value chemicals, among which vanillin and syringaldehyde are of prime interest. Vanillin with an approximate selling price of $5–9/pound finds application in different industries such as food, flavor, and pharmaceuticals. Syringaldehyde, with an even higher selling price is used in similar applications as vanillin due to the similar chemical structure and properties. Its prospective use includes the production of hair and fiber dye and drugs for obesity and breast cancer [72]. The extent of lignin oxidation may be improved by the introduction of catalysts. Lignin dissolved during HWE can be examined for oxidation in alkaline mediums in the presence of Fe^{3+} and Cu^{2+} catalysts, in the form of CuO, Fe_2O_3, mixture of CuO and Fe_2O_3, and in the presence of recently developed metal organic frameworks(MOFs) (Cu/Fe/benzenetricarboxylate) has demonstrated yields of up to 80% of the original lignin used [73]. In addition, lignin can be oxidatively depolymerized with polyoxometalates (POMs) under acid conditions. Oxidation in the presence of POMs can be performed anaerobically when POMs perform the oxidizing function, in which case reduced POMs are reoxidized in a separate step with oxygen. Alternatively, in aerobic conditions POMs perform a catalytic function [74].

4.4.9 Isolation and Potential Use of the Sol Fraction

The permeate obtained in ultrafiltration of hot-water extract (from which the Insol fraction has been removed) was extracted in laboratory conditions with chloroform and the Sol fraction was recovered after chloroform evaporation. This method of isolation of the Sol fraction is unacceptable in industrial conditions (a large amount of a chlorinated organic solvent, which is reasonably anticipated to be a human carcinogen, is used) and feasible ways of separation that are useful for industrial conditions are being evaluated. There are different promising approaches in this effort, including the use of hydrophobic membranes and a combination of extraction with an appropriate nonpolar solvent (hexane) and adsorption onto a neutral acrylic ester sorbent [18c,18d,75]. Recently, a new category of "switchable solvents" has been introduced. These solvents change from being hydrophobic, with very low miscibility with water when in air, to being hydrophilic and completely miscible in water under an atmosphere of CO_2 [76]. Alternatively, the Sol fraction of HWE might be recovered in a two-step process consisting of polymerization followed by membrane separation of the polymers. Polyphenol oxidases/laccases are able to oxidize phenols with a redox potential value lower than that of laccases (0.5–0.8 V/NHE) [77]. The laccase treatment of phenolic compounds present in the HWE would result in the production of phenoxy radicals, which would subsequently participate in coupling reactions forming aromatic oligomers/polymers. Polymerization of phenols to poly(phenylene oxide)s has been demonstrated using, for example, syringic acid and 2,6-dimethylphenol in the presence of laccases of different origins [78]. In the presence of hydrogen peroxide (H_2O_2) which acts as an electron acceptor, lignin peroxidase (LiP) can also catalyze the oxidative polymerization of phenols [78b,79]. Treatment of willow hydrolysate with laccase, LiP, and laccase and LiP combined led to the removal of monoaromatic phenolic compounds, which increased fermentability of the hydrolysate. This detoxification effect of enzymes corroborated a mechanism of oxidative polymerization [64b].

4.5 BIOMASS PYROLYSIS AND GASIFICATION IN WATER

Instead of working on fractionated biomass components, another straightforward strategy used to overcome the recalcitrance of biomass involves first deconstructing the solid biomass into smaller molecules through pyrolysis or gasification, then refining these products (e.g., bio-oil, bio-crude, and syngas) into biofuels and biochemicals. Both pyrolysis and gasification of biomass can be achieved in sub- or supercritical water.

4.5.1 Biomass Pyrolysis in Water

The incentive behind biomass pyrolysis is to densify the bulky biomass into a liquid form, such as so-called "bio-oil" or "bio-crude", to facilitate transportation. Some experts in the area pointed out the advantages of a distributed processing based on bio-crude, which are more economically transported to a centralized Fischer–Tropsch synthesis facility to produce bio-fuels and bio-chemicals [80]. Typical conditions for direct SCW liquefaction involves converting biomass to an oily liquid by contacting the biomass with water at elevated temperatures ($\sim$250–400 °C) and pressures ($\sim$10–30 MPa), with residence times of up to 60 min. Alkali catalysts were often added to promote organic conversion. The primary product of this process is an organic liquid with reduced oxygen content (about 10–20%) and a higher heating value of

30–36 MJ kg^{-1}, with the primary by-product water containing soluble organic compounds [81]. In contrast, biomass feedstocks used for liquefaction typically have higher heating values of 10–20 MJ kg^{-1} and oxygen contents of 30–50%, indicating a significant amount of upgrading from biomass to bio-crude [81b]. The primary goal of liquefaction is to remove oxygen heteroatoms; the oxygen is preferentially removed as CO_2 or H_2O, which themselves have no heating value, thus preserving as much of the feedstock's heating value as possible [81b]. Preferentially removing oxygen as CO_2 may be more desirable as this also has the advantage of increasing the H:C ratio in bio-crude, which leads to a more desirable fuel product.

A review on the engineering assessment of direct liquefaction of wood, mainly done in Canada from 1920 to 1980, was previously published [82]. In the early 1980s, Pittsburgh Energy Research Center (PERC) and Lawrence Berkeley National Laboratory (LBL) developed their own processes of biomass liquefaction using subcritical water. The process developed by PERC works with wood chips. The biomass and water mixture is pumped through a tube reactor with a residence time of 10–30 min at temperatures between 330 to 370 °C and a pressure of 20 MPa. The oil yield amounts to 45–55% of the employed dry matter of organic materials; the recycled oil serves as a hydrogen supplier [83]. Lawrence Berkeley Laboratory's process starts with the hydrolysis of biomass with sulfuric acid; this is followed by neutralization with sodium carbonate. The liquefaction occurs at pressures of 10–24 MPa and temperatures between 330 and 360 °C. Then the resultant mixture is homogenized in a refiner and later pumped through a tube reactor. The product of this direct liquefaction is a liquid material similar to bitumen (high viscosity, density 1.1–1.2 $kg{\cdot}m^{-3}$, composition: 15–19% oxygen, 6.8–8% hydrogen and 74–78% carbon). The higher heating value amounts to 34 $MJ{\cdot}kg^{-1}$ [83].

The Shell Research Laboratory in Amsterdam, Netherlands started the fundamental investigation of HydroThermal Upgrading (HTU) of biomass for fuel production and its process development in 1982. The HTU process converts a large variety of biomass feedstocks into a liquid fuel that can be upgraded to a high quality diesel fuel. In the SCW upgrading process, different biomasses (with high moisture content) can be liquefied under high pressure. A high-pressure slurry pump is used to pump the biomass suspension into a reactor. The reaction takes place under conditions of a temperature range of 300–350 °C, pressures between 12 and 18 MPa, and a residence time of 5–20 min. The oxygen in the biomass is removed as water and CO_2, and bio-crude with oxygen content between 10–18% can be produced. The product distribution is around 45% bio-crude (wt% of dry input materials without ash), 25% gas (>90% CO_2), 20% H_2O, and 10% dissolved organic materials (e.g., acetic acid, ethanol etc.). The bio-crude is a heavy organic fluid with a heating value between 30 and 35 $MJ{\cdot}kg^{-1}$ and the H:C ratio is 1:1. The thermal efficiency for one variant of this process amounts to 74.9%, and theoretically a maximum of 78.6% could be reached [84]. It was claimed that an economically viable process of biofuel production using HTU routes could be possible if fossil fuels costs stand above 30$/bbl and biomass feedstock cost is <2€/GJ in 2015 [84].

There are also a few commercial-scale applications in process now. Developers using subcritical water technology for bio-crude production include Changing World

Technologies Inc. [85], EnerTech Environmental Inc. (Atlanta, GA) [86], and Biofuel B.V. (Heemskerk, Netherlands) [84]. In 1998, Changing World Technologies Inc. (CWT) started a subsidiary of Thermo-Depolymerization Process LLC (TDP), which is now running a pilot that can convert approximately 250 tons day^{-1} of turkey offal and fats into approximately 500 barrels of renewable diesel fuel oil [85]. Recently, EnerTech Environmental Inc. announced the ceremonial commissioning of its first biosolids-to-renewable energy facility, designed to process over 270,000 wet tons of biosolids per year; the SlurryCarb^{TM} facility in Rialto, CA, will generate over 60,000 tons of renewable fuel annually for the Southern California area [86].

SCW liquefaction for bio-crude production is able to create relatively lower oxygen content in the bio-crude than fast pyrolysis which is another promising route for bio-oil production. Part of the oxygen, as well as part of the heating content of the feedstock, exists as small organic compounds that partition into the aqueous phase after SCW reaction. The yield of oil highly depends on the reaction conditions, such as reaction parameters, reaction atmosphere, catalysts, as well as properties of the feedstock. A comparison of major chemicals in switchgrass bio-oils produced via SCW conversion versus fast pyrolysis is presented in Table 4.1 [87]. Oils produced

TABLE 4.1 Comparison of Major Chemicals in Switchgrass Bio-Oil Produced via SCW Liquefaction vs. Fast Pyrolysis

Category	Origin	SCW Liquid Products	Fast Pyrolysis Crude Oil
Saccharides and derivatives	Cellulose and hemicellulose	Oligosaccharides	Levoglucosan
		Cellobiose	Pentopyranose
		Glucose	Furfural
		Xylose	2-(Hydroxymethyl)furan
		5-HMF	Hydroxyacetaldehyde
		Furfural	Acetol
Phenol-related derivatives	Lignin	3-ethylphenol	Phenol
		Ethylguaiacol	Catechol
		Guaiacol	2-Methyl-1,2-benzenediol
		4-Ethyl-phenol	3-Methyl-phenol
		2,4-Dimethoxyphenol	4-(Hydroxymethyl)phenol
		3-Methyl-2-methoxyphenol	4-Ethyl-phenol
		4-Methyl-2-methoxyphenol	2-Methoxy-4-methyl –phenol
		4-Ethyl-2-methoxyphenol	2,4-Dimethoxyphenol
			2-Methoxy-4-vinylphenol
			4-Ethyl-2-methoxy-Phenol
			2-Methoxy-3-(2-propenyl)-phenol
Other main products	Fragmentation of saccharides	Acetic acid	Acetic acid,
		1-Hydroxy-2 -propanone	Propanoic acid
			1-Hydroxy-2-propanone

Source: Ref. [87].

from subcritical water liquefaction typically have more desirable quantities than fast pyrolysis oils, and can be made with higher energetic efficiency when using biomass with high moisture content (by avoiding evaporating water); however, fast pyrolysis route has the advantage of short residence times and lower capital costs [81b]. Elliott et al. [81a] gave a review on hydrothermal liquefaction technologies, mainly focusing on the hydrogenation and heteroatoms removal (such as O, N, and S) of bio-crude using different catalysts. Detailed information of the process and catalyst system for upgrading can be found in this review [81a].

Producing renewable liquid fuels that are suitable for use in motor vehicles is perhaps the greatest challenge in the biofuel area. KaragÖz and co-workers [88] at Okayama University (Japan) performed a number of studies exploring the effects of various parameters on the liquefaction and conversion of biomass in SCW. Their conditions were generally at $\sim 280\,°C$ for 15 min. Alkali was found to be effective on the liquefaction of biomass in terms of both oil yield and composition. Based on the conversion and yield of liquid products, the catalytic activity can be ranked as: $K_2CO_3 > KOH > Na_2CO_3 > NaOH$. The volatility distribution of hydrocarbons (ether extract) was characterized by using C-NP gram (C, carbon and NP, paraffin) and it showed that most of the hydrocarbons for all runs were distributed with boiling point range of n-C_{11}.

The hydrolysis and liquefaction characteristics of heterogeneous biomass are not very clear and data are lacking. Some have criticized that the bio-crude is too complex a product that resulted from intensive processing of complex biomass. Instead of focusing on bio-crude production using SCW conditions, recent research [87b,89] is aiming at real biomass phase behavior in SCW in order to achieve better understanding of the fraction separation using SCW technologies by controlling the properties of water and reaction parameters. One example is the work done by Fang and co-workers [89d,90] to study phase behavior of willow wood at different SCW conditions. Willow shows dramatically different solubility and conversion characteristics at different water properties. This means that the phase behavior of real biomass in SCW can be controlled by controlling the properties of water. It has also been found that different fractions of biomass components can be recovered using SCW or different organic solvents, such as methanol or phenol [89a–c,91], indicating a possible way to separate/collect different biomass fractions by gradually decreasing the temperature or pressure of water after reaction. Although these are very basic works, insight rendered by the research may result in high fraction separation using SCW as reaction media and the subsequent production of value-added chemicals or liquid fuels using SCW technologies.

4.5.2 Biomass Gasification in Water

Facing the complex, heterogeneous biomass and the recalcitrance, the philosophy behind biomass gasification is to deconstruct biomass at high temperature with the aid of oxygen into simple useful building blocks such as syngas and further build a synthetic platform based on syngas for the production of energy, fuels, and materials. A complex combination of reactions in all phases can occur during biomass gasification including pyrolysis, partial oxidation, steam reforming, water–gas shift, methanation, and possibly many catalytic reactions. [10,92]

Gasification may be carried out in air, oxygen, subcritical steam, or water near or above its critical point. This section concerns hydrothermal gasification of biomass above or close to the water's critical point to produce energy, fuels, and/or chemicals. Conventional thermal gasification faces major problems from the formation of undesired tar and char. The tar can condense on downstream equipment, causing serious operational problems, or it may polymerize to a more complex structure, which is undesirable for hydrogen production. Char residues contribute to energy loss and operational difficulties. Furthermore, very wet biomass can be a major challenge to conventional thermal gasification because it is difficult to economically convert if the biomass contains more than 70% moisture. The energy used in evaporating biomass moisture, which effectively remains unrecovered, consumes a large part of the energy in the product gas. Biomass in general contains substantially more moisture than do fossil fuels such as coal. Some aquatic species such as algae and municipal sewage can have water content exceeding 90%. The efficiency of the conventional steam gasification of a biomass containing 80% water is only 10%, while that of hydrothermal gasification in SCW can be as high as 70% [10]. Yoshida et al. found the SCW gasification efficiency is insensitive to biomass moisture content, at 31% and 51%, respectively, even when the moisture in the biomass increased from 5% to 75% [89a,93].

The solubility of gases in water is usually thought to decrease with temperature, but this only occurs to a certain temperature, then solubility increases again. Therefore, gases are quite soluble in superheated water at elevated pressures. Above the critical temperature, water is completely miscible with all gasses. The increasing solubility of oxygen in particular allows SCW to be used for the wet oxidation processes, so that biomass with the natural water content ("green biomass") can be converted completely and energetically to gases. Gasification in near- or supercritical water therefore offers the following benefits:

- SCW offers rapid hydrolysis of biomass, high solubility of intermediate products including gases, and high ion products near the critical point that helps ionic reactions.
- The production of tar or char is suppressed in SCW gasification; the tar and char precursors, such as phenolic molecules, are completely soluble in SCW and so can be efficiently reformed in SCW gasification.
- Water–gas shift reaction and methane formation can be combined in SCW gasification to produce hydrogen-rich gas or methane-rich gas in one step with low carbon monoxide, eliminating the need for an additional shift or methane synthesis reactor downstream.
- Carbon dioxide can be easily separated because of its much higher solubility in high-pressure water; meanwhile, hydrogen or methane is produced at high pressure, making it ready for downstream commercial use.
- Heteroatoms such as S, N, and halogens leave the process with aqueous effluent, avoiding expensive gas cleaning; inorganic impurities, being insoluble in SCW, can also be removed easily.

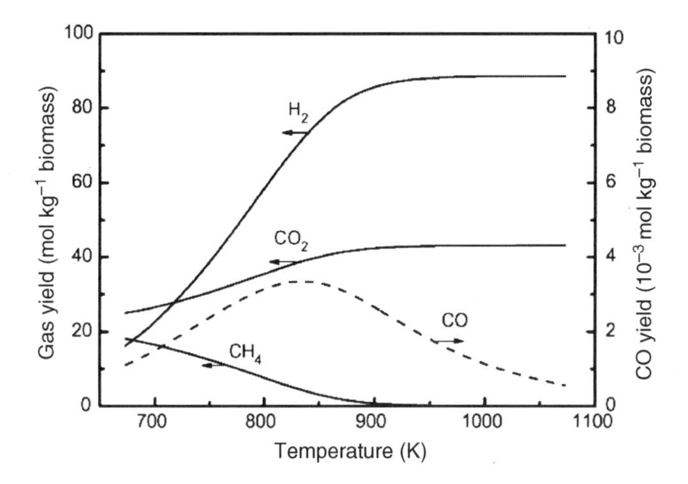

FIGURE 4.4 Equilibrium gas yields as a function of temperature for biomass gasification at 25 MPa with 5 wt% dry biomass content. (Reproduced from Ref. 94 with permission from Elsevier © 2007.)

Thermodynamic equilibrium calculations teach us that methane is preferably formed at lower temperatures, whereas at higher temperatures, hydrogen is the main product, besides CO_2 (Fig. 4.4) [94]. Depending on the reaction conditions and gaseous product desired, SCW gasification of biomass has been divided into three general temperature categories: high, medium, and low with specifically desired products for each [6]. Relatively high temperature (>500 °C) favors the target production of hydrogen; the target production of methane is usually conducted at temperatures just above the critical temperature (~374 °C) but below 500 °C; and the third category gasifies at subcritical temperatures, using simple organic compounds, such as glucose that can be derived from biomass, as its feedstock (see Section 4.6.2). The last two groups, because of their low-temperature operation, need heterogeneous catalyst (Pt, Pd, Rh, Ni, Fe, Cr, Ru, and Ir, etc.) to speed up the reactions. Here the focus is on the first two principle ways of SCW gasification that directly use real biomass instead of model compounds for fuel and energy production.

4.5.2.1 SCW Gasification for Hydrogen Production The high hydrogen yield predicted by thermodynamic calculations and the special properties of supercritical water were the main reasons why the process of supercritical water gasification was investigated. The energy required for heating up the relatively high water amount can be recovered by a compact heat exchanger, which is very important for the overall energy balance. The chemistry of biomass degradation in supercritical water is rather complex, but fortunately knowledge acquired from experiments with model compounds about the main reaction pathways and their dependencies on reaction conditions can be applied to the studies of real biomass conversion. Biomass may include proteins and salts, which have a significant influence on the gasification:

salts increase and proteins decrease the gas yield at comparable reactions conditions. In addition, the heating-up rate and the reactor type used influence the results. Good reviews and summaries exist in literature [6,94,95] on the high-temperature SCW gasification to hydrogen. Here a pilot-scale plant in Germany operating at 100 kg h^{-1} sludge [92a] is used to illustrate the technological aspects of SCW gasification. A typical SCW gasification plant includes the following key components:

- Feedstock pumping system
- Heat exchanger for feed preheating
- Gasification reactor
- Heat-recovery (product-cooling) exchanger
- Gas–liquid separator
- Optional product-upgrading equipment

The feed preheating system is very elaborate and accounts for the majority ($\sim$60%) of the capital investment in a SCW gasification plant. As depicted in Figure 4.5, biomass is made into slurry for feeding. It is then pumped to the required supercritical pressure. Alternatively, water may be pressurized separately and the biomass fed into it. In any case, the water needs to be heated to the designed inlet temperature for the gasifier, which must be above the critical temperature and well above the designed gasification temperature because the enthalpy of the water provides the energy required for the endothermic gasification reactions. This temperature is a critical design parameter. Sensible heat of the gasification products

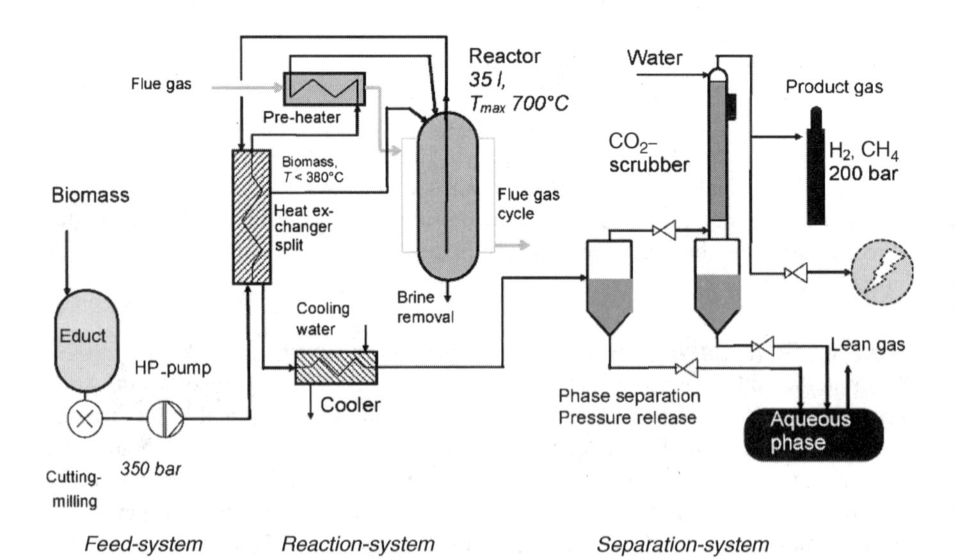

FIGURE 4.5 Flowsheet of a pilot plant for hydrothermal biomass gasification. (Reproduced from Ref. 92a with permission from Elsevier © 2009.)

may be partially recovered in a waste heat-recovery exchanger and used for preheating the feed. For complete preheating, additional heat may be obtained from either burning part of the fuel gas produced to supplement the external fuel, or controlled burning of unconverted char in the reactor system. After gasification, the products are first cooled in the waste heat-recovery unit. Thereafter, it cools to room temperature in a separate heat exchanger by giving off heat to an external coolant. The next step involves separation of the reaction products. The solubility of hydrogen and methane in water at low temperature but high pressure is considerably low, so they are separated from the water after cooling, while the carbon dioxide, because of its high solubility in water, remains in the liquid phase. The gaseous hydrogen is separated from the methane in a pressure swing adsorption unit. The CO_2-rich liquid is depressurized to the atmospheric pressure, separating the carbon dioxide from the water and unconverted salts.

4.5.2.2 SCW Gasification for Methane Production Through the use of metal catalysts, biomass gasification can be accomplished with high levels of carbon conversion to methane-rich gas at relatively low temperature. Prominent research groups in this area include Massachusetts Institute of Technology [96], U.S. Pacific Northwest National Laboratory [97], University of Hawaii [95,98], State Key Laboratory of Multiphase Flow in Power Engineering (Xi'an Jiaotong University, China) [94], Forschungszentrum Karlsruhe in Germany [92a,99], Paul Scherrer Institut in Switzerland [100], and multiple research groups in Japan [8,101]. Although active heterogeneous catalysts using supported noble metals, such as Rh and Pt, can be used to completely gasify solid lignocellulosic biomass to a methane-rich gas around 400 °C and 30 MPa, continuous efforts have been put on finding less expensive catalysts.

Elliott et al. at U.S. Pacific Northwest National Laboratory are representative of CH_4 formation using less expensive Ni-based catalysts [97b,102]; their early catalysts were hardly stable in the long term. Investigations of the stability of support materials in SCW by Elliott and coworkers found useful support materials, for example, carbon, monoclinic zirconia or titania (TiO_2), and α-alumina, and they reported that nickel had a significant effect on hydrothermal gasification only in the reduced form. A demonstration unit of continuous SCW gasification of biomass to methane-rich gas was established at the U.S. Pacific Northwest National Laboratory, confirming their earlier batch test results that high conversion of biomass solids to gas could be achieved with high concentration of methane in the product gas using a number of biomass feedstocks, such as sorghum, spent grain, and cheese whey. The initial problem of plugging of the catalyst bed by the biomass was solved by a liquefaction step in a CSTR prior to gasification by the catalyst bed reactor. However, also seen in these tests was the rapid deactivation of the nickel catalysts used, although it could be stabilized by doping with another metal.

Ruthenium catalysts of various formulations were also investigated. Some were found to be promising with regards to the activity and selectivity. Often, the Ru catalyst lifetime was increased compared to Ni. Osada et al. [101] have investigated a range of catalysts and biomass feedstocks in a small batch reactor. After 15 min at

400 °C and 37.1 MPa, the carbon gas yields from lignin range from 3.7% without catalyst to 5.5% with a nickel-on-alumina catalyst to 31.1% with a ruthenium-on-titania catalyst. For cellulose, the same pattern emerged: 11.3% without catalyst, 17.3% with Ni, and 74.4% with Ru. The titania support was not specified, but its surface area was listed at 24.9 m^2 g^{-1}, suggesting that it was a mixture of anatase (unstable, high-surface area) and rutile (stable, low-surface area). At this supercritical condition, it was also found that the rate of gasification was dependent on the water density, increasing over the range of 0.1–0.5 g cm^{-3}.

In a continuous SCW gasification study for methane formation from selected mixture of five organic compounds, Vogel et al. [103] found that after a few hours on stream, both skeletal nickel and a ruthenium-stabilized variant were deactivated due to sintering of the nickel crystallites. Further screening work in their batch reactor using other skeletal type nickel catalysts and Ru/C catalysts resulted in the identification of an active and selective catalyst that exhibited no signs of deactivation after the batch test. A long-term test performed with this catalyst using a synthetic liquefied wood mixture confirmed that Ru-on-carbon is a hydrothermally stable catalyst for the production of methane. However, details of the Ru/C catalyst are proprietary to the manufacturer. The authors claimed that this process is a good example for the application of green chemistry and green engineering, using a green feedstock (biomass), a green solvent and coreactant (water), exhibiting a high thermal efficiency, and producing a green fuel (synthetic natural gas) with a low environmental impact.

Waldner and Vogel [104] have generated some process economic estimates for an application of the SCW technology and concluded that using a wood feedstock at \$67 $tonne^{-1}$ would yield a gas product valued at \$10.3/GJ (\$9.8/million Btu). The wet tonne day^{-1} plant (39,467 L h^{-1}), operating at 420 °C and 30 MPa, would have an installation capital cost of \$5.9 million. Ro and Cantrell et al. [102a] completed a system analysis for use of subcritical gasification with hog manure feedstock, suggesting that catalytic hydrothermal gasification of flushed swine manure feedstocks with solids concentration greater than 0.8 wt% can be a net energy producer. The installed capital cost for the 1580 L h^{-1} unit (serving 4400 head swine) operating at 350 °C and 22 MPa, would be \$0.99 million. The net product gas would have a value of \$47,006 per year at \$8/million Btu. The authors concluded that the costs are higher than a conventional anaerobic digestion lagoon system; however, the high rate of conversion of organic matters into gas drastically decreases the land requirement for manure application and costs for transportation and tipping fees. In addition, the catalytic gasification process would destroy pathogens and bioactive organic compounds, producing relatively clean water for reuse, and the ammonia and phosphate byproducts have potential value in the fertilizer market.

Challenges still exist for a practical application of SCW gasification, including corrosiveness of water at supercritical conditions with salts, pumping system handling biomass slurry with high solid loading, and optimizing dry matter loading to ensure that the energy content of the feedstock is higher than the energy loss of the processing plant. SCW gasification appears to be a unique technology, which requires further development. Some areas of future research include the development of highly active, stable, contaminant-resistant (especially to sulfur), and selective novel catalysts, reaction chemistry studies, and reactor designs [92b].

4.6 CHEMICAL CONVERSION OF BIOMASS IN SCW

The majority of previous research has taken a reductionist approach by studying model compounds of individual biochemicals contained in heterogeneous biomass, leaving the kinetics and interactions among many components under SCW conditions largely unexplored. However, following biomass pretreatment and fractionation, knowledge of the reaction characteristics of these model compounds/components is important to the understanding of real biomass conversion in SCW. Biomass species (and the contained compounds/components) covered in this section include lignocellulosic biomass (mainly forest/agricultural residues containing hemicellulose, cellulose, and lignin), protein and amino acids, vegetable oils, and algae/microalgae (triglycerides, fatty acids, glycerol). Different structural matrices of the biomass species result in different reaction characteristics in SCW media.

4.6.1 Hemicelluloses and Cellulose

The hydrolysis reaction is an initial step in biomass conversion in SCW. Hydrolysis plays a critical role in forming glucose and its oligomers from lignocellulosic biomass, which can then be decomposed quickly to nonglucose aqueous products, oil, char, and gases. The monosaccharides produced make suitable sugars for fermentative processes, such as those used in the production of cellulosic-ethanol and other biofuels and materials. However, it has been suggested that some aromatic compounds formed in hydrothermolysis must be removed because they may inhibit some fermentation processes [81b]. Hydrolysis, at the very initial stage of biomass conversion in SCW, is the most different reaction from those in "dry" processes such as fast pyrolysis or gasification.

Hemicelluloses can be any of several heteropolymers (matrix polysaccharides) present in almost all plant cell walls along with cellulose. Hemicelluloses consist of shorter chains of $\sim$500–3000 monosugar units, and have a random, amorphous structure with little strength. Generally, hemicelluloses are easily hydrolyzed in the presence of catalysis of dilute acids, bases, as well as a great number of hemicellulase enzymes; and it can also be easily removed in subcritical water without catalysts [105]. In this way, investigation of hemicelluloses conversion in subcritical water is rather rare compared with those of cellulose.

Cellulose is a homogenous polymer of D-glucose units linked through $\beta(1\rightarrow4)$-glycosidic bonds. Peterson et al. [81b] gave a detailed review of cellulose hydrolysis kinetics in SCW media; the kinetics appear to show a change in activation energy compared with the "dry" pyrolysis process. An activation energy of 215 kJ mol^{-1} is obtained in SCW, while the activation energy for cellulose pyrolysis in the absence of condensed water is about 228–238 kJ mol^{-1}. Below 230 °C with no additional pressure, no dramatic change will happen in cellulose, but liquid water at elevated temperatures and pressures can both break up the hydrogen-bound crystalline structure and hydrolyze the β(1→4)-glycosidic bond, resulting in the production of glucose monomers.

Several factors have significant influences on cellulose hydrolysis in SCW. The most important factors are: temperature and pressure (which also determine the properties of water), the time–temperature history experienced by reacting particles, and catalysts and other additives. Temperature markedly affects the rate of the hydrolysis reaction (as well as side reactions such as the fragmentation reactions), while the operating pressure has a less significant influence on the degree of cellulose decomposition than the temperature does [106].

The heating rate also affects the cellulose hydrolysis reaction. A high heating rate shortens the residence time of biomass, resulting in a reduced degradation of glucose product, therefore leading to a high production yield of glucose. Fang et al. studied the cellulose decomposition mechanism without a catalyst in a diamond anvil cell coupled with optical and infrared microscopy in subcritical water [107]. A phase difference between high heating rate and low heating rate behavior was observed. Using low heating rates (e.g. $0.18\,°C\ s^{-1}$), the reactions occur mostly under heterogeneous conditions; but it was seen that when higher heating rates are used ($>2.2\,°C\ s^{-1}$), the reactions (hydrolysis and decomposition) can occur in a homogeneous phase. The reaction mechanisms of cellulose under homogeneous and heterogeneous conditions are very different as is evident by the formation of "glucose char" in the former or the formation of "cellulose char" in the latter. They also visually observed breakup of the cellulose fibers very shortly after the loss of crystallinity, suggesting that the crystallinity was preventing breakdown of the cellulose [107]. This conclusion is further confirmed by a later study [11].

Generally, the decomposition of hemicelluloses and cellulose leads to the formation of oligosaccharides, cellobiose, monomer sugars, as well as dehydration or fragmentation derivatives of those sugars produced in SCW, such as furfural and 5-hydroxymethylfurfural (HMF). Huber et al. summarized the chemistry for the decomposition pathways of cellulose and glucose in supercritical water, shown in Figure 4.6 [92b]. Glucose obtained from cellulose hydrolysis undergoes isomerization to form fructose, which can then undergo dehydration to form 5-hydroxymethylfurfural. Further dehydration of 5-hydroxymethylfurfural yields a 1:1 mixture of levulinic and formic acids. Angelica lactones are formed by dehydration of levulinic acid. Retroaldol reactions produce glycolaldehyde, dihydroxyacetone, glycolaldehyde, and erythrose from fructose and glucose. These intermediates react further to form pyruvaldehyde, glycolaldehyde, and acids. Glucose can also form 1,6-anhydroglucose by dehydration. Hemicelluloses undergo analogous reaction pathways to those shown in Figure 4.6 [92b].

4.6.2 Monosaccharides

Monosaccharides are the most important hydrolysis products from biomass for subsequent conversion to produce transportation fuels such as ethanol. However,

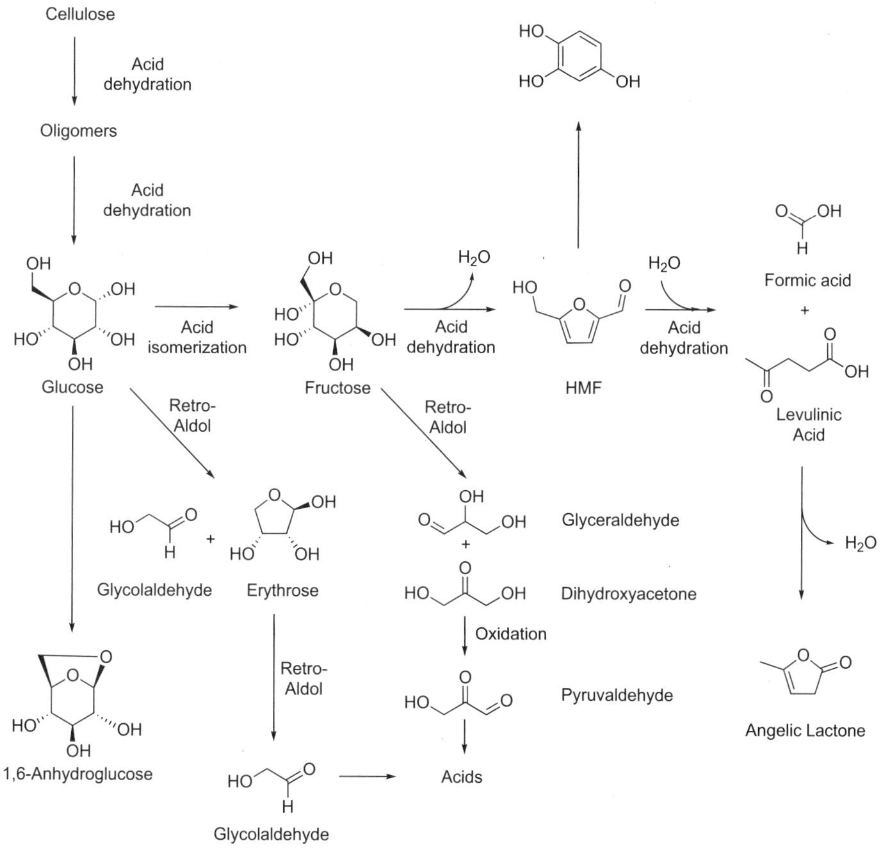

FIGURE 4.6 Cellulose decomposition pathways in supercritical water. (Reproduced from Ref. 92b with permission from American Chemical Society © 2006.)

monosaccharides, (e.g., glucose) may be easily decomposed in SCW to form other products, such as 5-hydroxymethylfurfural (Fig. 4.6) [108]. Mechanisms of monosaccharide decomposition in SCW have been studied previously, with or without a catalyst. As demonstrated by Watanabe et al., [109] at higher temperature and low water density, radical reactions are likely to occur in SCW; on the other hand, ionic reactions mainly occur at lower temperature and high water density. In addition, the reaction conditions (temperature, pressure, or water density) or the presence of additives (acid or base) may also influence glucose reactions, especially at low temperatures.

Homogeneous acid and base additives, such as sulfuric acid (H_2SO_4) and sodium hydroxide (NaOH), are commonly used to change the acidic or basic environment in SCW. The addition of alkali promotes isomerization and retroaldol condensation reactions involving glucose, while the addition of acid promotes the dehydration of

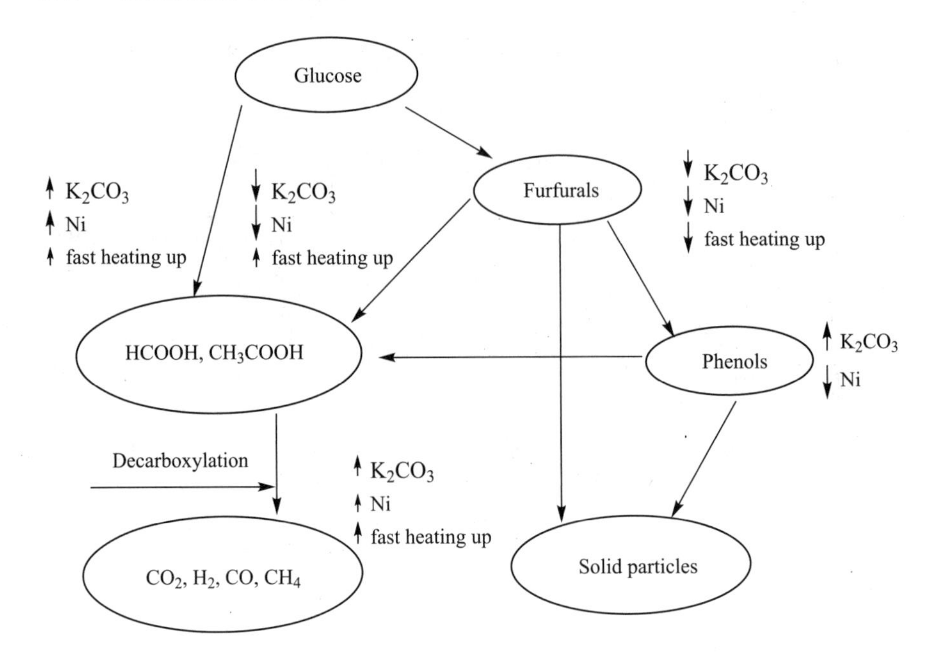

FIGURE 4.7 Glucose decomposition mechanism in sub- or supercritical water in the presence of catalysts. (Reproduced from Ref. 111a with permission from American Chemical Society © 2003.)

glucose. Watanabe and coworkers [109a,109c,109d,110] studied the influence of heterogeneous acid and base additives, such as metal oxides, on the glucose reactions. They found that the addition of ZrO_2 promotes isomerization of glucose and fructose, and thus, ZrO_2 can be considered as a base catalyst for glucose; however, anatase TiO_2 promotes isomerization and dehydration of glucose into 5-hydroxymethylfurfural, indicating the existence of both base and acid sites on the surface. In SCW, a nickel catalyst can catalyze the steam reforming of aqueous products as intermediates of biomass conversion, forming methane [97a].

It has been reported that retroaldol condensation is favored under alkali conditions, and that an acid catalyst is effective to enhance dehydration reactions. Sinag and coworkers [99a,111] studied glucose decomposition in supercritical water in the range of 400–500 °C and 30–50 MPa, with the presence of alkali or Ni catalysts; the influences of heating rate and catalysts on the formation of intermediates were summarized as shown in Figure 4.7 [111a]. Glucose converts into furfurals and acids/aldehydes in parallel. Further conversion of furfurals leads to acids/aldehydes and phenol formation. Acids and aldehydes are precursors of the gaseous compounds. Furfural is formed through dehydration, and further dehydration of furfural gives phenols. In contrast, acids and aldehydes are produced by bond-breaking reactions. Such a simple mechanism is very useful for estimating the predominance of the two pathways (dehydration and bond breaking) from the yields of the key compounds, such as furfurals, acids, aldehydes, phenols, and gaseous products. The influences of heating rate and catalysts on the formation of intermediates are also indicated in Figure 4.7.

In pure SCW medium, only a small amount of lactic acid was obtained from the degradation of carbohydrates and the derivatives, such as fructose and glucose; however, by adding small quantities of metal ions such as Co(II), Ni(II), Cu(II), and Zn(II) to that reaction media, lactic acid yield could be increased up to 42% ($g\,g^{-1}$) starting from sucrose and to 86% ($g\,g^{-1}$) starting from dihydroxyacetone at 300 °C and 25 MPa; Zn(II) gave the best results with regard to the lactic-acid yield [112].

Jin et al. [113] reported a two-step SCW process to improve the production of acetic acid from sugars derived from cellulose and starch. The first step was to accelerate the formation of 5-hydroxymethylfurfural, 2-furaldehyde, and lactic acid, and the second step was to further convert the furans (5-hydroxymethylfurfural and 2-furaldehyde) and lactic acid produced in the first step to acetic acid by oxidation with newly supplied oxygen. The acetic acid obtained by the two-step process had not only a high yield but also high purity. The contribution of two pathways via furans and lactic acid in the two-step process to convert carbohydrates into acetic acid was roughly estimated as 85–90%, and the ratio of contributions of furans to the lactic acid in the acetic acid yield was estimated at 2:1.

Using relatively lower temperatures (200–260 °C) in aqueous phase (1–5 MPa), Dumesic and coworkers [114] recently developed aqueous-phase catalytic processes for the conversion of monosaccharides or sugar-derived feed with liquid water and heterogeneous (solid) catalysts to hydrogen, chemicals, and alkanes ranging from C_1 to C_{15}. Roman et al. developed a two-phase reactor system for the selective dehydration of fructose to 5-hydroxymethyl furfural that operated at high fructose concentrations (10–50 wt%), achieved high yields (80% selectivity to 5-hydroxymethyl furfural at 90% fructose conversion), and delivered 5-hydroxymethyl furfural in a separation-friendly solvent (methylisobutylketone) in the presence of an acid catalyst (hydrochloric acid or an acidic ion-exchange resin) [114a,114b]. Huber et al. [114f–h] demonstrated that it is possible to produce light alkanes by aqueous-phase reforming (APR) of biomass-derived oxygenates such as sorbitol, which can be obtained from glucose by hydrogenation. They pointed out that the production of alkanes from aqueous carbohydrate solutions would be advantageous because of the easy separation of the alkanes from water. Accordingly, it was estimated that the overall energy efficiency for alkane production from corn would be greatly increased, assuming that this process eliminates the energy-intensive distillation step but still requires all of the remaining energy input needed for the production of ethanol from corn [114f]. The alkane selectivity depends on relative rates of C–C bond cleavage, dehydration, and hydrogenation reactions, and it can be varied by changing the catalyst composition and reaction conditions, and by modifying the reactor design [92b].

Further results of the production of higher carbon number alkanes from carbohydrates have been reported by Huber et al. using a four-phase dehydration/hydrogenation unit (4-PD/H); the large liquid alkanes (ranging from C_7 to C_{15}) were obtained by dehydration and hydrogenation over a bifunctional catalyst containing

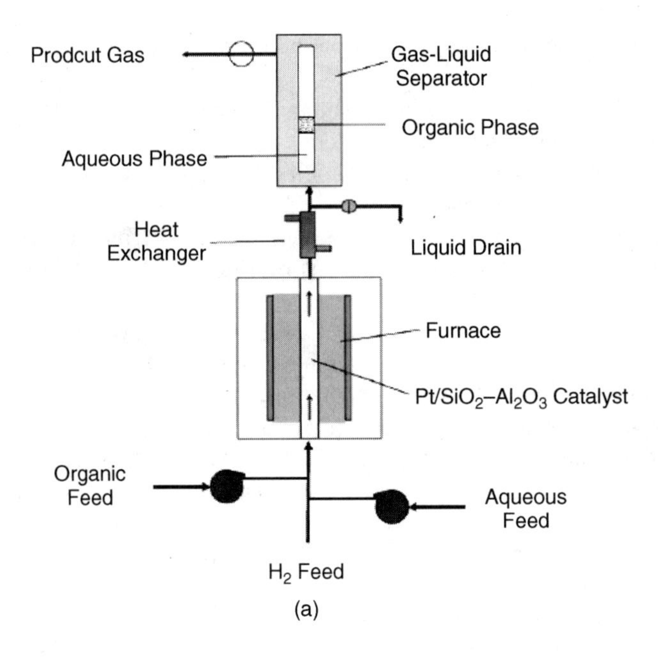

FIGURE 4.8 Four-phase dehydration/hydrogenation unit and proposed mechanism of alkanes production from monosaccharides (a) Four-phase dehydration/hydrogenation unit. (Reproduced from Ref. 116b with permission from IOP Science © 2004.) (b) Proposed mechanism of alkanes production from monosaccharides. (Reproduced from Ref. 115 with permission from Elsevier © 2006.)

metal and acid sites and the process is depicted in Figure 4.8 (a) [115]. The main reaction pathway includes acid-catalyzed dehydration, which was followed by aldol condensation with ketones (e.g., acetone) over solid base catalysts to form large organic compounds, as shown in Figure 4.8(b) [114f,114h,115]. The reason for utilizing the organic phase is that the aqueous organic reactant becomes more hydrophobic and a hexadecane alkane stream can remove hydrophobic species from the catalyst before they go on further to form coke. These liquid alkanes have the appropriate molecular weight to be used as transportation fuel components, and they contain 90% of the energy of the carbohydrate and H_2 feeds [114h]. It is also reported recently that the ketones for aldol condensation can also be produced from sugars and polyols over a Pt–Re catalyst [114i].

4.6.3 Lignin

Lignin is an integral cell-wall constituent in all vascular plants, including herbaceous varieties. Lignin provides rigidity, water-impermeability, and resistance against microbial attack. Its amount in lignified plants ranges from 15 to 36% by mass. Lignin is an aromatic polymer consisting of guaiacyl- (G), syringyl- (S), and *p*-hydroxyphenyl- (H) phenylpropanoid units whose proportions differ with the

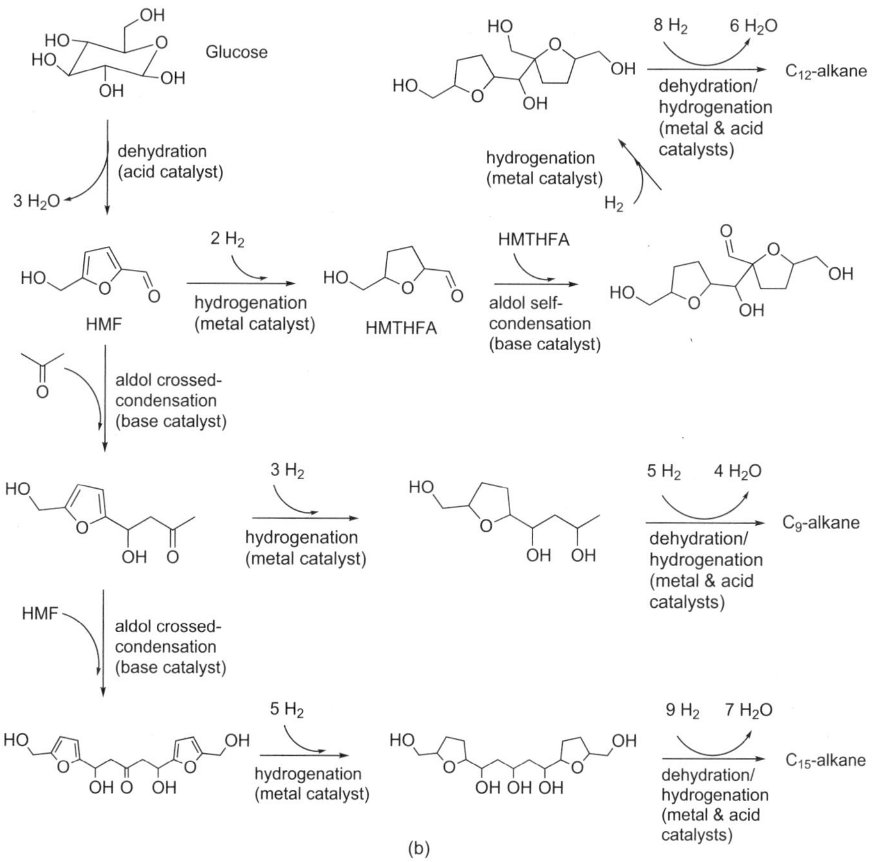

(b)

FIGURE 4.8 (*Continued*)

botanical origin of the lignin. The phenylpropanoid units are attached to each other by a series of C–O–C and C–C bonds such as β-O-4, β-5, α-O-4, β–β, and 5-5. The polymer is branched, and cross-linking occurs [41a–c]. Lignin polymer models are well illustrated elsewhere [65].

Recent research on lignin conversion in SCW is summarized in Table 4.2 [116]. Lignin conversion in pure SCW medium without other cosolvents or catalysts will lead to the formation of chars during reactions [116e]. The proposed mechanism is that the intermediates formed by hydrolysis and pyrolysis of lignin can be recombined, leading to char formation. However, with the addition of phenol-related compounds, good solubility of lignin can be obtained, as shown in Figure 4.9 (a) [116c] and summarized in Table 4.2. Okuda et al. proposed that the formation of char can be greatly restrained by the addition of phenol [116c] or *p*-cresol [117] as radical capture agents. Okuda et al. [117] have recently reported that lignin was selectively converted into a single chemical species in water/*p*-cresol mixtures at 400 °C within a

TABLE 4.2 Summary of Recent Research on Lignin Reaction Characteristics in Sub- or Supercritical Water

Feedstock	Reactor	Reaction Condition	Key Results	Literature
Sulfuric acid lignin (byproduct of wood hydrolysis)	12 mL micro reactor	175–280°C, 4 h, 0.1 M NaOH	Hydrophilic groups were introduced to the lignin	[118a]
Organosolv lignin	5 mL micro-reactor	Phenol water mixture, 400°C, water density: 0.36–0.5 $cm^3 g^{-1}$	99% is THF soluble; no char formation	[118b–d]
Organosolv lignin	10 mL batch reactor	400°C, water density: 0–0.5 $g cm^{-3}$, 0–1 h	Phenol prevents char formation	[116,117]
Organosolv lignin	Diamond anvil cell micro-chamber (ca. 50 nL)	400°C–600°C water density: 0.428–0.683 $cm^3 g^{-1}$	Mixtures of phenol and lignin become homogeneous in SCW	[118d]

reaction time of 4 min (as shown in Figure 4.9(b)); complete depression of char formation was realized in a mixture of 1.8 g of water and 2.5 g of *p*-cresol. The chemical species obtained had a molecular weight of 214 (M^+) assigned by gas chromatography–mass spectroscopy (GC-MS) and was identified as hydroxylphenyl-(hydroxyltolyl)-methane (HPHTM) by ^{1}H and ^{13}C nuclear magnetic resonance. Its yield approached the maximum of 80% based on carbon after 30 min of reaction time.

Fang et al. [116f] proposed a mechanism for the reaction paths of lignin in SCW, as shown in Figure 4.10. Dissolved products (e.g., lignin, oligomers, monomers) and nondissolved lignin have different reaction mechanisms towards four phases, which were: (i) oil phase (phenolics, polycyclic aromatic hydrocarbons, and heavy hydrocarbons), (ii) aqueous phase (acids, aldehydes, alcohols, catechol, and phenols), (iii) gas phase (CO_2, CO, H_2, and C_1–C_4 hydrocarbons), and (iv) solid residue phase. The dissolved samples with ether linkages were probably homogeneously hydrolyzed to single-ring phenolic oil (syringols, guaiacols, catechols, and phenols). Syringols and guaiacols in the oil phase were further hydrolyzed and dealkylated into aqueous products of methanol and catechols in the aqueous phase [116e]. Minor catechols were probably decomposed to phenols and aromatics, and aromatics subsequently changed to Polycyclic Aromatic Hydrocarbons (PAHs) and aromatic char. At high temperatures, stable C–C bonds in lignin and oligomers were most likely cleaved into single- and multiring phenolics. Dealkylation reactions probably led to the formation of gas, hydrocarbons (HCs), and aqueous products (alcohols, aldehydes, and acids). At higher temperatures and

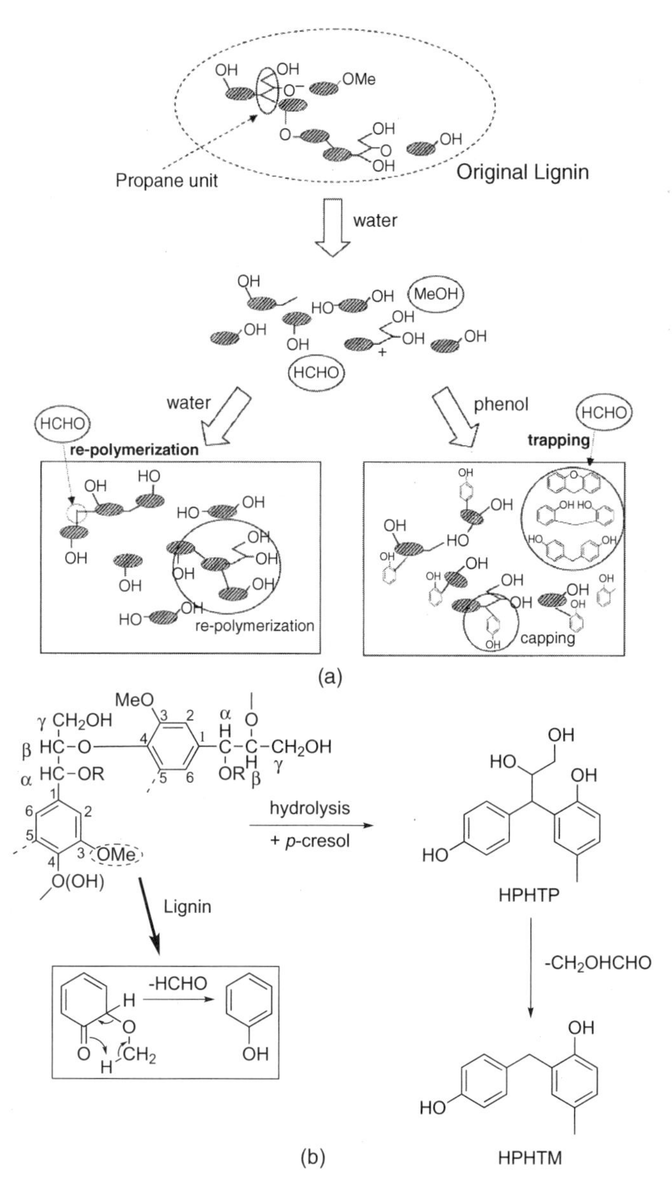

FIGURE 4.9 Mechanism of lignin conversion in SCW with phenol or *p*-cresol (a) Mechanism of lignin conversion in SCW with or without the presence of phenol (Reproduced from Ref. 116c with permission from Elsevier © 2004.) (b) Mechanism of lignin conversion into a single chemical species in SCW with the presence of *p*-cresol. (Reproduced from Ref. 117 with permission from IOP Science © 2004.)

long reaction times, phenolics (both monomers and oligomers) repolymerized with aldehydes to form heavier cross-linked phenolic char [116e], which was precipitated as residue. The nondissolved samples probably underwent pyrolysis to yield gas, HCs, a mixture of phenolics, water-soluble products (methanol, acids, and

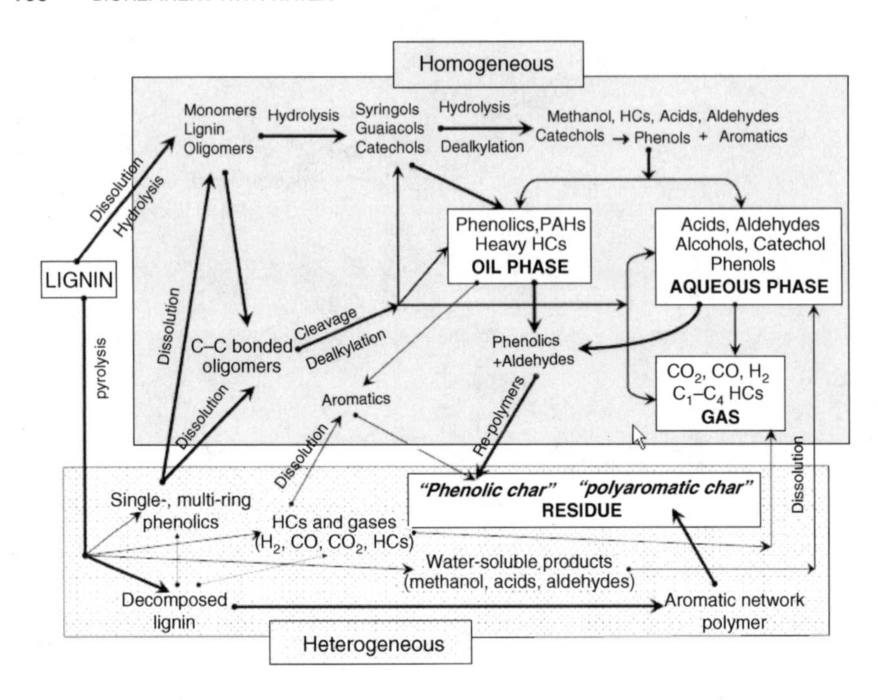

FIGURE 4.10 Reaction paths of lignin decomposition promoted in homogeneous and heterogeneous environments in supercritical water. (Reproduced from Ref. 116f with permission from Elsevier © 2008.)

aldehydes) and decomposed lignin via free-radical and concerted mechanisms or acid-catalyzed decomposition.

Interestingly, Chen et al. [118] reported a new catalytic route with bifunctional combination of a stable carbon-supported noble-metal catalyst and a mineral acid to convert phenolic bio-oil components (phenols, guaiacols, and syringols) with high selectivity to cycloalkanes and methanol. This new catalytic route for the production of alkanes by upgrading aqueous phenolic bio-oil was developed under the heat and pressure conditions of 250 °C and 5 MPa in the presence of H_2. A carbon-supported noble-metal catalyst (5 wt% Pt supported on carbon) in combination with the mineral acid H_3PO_4 acted as bifunctional catalysts for the one-pot hydrodeoxygenation conversion of bio-oil through multistep reactions consisting of hydrogenation, hydrolysis, and dehydration [118]. The mixture of alkane products formed a second phase, which was easily separated from the aqueous phase. This route provides a feasible approach for the direct use of crude aqueous bio-oil mixture in subcritical water, facilitating an energy-efficient process.

4.6.4 Triglycerides and Fatty Acids

Fats and oils in biological systems are typically in the form of triglycerides (TAGs), which consist of three fatty acids bound to a glycerol backbone. Another important source of triglycerides is aquatic biomass such as algae. While algae appear to provide natural raw materials in the form of a lipid-rich feedstock, we lack the requisite understanding of the details of lipid metabolism that would enable manipulation of the process physiologically and genetically. There is a need for concentrated research on the biosynthesis of algal lipids and hydrocarbons, if we are to better understand and manipulate algae for the production of biofuels.

As discussed in Section 4.2, the dielectric constant of SCW is drastically lower than that of room temperature water, causing the solvation properties of SCW to more closely resemble organic solvents. Under SCW conditions, hydrogen bonding between water molecules becomes weaker, allowing greater miscibility between lipids and water. The increase in temperature causes fats and oils to become increasingly soluble in water as the temperature rises under SCW conditions, ultimately becoming completely miscible by the time the water has reached its supercritical state [119].

Hydrolysis of TAGs occurs primarily in the oil phase, and proceeds to an increasing equilibrium level with increasing water-to-oil ratios. As the hydrolysis reaction proceeds, more fatty acids are generated, increasing the solubility of water in the oil phase and thus the observed reaction rate. Moquin and Temelli [120] provide a good overview of this phenomenon in the introduction of their recent articles on canola oil hydrolysis in supercritical water media. Tomoyuk et al. [121] have confirmed that a first-order reaction drives the degradation kinetics of monoacylglycerides in aqueous phase and showed a first-order reaction rate that follows Arrhenius kinetics, with activation energy of $77.5\,\mathrm{kJ\,mol^{-1}}$ and frequency factor of $1.01 \times 10^5\,\mathrm{s^{-1}}$.

Free fatty acids (FFAs) have been shown to degrade in sub- or supercritical water systems, producing long-chained hydrocarbons [122]. Decarboxylation is a very important reaction for alkane production from FFAs. Watanabe et al. [122] studied stearic acid ($C_{17}H_{35}COOH$) decomposition to a maximum processing temperature of $400\,^{\circ}\mathrm{C}$ in a batch reactor at a fixed density of water of $0.17\,\mathrm{g\,cm^{-3}}$; they found two major products, $C_{17}H_{36}$ and $C_{16}H_{32}$. It has been reported [122] that the sodium acetate forms of FFAs decarboxylate more rapidly than the acid form. When NaOH or KOH was added to a SCW system, decomposition of stearic acid increased significantly and alkanes again became the dominant products, while lower hydrocarbons were suppressed; KOH had a larger effect than NaOH. Watanabe et al. [122] also found that fragmentation reaction of stearic acid to short-chain hydrocarbons was suppressed in the SCW experiments compared to that of pyrolysis reaction without water.

4.6.5 Glycerol

The other product of triglycerides hydrolysis is glycerol. It is noteworthy to point out that a glycerol glut has occurred as a by-product of transesterification in recent years due to the large-scale production of biodiesel. Duane et al. [123] provided a detailed review of promising options for both the catalytic and biological conversion of glycerol into various value-added products, many of which are bio-based alternatives to petroleum-derived chemicals. Among those value-added products, acrolein [124] is gaining more attention because it is an important intermediate product that can be easily further converted to many fine chemical products, such as acrylic acid, polyester resin, polyurethane, propylene glycol, and acrylonitrile, etc.

The reactions of glycerol in SCW are mainly dehydration and fragmentations. Some studies aim at gas-phase production of acrolein in the presence of acid catalysts [124c–f,125]. However, the problem of severe coking on the catalysts has been an issue that restrained the long-term run and process economics. Nonetheless, it has been reported that glycerol dehydration for acrolein can take place in SCW with the aid of a mineral acid or salt as catalyst [124a,124b]. This is a promising way to achieve the production of valuable chemicals from glycerol. When protonation occurs at the secondary hydroxyl group of glycerol, a water molecule and a proton are eliminated from the protonated glycerol, and then 3-hydroxypropanal is produced by tautomerism; 3-hydroxypropanal was difficult to detect in the actual reaction because it is unstable, and it is readily dehydrated into acrolein. This mechanism is shown in Figure 4.11, which summarizes the chemical products and possible pathways using acid or base catalysts in SCW [124a,126]. Possible side-reaction pathways can also take place, leading to the formation of hydroxyacetone when protonation proceeds at a terminal hydroxyl group of glycerol.

SCW decomposition experiments of glycerol with an alkali have shown that glycerol could be converted into lactic acid with a high yield of about 90 mol% based on glycerol used [126]. Discussion on the pathway for the conversion of glycerol to

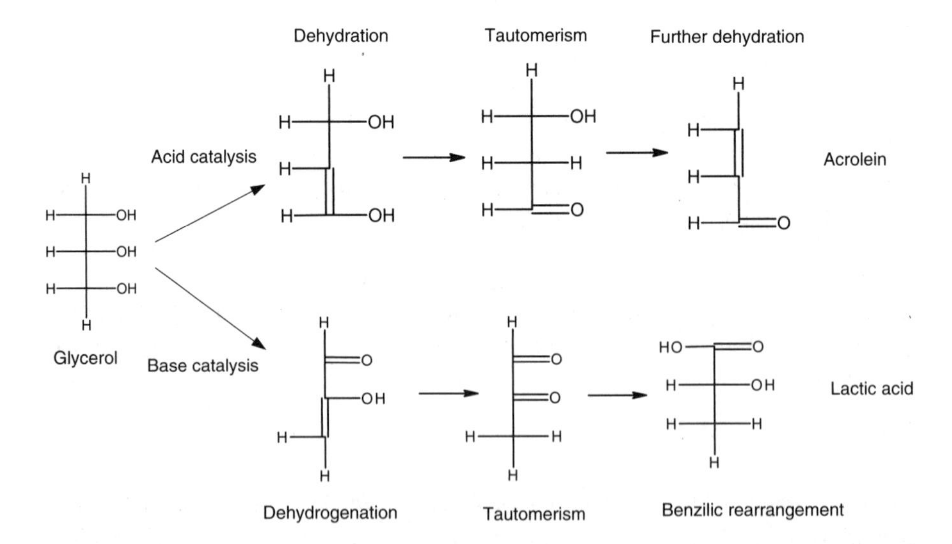

FIGURE 4.11 Glycerol conversion in SCW with the presence of acid or base catalysis. (Adapted from Ref. [124a and 126].)

lactic acid suggests that glycerol is first decomposed to pyruvaldehyde with the elimination of hydrogen. Pyruvaldehyde formed is then converted into lactic acid by the benzilic acid rearrangement (as shown in Fig. 4.11).

4.6.6 Proteins and Amino Acids

Proteins are another type of important biopolymers. They are also present in algae and food/agriculture waste. Reactions of proteins in SCW can be of interest for producing oligomers, and amino acids as the building blocks of the proteins. In protein hydrolysis, first a proton is attached to the nitrogen atom of the peptide bonding, leading to a split of the bonding and forming a carbocation and an amino group. In the next step, a hydroxide ion from a dissociated water molecule attaches to the carbocation, forming a carboxyl group [106].

Amino acids, the building blocks of proteins, have high commercial value (for use in feed, food, pharmaceuticals, and cosmetics) compared to most other fractions of biomass. Therefore, a number of studies [126,127] have explored the potential of SCW technologies to extract amino acids from various protein-rich feedstocks, particularly marine waste in Japan. Hydrolysis of a model protein (bovine serum albumin, BSA) and sklero-proteins like feathers and hair, carried out in a continuous plug-flow reactor, resulted in a total liquefaction of the proteins and in the formation of amino acids [127c]. Production of amino acids depends mainly on reaction temperature, with an optimum at 310 °C. Pressure in the range of 15–27 MPa had no significant effect on the reaction.

The 20 common amino acids have different chemical structures, and therefore react according to different pathways. However, all amino acids also have the same peptide backbone, and undergo similar decarboxylation and deamination reactions. Klinger et al. [128] recently studied glycine and alanine, two of the simplest amino acids, and found the primary mechanisms of degradation of these amino acids to be decarboxylation and deamination. They also found similar decomposition kinetics for these two compounds, with about 50% of their starting material degraded in 5–15 s in 350 °C water at 34 MPa; both compounds had decomposition activation energy of about 160 kJ mol^{-1}, which is similar to values reported earlier by Sato et al. [129]. Klinger et al. [128] also found no effect of pressure on the decomposition rate between 24 and 34 MPa at 300–350 °C. Li and Brill [130] reported decarboxylation kinetics for seven different amino acids at temperatures ranging from 310 to 330 °C, providing useful information about the stability/instability of amino acids at SCW conditions.

4.7 OPPORTUNITIES, CHALLENGES, AND OUTLOOK

SCW technologies have seen significant progress over the last 20 years and the supercritical water oxidation of wastes has emerged as an industrial application as a result. There exist strong incentives and good opportunities to develop "green" SCW technologies for the production of biofuels, and renewable chemicals and materials, since the evolution from lab-scale to pilot-scale facilities has provided data on reaction mechanisms, kinetics, modeling, and reactor technology [131].

The cost-competitive conversion of biomass into liquid fuels and renewable chemicals should involve the integration of operations on heterogeneous ligno-cellulosic, protein, and triglyceride feeds with processes that convert specific fractions of biomass as outlined in Figure 4.12. Much research has been done on specific biomass components and different aspects of biomass conversion, such as dewatering in liquid phase, selective solvation of hemicelluloses, modification of cellulose crystallinity, and lignin structure decomposition in SCW, leaving the kinetics of and interactions among heterogeneous biomass constituents largely unexplored. Insight into biomass phase behavior in SCW and reaction characteristics of biomass model compounds, achieved in previous studies, may lead to new process developments in the near future such that products generated from different biomass components can be efficiently separated and collected by controlling the properties of SCW with the aids of catalysts. Complementary to gasifying biomass or converting whole biomass into bio-crude and upgrading this crude oil, each fraction could be selectively extracted or converted to desired alkanes and chemicals. This concept (Fig. 4.12) could be used for future process development of advanced biofuels and renewable chemicals. A broad range of fundamental and exploratory research is needed to further advance the understanding of biomass conversion physics, chemistry, and engineering in SCW, such as real biomass phase behavior and reaction characteristics, clean fractionation of biomass components, effects of impurities in lignocellulosics and algae, selective hydrogenation and deoxygenation,

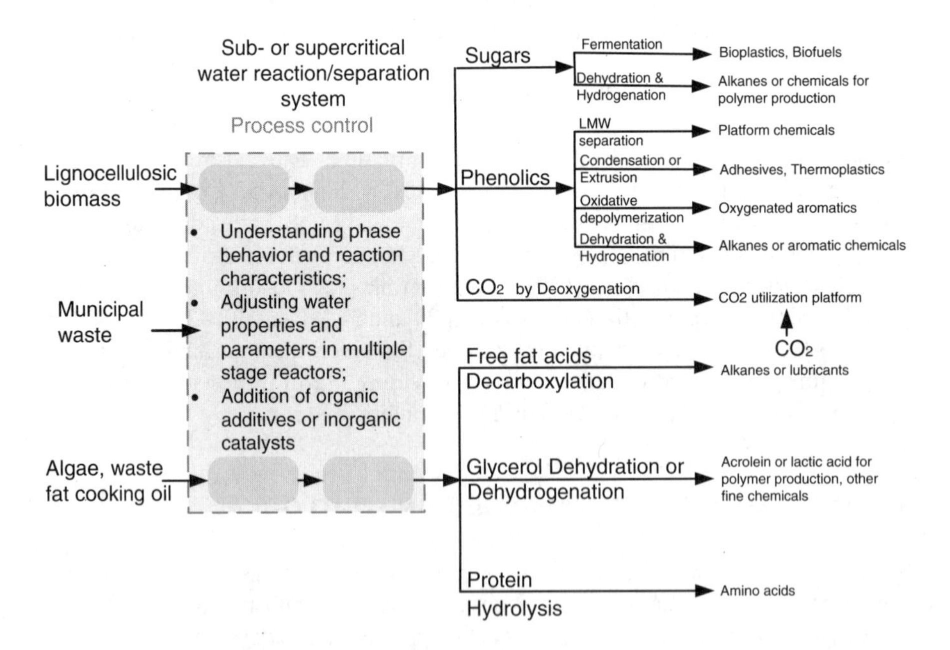

FIGURE 4.12 Future perspective of biorefinery with water.

catalyst chemistry and engineering in SCW media, coliquefaction of biomass with low-value carbonaceous materials, and the integration of direct biomass conversion with existing petroleum refineries.

Environmental benefits, including that SCW is an environmentally friendly reaction medium and that costs for carbon dioxide sequestration are low, should be fully considered in assessing biomass conversion in SCW. Good opportunities to utilize SCW technologies exist because of the increasing attention to microalgae as feedstock for the production of biofuels and renewable chemicals. These algae feedstocks have very high moisture content. SCW as reaction media may provide good solubility and straightforward chemistry in the production of liquid alkanes and value-added amine products. Furthermore, high-concentration carbon dioxide rejected from the SCW processes can be fed back for algae production.

Limited studies on the process economics can be found in scattered literature for supercritical water oxidation of wastes and SCW pyrolysis or gasification of biomass. Offsetting the advantage of being energy efficient, the high pressure requirement for SCW technologies may demand heavy initial investment. However, there exist good opportunities to develop green and economically competitive industrial processes in the utilization of biomass with high moisture content. The advantages of SCW technologies in terms of greenhouse gas emissions and sequestration, comparative air-pollutant emissions, and comparative water withdrawal and consumption should be taken into consideration. Economic evaluation of the emerging biorefinery with SCW should incorporate flow of mass and energy balance modeling, heat and power integration, together with life cycle cost analysis and environmental life cycle analysis, since this incorporation is essential to system efficiency, performance, and environmental impacts and credits.

REFERENCES

1. A. J. Ragauskas, C. K. Williams, B. H. Davison, G. Britovsek, J. Cairney, C. A. Eckert, W. J. Frederick, J. P. Hallett, D. J. Leak, C. L. Liotta, J. R. Mielenz, R. Murphy, R. Templer and T. Tschaplinski, *Science* **2006**, 311, 484–489.
2. U. M. Lindstrom, *Chem. Rev.* **2002**, 102, 2751–2722.
3. N. Akiya and P. E. Savage, *Chem. Rev.* **2002**, 102, 2725–2750.
4. M. H. Eikani, F. Golmohammad and S. Rowshanzamir, *J. Food Eng.* **2007**, 80, 735–740.
5. J. C. Johnson, Massachusetts Institute of Technology. **2006**.
6. A. A. Peterson, F. Vogel, R. P. Lachance, M. Fröling, J. M. J. Antal and J. W. Tester, *Energy Environmen. Sci.* **2008**, 1, 32.
7. *Engineers Steam Properties Database, National Institute of Standards and Technology/American Society of Mechanical Engineers*, **1996**.
8. H. Nakagawa, A. Namba, M. Böhlmann and K. Miura, *Fuel* **2004**, 83, 719–725.
9. Q. Cai, Z.-H. Huang, F. Kang and J.-B. Yang, *Carbon* **2004**, 42, 775–783.
10. E. Dinjus and A. Kruse, *J. Phys. Condens. Matter* **2004**, 16, S1161–S1169.
11. S. Deguchi, K. Tsujii and K. Horikoshi, *Chem. Commun.* **2006**, 31, 3293–3295.

12. A. Demirbaş, *Energy Conv. Manag.* **2001**, 183–188.

13. J. J. Meister, *J. Macromol. Sci. Polym. Rev.* **2002**, 42, 235–289.

14. (a) Y. Sun and J. Cheng, *Biores. Technol.* **2002**, 83, 1–11; (b) S. Liu, T. E. Amidon, R. C. Francis, B. V. Ramarao, Y. Z. Lai and G. M. Scott, *Ind. Biotechnol.* **2006**, 2, 113–120.

15. G. Garrote, H. Dominguez and J. C. Parajó, *J. Chem. Technol. Biotechnol.* **1999**, 74, 1101–1109.

16. (a) A. J. Stipanovic, J. S. Haghpanah, T. E. Amidon, G. M. Scott, V. Barber and K. Mishra, *Mater. Chem. Energy Forest Biomass* Vol. 954, **2007**, 107–120; (b) T. E. Amidon, C. D. Wood, A. M. Shupe, Y. Wang, M. Graves and S. Liu, *J. Biobased Mater. Bioenergy* **2008**, 2, 100–120.

17. S. Fernando, S. Adhikari, C. Chandrapal and N. Murali, *Energy Fuels* **2006**, 20, 1727–1737.

18. (a) S. Kubo, R. D. Gilbert and J. E. Kadla, in *Natural Fibers, Biopolymers, and Biocomposites* (Eds.: M. M. A.K. Mohanty, L.T. Drzal), CRC Taylor & Francis Group, Boca Raton, **2005**, 671–697; (b) D. Stewart, *Ind. Crops Prod.* **2008**, 27, 202–207; (c) H. D. Embree, T. Chen and G. F. Payene, *Chem. Eng* **2001**, 84, 133–147; (d) J. A. Koehler, B. J. Brune, T. Chen, A. J. Glemza, P. Vishwanath, P. J. Smith and G. F. Payne, *Ind. Eng. Chem. Res.* **2000**, 39, 3347–3355.

19. (a) P. Alvira, E. Tomás-Pejó, M. Ballesteros and M. J. Negro, *Biores. Technol.* **2010**, 101, 4851–4861; (b) N. Mosier, C. Wyman, B. Dale, R. Elander, Y. Y. Lee, M. Holtzapple and M. Ladisch, *Biores. Technol.* **2005**, 96, 673–686; (c) C. E. Wyman, B. E. Dale, R. T. Elander, M. Holtzapple, M. R. Ladisch and Y. Y. Lee, *Biores. Technol.* **2005**, 96, 1959–1966.

20. P. Alvira, E. Tomás-Pejó, M. Ballesteros and M. J. Negro, *Biores. Technol.* **2010**, 101, 4851–4861.

21. S. Liu, *Biotechnol. Adv.* **2010**, 28, 563–582.

22. (a) D. J. Brasch and K. W. Free, *TAPPI* **1965**, 48, 245–248; (b) R. L. Casebier, J. K. Hamilton and H. L. Hergert, *TAPPI* **1969**, 52, 2369–2377; (c) J. H. Lora and M. Wayman, *TAPPI* **1978**, 61, 47–50; (d) S. H. Yoon and A. van Heiningen, *TAPPI J.* **2008**, 7, 22–27; (e) S. H. Yoon, K. Macewan and A. van Heiningen, *TAPPI J.* **2008**, 7, 27–31; (f) W. W. Al-Dajani, U. Tschirner and T. Jensen, *TAPPI J.* **2009**, 8, 30–37.

23. (a) S. I. Aronovsky and R. A. Gortner, *Ind. Eng. Chem.* **1930**, 22, 264–274; (b) O. Goldschmid, *TAPPI J.* **1955**, 38, 728–732; (c) M. G. S. Chua and M. Wayman, *Can. J. Chem.* **1979**, 57, 1141–1149; (d) M. G. S. Chua and M. Wayman, *Can. J. Chem.* **1979**, 57, 2603–2611; (e) J. H. Lora and M. Wayman, *Can. J. Chem.* **1980**, 58, 668–676; (f) G. Garrote, H. Dominguez and J. C. Parajó *Holz Roh Werk* **1999**, 57, 191–202; (g) G. Garrote, H. Domínguez and J. C. Parajó, *Holz Roh Werk* **2001**, 57, 191–202; (h) H. Ando, T. Sakaki, T. Kokusho, M. Shibata, Y. Uemura and Y. Hatate, *Ind. Eng. Chem. Res.* **2000**, 39, 3688–3693; (i) J. Li and G. Gellerstedt, *Ind. Crops Prod.* **2008**, 27, 175–181; (j) T. Song, A. Pranovich, I. Sumersky and B. Holmbom, *Holzforschung* **2008**, 62, 659–666; (k) M. Leschinsky, G. Zuckerstätter, H. K. Weber, R. Patt and H. Sixta, *Holzforschung* **2008**, 62, 645–652; (l) M. Leschinsky, G. Zuckerstätter, H. K. Weber, R. Patt and H. Sixta, *Holzforschung* **2008**, 62, 653–658; (m) M. Leschinsky, H. K. Weber and R. S. Patt, *Lenzinger Berichte* **2009**, 87, 16–25; (n) K. Leppännen, P. Spetz, A. Pranovich, K. Hartonen, V. Kitunen and H. Ilvesniemi, *Wood Sci. Technol.* **2010**.

24. R. P. Overend and E. Chornet, *Phil. Trans. R. Soc. Lond. A*, 321, 523–536.

25. (a) M. S. Tunc and A. R. P. van Heiningen, *Nord. Pulp Paper Res. J.* **2009**, 24, 46–51; (b) A. Mittal, S. G. Chatterjee, G. M. Scott and T. E. Amidon, *Holzforschung* **2009**, 63, 307–314.
26. (a) S. G. Allen, L. C. Kam, A. J. Zemann and M. J. Antal, Jr., *Ind. Eng. Chem. Res.* **1996**, 35, 2709–2715; (b) Y. Yu, X. Lou and H. Wu, *Energy Fuels* **2007**, 22, 46–60.
27. G. P. van Walsum, S. G. Allen, M. J. Spences, M. S. Laser, M. J. Antal and L. R. Lynd, *Appl. Biochem. Biotechnol.* **1996**, 57/58, 157–170.
28. I. Hasegawa, K. Tabata, O. Okuma and K. Mae, *Energy Fuels* **2004**, 18, 755–760.
29. (a) K. Hartonen, J. Parshintev, K. Sandberg, E. Bergelin, L. Nisula and M. L. Riekkola, *Talanta* **2007**, 74, 32–38; (b) C. C. Teo, S. N. Tan, J. W. H. Yong, C. S. Hew and E. S. Ong, *J. Chromatogr. A* **2010**, 1217, 2484–2494.
30. M. S. Tunc and A. R. P. van Heiningen, *Carbohydr. Polym.* **2010**, 83, 8–13.
31. J. Xu and S. Liu, *Renew. Energ.* **2009**, 34, 2353–2356.
32. Sun, Z., Shupe, A., Liu, T., Hu, R., Amidon, T.E. and Liu, S. Biores. *Technol.* **2011**, 102, 2133–2136.
33. Sun, Z., Fitzgerald, L., Mukherjee, S.S., and Liu, S. The Annual Conference of the American Institute of Chemical Engineers (AIChE), 2009, Nashville, TN.
34. T. M. Keenan, S. W. Tanenbaum, A. J. Stipanovic and J. P. Nakas, *Biotechnol. Prog.* **2004**, 20, 1697–1704.
35. A. Hasan, B. Bujanovic and T. Amidon, *J. Biobased Mater. Bioenergy*, **2010**, 4, 46–52.
36. C. Gong, M. J. Goundalkar, B. Bujanovic and T. Amidon, *J. Wood Chem. Technol.* **2012**, 32, 93–104.
37. T. Amidon and J. Howard, Int. Biorefinery Conf. IBC 09, October 6–7, 2009, Syracuse, NY; oral presentation; IBC Book of Abstracts:2C.
38. V. Yadama, T. E. Amidon, *et al.*, International Biorefinery Conference, IBC 09, October 8–9, 2009, Syracuse, NY: Abstracts of Presentations: 7A, 25.
39. K. E. Vroom, *Pulp Paper Mag. Can.* **1957**, 58, 228–231.
40. H. Sixta, *Handbook of Pulp*, Wiley-VCH, Weinheim, Germany, **2006**.
41. (a) D. Fengel and G. Wegener, *Wood: Chemistry, Ultrastructure, Reactions*, Walter de Gruyter, Berlin, **1989**; (b) E. Sjöström, *Wood Chemistry. Fundamentals and Applications*. 2nd ed., Academic Press, San Diego, **1993**; (c) Analytical Methods in Wood Chemistry, Pulping, and Papermaking, Springer Series in Wood Science, 1999, Springer-Verlag, Berlin Heidelberg; (d) K. Yates and R. A. McClelland, *J. Am. Chem. Soc.* **1967**, 89, 2686–2692; (e) P. Krammer and H. Vogel, *J. Supercrit. Fluids* **2000**, 16, 189–206.
42. S. Caparrós, G. Garrote, J. Ariza and F. López, *Ind. Eng. Chem. Res.* **2006**, 45, 8909–8920.
43. H. M. Wang, D. Loganathan and R. J. Linhardi, *Biochem. J.* **1991**, 278, 689–695.
44. I. Fraeye, T. Duvetter, I. Verlent, D. N. Sila, M. Hendrickx and L. A. Van, *Inn. Food Sci. Emerg. Technol.* **2007**, 8, 93–101.
45. J. Hafrén and U. Westermark, *Nord. Pulp Paper Res. J.* **2001**, 16, 284–289.
46. (a) B. Bujanovic, Goundalkar, M. J. and T. E. Amidon, Non-carbohydrate-based products extracted during hot-water extraction of sugar maple, *30th SETAC North America Annual Meeting: Green Biorefinery*, November 19–23, New Orleans, LA, platform presentation: 425, **2009**. (b) M. J. Goundalkar, B. Bujanovic and T. Amidon,

Lignin in the Hot-Water Extract of Sugar Maple—Isolation, Characterization and Potential Use, *Int. Bioref. Conf. IBC*, October 6–7 Syracuse, NY, IBC Book of Abstracts: 3A, **2009**. (c) M. J. Goundalkar, B. Bujanovic, C. Gong, and T. Amidon, Characterization of Organic Precipitate from Hot-Water Extraction of Hardwoods, *37th Northeast Regional Meeting of the ACS, NERM 2010:* "Chemistry for a Sustainable World," SUNY-Potsdam, NY, June 2–5, Poster 39, **2010**. (d) M. J. Goundalkar, B. Bujanovic, and T. Amidon, Analysis of non-carbohydrate based low-molecular weight organic compounds dissolved during hot-water extraction of sugar maple, *Cell. Chem. Technol.* **2010**, 44, 27–33.

47. S. Y. Lin and C. W. Dence, *Methods in Lignin Chemistry*, Springer Series in Wood Science, Springer, Berlin, Heidelberg, **1992**.
48. K. Lundquist and R. Lundgren, *Acta Chem. Scand.* **1972**, 26, 2005–2023.
49. S. Li and K. Lundquist, *Nord. Pulp Paper Res. J.* **2000**, 15, 292–299.
50. G. H. van der Klashorst, Lignin Properties and Materials, Vol. W. G. Glasser S. Sarkanen (Ed.: W. G. S. S.Glasser), *ACS Symposium Series 397*, Washington DC, **1989**, 346–360.
51. T. Koshijima and T. Watanabe, *Association Between Lignin and Carbohydrates in Wood and Other Plant Tissues. Springer Series in Wood Science*, Springer Berlin, Heidelberg, **2003**.
52. (a) B. Košikova, D. Joniak and L. Kosakova, *Holzforschung* **1979**, 33, 11–14; (b) M. Poláková, D. Joniak and M. Ďuriš, *Monatschefte für Chemie* **2000**, 131, 1197–1205.
53. W. Vermerris and R. Nicholson, *Phenolic Compound Biochemistry*, Springers, Dordrecht, The Netherlands, **2006**, 1–34.
54. S. P. Pietarinen, S. M. Willför, M. O. Ahotupa, J. E. Hemming and B. R. Holmbom, *J. Wood Sci.* **2006**, 52, 436–444.
55. S. P. Pietarinen, S. M. Willför, F. A. Vikström and B. R. Holmbom, *J. Wood Chem. Technol.* **2006**, 26, 245–258.
56. S. P. Pohjamo, J. E. Hemming, S. M. Willför, M. H. T. Reunanen and B. R. Holmbom, *Phytochemistry* **2003**, 63, 165–169.
57. E. S. Ong, J. S. H. Cheong and D. Goh, *J. Chromatogr. A.* **2006**, 1112, 92–102.
58. K. Zygmunt, B. Faubert, J. MacNeil and E. Tsiani, *Biochem. Biophys. Res. Co.* **2010**, 398, 178–183.
59. N. M. Saarinen, A. Wärri, S. I. Mäkelä, C. Eckerman, M. Reunanen, M. Ahotupa, S. M. Salmi, A. A. Franke, L. Kangas and R. Santti, *Nutr. Cancer* **2002**, 36, 216.
60. P. C. Eklund, A. I. Riska and R. E. Sjöholm, *J. Org. Chem.* **2002**, 67, 7544–7546.
61. P. Eklund, R. Sillanpää and R. E. Sjöholm, *J. Chem. Soc., Perkin Trans. 1* **2002**, 16, 1906–1910.
62. P. Eklund, A. Lindholm, J. P. Mikkola, A. Smeds R. Lehtilä and R. E. Sjöholm, *Org. Lett.* **2003**, 5, 491–493.
63. P. C. Eklund and R. E. Sjoholm, *Tetrahedron* **2003**, 59, 4515–4523.
64. (a) S. I. Mussatto and I. C. Roberto, *Biores. Technol.* **2004**, 93, 1–10; (b) L. J. Jönsson, E. Palmqvist, N. O. Nilverbrant and B. Hahn-Hägerdal, *Appl. Microbiol. Biotechnol.* **1998**, 49, 691–607.
65. J. Ralph, G. Brunow and W. Boerjan, *Encyclopedia of Life Sciences* (Eds. F. Rose, K. Osborne), Wiley **2007**, 1–10.

66. A. Mittal, G. M. Scott, T. E. Amidon, D. J. Kiemle and A. J. Stipanovic, *Biores. Technol.* **2009**, 100, 6398–6406.
67. S. K. Bose, R. C. Francis, M. Govender, T. Bush and A. Spark, *Biores. Technol.* **2009**, 100, 1628–1633.
68. F. Z. Erdemgil, S. Şanli, N. Şanli, G. Özkan, J. Barbosa, J. Guiteras and J. L. Beltrán, *Talanta* **2007**, 72, 496.
69. U. Westermark, B. Samuelsson and K. Lunquist, *Res. Chem. Intermed.* **1995**, 21, 343–352.
70. Y. Li and S. Sarkanen, *Macromolecules* **2002**, 35, 9707–9715.
71. E. K. Pye, *Biorefineries—Industrial Processes and Products*, Vol. 2 (Eds. B. Kamm, P. R. Gruber, M. Kamm), Wiley-VCH Verlag GmbH & Co. KGaA, Weinheim, **2006**, 165–200.
72. C. Eckert, C. Liotta, A. Ragauskas, J. Hallett, C. Kitchens, E. Hill and L. Draucker, *Green Chem.* **2007**, 9, 545–548.
73. (a) Q. Xiang and Y. Y. Lee, *Appl. Biochem. Biotechnol.* **2001**, 71, 91–93; (b) M. P. Masingale, E. F. Alves, T. N. Korbieh, S. K. Bose and R. C. Francis, *BioRes.* **2009**, 4, 1139–1146.
74. B. Bujanovic, S. A. Ralph, R. S. Reiner, K. Hirth and R. H. Atalla, *Materials* **2010**, 3, 1888–1903.
75. N. J. Walton, A. Narbad, C. B. Faulds and G. Williamson, *Curr. Opin. Biotechnol.* **2000**, 11(5):490–496.
76. P. G. Jessop, L. Phan, A. Carrier, S. Robinson, C. J. Dürr and J. R. Harjani, *Green Chem.* **2010**, 12, 809–814.
77. F. Xu, J. S. Kulys, K. Li, *et al.*, *Appl. Env. Microbiol.* **2000**, 66, 2052–2056.
78. (a) A. Marjasvaara, M. Torvinen, H. Kinnunen and P. Vainiotalo, *Biomacromolecules* **2006**, 7, 1604–1609; (b) R. Ikeda, J. Sugihara, H. Uyama and S. Kobayashi, *Macromolecules* **1996**, 29, 8705.
79. C. Regalado, B. E. Garcia-Almendarez and M. A. Duarte-Vazquez, *Phytochem. Rev.* **2004**, 3, 256.
80. M. M. Wright, R. C. Brown and A. A. Boateng, *Biofuels Bioprod. Biorefin.-Biofpr* **2008**, 2, 229–238.
81. (a) D. C. Elliott, *Energy Fuels* **2007** 21, 1792–1815; (b) A. A. Peterson, F. Vogel, R. P. Lachance, M. Fröling, J. Antal and J. W. Tester, *Energy Environ. Sci.* **2008**, 1, 33–65.
82. J. M. Moffatt and R. P. Overend, *Biomass* **1985**, 7, 99–123.
83. F. Behrendt, Y. Neubauer, M. Oevermann, B. Wilmes and N. Zobel, *Chem. Eng. Technol.* **2008**, 31, 667–677.
84. J. E. Naber and F. Goudriaan, ACS Division of Fuel Chemistry, **2005**.
85. Changing World Technologies: http://www.changingworldtech.com/what/index.asp, accessed on May 30, **2011**.
86. EnerTech Environmental Inc.:, http://enertech.com/about/mediarelations/pdfs/Ribbon-Cutting_6-11-2009.pdf, accessed on May 30, **2011**.
87. (a) L. Cheng, L. Liu and X. P. Ye, *Proceedings of AIChE 2009 Annual Meeting* **2009**; (b) L. Cheng, X. P. Ye, R. He and S. Liu, *Fuel Process. Technol.* **2009**, 90, 301–311; (c) R. He, X. P. Ye, B. C. English and J. A. Satrio, *Biores. Technol.* **2009**, 100, 5305–5311.

88. (a) S. Karagöz, T. Bhaskar, A. Muto, Y. Sakata, T. Oshiki and T. Kishimoto, *Chem. Eng. J.* **2005**, 108, 127–137; (b) S. Karagoz, T. Bhaskar, A. Muto, Y. Sakata and M. A. Uddin, *Energy Fuels* **2003**, 18, 234–241.
89. (a) K. Yoshida, J. Kusaki, K. Ehara and S. Saka, *Appl. Biochem. Biotechnol.* **2005**, 123, 11; (b) D. Takada, K. Ehara and S. Saka, *J. Wood Sci.* **2004**, 50, 253–261; (c) K. Ehara, D. Takada and S. Saka, *J. Wood Sci.* **2005**, 51, 256–261; (d) Z. Fang and C. Fang, *AIChE J.* **2008**, 54, 2751–2758.
90. R. Hashaikeh, Z. Fang, I. S. Butler, J. Hawari and J. A. Kozinski, *Fuel* **2006**, 86, 1614–1622.
91. M. Wang, C. Xu and M. Leitch, *Biores. Technol.* **2009**, 100, 2305–2307.
92. (a) A. Kruse, *J. Supercrit. Fluids* **2009**, 47, 391–399; (b) G. W. Huber, S. Iborra and A. Corma, *Chem. Rev.* **2006**, 106, 4044–4098.
93. (a) H. Yoshida, H. Tokumoto, K. Ishii and R. Ishii, *Biores. Technol.* **2009** 100, 2933–2939; (b) K. Yoshida, J. Kusaki, K. Ehara and S. Saka, *Twenty-Sixth Symposium on Biotechnology for Fuels and Chemicals* (Eds. B. H. Davison, B. R. Evans, M. Finkelstein, J. D. McMillan), Humana Press, Inc., New York, **2005**, 795–806.
94. L. J. Guo, Y. J. Lu, X. M. Zhang, C. M. Ji, Y. Guan and A. X. Pei, *Catal. Today* **2007**, 129, 275–286.
95. Y. Matsumura, T. Minowa, B. Potic, S. R. A. Kersten, W. Prins, W. P. M. van Swaaij, B. van de Beld, D. C. Elliott, G. G. Neuenschwander, A. Kruse and M. Jerry Antal Jr, *Biomass Bioenergy* **2005**, 29, 269–292.
96. M. Modell, R. C. Reid and S. I. Amin, *US Patent 4,113,446*, **1978**.
97. (a) D. C. Elliott, *Biofuels, Bioprod. Biorefin.* **2008**, 2, 254–265; (b) D. C. Elliott, T. R. Hart and G. G. Neuenschwander, *Ind. Eng. Chem. Res.* **2006**, 45, 3776–3781.
98. M. J. Antal, S. G. Allen, D. Schulman, X. Xu and R. J. Divilio, *Ind. Eng. Chem. Res.* **2000**, 39, 4040–4053.
99. (a) A. Sinag, A. Kruse and V. Schwarzkopf, *Eng. Life Sci.* **2003**, 3, 469–473; (b) A. Kruse, *Biofuels, Bioprod. Biorefin.* **2008**, 2, 415–437.
100. S. Rabe, M. Nachtegaal, T. Ulrich and F. Vogel, *Angew. Chem. Int. Ed.* **2010**, 49, 6434–6437.
101. M. Osada, T. Sato, M. Watanabe, M. Shirai and K. Arai, *Combustion Sci. Technol.* **2006**, 178, 537–552.
102. (a) K. S. Ro, K. Cantrell, D. Elliott and P. G. Hunt, *Ind. Eng. Chem. Res.* **2007**, 46, 8839–8845; (b) D. C. Elliott, L. J. Sealock and E. G. Baker, *Ind. Eng. Chem. Res.* **2002**, 32, 1542–1548; (c) D. C. Elliott, G. G. Neuenschwander, M. R. Phelps, T. R. Hart, A. H. Zacher and L. J. Silva, *Ind. Eng. Chem. Res.* **1999**, 38, 879–883.
103. F. Vogel, M. H. Waldner, A. A. Rouff and S. Rabe, *Green Chem.* **2007**, 9, 616–619.
104. M. H. Waldner and F. Vogel, *Ind. Eng. Chem. Res.* **2005**, 44, 4543–4551.
105. S. E. Jacobsen and C. E. Wyman, *Ind. Eng. Chem. Res.* **2002**, 41, 1454–1461.
106. G. Brunner, *J. Supercrit. Fluids* **2009**, 47, 373–381.
107. Z. Fang, T. Minowa, R. L. Smith, T. Ogi and J. A. Kozinski, *Ind. Eng. Chem. Res.* **2004**, 43, 2454–2463.
108. Z. Srokol, A.-G. Bouche, A. van Estrik, R. C. J. Strik, T. Maschmeyer and J. A. Peters, *Carbohydr. Res.* **2004**, 339, 1717–1726.

109. (a) M. Watanabe, M. Osada, H. Inomata, K. Arai and A. Kruse, *Appl. Catal. A: Gen.* **2003**, 245, 333–341; (b) M. Watanabe, T. Iida, Y. Aizawa, H. Ura, H. Inomata and K. Arai, *Green Chem.* **2003**, 5, 5; (c) M. Watanabe, Y. Aizawa, T. Iida, T. M. Aida, C. Levy, K. Sue and H. Inomata, *Carbohydr. Res.* **2005**, 340, 1925–1930; (d) M. Watanabe, Y. Aizawa, T. Iida, C. Levy, T. M. Aida and H. Inomata, *Carbohydr. Res.* **2005**, 340, 1931–1939.

110. M. Watanabe, Y. Aizawa, T. Iida, R. Nishimura and H. Inomata, *Appl. Catal. A: Gen.* **2005**, 295, 150–156.

111. (a) A. Sınag, A. Kruse and J. Rathert, *Ind. Eng. Chem. Res.* **2003**, 43, 502–508; (b) A. Sınag, A. Kruse and V. Schwarzkopf, *Ind. Eng. Chem. Res.* **2003**, 42, 3516–3521.

112. M. Bicker, S. Endres, L. Ott and H. Vogel, *J. Molec. Catal. A : Chem.* **2005**, 239, 151–157.

113. F. Jin, Z. Zhou, T. Moriya, H. Kishida, H. Higashijima and H. Enomoto, *Environ. Sci. Technol.* **2005**, 39, 1893–1902.

114. (a) Y. Roman-Leshkov, C. J. Barrett, Z. Y. Liu and J. A. Dumesic, *Nature* **2007**, 447, 982–985; (b) Y. Roman-Leshkov, J. N. Chheda and J. A. Dumesic, *Science* **2006**, 312, 1933–1937; (c) Juben N. Chheda, George W. Huber and James A. Dumesic, *Angew. Chem. Int. Ed.* **2007**, 46, 7164–7183; (d) R. D. Cortright, R. R. Davda and J. A. Dumesic, *Nature* **2002**, 418, 964–967; (e) R. R. Davda, J. W. Shabaker, G. W. Huber, R. D. Cortright and J. A. Dumesic, *Appl. Catal. B: Environ.* **2005**, 56, 171–186; (f) G. W. Huber, J. N. Chheda, C. J. Barrett and J. A. Dumesic, *Science* **2005**, 308, 1446–1450; (g) G. W. Huber, R. D. Cortright and J. A. Dumesic, *Angew. Chem. Int. Ed.* **2004**, 43, 1549–1551; (h) G. W. Huber and J. A. Dumesic, *Catal. Today* **2006**, 111, 119–132; (i) E. L. Kunkes, D. A. Simonetti, R. M. West, J. C. Serrano-Ruiz, C. A. Gartner and J. A. Dumesic, *Science* **2008**, 322, 417–421.

115. G. Huber and J. Dumesic, *Catal. Today* **2006**, 111, 119–132.

116. (a) Y. Matsushita, T. Inomata, T. Hasegawa and K. Fukushima, *Biores. Technol.* **2009**, 100, 1024–1026; (b) K. Okuda, X. Man, M. Umetsu, S. Takami and T. Adschiri, *J. Phys. Condens. Matter* **2004**, 16, S1325–S1330; (c) K. Okuda, M. Umetsu, S. Takami and T. Adschiri, *Fuel Process. Technol.* **2004**, 85, 803–813; (d) K. Okuda, S. Ohara, M. Umetsu, S. Takami and T. Adschiri, *Biores. Technol.* **2008**, 99, 1846–1852; (e) M. Saisu, T. Sato, M. Watanabe, T. Adschiri and K. Arai, *Energy Fuels* **2003**, 17, 922–928; (f) Z. Fang, T. Sato, R. L. Smith Jr, H. Inomata, K. Arai and J. A. Kozinski, *Biores. Technol.* **2008**, 99, 3424–3430.

117. K. Okuda, X. Man, M. Umetsu, S. Takami and T. Adschiri, *J. Phys. : Condens. Matter* **2004**, 16, S1325–1330.

118. Z. Chen, K. Yuan, A. L. Angeliki, L. Xuebing and A.J. Lercher, *Angew. Chem.* **2009**, 121, 4047–4050.

119. (a) K. Pramote, F. Tomoyuki, A. Shuji, K. Yukitaka and M. Ryuichi, *Chem. Eng. J.* **2004**, 99, 1–4; (b) K. Pramote, A. Shuji and M. Ryuichi, *Biosci. Biotechnol. Biochem.* **2002**, 66, 1723–1726.

120. (a) P. Moquin and F. Temelli, *J. Supercrit. Fluids* **2008**, 45, 94–101; (b) F. Temelli, *J. Supercrit. Fluids* **2009**, 47, 583–590.

121. F. Tomoyuk, K. Pramote, K. Yukitaka and A. Shuji, *Food Chem.* **2006**, 94, 341–347.

122. M. Watanabe, T. Iida and H. Inomata, *Energy Conv. Manag.* **2006**, 47, 3344–3350.

123. T. J. Duane and A. T. Katherine, *Environ. Prog.* **2007**, 26, 338–348.

124. (a) M. Watanabe, T. Iida, Y. Aizawa, T. M. Aida and H. Inomata, *Biores. Technol.* **2007**, 98, 1285–1290; (b) L. Ott, M. Bicker and H. Vogel, *Green Chem.* **2006**, 8, 214–220; (c) H. Atia, U. Armbruster and A. Martin, *J. Catal.* **2008**, 258, 71–82; (d) E. Tsukuda, S. Sato, R. Takahashi and T. Sodesawa, *Catal. Commun.* **2007**, 8, 1349–1353; (e) S. Chai, H. Wang, Y. Liang and B.-Q. Xu, *J. Catal.* **2007**, 250, 342–349; (f) S. Chai, H. Wang, Y. Liang and B. Xu, *Green Chem.* **2007**, 9, 1130–1136.

125. L. Cheng and X. P. Ye, *Catal. Lett.* **2009**, 130, 100–107.

126. H. Kishida, F. Jin, Z. Zhou, T. Moriya and H. Enomoto, *Chem. Lett.* **2005**, 34, 1560–1562.

127. (a) X. Zhu, C. Zhu, L. Zhao and H. Cheng, *Chin. J. Chem. Eng.* **2008**, 16, 456–460; (b) A. T. Quitain, H. Daimon, K. Fujie, S. Katoh and T. Moriyoshi, *Ind. Eng. Chem. Res.* **2006**, 45, 4471–4474; (c) T. Rogalinski, S. Herrmann and G. Brunner, *J. Supercrit. Fluids* **2005**, 36, 49–58.

128. D. Klingler, J. Berg and H. Vogel, *J.Supercrit. Fluids* **2007**, 43, 112–119.

129. N. Sato, A. T. Quitain, K. Kang, H. Daimon and K. Fujie, *Ind. Eng. Chem. Res.* **2004**, 43, 3217–3222.

130. J. Li and T. B. Brill, *Int. J. Chem. Kinet.* **2003**, 35, 602–610.

131. A. Loppinet-Serani, C. Aymonier and F. Cansell, *ChemSusChem* **2008**, 1, 486–503.

CHAPTER 5

Supercritical CO_2 as an Environmentally Benign Medium for Biorefinery

RAY MARRIOTT and EMILY SIN

5.1 INTRODUCTION

The biorefinery involves a concept where renewable feedstocks are processed to provide a wide spectrum of products [1]. The main driver in the development of new biorefineries is the decline in availability of nonrenewable resources such as natural gas, petroleum, and coal and to a lesser extent the impact of greenhouse gases on climate change. Much of the biorefinery development has been centered in Europe and this has been galvanized by the European Union 20-20-20 agreement to reduce climate change under which energy consumption and carbon emissions are to be reduced by 20% by 2020. In addition, a 20% share of the energy generated should be sourced from a renewable source [2–4]. As part of this strategy, the EU has set a target for a minimum biofuel content in road-transport fuels of 2% by 2005 and 5.75% by 2010. This will be calculated on the basis of the energy content of all the transport fuels [5]. The US Environmental Protection Agency (EPA) has stated the legal requirement for all road-transport fuels to be a minimum of 8.01% by 2011 [6].

In a direct parallel to petroleum refining, the production of biofuels presents an opportunity to produce coproducts from renewable resources that can be converted into industrial products and intermediates [7]. The production of bioethanol in particular generates significant quantities of CO_2 and utilization of this should be considered in preference to capture and storage in oceans or terrestrial sinks [8]. CO_2 generated by fermentation is relatively pure and can be easily dried, recompressed, and stored before being potentially used to extract valuable metabolites or used as a reaction solvent.

The Role of Green Chemistry in Biomass Processing and Conversion, First Edition.
Edited by Haibo Xie and Nicholas Gathergood.

5.2 PROPERTIES OF CO_2

CO_2 is a valuable renewable resource for use as an alternative medium for extractions and reactions, however, its stability makes its economical utilization as a feedstock for chemicals or fuels a formidable challenge. Plants are able to convert CO_2 and water to carbohydrates and oxygen through the use of sunlight (photosynthesis). CO_2 reduction is a thermodynamically uphill process requiring significant amounts of energy for the transformation of CO_2 into carbon-containing products [9]. This energy can be sustainably and economically produced through renewable energy, such as solar power. Typically such systems would use homogeneous and heterogeneous catalysts for artificial photosynthesis, electrochemical reduction or hydrogenation of CO_2 using solar generated hydrogen and energy. Work by Traynor and Jensen has demonstrated direct solar reduction of CO_2 to CO and oxygen using only solar energy [10], Although this work demonstrates promise, further work is needed to realize the full potential of CO_2 reduction.

A more immediate opportunity is the replacement of fossil fuel derived solvents with CO_2, either as a liquid or supercritical fluid, and this is often perceived as a greener alternative to organic solvents. A supercritical fluid (SCF) is a substance above its critical temperature (T_c) and pressure (P_c) and the T_c for CO_2 is 304 K (31 °C) and P_c is 72.8 bar (7.3 MPa) as indicated in Figure 5.1, making the supercritical fluid phase easily accessible [11, 12]. Above the critical point CO_2 exists as a supercritical fluid and the manipulation of temperature and pressure produces a highly tunable solvent that can be used to selectively extract a wide range of molecules. Most commercial extraction plants are designed to operate using supercritical CO_2 rather than liquid CO_2 for this reason.

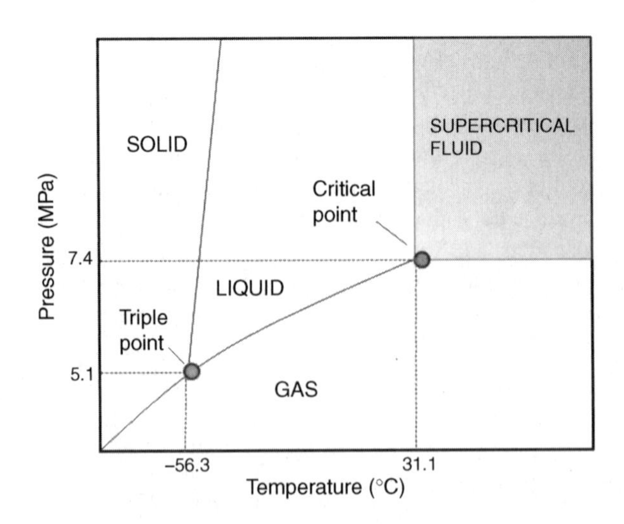

FIGURE 5.1 Phase diagram for CO_2.

TABLE 5.1 Physical Properties of Gas, SCF and Liquid CO_2

Property	Gas	SCF	Liquid
Density ($g\ cm^{-1}$)	10^{-3}	0.1–0.9	1
Viscosity (Pa s)	10^{-5}	10^{-4}	10^{-3}
Diffusivity ($cm^2\ s^{-1}$)	0.1	10^{-3}	10^{-5}–10^{-6}

Source: Ref. [10].

The properties of SCFs have been described as an intermediate between gas and liquid and have the advantages of each phase, high diffusivity and low viscosity. The physical properties of gaseous, liquid, and supercritical CO_2 are shown in Table 5.1.

Supercritical CO_2 is an ideal solvent for extraction and reactions as the solvent is easily removed owing to its "zero" surface tension and extraction products are recovered in a solvent-free state. Its rapid diffusion can enhance the extraction rates and after the extraction, the solvent can be recycled, resulting in nearly zero waste. It is nontoxic, nonflammable, colorless, odorless, and tasteless which makes it a good extraction solvent for natural products particularly where the extract or residual products are ultimately to be used in consumer products. Another distinct property of supercritical CO_2 is that all gases are miscible in $scCO_2$ and it is itself highly soluble in gases, which can enhance the transport rates in reactions involving gases such as hydrogenation [11]. Under supercritical conditions, the concentration of hydrogen in a mixture of hydrogen and CO_2 at 120 bar and 50 °C is 3.2 M, however, under the same condition in THF, the concentration is only 0.4 M [13].

The solvent properties and strength can be manipulated by pressure and temperature [11,12] and this can be expressed using a number of different parameters such as dipole moment, dielectric constant, refraction index, or degree of solubility [14]. Solvent properties can be measured by considering the solvent–solute interaction forces and these can be observed using various techniques such as IR, NMR, or UV–VIS spectroscopy [15, 16]. Heller compared the solubility parameters of CO_2 to *n*-alkanes in 1980s [17]. In 1991, Ikushima used an infrared spectroscopic technique to measure the solvent polarity parameters of $scCO_2$. The results showed close correlation to nonpolar solvents *n*-hexane and *n*-heptane [14]. Table 5.2 shows the comparison of some physical properties of CO_2 with a selection of common organic solvents [18].

Apart from environmental advantages, there are many chemical advantages in using CO_2 as a solvent. CO_2 is an aprotic solvent and can carry out reactions where a labile proton may impede reaction processes [19]. CO_2 cannot be oxidized and can be used in oxidation reactions without the formation of by-products [20,21]. CO_2 exhibits a low dielectric constant suggesting low solvent power [18,21]. The dielectric constant for liquid CO_2 is about 1.5 and for $scCO_2$ is between 1.1 and 1.5 compared to organic solvents such as dichloromethane (9.1), ethanol (30), and hexane (2) [19]. In contrast, the solvent powers measured by Reichardts dye suggests that liquid CO_2 (LCO_2) and supercritical CO_2 ($scCO_2$) are 0.081 units more polar than hexane (Table 5.2).

When used as a reaction solvent a number of properties of CO_2 have to be considered. CO_2 can dissolve in water at moderate pressures of <100 bar, carbonic

TABLE 5.2 Physical Properties of CO_2 Compared with Organic Solvents

Solvent	Density ($g\ mL^{-1}$)	Viscosity (Pa s)	Flash Point (°C)	Heat Capacity ($kJ\ kg^{-1}\ K^{-1}$)^d	Reichardt's Polarity Scale E_T^{Nd}
DCM	1.326	4.06×10^{-4}	N/A	1.19	0.309
Hexane	0.655	2.95×10^{-4}	−23	2.27	0.009
Water	1	8.94×10^{-4}	N/A	4.18	1
Ethanol	0.789	1.074×10^{-3}	12	2.44	0.654
Ethyl Acetate	0.894	4.31×10^{-4}	61	1.9	0.228
$scCO_2$	0.956^a	1.06×10^{-4c}	N/A	0.846^c	0.09 (variable)
LCO_2	1^b	1.2×10^{-4d}	N/A	3.14^e	0.09 (variable)

Note: Table adapted from data in reference [17].
^{a}Under supercritical conditions @ 40°C and 400 bar.
^{b}Under liquid CO_2 conditions @ 25°C and 400 bar.
^{c}@ 40°C.
^{d}@ 25°C.
^{e}@ 10°C.

acid formation under these conditions can result in a pH drop to 2.85, a significant problem in pH-dependent reactions. CO can also be produced as a by-product in hydrogenation reactions over a noble-metal catalyst and thus poison such a catalyst [13]. Primary and secondary amines can also react with CO_2 to form carbamates, which can be advantageous or harmful depending upon application [22] and $scCO_2$ is a Lewis acid and can react with strong bases [23]. Anilines and tertiary amines do react with CO_2 but to a lesser extent [22].

Liquid and supercritical CO_2 are now accepted as alternative green solvents although their adoption has been relatively slow except in areas such as extraction of food, cosmetic, and pharmaceutical ingredients from botanical materials.

5.3 USING CO_2 WITHIN THE BIOREFINERY

The very definition of a biorefinery implies that the feedstock will be a renewable material of plant origin containing a complex mixture of primary and secondary metabolites. For most biorefinery schemes, it is the primary metabolites, cellulose, hemicellulose, starch, or lipids from which the principle products will be produced. Protein is recovered for use as animal feed and the lignin is either used as a derivative such as lignosulfonate or depolymerized and used as a feedstock for smaller phenolic molecules, such as the production of vanillin. In most current biorefineries, the secondary metabolites are not considered in the overall process scheme as their concentration is relatively low and the isolation of these molecules complex and costly.

For bioethanol-based biorefineries CO_2 is an abundant by-product and most already capture, purify, and sell the CO_2 for a variety of applications outside the biorefinery as part of the overall economic mix. Indeed most new installations subcontract this activity to a specialist gas company and an example of this is the

FIGURE 5.2 CO_2 storage at the ENSUS/Yara site on Teeside (UK). (Photograph courtesy of Yara (UK) Ltd.)

YARA installation at the ENSUS bioethanol plant on Teeside (UK) (Fig. 5.2). It is not therefore a large step to use this solvent for processes within the biorefinery and this could include extraction of secondary metabolites such as essential oils, phytosterols, or cuticle waxes or the recovery of lipids for use as food [24–27] or cosmetic ingredients [26,27] or in the production of biodiesel [28,29]. As the production of liquid biofuels such as ethanol increases, capturing and utilizing the CO_2 generated as part of an integrated biorefinery will further enhance the benefit from using these renewable feedstocks.

There are currently no industrial biorefineries that are using an integrated supercritical fluid extraction (SFE) step as part of their processes. However, there are recent patents that demonstrate that such integration is being considered as an addition to existing processes. The most advanced proposed process [30] envisages capturing the CO_2 generated from the production of ethanol and using this to extract the lipids from the corn germ as a raw material for biodiesel. Figure 5.3 shows how carbon dioxide created during corn ethanol production is captured, purified, and used as an extraction solvent in the corn biorefinery for the recovery of corn oil.

This could be further extended to use $scCO_2$ as an extraction solvent for secondary metabolites when ligno-cellulosic sources are used as a biomass source for ethanol production [31] as shown in Figure 5.4. This approach has numerous applications as future ethanol production will rely more on second-generation nonfood sources such as ligno-cellulose and less on first-generation feedstocks such as grain. The range of

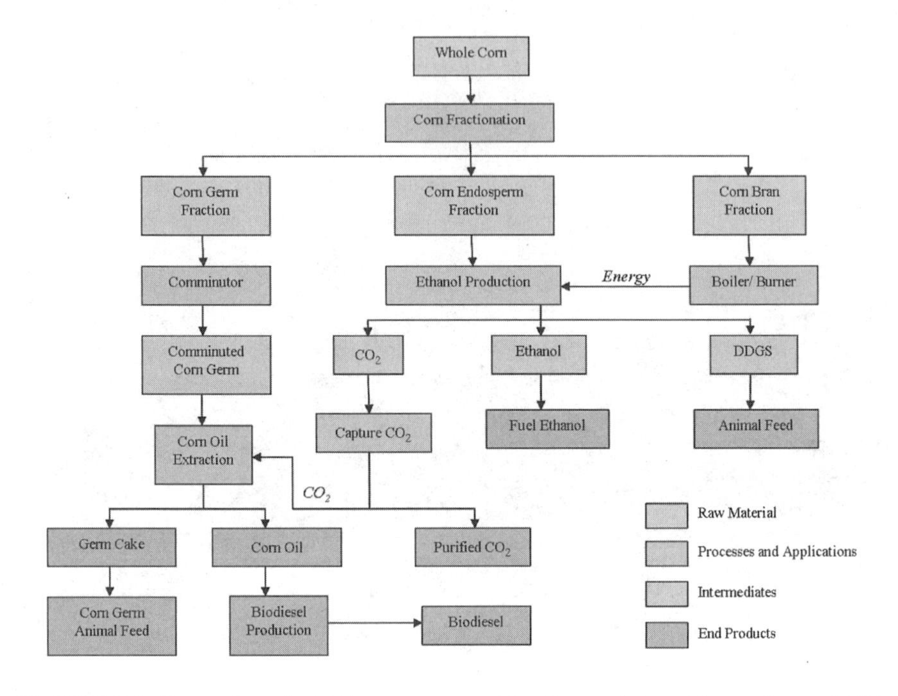

FIGURE 5.3 Integration of an SFE step into a corn biorefinery. (Adapted from reference [30].)

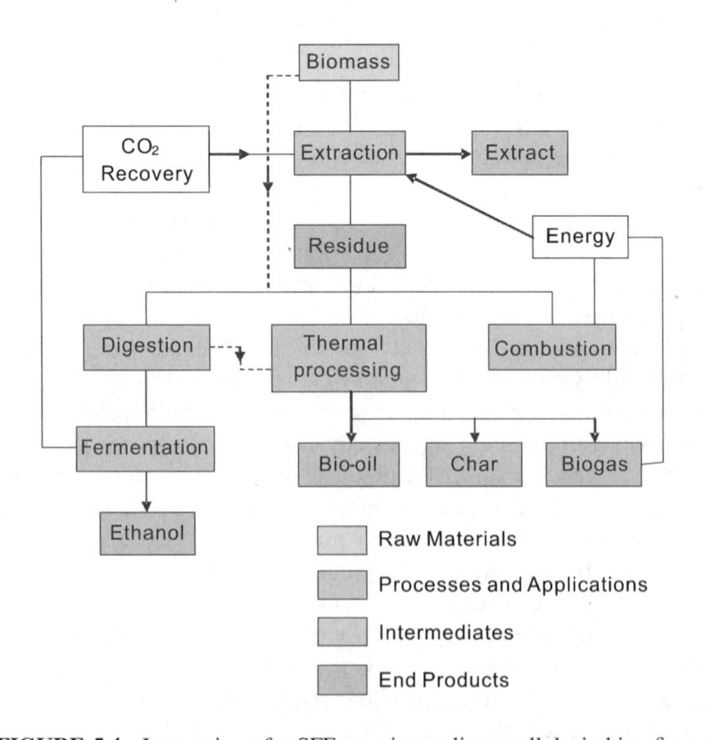

FIGURE 5.4 Integration of a SFE step into a ligno-cellulosic biorefinery.

extractable molecules is much greater from these sources, forestry products are a source of terpenoids, resins, sterols, and phenolics and almost all plant surfaces including cereal straw have a protective cuticle layer comprising a complex mixture of alkanes, wax esters, sterols, fatty acids, alkanols, and ketones. The process of extraction even at 40–60 MPa has little impact on the primary composition and structure and if the residue is going for co-firing as opposed to digestion for liquid biofuels the calorific value is not compromised. Any reduction in calorific value from removal of the extracts is compensated by a simultaneous reduction in water content, the residue normally have only 3–4% moisture after extraction due to the solubility of water in supercritical CO_2. The water content of the extracting biomass is $<10\%$ as large amounts of water in biomass can act as an entrainer during the supercritical CO_2 extraction affecting the selectivity and quality of the extracts [32].

CO_2 can also be used as a benign solvent for the production of derivatives of natural products that originate in the biorefinery. With an increasing demand for "natural ingredients" in many consumer products using a green solvent such as CO_2 with an acceptable biocatalyst such as an isolated enzyme or whole-cell preparation allows the end product to be labeled as a natural and often "organic" ingredient. This application and the development of future integrated extraction–reaction systems are discussed later.

5.4 EXTRACTION WITH CO_2

The advantages of using CO_2 as a benign and green solvent are to some extent counteracted by the capital investment required to capture and utilize this by-product. The main cost is in the construction of the equipment with a suitably high working pressure (typically 30–55 MPa) and the provision of compressors and pumps to deliver this pressure. However, there are now over 150 installations worldwide using liquid and supercritical CO_2 to extract commodity products such as edible oils as well as high-value products. CO_2 is widely used as an extraction solvent for high-volume applications such as decaffeination of coffee and tea [33,34], production of hop extracts for the brewing industry [35, 36], recovery of flavors and aromas from herbs and spices [37, 38], nicotine extraction from tobacco [39,40], extraction and fractionation of edible oils [41–43], and removal of contaminants as a cleaning procedure [44, 45]. The effectiveness of the extraction is highly dependent on the kinetic extraction properties of the raw materials, solvent strength, and the operating conditions [46].

Industrial-scale supercritical CO_2 extractions are currently all carried out as a batch process as shown in Figure 5.5. The quick loading and unloading of materials into the extractors minimizes equipment downtime. An extractor with an internal basket or hydraulically operated top and bottom lids, allow for the gravity discharge of materials into hoppers or conveyors which can be used to aid the loading and unloading process. A typical supercritical CO_2 extractor operates as a pumped closed loop. Liquid CO_2 is taken from a storage vessel normally held refrigerated at 2.4 MPa and then further cooled before entering the primary pump. The compressed CO_2 is then heated to the desired extraction temperature and then passed through the

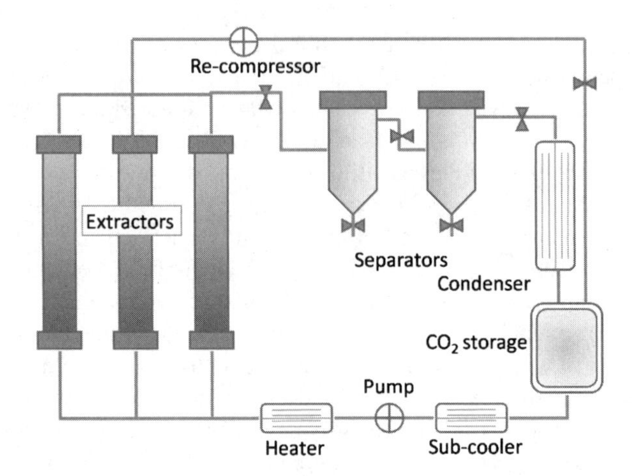

FIGURE 5.5 Typical supercritical CO_2-extraction equipment components showing optional recompression stage.

extractors which may be arranged either in parallel or in series. The extract is recovered in a series of separators that have sequentially lower pressures resulting in precipitation of the increasingly insoluble extracted components and the CO_2 is finally condensed and returned to the storage tank. However, CO_2 will be lost in this closed circuit when the extractors are emptied after decompression, although this CO_2 can be recompressed and returned to the storage tank. In cases when liquid CO_2 is used as the extraction solvent, separators are replaced with evaporators.

Solubility in supercritical CO_2 is essentially proportional to density [47,48] and this in turn is proportional to pressure and inversely proportional to temperature. Figure 5.6

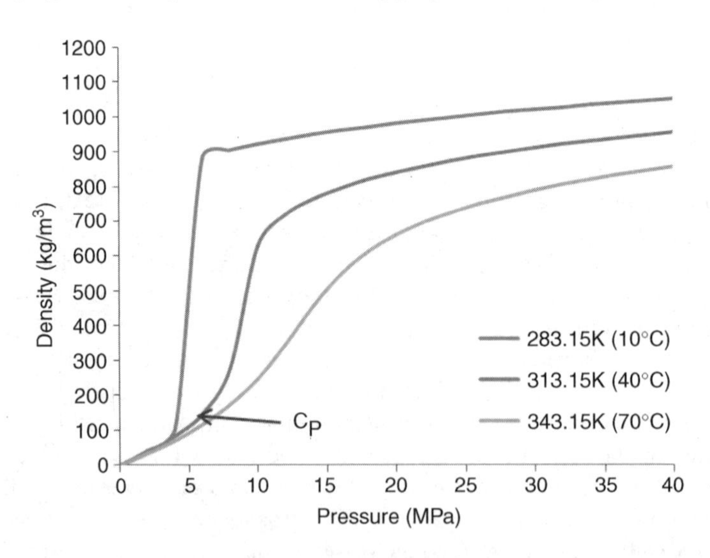

FIGURE 5.6 Supercritical CO_2 density as a function of pressure and temperature [49].

shows the relationship between density and pressure at selected temperatures and it can be seen that the density rises rapidly once past the critical point (C_p).

In addition to the temperature and pressure of supercritical CO_2, extraction efficiency is also directly influenced by other process parameters.

(1) Extraction time—equates to total kg CO_2/kg extract
(2) Extractor design
(3) Raw material pretreatment
  a. Particle size
  b. Particle porosity
  c. Bulk density

Extraction time is largely determined at the design stage of any equipment and the circulation pumps sized to meet a predetermined batch-extraction time. Extractor design is also optimized at the same time.

Extraction of molecules that have very limited solubility in supercritical CO_2 can be enhanced by the addition of entrainers. These are cosolvents added as a percentage of the supercritical CO_2 flow prior to the extractor to increase the solvating power of the extraction solvent. Ethanol is commonly used as a cosolvent and this can of course be obtained within the biorefinery.

5.5 EXTRACTION OF LIPIDS

The application of CO_2 in maximizing lipid extraction for biodiesel production either by replacing hexane, enhancing the mechanical pressing process, or for recovering lipids from unused parts of the feedstock, such as the bran, offers an opportunity to integrate the use of supercritical CO_2 into the biorefinery.

Supercritical CO_2 is often compared to hexane due to their similar polarity [14] and significant work has been carried out to transfer the extraction of lipids using hexane to $scCO_2$, so that this toxic and highly flammable solvent can be replaced. In the United States, most of the 8 million tons of oil isolated each year from seeds and grain commodities are extracted using hexane. Hexane had been recognized as a hazardous air pollutant by the US EPA and it has been reported by the EPA Toxic Release Inventory (TRI) that more than 20,000 tonnes of hexane are released to the atmosphere each year from the extraction of vegetable oils [50]. The US EPA also identified hexane as a potential carcinogen in the national emission standards for hazardous air pollutants [51]. Research also suggested that hexane is neurotoxic and can affect the nervous system by enhancing the development of peripheral neuropathy [52,53].

More than 30 million tonnes of soybeans are crushed and extracted annually for oil production in the United States [41,42]. Hexane is typically used for extraction of soybean oil resulting in a mixture of triglycerides, free fatty acids, phospholipids, pigments, and unsaponifiable components [41]. Although extraction with hexane has a lower unit cost than extraction with scCO_2, production costs increase as extra refining steps and energy input are needed to reduce solvent residues below legislated requirements and to recover the solvent. Due to its flammability, uncertain future availability, and the increase in petroleum costs, the drive for an alternative to organic solvents is high [54]. Extraction yields and properties of scCO_2 extracted oil (odor and color) are comparable to or better than hexane [41,54] and extraction of soybean triglycerides using scCO_2 has been demonstrated [55]. Other materials such as corn, wheat, germ, sunflower seeds, safflower seeds, and peanuts have also been successfully extracted using scCO_2 [55] but the adoption of this technology on a large scale has been slow largely due to the significant existing investment in hexane-extraction plant. More recently one of the biggest worldwide industrial scCO_2 plants was built in South Korea for the extraction of sesame oil with extractor volumes of 2 – 3800 L and operating pressure of up to 55 MPa [56,57].

Cold pressing is also a popular way of processing oil and the quality of cold pressed oil is considered to be higher than oil produced by solvent extraction [58,59]. Oil recovery from cold pressing is lower than hexane extraction and a second extraction of the pressed meal is often carried out using hexane as the solvent and there is no reason why this could not also be carried out using scCO_2. An intermediate solution is the use of Gas Assisted Mechanical Extraction (GAME) which uses an injection of liquid or supercritical CO_2 into a mechanical press to solubilize the oil more effectively. In a recent study [60] yields obtained with GAME were found to be 30% higher than conventional expression under the same conditions (40 °C, 10–30 MPa effective mechanical pressure and 10 MPa CO_2 for GAME experiments). The displacement of oil by dissolved CO_2 was identified as the major cause of increased oil yields in dehulled seeds and for hulled seeds the entrainment of oil by CO_2 during depressurization also played a significant role in the increase of the oil yields. This technology has now been developed commercially and offers a method of increasing seed oil yields without using solvents such as hexane.

5.6 EXTRACTION OF SECONDARY METABOLITES FROM LIGNO-CELLULOSIC FEEDSTOCKS

The range of potential extractives from ligno-cellulosic feedstocks is only limited by the availability of low-cost sources of biomass and as the production of ethanol from ligno-cellulosic sources becomes more economic a greater range of products will become available. At present cereal by-products, forestry residues, and sugarcane bagasse are the largest volumes available.

Plant leaf and some stem surfaces are coated with a thin layer of waxy material that has a myriad of functions. This layer is microcrystalline in structure and forms the outer boundary of the cuticle membrane; it is the interface between the plant and the atmosphere. It serves many purposes, for example, to limit the diffusion of water and solutes, while permitting a controlled release of volatiles that may deter pests or attract pollinating insects. The wax provides protection from disease and insects, and

TABLE 5.3 The Major Constituents of Plant Leaf Waxes

Group	Generic Formulae	Number of Carbons	Odd or Even
Hydrocarbons	$CH_3(CH_2)_nCH_3$	21–35	Odd
Wax esters	$CH_3(CH_2)_xCOO(CH_2)_yCH_3$	34–62	Even
Fatty acids	$CH_3(CH_2)_nCOOH$	16–32	Even
Primary alcohols	$CH_3(CH_2)_nCH_2OH$	22–32	Even
Aldehydes	$CH_3(CH_2)_nCHO$	22–32	Even
Ketones	$CH_3(CH_2)_xCO(CH_2)_yCH_3$	23–33	Odd
Secondary alcohols	$CH_3(CH_2)_xCHOH(CH_2)_yCH_3$	23–33	Odd
β-diketones	$CH_3(CH_2)_xCOCH_2CO(CH_2)_yCH_3$	27–33	Odd

helps the plants resist drought. As plants cover much of the earth's surface, it seems likely that plant waxes are the most abundant of all natural lipids. The range of lipid types in plant waxes is highly variable, both in nature and in composition. Table 5.3 illustrates some of this diversity in some of the main components.

In addition, there may be hydroxy-β-diketones, oxo-β-diketones, alkenes, branched alkanes, acids, esters, acetates and benzoates of aliphatic alcohols, methyl, phenylethyl and triterpenoid esters and acids, and many more. The amount of each lipid class and the nature and proportions of the various molecular species within each class vary greatly according to the plant species and the site of wax deposition (leaf, flower, fruit, etc.) and some data for well-studied species is listed in Table 5.4.

The composition of straw waxes has been extensively studied; however, most work has been carried on the *Triticeae* and in particular wheat (*Triticum aestivum*), barley (*Hordeum vulgare*), and rye (*Secale cereale*). The key groups of compounds found in the straw waxes are listed in Table 5.5. In most studies, the level of total "waxes" varied from 1% to 3% depending on the variety, straw parts used (leaves, node, or internode), crop year, and method of extraction. The composition of "waxes" from wheat straw was first described in 1969 [61] and fractionation of the wax carried out using classical separation methods. More recent studies [62,63] have identified a wider range of compounds but extraction was carried out with organic solvent mixtures. The "waxes" contain a wide range of compounds including alkanes, alkanols, fatty acids, sterols, triglycerides, and waxes and these have all

TABLE 5.4 Relative Proportions (wt%) of the Common Wax Constituents in Some Plant Species

	Grape Leaf	Rape Leaf	Apple Fruit	Rose Flower	Pea Leaf	Sugarcane
Hydrocarbons	2	33	20	58	40–50	2–8
Wax esters	6	16	18	11	5–10	6
Aldehydes	6	3	2	–	5	50
Ketones	–	20	3		–	–
Secondary alcohols	–	8	20	9	7	–
Primary alcohols	60	12	6	4	20	5–25
Fatty acids	8	8	20	5	6	3–8

TABLE 5.5 Key Groups of Compounds in Cereal Straw Waxes

Group	Basic Structure	Examples Found
Alkane	R	Heptacosane (C_{27}), Nonacosane (C_{29}), Hentriacontane (C_{31})
Fatty alcohol	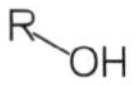 R–OH	Hexacosanol (C_{26}), Octacosanol (C_{28})
Fatty acid	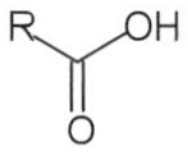 R, OH, O	Palmitic acid ($C_{16:0}$), Linoleic acid ($C_{18:2}$), Oleic acid ($C_{18:1}$), Stearic acid ($C_{18:0}$)
Aldehyde	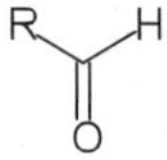 R, H, O	Octacosanal (C_{28})
Ketone	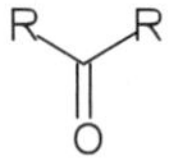 R, R, O	6,10,14-trimethyl 2-pentadecanone
β-diketone	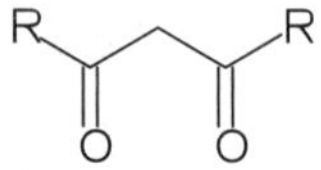 R, R, O, O	14,16-hentriacontanedione (C_{31}), 14,18-triatriacontanedione (C_{33})
Wax ester	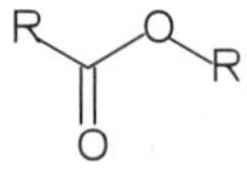 R, O, R, O	Hexacosanyl hexadecanoate (C_{26}:C_{16}), Octacosanyl hexadecanoate (C_{28}:C_{16}), Octacosanyl octadecanoate (C_{28}:C_{18})
Sterol	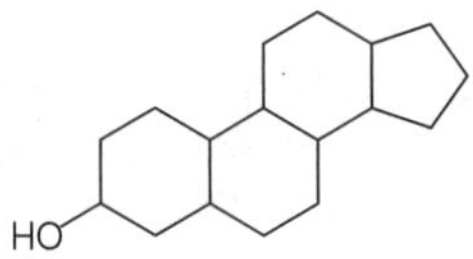 HO	Stigmasterol, Sitosterol

been characterized. The wax fraction from wheat straw has also been extracted using supercritical CO_2 [64] and this has demonstrated that the use of this solvent leads to a highly selective extraction of the lipid and wax fraction with minimal content of other compounds such as pigments. The complete breakdown of the composition in the supercritical CO_2 extract of wheat straw wax is shown in Figure 5.7.

The composition of barley straw wax has also been studied [66] and found to be very similar to that of wheat straw wax. These cuticular "waxes" protect the plant against microbial attack and have been found to exist in three distinct layers in barley

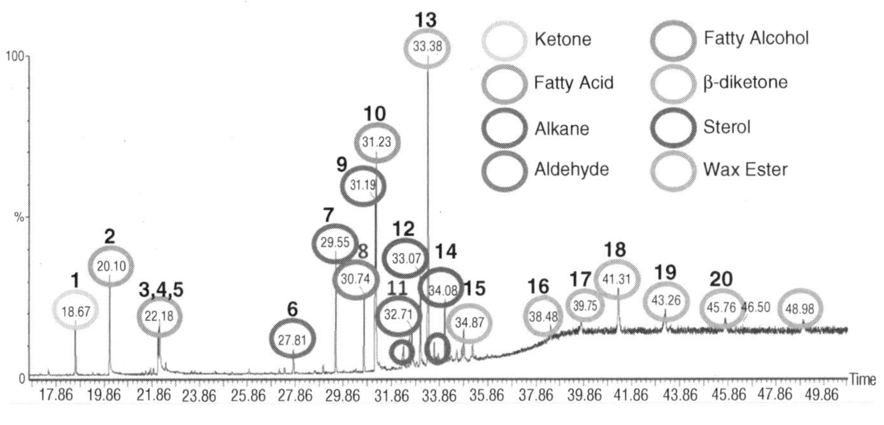

1) 6,10,14-trimethyl-2-pentadecanone
2) Palmitic acid (Hexadecanoic acid)
3) Linoleic acid (9,12-Octadienoic acid)
4) Oleic acid (9-octadecenoic acid)
5) Stearic acid (Octadecanoic acid)
6) Heptacosane
7) Nonacosane
8) Octacosanal
9) Hentriacontane
10) Octacosanol
11) Triatriacontane
12) Stigmasterol
13) 14,16-hentriacontanedione
14) Sitosterol
15) 16,18-triatriacontanedione
16) Tetracosanyl hexadecanoate
17) Hexacosanyl hexadecanoate
18) Octacosanyl hexadecanoate
19) Octacosanyl octadecanoate
20) Octacosanyl eicosanoate

FIGURE 5.7 Composition of wheat straw wax extracted with supercritical CO_2 [65].

straw [67] each having different degrees of order and composition. The composition of "waxes" from the cuticular wax layers of rye leaves has also been reported [68]. The total wax mixture from both sides of the leaves was found to contain primary alcohols (71%), alkyl esters (11%), aldehydes (5%), and small amounts (<3%) of alkanes, sterols, secondary alcohols, fatty acids, and unknowns. A homologous series of alkyl resorcinols was also identified.

Straw "waxes" from other plant families have also been examined. Flax (*Linum usitatissimum*) straw was first examined in 1931 [69] and a more detailed study has recently been reported [70]. However, there has been very little work carried out on oilseed rape (*Brassica napus*), or sunflower straw (*Helianthus annuus*) other than examination of the structural components. The total benzene/ethanol extractable components from rape straw have been reported [71] and this indicates a yield of 1–2%. Analysis of surface leaf waxes of oilseed rape [72] has indicated that they may be particularly rich in alkanes. Recent extraction trials [73] have indicated that yields from sunflower and rape straw are in the range 0.3–2% of the biomass depending on the extraction solvent used (Table 5.6).

Analysis of these extracts indicates that both contain complex mixtures of fatty acids, alkanes, and sterols with other minor components. The rape straw extract contained higher levels of odd-numbered alkanes than the sunflower extracts. The extraction yield and composition of the hexane extract is almost identical to that obtained when using supercritical CO_2 (30 MPa and 50 °C) and this is not

TABLE 5.6 Comparison of Crude Yields on Solvent and CO_2 Extracts

Raw Material/Solvent	Hexane	Ethanol	$scCO_2$
Sunflower straw	0.33%	1.99%	0.33%
Rape straw	0.71%	1.83%	–

surprising given that the polarity of both solvents is almost identical [74]. These yields provide only a "snapshot" of the potential yields as significant variation in both yield and composition can be expected between varieties, crop year, and harvest conditions.

It should be recognized that the extraction of these "waxes" is largely from the surface of the straw and as these are commercial crops there may be plant-protection residues present on the straw. Nonpolar residues would be soluble in supercritical CO_2 and given that the extract will represent a 30–50 fold concentration the levels might be significant. A survey of all possible residues is outside the scope of this study but should be considered in future work.

In addition to crop residues from cereal or oilseed crops, there are dedicated energy crops that are being developed either for direct cofiring or as a second-generation bioethanol feed stock and these are also amenable to supercritical CO_2 pretreatment to capture lipophilic molecules. Miscanthus, reed canary grass, and hemp are all being developed as potential energy crops and the lipophilic extractives of these have all been studied [75–79] and show a mixture of alkanes, fatty alcohols, fatty acids, aldehydes, and sterols that are common to all three but vary in relative concentration according to species. Giant reed (*Arundo donax*) has also been studied [78,80] and this shows a very different composition as being predominantly fatty acids and sterols with only low levels of alkanes, fatty alcohols, and aldehydes.

Hardwood and softwood species are not normally considered as feedstocks for biorefineries but development of more integrated processing in Sweden for example [81] has opened up the opportunity for the use of supercritical CO_2 to extract molecules from the bark and leaves of the woody raw material. Softwoods such as spruce and pine [82,83] have been extracted with CO_2 and show a diverse range of terpenoids, sterols, and phenolics. Hardwood species such as eucalyptus, which are now widely grown in South America also contain high levels of sterols and sterol glycosides, aliphatic alcohols, and aromatic compounds [84].

5.6.1 Potential Applications of Cuticle "Waxes"

Waxes are produced commercially in large amounts for use in cosmetics, lubricants, polishes, surface coatings, inks, and many other applications. Many of these are of mineral origin but four in particular are derived from plant sources: beeswax, carnauba wax, candelilla wax, and jojoba wax. In addition there are also other plant waxes that are commercially produced in smaller quantities such as citrus wax and apple wax. These are mostly used in cosmetic or personal care applications [85]. There is so far no commercial production of straw "waxes"; however, within the groups of molecules so far reported in wheat and barley straw, potential applications have been identified [86] for the, alkanes, alkanols, and sterol wax fraction.

Waxes are currently extracted with organic solvents and can be derived from plant or animal sources. Using supercritical CO_2 as the extraction solvent, extracts can be produced with no organic solvent residues and furthermore can be certified as organic in many countries [87]. These waxes can be used as replacements for a wide range of existing products, such as cosmetics, as their physical properties can be varied by selecting the temperature and pressure of the supercritical CO_2 used in the extraction [64]. Importantly, the waxes from straws have been shown to have a microcrystalline structure and this property is important for many cosmetic uses.

Most straw waxes appear to contain a range of straight-chain and branched alkanes and it has recently been shown [88] that similar odd-numbered alkanes are important semiochemical molecules. Ladybirds secrete these molecules as they inhabit plant surfaces and aphids are able to detect these and avoid these surfaces. Application of wheat wax extracts has been shown [89] to reduce aphid foraging on important food crops. Extraction of these molecules from straw offers a low-cost and sustainable source of these semiochemicals.

Polycosanols are used in the treatment of various chronic diseases such as diabetes and hypercholesterolelia and work by inhibiting the production of cholesterol in the liver. Recent work [90] has shown that the polycosanol content of wheat straw (164 mg kg^{-1}) and sugar cane leaves (181 mg kg^{-1}) are remarkably similar reinforcing the potential of wheat straw as an alternative source of this important group of molecules. Sterols are also used to reduce plasma cholesterol and LDL by interfering with the intestinal absorption of cholesterol originating in the diet. Combinations of polycosanols and sterols are claimed to act synergistically [91] and development of the use of these molecules from straw extracts could be delivered as either dietary supplements or incorporated into high-fat foods such oils or emulsified spreads.

5.6.2 Economic Considerations when Extracting with Supercritical CO_2

Almost all CO_2 -extraction plants are now individually designed to deliver optimum performance for their intended application with the lowest possible energy consumption. In designing an extraction plant that would be suitable for a biorefinery application a number of important considerations need to be taken into account.

(1) Although continuous extraction would offer significant cost savings no commercial plant exists for this process therefore a high-throughput batch process will need to be considered.

(2) Volumes of feedstocks will be high and the loading and unloading of the extraction vessels will be rate limiting.

(3) Extract yields will be relatively low ($<3\%$ of feed volume) so separator sizes can be scaled down.

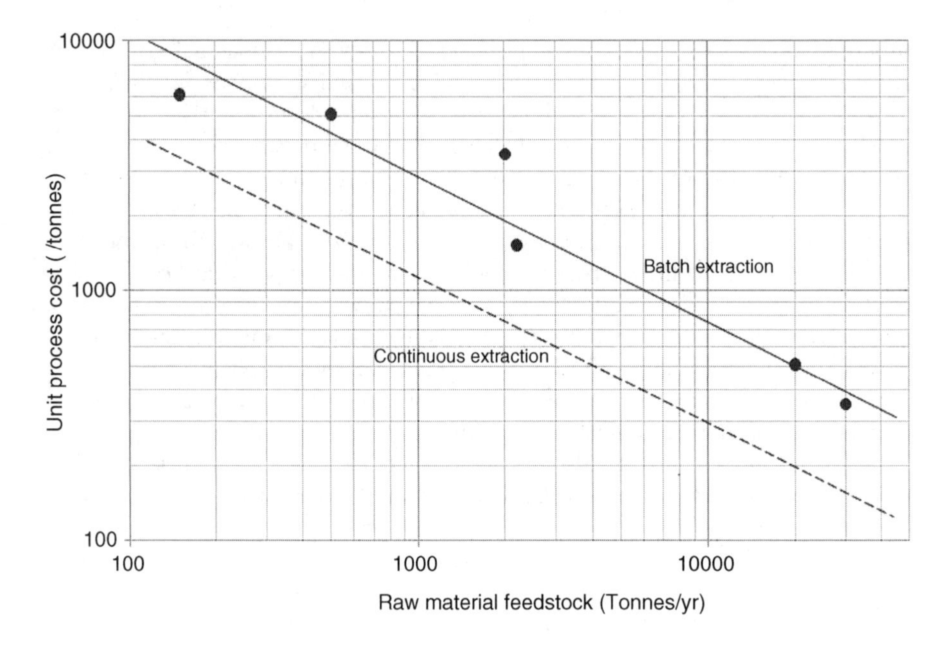

FIGURE 5.8 Extraction costs (€/ton) relative to throughput. (Adapted from reference [46].)

(4) Maximum load for each extraction needs to be achieved so bulk density of feedstock needs to be maximized. However this should not be so excessive that diffusion of the CO_2 is compromised.

(5) Post-extraction processing may need to be considered if the residue needs to be delivered in specific format, for example, cofiring. However, a pelleted form is often desirable to reduce dust.

The extraction of biomass prior to its use as a fuel could potentially require extraction plants of a much larger scale than has been built so far. One of the largest commercial CO_2-extraction plants for oil extraction has a extractor volume of $2 \times 3{,}8\,m^3$ operating at 550 bar and the price for such a plant is currently approximately €5 million without utilities or building. This would have a processing capacity is about 15 tonnes of seed per day. Larger plants for specific applications such as the decaffeination of coffee are in widespread use.

The operating costs of a supercritical CO_2-extraction plant are inversely proportional to the size of the plant. Figure 5.8 is a compilation of published data [46] reflecting the unit costs of batch and, in theory, continuous extraction processes; however, such cost estimates have an agreed variability of $\pm 30\%$.

Based on an extraction plant that would match the output of a typical straw pelleting plant (30,000 t $year^{-1}$ or 80 t day^{-1}) we would expect an extraction cost of approximately €350 t^{-1}. The lower line in Figure 5.8 represents continuous extraction and an anticipated threefold increase in productivity (mostly due to shorter use

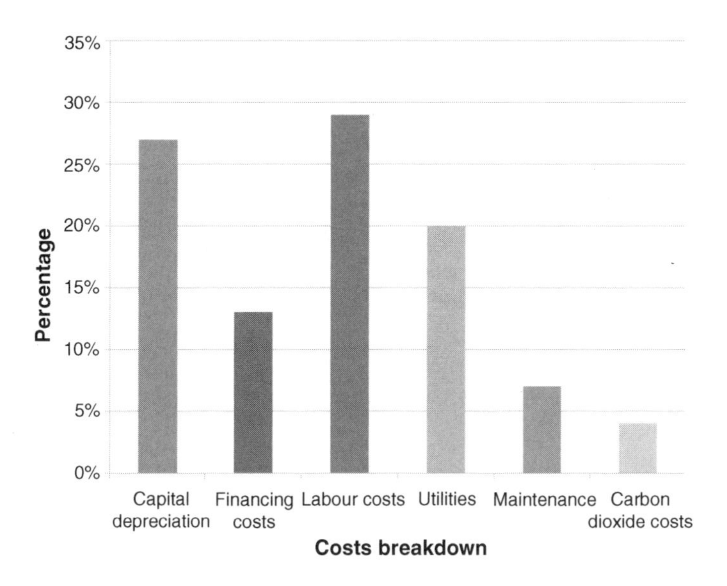

FIGURE 5.9 Typical operating costs for CO_2 extraction. (Adapted from reference [64].)

of high-pressure volume for extraction) reducing operating costs in the same proportion. However, continuous extraction has so far only been carried out on laboratory scale.

The unit cost of extraction includes all direct and indirect costs and the breakdown of these will again depend on the size of the extraction plant. Typical figures are provided by a number of extraction plant manufacturers as indicated in Figure 5.9 [56] and clearly show that the initial capital costs can present a significant barrier to adopting this technology and that labor and utilities are the largest contributors to operating costs.

5.6.3 Biomass Densification

The operating costs of extraction are influenced predominantly by plant size/capacity and by the bulk density of the material to be extracted. The impact of bulk density on daily capacity of medium-sized plants demonstrates the necessity for straw densification prior to extraction. For example, a plant with extraction capacity of 3 m^3 will have a daily capacity of 6300 kg if the feed material has a bulk density of 350 kg m^{-3} and 9900 kg day^{-1} if the bulk density is 550 kg m^{-3}. It is anticipated that straw pellet density will be in excess of 500 kg m^{-3} and at this density the pellets will still retain sufficient porosity (>300 cm^3 dm^{-3}) for efficient extraction.

Biomass pelletizing is a commercially established technology with many operational sites across Europe producing pellets for domestic and industrial boilers or for cofiring. There are two main types of pellet press: flat-die and ring-die pelletizers the latter being the most common. Straw is normally passed through a hammer mill to achieve the desired particle size before pelleting. The milled biomass is then fed to a pellet press where the combined action of die rotation and roller pressure forces material through the die to produce pellets that are normally 6–8 mm diameter and 15–20 mm long. The pellet length is controlled by adjustable knives. During the pelleting process, the temperature is raised to 60–120 °C depending on the pelleting pressure applied and this softens the lignin that acts to bind the straw fibers as the pellet is extruded [92]. The pellets are cooled immediately after pressing to harden them prior to conveying and packaging if appropriate. There is almost no mass loss during milling and pelleting; however, it is prudent to apply a 98% yield to this process.

Pelleting can be used to increase bulk density to a maximum of 700 kg m^{-3} but although this would maximize the extractor capacity it would be undesirable as the pellet porosity would be too low. A density of 500–550 kg m^{-3} would enable the pellets to be charged directly to the extractor without further milling and at this density the low viscosity of the $scCO_2$ would be able to penetrate the pellet. The pellets would remain intact during and after the extraction process thereby minimizing dust production during subsequent handling.

5.7 REACTIONS IN SUPERCRITICAL CO_2

Supercritical CO_2 offers an alternative nonpolar green solvent for carrying out reactions using heterogeneous catalysts, which can be transition-metal catalysts or biocatalysts. These conversions can be carried out as a separate off-line process or potentially included within the extraction system if reaction rates can be synchronized with extraction rates. There are many examples of successful catalysis in $scCO_2$ and several research groups have now demonstrated that supercritical CO_2, is a highly effective medium for continuous catalytic reactions [93], including alkylation [94], etherification [95], hydroformylation [96], oxidation [97], and particularly hydrogenation [98]. Some of these reactions have been successfully optimized and transferred to commercial production often producing products of sufficient purity to not require further downstream purification [98].

Hydrogenation of various aliphatic and aromatic substances had been extensively studied by Poliakoff et al [98,99]. A pilot plant for the Thomas Swan Company in Durham (UK) by Chemateur Engineering was constructed due to the success of the reaction in supercritical CO_2 [98]. Poliakoff et al. also investigated etherification with different combinations of aliphatic alcohols and showed high selectivity [98,100]. Friedel–Crafts alkylation was also demonstrated by Poliakoff et al. on anisole and *m*-cresol in different acid catalysts [101,102]. An example of hydroformylation in supercritical CO_2 lead to 85% aldehyde being formed with the reaction benefited from reduction of waste product and an improved selectivity [19].

One example of the application to biorefinery schemes has been demonstrated by the continuous conversion of levulinic acid to γ-valerolactone [103] using water as a cosolvent and a Ru–SiO_2 catalyst (Fig. 5.10). After optimization of the process, a yield of >99% was achieved at 10 MPa and 200 °C. Levulinic acid is considered one of the 12 most important platform molecules [104,105] and the cyclization to

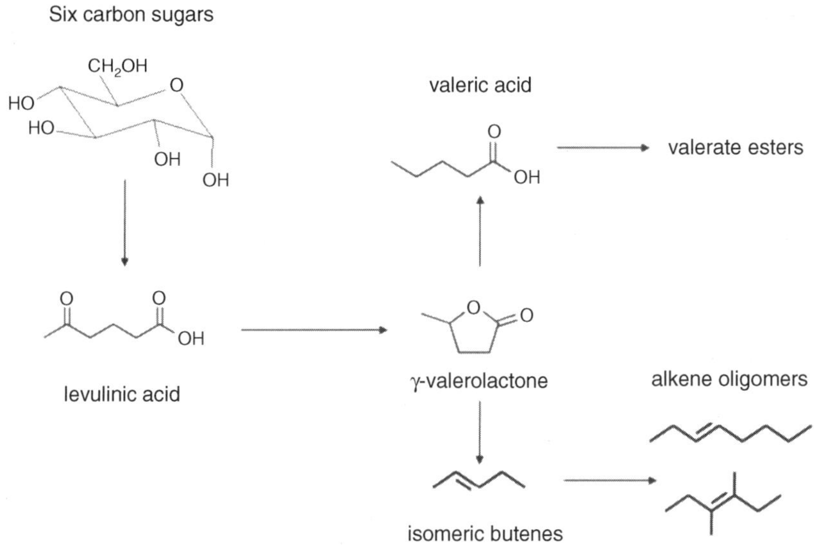

FIGURE 5.10 γ-Valerolactone as an intermediate for liquid transport fuels.

γ-valerolactone represents an important intermediate that can be converted to liquid alkenes of a molecular weight range suitable for liquid transport fuels as shown below [106].

Supercritical CO_2 also offers an alternative solvent for biocatalysis where reactions in nonaqueous conditions are required and the tunable solvent properties of $scCO_2$ can be exploited together with easy recovery of both product and catalyst free of solvent residues. The stability of enzymes in $scCO_2$ and reaction rates are similar to those observed in organic solvents such as *n*-hexane and cyclohexane [107,108]. Supercritical fluids were first used as a medium for enzyme reactions in 1985 and many examples of reactions in this solvent have been described since [109]. The temperature range used to establish supercritical CO_2 conditions is similar to the optimum temperature for many enzymes and with high diffusivity and low viscosity compared to liquid solvents mass-transfer rates of substrates to active sites of enzymes is high. However, $scCO_2$ as a solvent has one major limitation; it is highly nonpolar and as such the solubility of biomolecules in the solvent is limited to predominantly small hydrophobic molecules. To increase the polarity of the media, cosolvents and surfactants have been applied in order to dissolve both hydrophilic and hydrophobic molecules.

The limited water availability in $scCO_2$ reactions reduces denaturation of the enzymes; however, variation in operating conditions (temperature, pressure, and pH) can lead to conformational changes in the enzymatic structure leading to deactivation or reduced reaction rates [110–117]. The effect of compression and decompression of CO_2 may also affect the long-term stability of the enzyme. Kasche et al. found that α-chymotrypsin and trypsin were partially denatured by the decompression step of $scCO_2$ [118]. Immobilized enzymes have been shown to be more resistant to changes in the physical state of the $scCO_2$; the degree of increased stability depending on which physical or chemical immobilization method has been used and the structure of the enzyme itself [117].

In terms of process optimization, increased pressure, for example, from 8 to 10 MPa, increases the solubility of the substrate and substrate aggregation resulting in an overall increase in the rate of reaction [85]. However, these observations are specific to certain reactions and it has been reported that the rate of esterification of stearic acid increases when the pressure increases but the hydrolysis rate of ethyl stearate is optimum near the critical point [95]. In another example, a lipase-catalyzed acylation of methyl 6-*O*-trityl β-D-glucopyranose in subcritical CO_2 decreased with the increase of pressure; however, the rate was observed to increase when approaching the critical point [119]. A number of studies have also demonstrated that optimizing temperature and pressure of the $scCO_2$ enhances stereoselectivity through changes in the active site configuration and in many cases the optimum selectivity reached a peak before declining at higher pressures [120–123].

Significant research has demonstrated that $scCO_2$ is a promising alternative solvent for catalysis as part of a biorefinery scheme, but so far there are no industrial-scale examples of this technology. It is most likely that the development of this as part of an integrated biorefinery will be advanced when applications for simultaneous extraction and reactions are identified that result in a commercially viable route to high-value molecules. An example of this would be the extraction of sterols and subsequent esterification [124,125] of these using a lipase preparation.

5.8 CONCLUSIONS AND PROSPECTIVES

Supercritical CO_2 has been shown to be a suitable solvent for the extraction of useful molecules from biomass and as an extraction technique is already commercially accepted. Expanding its use into integrated biorefineries as a pretreatment step requires the development of large-scale facilities that are currently difficult to operate because of the restraints of batch processing. A key advance in the use of supercritical CO_2 would be the development of a large-scale continuous extractor that could be integrated into existing biorefinery infrastructure. Extending the use of supercritical CO_2 as a reaction solvent to produce valuable molecules from renewable feedstocks is closer to commercial reality as continuous reactions have already been demonstrated using both transition-metal catalysts and biocatalysts.

REFERENCES

1. B. Kamm and M. Kamm, *Appl. Microbiol. Biotechnol.*, **2004**, 64, 137–145.
2. B. Kamm and M. Kamm, *Chemie Ingenieur Technik*, **2007**, 79, 592–603.
3. B. Kamm and M. Kamm, *Nachrichten Aus Chemie Technik Und Laboratorium*, **1998**, 46, 342–343.

4. J. Rohn, *Chem. Ind.-London*, **2006**, 11–11.
5. European Parliament and Council, Towards a European strategy for the security of energy supply, **2002**, (Green Paper KOM2002/321) European Parliament, Strasbourg.
6. http://www.epa.gov/otaq/fuels/renewablefuels/420f10056.htm Last accessed: 16th May **2011**.
7. B. Kamm, P. R. Gruber and M. Kamm, (Eds.), *Biorefineries—Industrial Processes and Products Volume 1* **2006**, Wiley VCH, Weinheim, Germany, 3–40.
8. B. Metz, O. Davidson, H. de Coninck, M. Loos and L. Meyer (Eds.), IPCC Special Report, **2005**, 280–318.
9. M. M. Halmann, *Chemical Fixation of Carbon Dioxide: Methods for Recycling CO_2 into Useful Products*, CRC Press, Boca Raton, FL, **1993**.
10. A. J. Traynor and R. J. Jensen, *Ind. Eng. Chem. Res.*, **2002**, 41, 1935–1939.
11. (a) F. M. Kerton, *Alternative Solvents for Green Chemistry;* RSC Publishing, Cambridge, **2009**, 60–70. (b) P. G Jessop and W. Leitner, *Chemical Synthesis Using Supercritical Fluids;* Wiley-VCH, Weinheim, **1999**, 1–5.
12. (a) A. A. Clifford, *Fundamentals of Supercritical Fluids;* Oxford University Press: Oxford, **1998**, 1–5. (b) Y. Arai, T. Sako and Y. Takebayashi, *Supercritical Fluids: Molecular Interactions, Physical Properties, and New Applications;* Springer, New York, **2002**, 1–4. (c) H. Brogle, *Chem. Ind-London.*, **1982**, 12, 385–390.
13. P. G. Jessop, Y. Hsiao, T. Ikariya and R. Noyori, *J. Am. Chem. Soc.*, **1996**, 118, 344.
14. Y. Ikushima, N. Saito, M. Arai and K. Arai, *Bull. Chem. Soc. Jpn*, **1991**, 64, 2224–2229.
15. M. C. R. Symons, *Chem. Soc. Rev.* **1983**, 12, 12.
16. C. N. R. Rao, S. Singh and V. P. Senthilnathan, *Chem. Soc. Rev.* **1976**, 5, 297.
17. F. M. Orr, J. P. Heller, J. T. Taber and R. J. Card, *ChemTech.*, **1983**, 13(8), 482.
18. J. H. Clark and S. J. Tavener, *Org. Process Res. Dev.*, **2007**, 11, 149–155.
19. E. J. Beckham, *J. Supercrit. Fluid.*, **2004**, 28, 121–191.
20. E. R. Birnbaum, R. M. Le Lacheur, A. Horton and W. Tumas, *J. Mol. Cat. A: Chem.*, **1999**, 139, 11.
21. F. Loeker and W. Leitner, *Chem. Eur. J.*, **2000**, 6, 2011.
22. R. Wiebe and V. L. Gaddy, *J. Am. Chem. Soc.*, **1940**, 62, 815–817.
23. K. N. West, C. Wheeler, J. P. McCarney, K. N. Griffith, D. Bush, C. L. Liotta and C. A. Eckert, *J. Phys. Chem. A*, **2001**, 105, 3947.
24. E. Stahl, E. Schuetz and H. K. Mangold, *J. Agric Food.*, **1980**, 28(6), 1153–1157.
25. T. Tsuda, K. Mizuno, K. Ohshima, S. Kawakishi and T. Osawa, *J. Agric. Food. Chem.*, **1995**, 43(11), 2803–2806.
26. E. Vági, B. Simándi, Á. Suhajda and É. Héthelyi, *Food Res. Int.*, **2005**, 28(1), 51–57.
27. B. Simándi, A. Deák, E. Rónyai and G. Yanxiang, *J. Agric. Food. Chem.*, **1999**, 47(4) 1635–1640.
28. A. Demirbaç, *Energ. Source.*, Part A, **2009**, 31(2), 163–168.
29. M. Aresta, A. Dibenedetto, M. Carone, T. Colonna and C. Fragale, *Env. Chem. Lett.*, **2005**, 3(3), 136–139.
30. K. Deline, D.L Claycamp, D. Fetherston and R.T. Marentis. World Patent No. 2008/020865, 2008.
31. F. E. I. Deswarte, J. H. Clark, J.J.E. Hardy and P. M. Rose, *Green Chem.*, **2006**, 8, 39–42.

32. M. G. Sajilata, M. V. Bule, P. Chavan, R. S. Singhal and M. Y. Kamat, *Sep. Purif. Technol.*, **2010**, 71, 173–177.
33. K. Ramalakshmi and B. Raghavan, *Crit. Rev. Food. Sci.*, **1999**, 39(5), 441–456.
34. C. J. Chang, K.-L. Chiu, Y.-L. Chen and C.-Y. Chang, *Food Chem.*, **2000**, 68, 109–113.
35. Z. Zekovic, I. Pfaf-Sovljanski and O. Grujic, *J. Serb. Chem. Soc.*, **2007**, 72, 81–87.
36. R. Vollbrecht, *Chem. Ind.-London*, **1982**, 19, 397–399.
37. G. Leeke, F. Gaspar and R. Santos, *Ind. Eng. Chem. Res.*, **2002**, 41, 2033–2039.
38. D. Ehlers, E. Czech, K. W. Quirin and R. Weber, *Phytochem. Analysis*, **2006**, 17, 114–120.
39. J. RincóN, A. D. Lucas, M. A. García, A. García, A. Alvarez and A. Carnicer, *Separ. Sci. Technol.*, **1998**, 33, 411–423.
40. W. Huang, Z. S. Li, H. Niu, J. W. Wang and Y. Qin, *Biochem. Eng. J.*, **2008**, 42, 92–96.
41. J. P. Friedrich and G. R. List, *J. Agric. Food Chem.*, **1982**, 30, 192–193.
42. J. P. Friedrich, G. R. List and A. J. Heakin, *J. Am. Oil Chem. Soc.*, **1982**, 59, 288–292.
43. N. A. N. Norulaini, W. B. Setianto, I. S. M. Zaidul, A. H. Nawi, C. Y. M. Azizi and A. K. M. Omar, *Food Chem.*, **2009**, 116, 193–197.
44. W. A. Den Otter, *Tekstil*, **2007**, 56, 514–515.
45. M. J. E. van Roosmalen, G. F. Woerlee and G. J. Witkamp, *J. Supercrit. Fluids*, **2003**, 27, 337–344.
46. G. Brunner, *J. Food Eng.*, **2005**, 67, 21–33.
47. J. Chrastil, *J. Phys. Chem.*, **1982**, 86(15), 3016–3021.
48. R. Hartono, G. A. Mansoori and A. Suwono, *Sep. Sci. Technol.*, **2001**, 56, 6949–6958.
49. Nist web book http://webbook.nist.gov/chemistry/Last accessed: 16th May **2011**.
50. J. M. DeSimone, *Science*, **2002**, 297, 799–803.
51. US Environmental Protection Agency , *National Emission Standards for Hazardous Air Pollutants: Solvent Extraction for Vegetable Oil Production, Fed. Reg.*, **2001**, 66, 71.
52. H. H. Schumburg and P. S. Spencer, *Brain*, **1976**, 99, 183–192.
53. P. S. Spencer and H. H. Schumburg, *Proc. R. Soc. Med.*, **1977**, 70(1), 37–38.
54. A. P. Gandhi, K. C. Joshi, K. Jha, V. S. Parihar, D. C. Srivastav, P. Raghunadh, J. Kawalkar, S. K. Jain and R. N. Tripathi, *Int. J. Food Sci. Technol.*, **2003**, 38, 369–375.
55. M. McHugh and V. Krukonis, *Biotechnology and Food Process Engineering*; Basic Symposium Series, **1990**, 203–212.
56. Natex Prozesstechnologie http://www.natex.at Cited: 6th November **2010**.
57. O. Fragner and H. Seidlitz, Personal communication December **2008**.
58. R. Usuki, T. Suzuli, Y. Endo and T. Kaneda, *J. Am. Oil Chem. Soc.*, **1984**, 61, 785.
59. C. Franzke, K. S. Grunert and H. Kroshel, *Nahrung*, **1972**, 16, 859.
60. P. Willems, N. J. M. Kuipers and A. B. de Haan, *J. Supercrit. Fluids*, **2008**, 45, 298–305.
61. A. P. Tulloch and R. O. Weenink, *Can. J. Chem.*, **1969**, 47, 3119.
62. R. C. Sun and J. Tompkinson, *J. Wood. Sci.*, **2003**, 49(1), 1611–4663.
63. R. C. Sun and J. Tompkinson, *Cell. Chem. Technol.*, **2001**, 35(5), 471–485.
64. F. E. I. Deswarte, J. H. Clark, J. J. E. Hardy and P. M. Rose, *Green Chem.*, **2006**, 8, 39–42.
65. E. H. K. Sin, original work, University of York, **2010**.
66. R. Sun and X-F. Sun, *Chem. Eng. Sci.*, **2001**, 36(13), 3027–3048.

67. S. K. Wiśniewska, J. Nalaskowski, E. Witka-Jeżewska, J. Hupka and J. D. Miller, *Coll. Surf. B*, **2003**, 29(2), 131–142.
68. J. Xiufeng and J. Reinhard, *Phytochemistry*, **2008**, 69, 1197–1207.
69. W. H. Gibson, *Chem. Eng. Res. Des.*, **1931**, 9a, 30–35.
70. A. P. Tulloch and L. L. Hoffman, *J. Am. Oil Chem. Soc.*, **1977**, 54(12), 587–588.
71. F. Karaoosmanogalu, E. Tetick, B. Guerboy and A. Seanli, *Energ. Source.*, **1999**, 21, 801–810.
72. http://www.lipidlibrary.co.uk/Lipids/waxes Cited: December 18th **2008**.
73. E. H. K. Sin, original work, University of York, **2010**.
74. Y. Ikushima, N. Saito, M. Arai and K. Arai, *Bull. Chem. Soc. Jpn.*, **1991**, 64(7), 2224–2229.
75. J. J. Villaverde, R. M. A. Domingues, C. S. R. Freire, A. J. D. Silvestre, C. P. Neto, P. Ligero, and A. Veja, *J. Agric. Food Chem.*, **2009**, 57, 3626–3631.
76. D. A. Boadi, S. A. N. Moshtaghi, K. M. Wttenberg and W. P. McCaughey, *Can. J. Anim. Sci.*, **2002**, 82, 465–469.
77. J. J. Majnarich, US patent 3420935, 1969.
78. D. Coelho, G. Marques, A. Gutierrez, A. J. D. Silvestre and J. C. del Rio, *Ind. Crops Prod.*, **2007**, 26, 229–236.
79. A. Gutiérrez, I. M. Rodriaguez and J. C. del Rio., *J. Agric. Food Chem.*, **2006**, 54, 2138–2144.
80. G. Marques, J. Rencoret, A. Gutiérrez and J. C. del Río., *Open Agric. J.*, **2010**, 3, 1–9.
81. www.processum.se Last accessed: 16th May **2011**.
82. O. Yesil-Celiktas, F. Otto and H. Parlar., *Eur. Food Res. Technol.*, **2009**, 229, 671–677.
83. M. E. M. Braga R. M. S. Santos, I. J. Seabra, R. Facanali, M. O. M. Marques and H. C. de Sousa, *J. Supercrit. Fluid.*, **2008**, 47, 37–48.
84. C. S. R. Freire, P. C. R. Pinto, A. S. Santiago, A. J. D. Silvestre, D. V. Evtuguin and C. P. Neto, *BioResources*, **2006**, 1(1), 3–17.
85. E. Flemming and U. Hehner, US patent 5885561, 1999.
86. F. E. I. Deswarte, J. H. Clark, A. J. Wilson, J. J. E. Hardy, R. Marriott, S. P. Chahal, C. Jackson, G. Heslop, M. Birkett, T. J. Bruce and G. Whiteley, *Biofuels, Bioprod. Bioref.*, **2007**, 1, 245–257.
87. Soil Association Certification—clause 8.03.07a
88. Y. Nakashima, M. A. Birkett, B. J. Pye, J. A. Pickett and W. Powell, *J. Chem. Ecol.*, **2004**, 30, 1103–1116.
89. G. Powell, J. Hardie and J. A. Pickett, *Entomol. Exp. Appl.*, **1997**, 84, 189–193.
90. S. Irmak and N. T. Dunford, *J. Agric. Food Chem.*, **2005**, 53, 5583–5586.
91. C. K. Dartey, European Patent 1108364, 2001.
92. G. Evans, Techno-Economic Assessment of Biomass "Densification" Technologies NFCC Project 08-015, **2008**, 38.
93. E. J. Beckman, *J. Supercrit. Fluids*, **2004**, 28, 121–191.
94. R. Amandi, J. R. Hyde, S. K. Ross, T. J. Lotz and M. Poliakoff, *Green Chem.*, **2005**, 7, 288–293.
95. R. P. J. Bronger, J. P. Bermon, J. N. H. Reek, P. C. J. Kamer, P. Van Leeuwen, D. N. Carter, P. Licence and M. Poliakoff, *J. Mol. Catal. A: Chem.*, **2004**, 224, 145–152.

96. S. Dharmidhikari and M. A. Abraham, *J. Supercrit. Fluids*, **2000**, 18, 1–10.
97. G. Jenzer, M. S. Schneider, R. Wandeler, T. Mallat and A. Baiker, *J. Catal.*, **2001**, 199, 141–148.
98. P. Licence, J. Ke, M. Sokolova, S. K. Ross and M. Poliakoff, *Green Chem.*, **2003**, 5, 99–104.
99. M. Poliakoff, T. M. Swan, T. Tache, M. G. Hitzler, S. K. Ross and S. Wieland, US Patent No. 6, 126933, 2000.
100. W. K. Gray, F. R. Smail, M. G. Hitzler, S. K. Ross and M. Poliakoff, *J. Am. Chem. Soc.*, **1999**, 121, 10711–10718.
101. R. Amandi, P. Licence, S. K. Ross, O. Aaltonen and M. Poliakoff, *Org. Process Res. Dev.*, **2005**, 9(4), 451–456.
102. R. Amandi, J. R. Hude, S. K. Ross, T. J. Lotz and M. Poliakoff, *Green Chem.*, **2005**, 7, 288–295.
103. R. A. Bourne, J. G. Stevens, J. Ke and M. Poliakoff., *Chem. Commun.*, **2007**, 4632–4634.
104. P. Gallezot, *Catal. Today*, **2007**, 121, 76–91.
105. J. Q. Bond, D. M. Alonso, D. Wang, R. M. West and J. A. Dumesic, *Science*, **2010**, 327, 1110.
106. J. J. Bozell., *Science*, **2010**, 329, 522–523.
107. D. A. Miller, H. M. Blanch and J. M. Prausnitz, *Ind. Eng. Chem. Res.*, **1991**, 30, 939.
108. E. Castillo, A. Marty, D. Combes and J. S. Condoret, *Biotechnol. Lett.*, **1994**, 16, 967.
109. T. W. Randolph, H. W. Blanch, J. M. Prausnitz and C. R. Wilke, *Biotechnol. Lett.*, **1985**, 7, 325.
110. M. Habulin and Z. Knez, *J. Chem. Technol. Biotechnol.*, **2001**, 76, 1260.
111. M. Habulin, M. Primozic and Z. Knez, *Acta Chim. Slov.*, **2007**, 54, 667.
112. H. R. Hobbs and N. R. Thomas, *Chem. Rev.*, **2007**, 107, 2786.
113. S. Kamat, G. Critchley, E. J. Beckman and A. J. Russell, *Biotechnol. Bioeng.*, **1995**, 46, 610.
114. M. Habulin and Z. Knez, *Acta. Chim. Slov.*, **2001**, 48, 521.
115. S. V. Kamat, E. J. Beckman and A. Russell, *J. Crit. Rev. Biotechnol.*, **1995**, 15, 41.
116. S. V. Kamat, B. Iwaskewycz, E. J. Beckman and A. J. Russell, *Proc. Natl. Acad. Sci. USA*, **1993**, 90, 2940.
117. E. Knez, *J. Supercrit. Fluid.*, **2009**, 47, 357.
118. A. Overmeyer, S. Schrader-Lippelt, V. Kasche and G. Brunner, *Biotechnol. Lett.*, **1999**, 21, 65.
119. C. Palocci, M. Falconi, L. Chronopoulou and E. Cernia, *J. Supercrit. Fluid.*, **2008**, 45, 88.
120. Y. Ikushima, N. Saito, M. Arai and H. W. Blanch, *J. Phys. Chem.*, **1995**, 99, 8941.
121. Y. Ikushima, *Adv. Coll. Interf. Sci.*, **1997**, 259, 71.
122. N. Mase, T. Sako, Y. Horikawa and K. Takabe, *Tetrahedron Lett.*, **2003**, 44, 5175.
123. M. Albrycht, P. Kiełbasiński, J. Drabowicz, M. Mikołajczyk, T. Matsuda, T. Harada and K. Nakamura, *Tetrahedron: Asymmetr.*, **2005**, 16, 2015.
124. J. W. King, J. M. Snyder, B. Frykman and A. Neese, *Eur. Food Res. Technol.*, **2001**, 212, 566–569.
125. N. Weber, P. Weitkamp and K. D. Mukherjee, *Food Res. Int.*, **2002**, 35, 177–181.

CHAPTER 6

Dissolution and Application of Cellulose in NaOH/Urea Aqueous Solution

XIAOPENG XIONG and JIANGJIANG DUAN

6.1 INTRODUCTION

In the 21st century, the wide utilization of renewable resources to produce environment-friendly materials, which avoid using or producing any hazardous substances, has become a favorable international frontier. Cellulose, a fascinating and sustainable natural biopolymer from photosynthesis, is the most abundant, and can be the inexhaustible polymeric raw material [1,2].

The industrial-scale chemical processing of cellulose started in 1870, which is the well-known production of celluloid. Though great progress has been made, the full potential of cellulose is far from being achieved. Because of the stiff molecular nature and the dense packing of numerous intra- or intermolecular hydrogen bonds, cellulose is nonthermoplastic and insoluble in many common solvents. Therefore, great energy has been focused on seeking efficient solvents of cellulose in order to process the fascinating material. Direct dissolution of cellulose was first discovered by Schweitzer [3] who found that cellulose solution could be prepared in a mixture of copper (II) hydroxide and aqueous ammonia, in which tetraamminecopper (II) hydroxide (cuprammonium hydroxide) $[Cu(NH_3)_4](OH)_2$ was formed. An indirect way of chemically processing cellulose is to obtain cellulose derivatives first, [4–6] which are then made soluble to be handled in large-scale industrial production. For example, the present industrial production of regenerated cellulose fiber is based on the viscose process, in which cellulose is transferred into cellulose xanthogenate to be soluble in alkali medium [4,5]. Another important cellulose derivative for large-scale fiber production is cellulose carbamate, which is soluble in sodium

The Role of Green Chemistry in Biomass Processing and Conversion, First Edition.
Edited by Haibo Xie and Nicholas Gathergood.

hydroxide [6]. However, those methods of dissolving cellulose pose environmental hazards because of the use of CS_2 and heavy-metal compounds.

So far, many other solvents for direct dissolution of cellulose have been developed which include salt complexes, [7–9] aqueous solvents, [10–16] and organic systems [17–20]. Among the many solvents, only the *N*-methyl-morpholine-*N*-oxide (NMMO) solvent was commercialized by Courtaulds and Lenzing to dissolve cellulose to prepare regenerate cellulose fibers and films, which is known as the Lyocell process [17]. Recently, room temperature ionic liquids (ILs) [21,22] have been synthesized to directly dissolve cellulose. It has been claimed that solvents such as NMMO and ILs could be efficiently recovered, although with high cost, which is a major breakthrough in cellulose processing.

Nevertheless, increasing attention has been focused on a new solvent system of NaOH/urea aqueous solution due to its rapid, relatively simple, low cost, and environment-friendly process of cellulose dissolution [14,15,23–29]. This novel method uses reagents of NaOH and urea, both of which are inexpensive ($\sim$310 US dollar ton^{-1}) [30]. It is worth noting that urea is normally produced by consuming CO_2 [$CO_2 + 2NH_3 = CO(NH_2)_2 + H_2O$], and NaOH is mainly produced by electrolyzing aqueous solution of NaCl (meaning sustainable resources of the reagents). The dissolution of cellulose by using the NaOH/urea aqueous solution takes place at low temperature ($\sim -12.6\,°C$), so that there is no evaporation of any chemical reagents during dissolution as well as during post-treatment of the obtained cellulose solution from like regeneration. Moreover, the coagulation bath after the regeneration of cellulose solution can be neutralized to have the by-products of the technology, which are mainly Na_2SO_4 and urea. It has been suggested that the Na_2SO_4 and the urea in the coagulation bath can be easily separated by flash evaporation, and then crystallization based on the large difference in the solubility between Na_2SO_4 and urea [28,29]. The recovery of Na_2SO_4 is almost 100% and that of urea is over 90%, and they can be reutilized [28,29]. Compared to the currently dominant viscose process, the novel technology produces no toxic component during either the dissolution or the regeneration processes. More importantly, it is seen that both the dissolution and the regeneration of cellulose are carried out in aqueous media, displaying the characteristics of an environment-friendly process.

At the same time, this novel technology for the first time breaks the limitation of traditional costly heat-treatment methods to dissolve cellulose, and to create novel dissolution of cellulose at low temperature. Therefore, this low-cost, nonpolluting, and nonecotoxic pathway is a real "green" technology. A series of cellulose fibers, films, gels, and other functional materials were produced at the laboratory or mid-sized manufacturing scale based on the dissolution of cellulose in the novel solvent systems at low temperature. The correlation between the structure and the properties of the regenerate cellulose materials was intensively investigated and "unraveled." Based on the continuous contribution and the outstanding achievements in this field, Professor Lina Zhang, the inventor of the novel NaOH/urea solvent for cellulose, was presented the *Anselme Payen Award* by the American Chemical Society's Cellulose and Renewable Materials Division in 2011 [31].

6.2 DISSOLUTION OF CELLULOSE IN NaOH/UREA AQUEOUS SOLUTION AT LOW TEMPERATURE AND MECHANISM

At the beginning of the 21st century, Professor Lina Zhang and coworkers [14–16] discovered that cellulose could be dissolved in a series of aqueous-based new solvent systems such as NaOH/urea aqueous solution. At that time, cellulose was soaked in NaOH and urea mixture aqueous solution, and then the slurry was cooled down to be frozen for over 8 h. After being kept at ambient conditions, the frozen slurry thawed with the temperature increase. Then, the thawed slurry was stirred vigorously to obtain a transparent solution. More recently, [23–29] they cooled down the NaOH and urea mixture aqueous solution to $-12.6\,^{\circ}$C, then cellulose was added with rapid stirring. Within 2 minutes, cellulose with weight-average molecular weight (M_{w}) lower than 1.2×10^{5} g mol^{-1} could be dissolved on a typical laboratory scale. By the application of this method, cellulose solution at a 1000 L scale has been prepared in a tank within 5 min to meet midsized industrial manufacturing requirements [29]. In our experience, dissolution occurs rapidly on this scale and only several seconds may have been required, which would represent the most rapid dissolution of cellulose in history.

The dissolution of cellulose in the precooled NaOH and urea mixture aqueous solution is illustrated in Figure 6.1 [23–29]. The optimal concentration of NaOH in

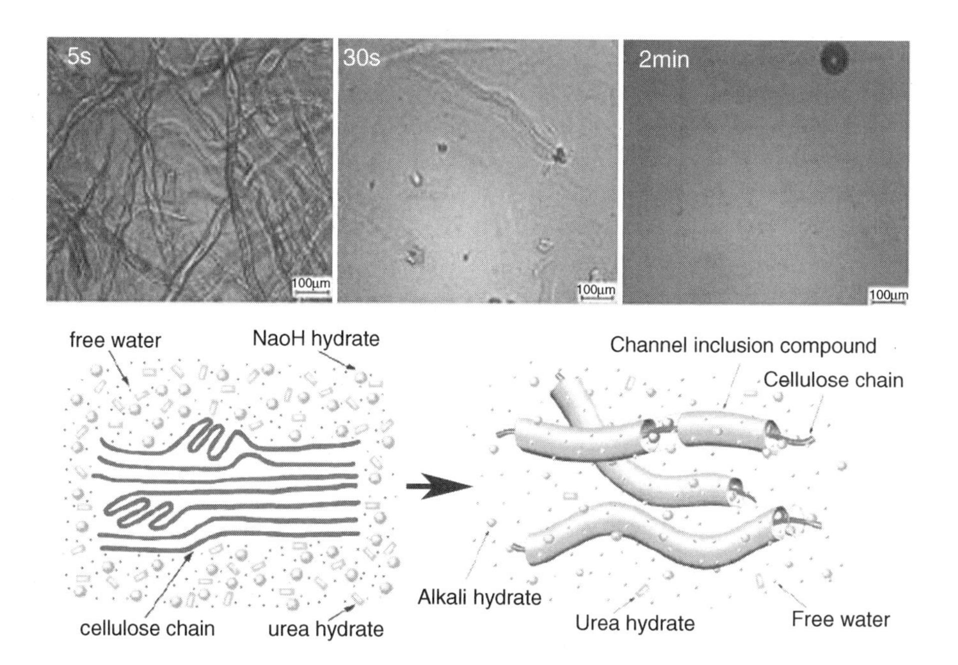

FIGURE 6.1 The OM images of cellulose in the NaOH/urea aqueous solution precooled to $-12\,^{\circ}$C (top) and the schematic diagrams of the dissolution mechanism at low temperature (bottom). (Reproduced from Refs. [23,25] with permission from Wiley © 2005.)

the mixture aqueous solution is found to be 7 wt%, and that of urea is 12 wt%. The NaOH and urea mixture aqueous solution is normally required to be below −12 °C in order to have satisfactory dissolution ability. After the cellulose sample was added into the precooled NaOH and urea mixture aqueous solution with stirring, the slurry was observed by optical microscope (OM). The OM images show that the cellulose fibers were swollen after 5 s, and the solid part was rare after 20 s, and homogeneous transparent solution was obtained within 2 min [23–29].

It has been suggested that alkali hydrates, urea hydrates, and free water were formed when the solvent was cooled [23,32,33]. Figure 6.2 shows the differential scanning calorimetry (DSC) thermograms of urea, NaOH, urea/NaOH, and 4% cellulose in urea/NaOH aqueous solutions [23]. The peaks at −10.5 °C (Fig. 6.2a) and −33.8 °C (Fig. 6.2b) are assigned to the melting of urea hydrates and NaOH hydrates, respectively. There are three peaks in the thermograms of NaOH/urea solution and the cellulose solution (Fig. 6.2c and d), corresponding to the melting of free water, urea hydrates, and NaOH hydrates, respectively. However, it is found that the melting temperatures of free water and urea hydrates in cellulose solution increase slightly when compared with those of NaOH/urea solution, while the

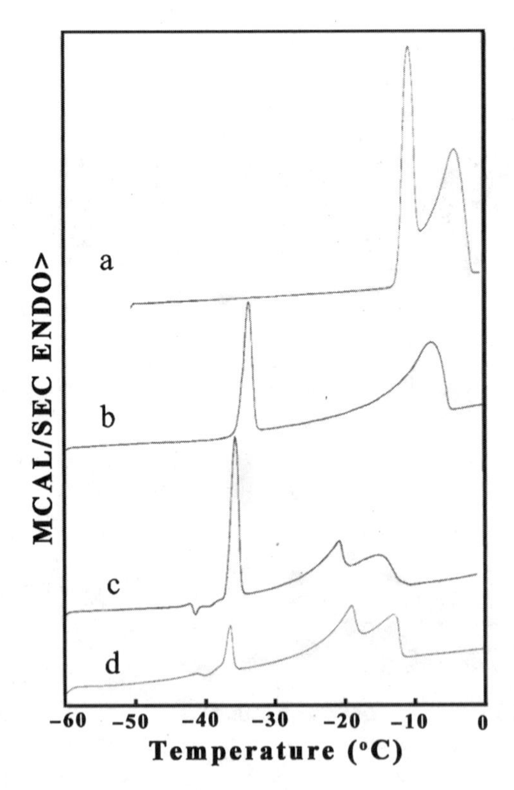

FIGURE 6.2 DSC thermograms of 12% urea (**a**), 7% NaOH (**b**), 12% urea/7% NaOH (**c**), and 4% cellulose in 12% urea/7% NaOH (**d**) aqueous solutions. (Reproduced from Ref. [23] with permission from Wiley © 2005.)

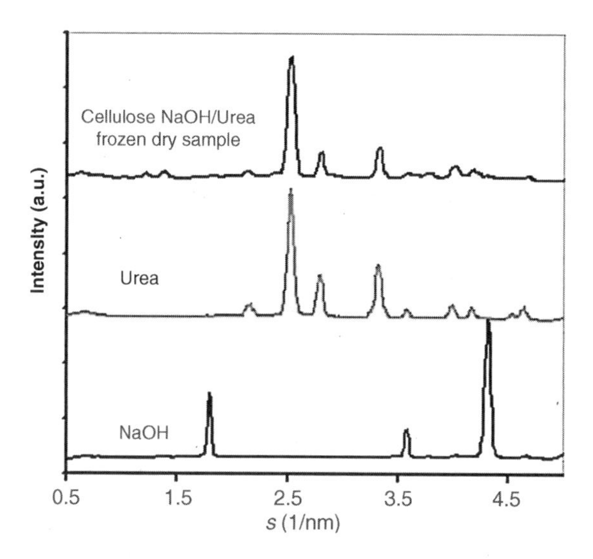

FIGURE 6.3 1D WAXD intensity profile of NaOH (powder), urea (powder), and cellulose in NaOH–urea solution subsequently frozen and dried. (Reproduced from Ref. [26] with permission from the American Chemical Society © 2008.)

enthalpy corresponding to the melting of NaOH hydrates decreases significantly. These results indicate that free water and urea hydrates are not bound to cellulose, but certain amounts of NaOH hydrates are. The wide angle X-ray diffraction (WAXD) intensity profiles of pure NaOH, urea, and freeze-dried sample from cellulose in 7 wt % NaOH–12 wt% urea aqueous solution are revealed in Figure 6.3 [26]. It is interesting to find that the peaks of NaOH and cellulose in the WAXD intensity profile of the cellulose in NaOH–urea solution disappear, and the profile is similar to that of pure urea (powder). This indicates that cellulose and NaOH were shielded, meaning that a complex of cellulose and NaOH hydrates was formed, and was encaged in urea hydrates that acted as host. Namely, the urea hydrates would aggregate at the surface of the new hydrogen-bonded networks associated with the cellulose chain and NaOH hydrates to create the cellulose inclusion complex.

Figure 6.4 shows the transmission electron microscope (TEM) images of the cellulose dilute solution dried at room temperature, which provide direct evidence on the cellulose inclusion complexes in the NaOH–urea aqueous solution [26]. The images display a wormlike pattern (Fig. 6.4a, b, and c), which is the cellulose inclusion complex rather than NaOH or urea. The kebab shape rather than a homogeneous surface is considered to be a result of the partial melting of urea on the surface of the cellulose inclusion complex due to the electron beam strike during TEM observation (Fig. 6.4b). Moreover, the electronic diffraction pattern of the crystallization structure on the surface of the cellulose was only observed in the sample prepared from NaOH aqueous solution (inset of Fig. 6.4d), which further confirmed that the surface of the cellulose bonded with NaOH hydrates was surrounded by urea hydrates.

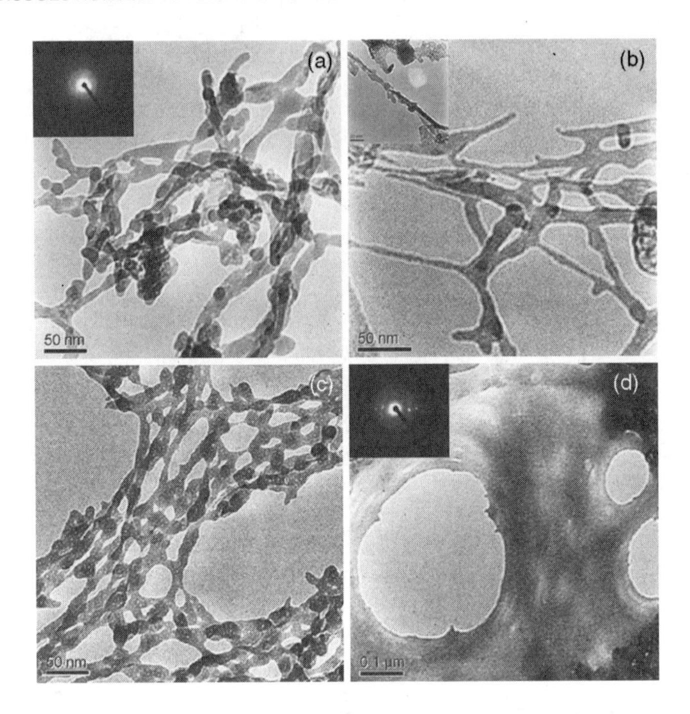

FIGURE 6.4 TEM images of cellulose solution at 5.0×10^{-4} g mL^{-1} in 12% urea/7% NaOH aqueous solution dried at room temperature, with the observation within 2 min (**a**), after 20 min (**b**), and after storing for 12 h at 5 °C (**c**). The TEM image shown in (**d**) is the cellulose solution at 5.0×10^{-4} g mL^{-1} in 7% NaOH aqueous solution. The insets (**a**, **d**) present their electron-diffraction pattern of the surface, and (**b**) shows melting urea on the surface of the cellulose-inclusion complex. (Reproduced from Ref. [26] with permission from the American Chemical Society © 2008.)

More recently, a new two-step method has been developed to efficiently dissolve cellulose by using precooled NaOH/urea aqueous solution, which is illustrated in Figure 1 of Ref [34]. First, cellulose was dispersed in 14 wt% NaOH aqueous solution precooled to 0 °C with stirring for 1 min (step 1), then 24 wt% urea aqueous solution precooled to 0 °C was added immediately with stirring vigorously for 2 min (step 2) to prepare transparent cellulose solution without any native fibers. It is pointed out that the first step involves formation and swelling of a cellulose–NaOH complex, and the second step is the dissolution of the cellulose–NaOH complex in the presence of urea. These experimental results suggest furthermore that NaOH hydrates can be more easily attracted to cellulose chains through the formation of new hydrogen-bonded networks, while urea hydrates can possibly be self-assembled at the surface of the NaOH hydrogen-bonded cellulose to form a transparent cellulose solution.

Moreover, the direct dissolution of cellulose in the precooled NaOH and urea mixture aqueous solution was confirmed by ^{13}C-NMR, as shown in Figure 6.5. The chemical shifts of the cellulose solution prepared in this way are summarized in Table 6.1, which also includes the chemical shifts of cellulose solutions in other nonderivatizing solvent systems. The chemical shifts of the carbon atoms of cellulose

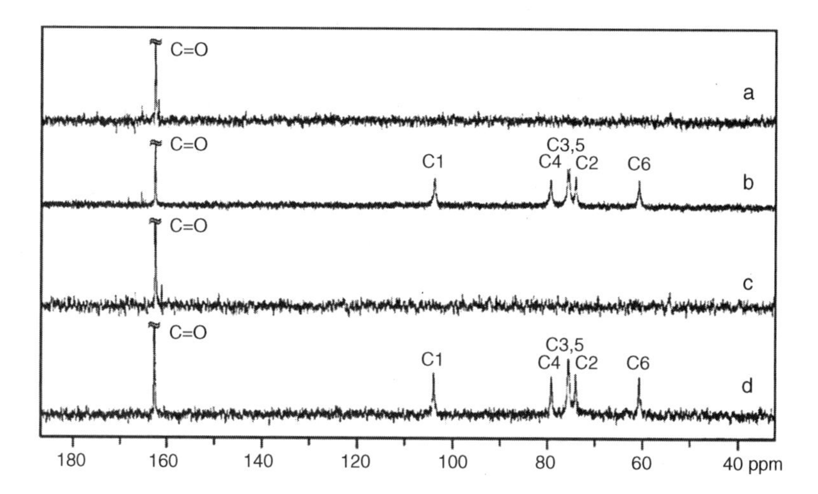

FIGURE 6.5 ^{13}C-NMR spectra of 4.2% LiOH/12% urea/D_2O (**a**), cellulose in 4.2% LiOH/12% urea/D_2O solution (**b**), 7% NaOH/12% urea/D_2O (**c**), and cellulose in 7% NaOH/12% urea/D_2O solution (**d**). (Reproduced from Ref. [23] with permission from Wiley © 2005.)

in NaOH/urea/D_2O are found to be 103.8 ppm for C1, 73.8–73.9 ppm for C2, 79.0–79.2 ppm for C4, and 60.5–60.6 ppm for C6, which are almost the same as those for cellulose in cadoxen or other good solvents [35–38]. Moreover, no new peak for cellulose derivatives in the ^{13}C-NMR spectra of the cellulose solution prepared in the novel solvent system has been observed. It is thus believed that the NaOH/urea aqueous solution is a good solvent for cellulose to form a true solution. Cuissinat and Navard [39] compared the solubility of cellulose in NaOH and in NaOH/additive systems and found out that the NaOH/urea aqueous solution had the strongest power to dissolve cellulose.

TABLE 6.1 ^{13}C-NMR Chemical Shifts of Cellulose Solution in Nonderivatizing Solvent Systems

	^{13}C-NMR Chemical Shift (ppm)						
Solvent System	C1	C2	C3	C4	C5	C6	Ref.
NaOH/urea/D_2O	103.8	73.8	75.6	79.0	75.5	60.5	[23]
	103.8	73.9	75.6	79.0	75.5	60.7	[34][a]
LiOH/urea/D_2O	103.8	73.9	75.7	79.2	75.3	60.6	[23]
NaOH/thiourea/D_2O	105.0	75.2	~76.7	80.3	~76.7	62.0	[39]
NaOH/urea/ZnO/D_2O	103.7	73.9	75.5	79.1	75.5	60.6	[44]
DMAc/LiCl	103.9	74.9	76.6	79.8	76.6	60.6	[35]
NaOH/D_2O	104.5	74.7	76.3	79.8	76.2	61.5	[36]
Cadoxen	103.8	74.9	76.6	78.9	76.4	61.8	[36]
NMMO/DMSO	102.5	73.3	75.4	79.2	74.7	60.2	[37]
NaSCN/D_2O	103.4	74.9	76.5	79.6	75.3	62.3	[38]

[a]Two-step method to dissolve cellulose [34].

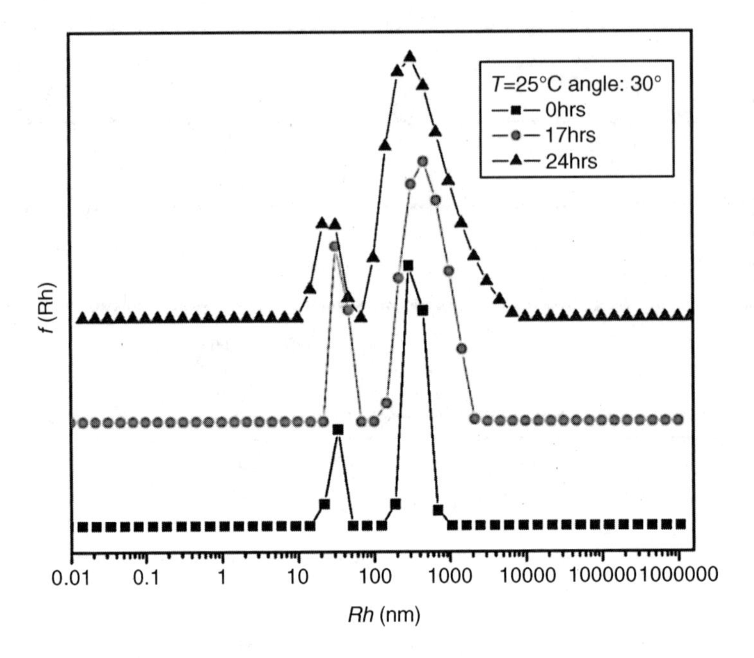

FIGURE 6.6 Hydrodynamic radius distributions $f(R_h)$ of the cellulose solution with concentration of $4.7 \times 10^{-4}\,g\,mL^{-1}$ in NaOH/urea solvent at different storage times. (Reproduced from Ref. [26] with permission from the American Chemical Society © 2008.)

In conclusion, the free water, urea hydrates, and NaOH hydrates could penetrate the macromolecule cluster of cellulose and destroy their intra- and inter-molecular hydrogen bonding at low temperature. A hydrogen-bonded network structure between cellulose macromolecules and the small molecules in the solvent was created rapidly, resulting in solvation of the cellulose chains. After absorbing sufficient such hydrates, cellulose chains would be enveloped by channels of inclusion complex hosted by urea, [25,26] so that cellulose chains could be dispersed in the solution homogeneously. The individual cellulose molecules encaged with urea and NaOH hydrates could be linked with each other to form aggregates as a result of the part dissociation of inclusion complex, which is evidenced in Figure 6.6 [26]. It can be seen that a peak with lower hydrodynamic radius (R_h) representing individual cellulose chains and a higher R_h peak attributed to its aggregates are always found in the dynamic light-scattering profiles of cellulose solutions stored for different times. It has been found that the aggregation was sensitive to temperature, polymer concentration, and storage time, indicating a fast dynamic self-assembly process of cellulose dissolved in the NaOH/urea solvent system.

In order to establish and to optimize the low-temperature solvent system, the dissolution behaviors of cellulose in NaOH/thiourea and LiOH/urea aqueous systems have also been investigated [16,23,40–42]. It has also been found that cellulose could be dissolved rapidly in 4.6 wt% LiOH/15 wt% urea and in 5 wt% NaOH/4.5 wt% thiourea aqueous solution at -12 and $-5\,°C$, respectively. The dissolubility of those

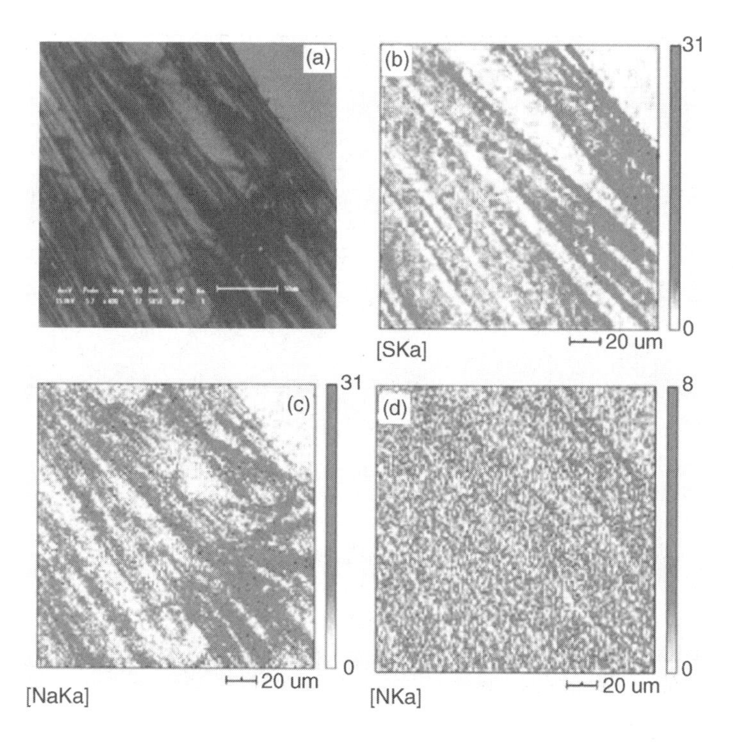

FIGURE 6.7 SEM image (**a**) and the S, Na, and N elemental distributions (**b**, **c**, and **d**) of the cellulose film on the surface of newly cleaved mica formed by depositing the dissolving cellulose NaOH/thiourea aqueous solution at room temperature. (Reproduced from Ref. [43] with permission from Wiley © 2007.)

two solvent systems is stronger than that of the NaOH/urea aqueous solution. The topochemistry investigation during the dissolving process of cellulose in the NaOH/thiourea aqueous solution has been carried out by employing SEM-EDX [43]. It has been discovered that both NaOH and thiourea were homogeneously distributed around the cellulose microfibers during dissolution (Fig. 6.7), suggesting their synergic interactions to dissolve cellulose. It was found that the dissolution of cellulose was a continuous process, and the ice-state formation promoted the dissolving process. Therefore, it is believed that the low-temperature plays a critical role in creating inclusion complex structures through hydrogen-bond networks between cellulose and solvent molecules, which leads to the dissolution of the cellulose macromolecules in the aqueous solution and is similar as the dissolution of cellulose in NaOH/urea solvent system [26,40].

It is interesting to note that ZnO was added to the NaOH/urea aqueous solution to increase the solubility of cellulose at low temperature [44]. The results show that the addition of a small amount of ZnO (about 0.5 wt%) can markedly improve the dissolution ability of cellulose for the NaOH/urea solvent (Fig. 6.8). It has been demonstrated that ZnO was transformed into $Zn(OH)_4^{2-}$ in the alkali system, which had stronger hydrogen bonds with cellulose than NaOH hydrates. Too much ZnO

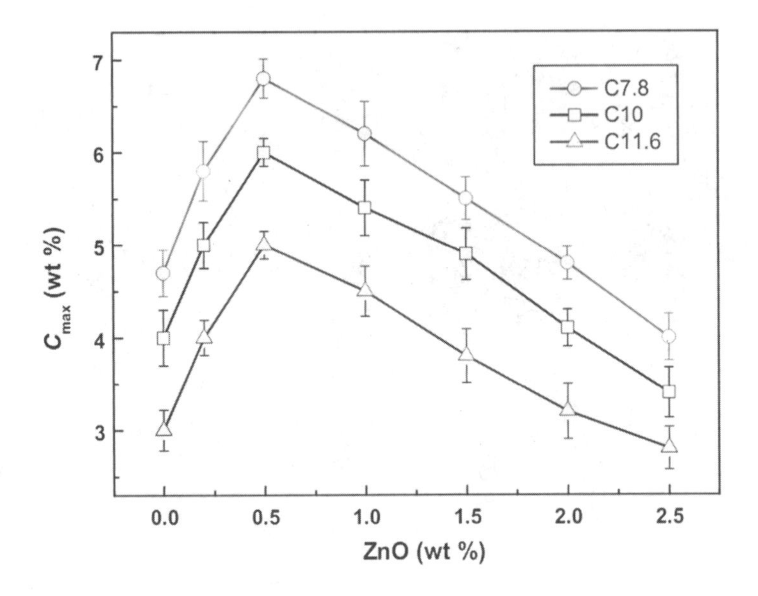

FIGURE 6.8 The dependence of the solubility values (C_{max}) of cellulose with M_η of 7.8×10^4 (C7.8), 1.0×10^5 (C10), and 1.16×10^5 (C11.6) on the content of ZnO addition. (Reproduced from Ref. [44] with permission from Elsevier © 2008.)

would, however, consume extra NaOH in the solvent system to decrease its contribution, which would inhibit the dissolution power of the solvent system.

6.3 SOLUTION PROPERTIES

Solubility of cellulose in NaOH/urea aqueous solution strongly depends on both temperature and the molecular weight of cellulose [29]. The solubility increased with a decrease in temperature, and cellulose having viscosity-average molecular weight (M_η) less than $10.0 \times 10^4\ \mathrm{g\,mol^{-1}}$ could be dissolved completely in NaOH/urea aqueous solution precooled to $-12.6\,°\mathrm{C}$. The apparent activation energy of dissolution was determined to be $-101\ \mathrm{kJ\,mol^{-1}}$. The concentrations of NaOH and urea also have important influences on the dissolution of cellulose. For example, the dissolution phase diagram demonstrated in Figure 6.9 shows the influences of the solvent components and the temperature on the dissolution of a cellulose sample with a M_η of $11.4 \times 10^4\ \mathrm{g\,mol^{-1}}$. By laser light scattering and viscometry measurements, the Mark–Houwink equation for cellulose in 6 wt% NaOH/4 wt% urea aqueous solution at 25 °C was established to be $[\eta] = 2.45 \times 10^{-2}\,(M_w)^{0.815}\ (\mathrm{mL\,g^{-1}})$ in the M_w region from 3.2×10^4 to $12.9 \times 10^4\ \mathrm{g\,mol^{-1}}$. The persistence length (q), the molar mass per unit contour length (M_L), and the characteristic ratio (C_∞) of cellulose in dilute solution were 6.0 nm, $350\ \mathrm{nm^{-1}}$, and 20.9, respectively, showing semiflexible chains of cellulose molecules in the solution [45].

However, aggregates were still found in the cellulose solution prepared from the NaOH/urea aqueous solvent system, [46] which is similar to the fringed-micelle like

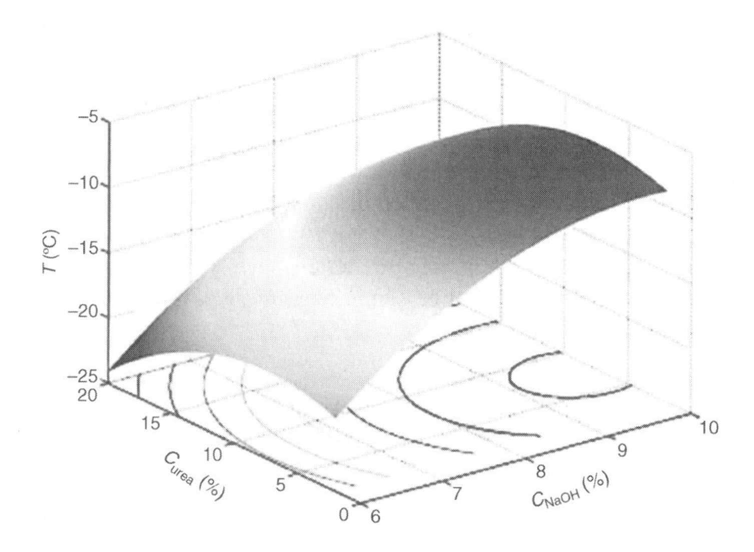

FIGURE 6.9 Three-dimensional phase diagram for solubility of cellulose in NaOH/urea aqueous solution at the precooled temperature of solvent (T), NaOH concentration (c_{NaOH}), and urea concentration (c_{urea}) relationship. The concentration of cellulose was 4 wt%. (Reproduced from Ref. [23] with permission from Wiley © 2005.)

molecular superstructure of polysaccharide solutions [47]. The radius of gyration (R_g) and the average R_h of cellulose in the NaOH/urea aqueous solvent was measured by light scattering to be 200~300 nm [46] and 160~170 nm [26,46] respectively, which is too large for a single cellulose chain ($R_g = 48\sim75$ nm, [26,45,46] $R_h = \sim25$ nm [26]). The R_g of cellulose in the NaOH/urea aqueous solvent is of a similar value to that of cellulose in viscose solution (266 nm) and in NMMO (200 nm), [47–50] indicating comparable solution behaviors of cellulose in those solvents. The dilute cellulose solution prepared in NaOH/urea aqueous solution is stable and can be deposited at room temperature for over 30 days, during which only slight decrease in viscosity of the solution or in the molecular weight of the recovered cellulose was observed [15,51,52].

The cellulose solution prepared in the precooled NaOH/urea aqueous solution is sensitive to temperature, time, molecular weight of cellulose, and cellulose concentration [24–26]. By changing one or more factors, cellulose solution will form a gel. It has been discovered that the sol–gel transition temperature decreased with increase of cellulose concentration and with its molecular weight. For moderate concentrations (3~5 wt% of cellulose), a longer time of deposition of cellulose solution could also lead to gelation at either high temperature (above 30 °C) or low temperature (below −3 °C). Once gelation had occurred, homogeneous cellulose solution could not be obtained even when the gel was cooled below −10 °C, the temperature at which cellulose had been dissolved previously. However, a moderate concentration cellulose solution remains a homogeneous liquid for over 24 h at temperatures ranging from 0 to 20 °C, indicating sufficient time for postprocessing of the cellulose solution.

6.4 NEW CELLULOSE MATERIALS PREPARED FROM THE NOVEL SOLVENT SYSTEMS

Based on the stability under ambient conditions, a sufficiently large "time window" exists, wherein cellulose solution prepared in the precooled NaOH/urea aqueous solution can be processed by various methods. As mentioned above, the cellulose solution is sensitive to many factors, so that corresponding actions can be performed to obtain various products or functional materials such as fibers, films, and gels. Moreover, cellulose derivatives can be synthesized in the novel solvent system by homogeneous reactions.

6.4.1 Novel Cellulose Fibers

By employing an extended laboratory-scale spinning apparatus, [14,46,53–56] the cellulose dope prepared from the NaOH/urea solvent system was spun into a coagulation bath composed of H_2SO_4/Na_2SO_4 aqueous solution at about 10 °C. After washing and drying, novel cellulose fibers have been prepared (Fig. 6.10). The obtained novel cellulose fibers have circular cross-sections (Fig. 6.11). The circular section of the novel fiber is similar to the round to oval cross-sectional shapes and to the homogeneous morphologies of carbamate and Lyocell fibers, [1] but is markedly different from the lobulate shape and the skin-core morphology of the viscose

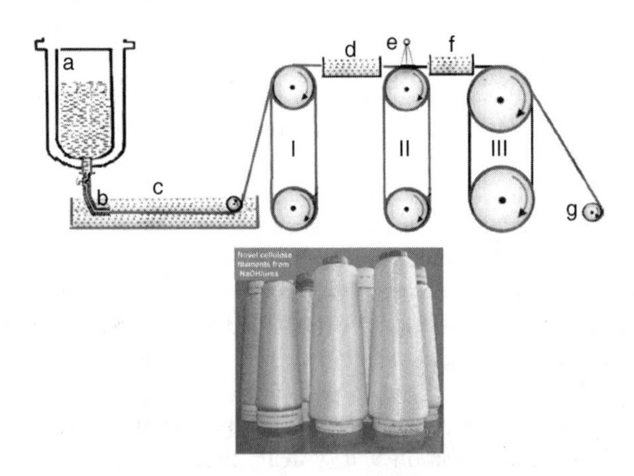

FIGURE 6.10 Schematic sketch (top) of the cellulose fiber prepared from the NaOH/urea solution at an extended laboratory scale: (**a**) a pressure extrude with a stainless cylinder having cooling jacket; (**b**) filter and spinneret (30 orifices, each of 100 mm diameter, and the ratio of nozzle length to diameter equal to 20); (**c**) first coagulation bath; (**d**) second coagulation bath; (**e**) hot water washing; (**f**) third post-treatment bath (plasticizing bath); (**g**) take-up device; (I) Nelson-type roller; (II) Nelson-type roller; (III) heated roller. Also shown is a photograph of the obtained products (bottom). [14, 46, 53–56]. (Reproduced from Ref. [55] with permission from Wiley © 2007.)

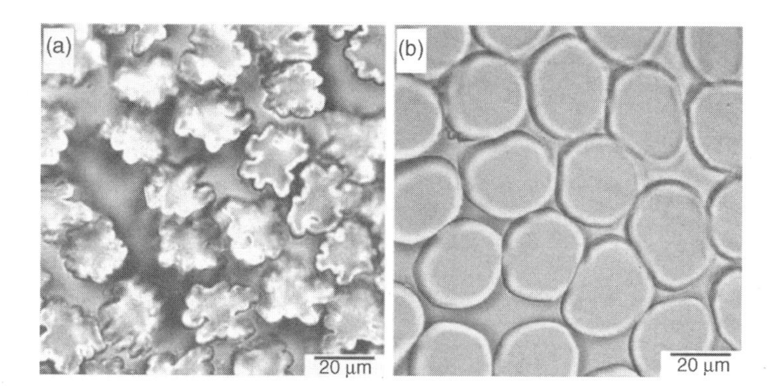

FIGURE 6.11 Optical polarizing microscope images of the viscose rayon (**a**) and of the multifilament (**b**) prepared from the cellulose dope in NaOH/urea solvent. (Reproduced from Ref. [53] with permission from Wiley © 2004.)

rayon [1,53]. The fiber prepared from the NaOH/thiourea aqueous solvent system also displays a circular section and smooth surface [57,58].

The multifilament fibers along the spinning process were at different degrees of drawing and orientation at the stages of roller I, II, and III. The scanning electron microscopy (SEM) images of the multifilament shown in Figure 6.12 reveal that the fibrils of the filaments could be oriented along the axis of the fibers to form a uniform and dense structure during drawing process. Meanwhile, cross-polarization/magic-angle spinning (CP/MAS) ^{13}C-NMR and 2D wide-angle X-Ray diffraction (WAXD) measurements indicate cellulose II of the crystal structure for the multifilament at all the stages [46,54,55]. The small-angle X-ray scattering (SAXS) patterns of the regenerate cellulose filaments show sharp and long equatorial streaks and very short meridional peaks, indicating the presence of needle-shape devoid or fibrillar structure aligned parallel to the fiber direction and with a periodic lamellar arrangement of crystalline and amorphous cellulose regions. The meridional peaks become stronger with drawing to induce stronger lamella arrangements, whose long period can reach 100 nm [55]. The results [14,46,53–56] indicate total crystallinity of 57~63% and crystal size of 4.78~5.47 nm of the novel cellulose fiber; the crystallinity of anisotropic part and its size were determined to be 58~62% and 5.30~5.87 nm, respectively; the orientation parameter π and the Hermans' orientation parameter $\bar{p}_{2,g}$ were measured to be 0.76~0.81 and 0.56~0.64, respectively. It has been reported that the crystallinity of the regenerate cellulose prepared from the novel solvent system was much higher than that of cellulose fibers regenerated from ferric chloride/sodium tartarate/sodium hydroxide (FeTNa) and NMMO systems [59]. Moreover, no obvious degradation of cellulose has been found during the dissolution and regeneration processes. As a result, the novel fiber possesses good mechanical properties to have a tensile strength of 1.7~2.2 cN dtex^{-1} and a break elongation of 9~13%, which are close to commercial viscose rayon (2.0~2.4 cN dtex^{-1}; ref. [60]). It has been claimed that better mechanical properties of the novel cellulose filaments could be expected by further optimization of the spinning process or by finishing treatment [61].

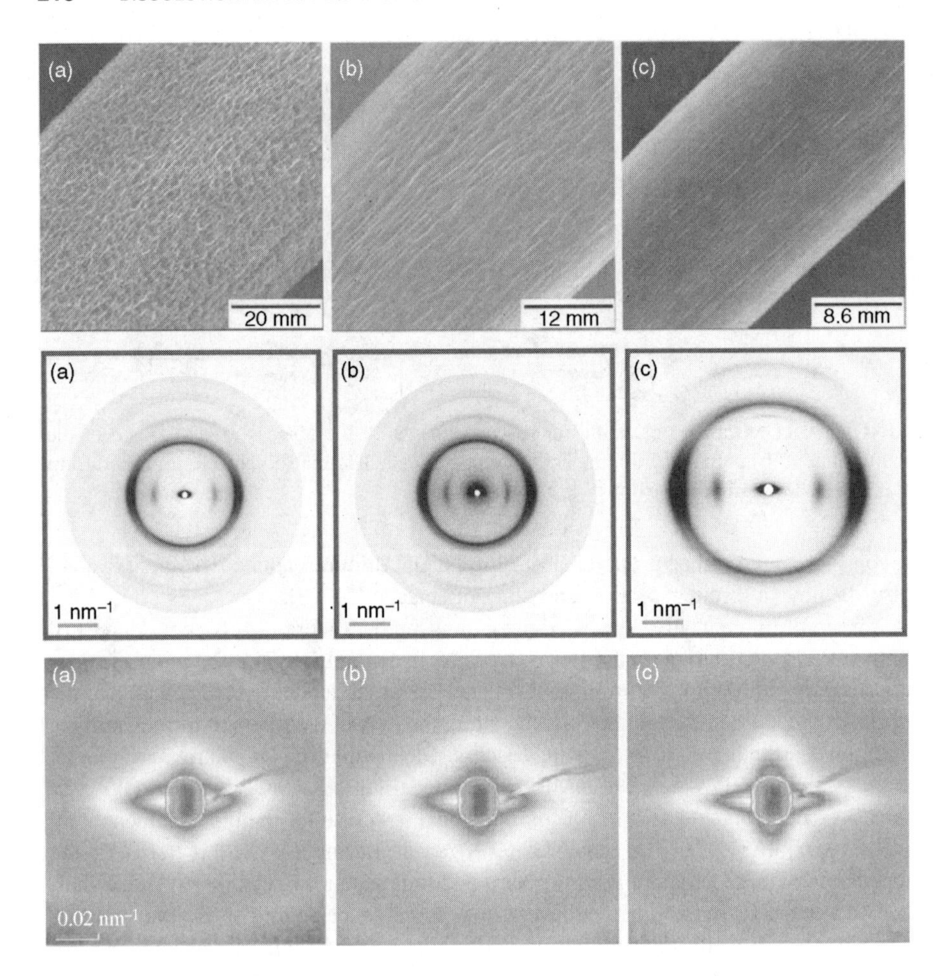

FIGURE 6.12 The SEM images (top), 2-D WAXD patterns (middle), and SAXS patterns (bottom) of the cellulose fiber at the stages of roller I (a), II (b), and III (c). (Reproduced from Ref. [55] with permission from Wiley © 2007.)

The first industrialized trial of wet-spinning the cellulose dope had been successfully carried out in 2010 [29]. As illustrated in Figure 6.13, cellulose has been dissolved completely in the precooled NaOH/urea aqueous solution within 5 min in a 1000 L tank, and then the transparent dope was spun into a coagulation bath of 15 wt% H_2SO_4/10 wt% Na_2SO_4 aqueous solution. The obtained fiber was further regenerated in a 5 wt% H_2SO_4 aqueous solution medium. The new cellulose fiber exhibits a bright surface and circular section, and its tensile strength reaches 1.7 cN $dtex^{-1}$. After neutralization, the main by-products of Na_2SO_4 and urea in the coagulation bath can be separated easily by flash evaporation, and then crystallization based on the large difference in the solubility between Na_2SO_4 and urea [28,29]. The recovery of Na_2SO_4 is almost 100% and that of urea is over 90%, and they can be reutilized [28,29]. It is seen

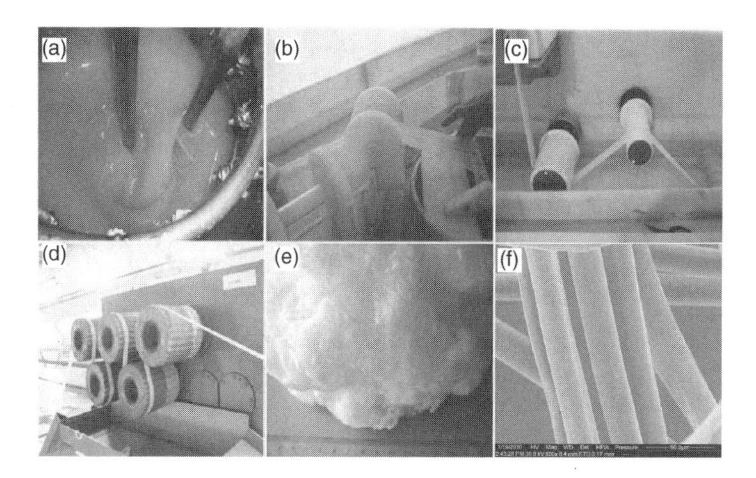

FIGURE 6.13 Photographs of the cellulose dissolution (**a**) and spinning process (**b**, **c**, **d**) during a mid-sized industrial manufacturing, and the photograph (**e**) and SEM image (**f**) of the novel cellulose fibers. (Reproduced from Ref. [29] with permission from the American Chemical Society © 2010.)

that the fiber-fabrication process is quick, nonpolluting and low cost, and the used reagents and the by-products are easily recycled, suggesting a truly "green" technology of the method. Moreover, the novel multifilament fibers are easy to dye to deep vibrant colors (Figure 9 of Ref. [61]), and the color strength (K/S) of the dyes on the novel cellulose fibers is higher than that of commercial viscose fibers [61]. Therefore, commercial products of the novel cellulose multifilament based on the "green" technology have high expectations, and no doubt will have great impact on the cellulose chemical industries.

The fiber in wet state exhibits nano-microporous structure matrix (Fig. 6.14a and b), in which metal ions can be absorbed [62,63]. After immersing the wet fiber in $FeCl_3$ aqueous solution and subsequent treatment in NaOH solution, iron oxide (Fe_2O_3) nanoparticles were synthesized in situ to locate in the pores (Fig. 6.14c, d, e, and f). The mean diameter of the introduced Fe_2O_3 nanoparticles was about 18 nm, endowing the obtained composite fiber with high mechanical strength, strong capability to absorb UV radiation, and superparamagnetic properties. If the composite fiber was subjected to calcinations, fiber-like Fe_2O_3 macroporous nanomaterials (Fig. 6.14g and h) were obtained, which have potential applications in the magnetic and electronic fields.

By the addition of other materials in the cellulose dope, functional fibers have been produced. For example, nanoscale fibrous materials other than spherical particles (Fig. 6.15) were obtained by electrospinning mixtures of cellulose and poly(ethylene glycol) (PEG) or poly(vinyl alcohol) (PVA) [64]. By blending cellulose with soy protein isolate in the novel solvent system, a new fiber with good mechanical characteristics has been fabricated which would have promising application in biomedical fields [65].

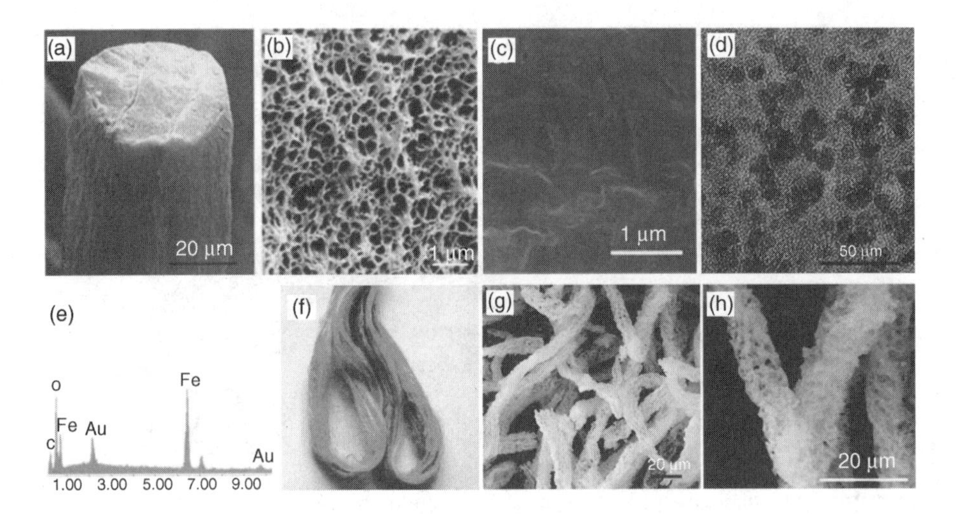

FIGURE 6.14 SEM images for the section of pure cellulose fiber in wet state (**a** and **b**), and the SEM (**c**) and transmission electron microscope (TEM) (**d**) images after introducing in situ synthesized iron oxide (Fe_2O_3) nanoparticles. The energy dispersive spectrum and the photograph of the composite fiber are shown in (**e**) and (**f**). After calcining the composite fiber, fiber-like Fe_2O_3 macroporous nano materials have been prepared (**g** and **h**). (Reproduced from Refs. [62,63] with permission from the American Chemical Society © 2008.)

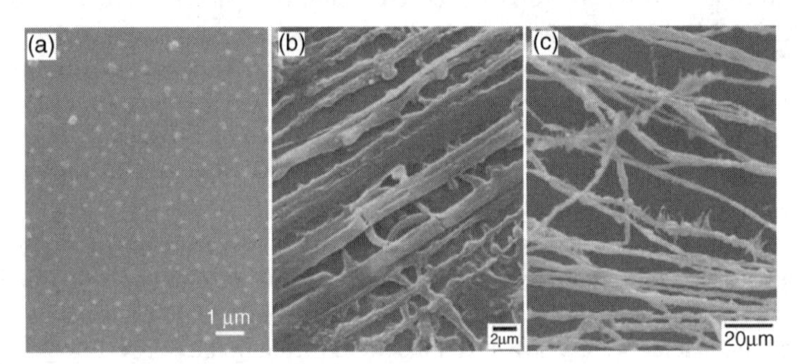

FIGURE 6.15 SEM images of materials obtained by electrospinning pure cellulose (**a**), cellulose/ PEG (**b**), and cellulose/PVA (**c**) solutions prepared in the NaOH/urea aqueous solvent. (Reproduced from Ref. [64] with permission from Wiley © 2010.)

6.4.2 Novel Cellulose Film

Various regenerate cellulose membranes and films have been prepared in the laboratory on the basis of dissolution of cellulose in the novel solvent systems. Usually, the cellulose dope is cast on support by spreading between thin wires with a glass stick. By adjusting the thickness of the wires and the length of the stick, the desired thickness and area of the spread cellulose solution can be obtained. By immersing the support with the spread cellulose solution in a coagulation bath for

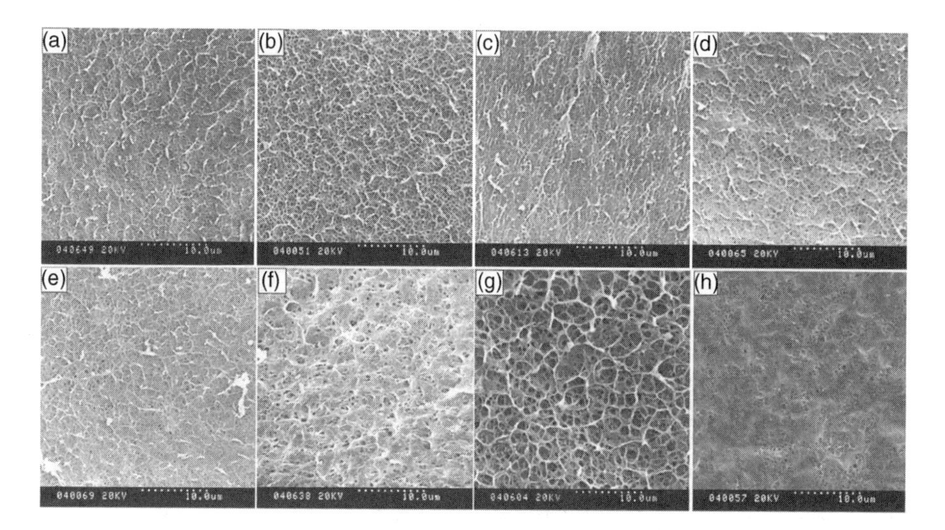

FIGURE 6.16 SEM images of the surface for the regenerated cellulose films prepared from the various coagulants: (**a**) 5 wt% H_2SO_4; (**b**) 3 wt% acetic acid; (**c**) 5 wt% H_2SO_4/5 wt% Na_2SO_4; (**d**) 5 wt% Na_2SO_4; (**e**) 5 wt% $(NH_4)_2SO_4$; (**f**) H_2O; (**g**) ethanol; (**h**) acetone. (Reproduced from Ref. [69] with permission from Elesvier © 2006.)

several minutes, cellulose can be regenerated to obtain the cellulose membrane or film after washing and drying.

Pure cellulose films were prepared by direct regeneration of the cellulose dope in coagulant baths [66–75]. The coagulants can be acidic aqueous solution, salt solution, organic nonsolvent of cellulose, or a complex combination. It has been found that the obtained films were of a microporous structure (Fig. 6.16) with porosity higher than 81%. The pore size of the films was determined by flow rate method to be in the range of 25~56 nm, [68] which was larger than that of membranes prepared from cuoxam (12 nm) [76] and viscose (5 nm) [77]. Consequently, the novel regenerate cellulose films with relatively high water permeability are postulated to have great potential applications in the separation field.

The porous properties and mechanical strength of the novel regenerate cellulose films are affected by many factors such as the molecular weight of cellulose, the coagulant, coagulation time, and temperature [66–75]. Zhang has suggested [69,70,78] that counter-diffusion and chemical neutralization between solvent in the cellulose solution and nonsolvent in the coagulant have occurred. During the coagulation, the removal of solvent from cellulose solution and the penetration of nonsolvent into the cellulose solution would result in formation of cellulose-rich and lean phases. Then, the phase separation would lead to the mesh porous structure of the novel regenerate cellulose film due to the hydroxyl junction of cellulose molecules as physical cross-links.

It is found that the mechanical properties such as tensile strength (σ_b) and elongation at break (ε_b) of the novel cellulose film increases with its molecular weight (Fig. 6.17) [28]. After drawing orientation, much stronger film can be obtained. Based on the temperature-dependent solubility of cellulose in the NaOH/urea aqueous solvent

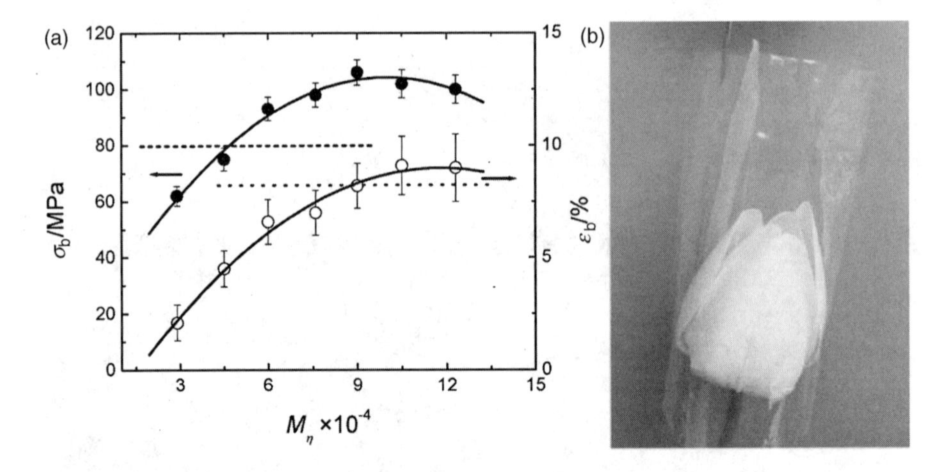

FIGURE 6.17 Dependence of the tensile strength (σ_b, ●) and elongation at break (ε_b, ○) of the regenerate cellulose film on its molecular weight (**a**). The mechanical properties of a commercial cellophane are also presented: (σ_b- - -) and (ε_b···) (**a**). (**b**) is a photograph of regenerate cellulose film with M_η of 7.8×10^4 g mol^{-1} packaging a flower. (Reproduced from Ref. [28] with permission from The Royal Society of Chemistry © 2008.)

system, an all-cellulose composite film has been prepared by the addition of native cellulose nano-whiskers to strengthen the pure cellulose film [79,80]. The cellulose whiskers retained their needlelike morphology with a mean length of 300 nm and diameter of 21 nm (Fig. 6.18, left) as well as their native crystallinity when added into the cellulose solution at ambient temperature. The tensile strength of the composite films could reach 157 MPa through a simple drawing process (Fig. 6.18, right). Moreover, the optical transmittance of the regenerate cellulose film prepared from

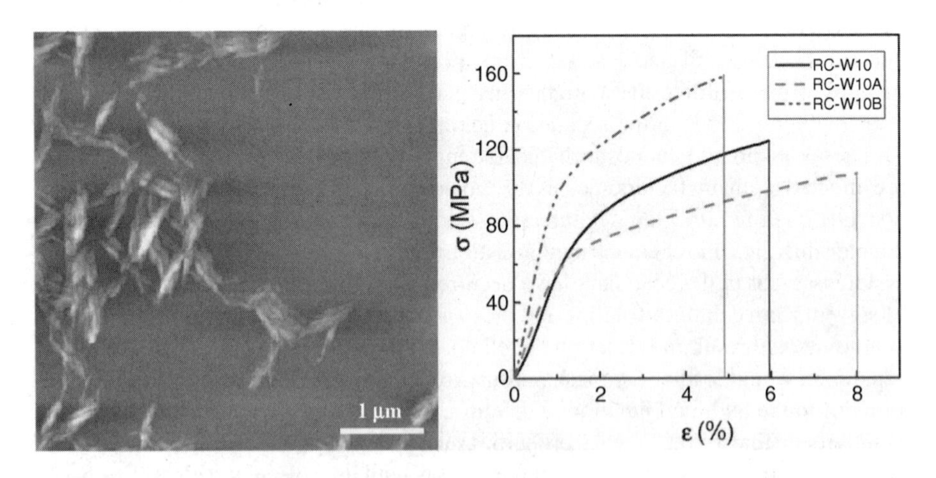

FIGURE 6.18 Atomic force microscopy (AFM) topography image of cellulose whiskers after drying on a mica surface (left) and the stress–strain curves of the all-cellulose composite films (right). (Reproduced from Ref. [79] with permission from the American Chemical Society © 2009.)

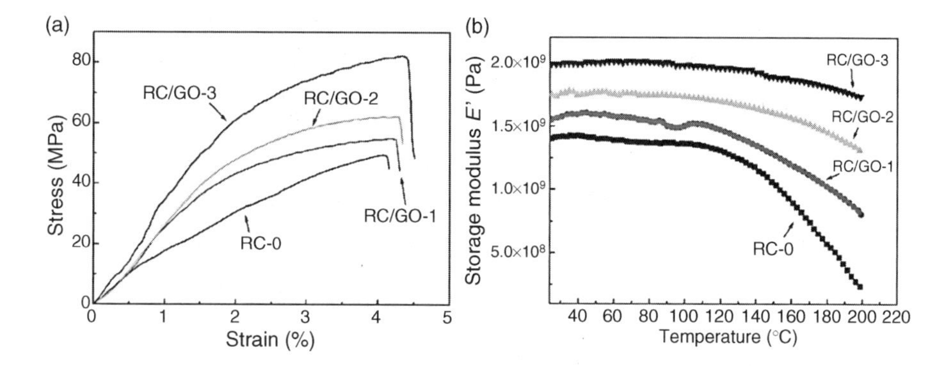

FIGURE 6.19 Stress–strain curves (**a**) and temperature dependencies of the dynamic storage modulus E' (**b**) of the regenerate cellulose films reinforced by addition of 0 (RC-0), 2.5 (RC/GO-1), 5.0 (RC/GO-2), and 7.5 wt% (RC/GO-0) GO. (Reproduced from Ref. [81] with permission from Wiley © 2011.)

cellulose with molecular weight lower than $11.1 \times 10^4\ g\ mol^{-1}$ is higher than that of commercial cellophane (85%) (Fig. 6.17b). Therefore, it is hoped that the cellulose film could be used as transparent packaging material and biomaterial because of its transparent, colorless, and biodegradable properties.

The reinforcement of the regenerate cellulose can be fulfilled by adding an inorganic filler such as graphite oxide (GO) [81]. It has been found that the GO could be homogeneously dispersed in the obtained regenerate cellulose film, which was prepared from the NaOH/urea solvent system. By the addition of a small amount of GO, the thermostability and mechanical properties of the regenerate cellulose films were greatly improved, especially at high temperatures, as shown in Figure 6.19.

By the dissolution of other natural polymers and cellulose in the same solvent of NaOH/urea aqueous solution, blend films have been prepared to have various functions [82–92]. For example, [88] blend films were prepared by mixing cellulose and soy protein isolate in NaOH/thiourea aqueous solution. The blend films were then hydrolyzed with NaOH aqueous solution to obtain a microporous structure. Due to the rough and microporous structure and the small amount of protein residue, the prepared film was suitable for culture of Vero cells (Fig. 6.20).

Cellulose can also be considered as an additive to strengthen other natural polymer-based materials. For example, cellulose nanoparticles were coagulated from the NaOH/urea aqueous solution to improve the thermal stability of starch-based films [93].

Hybrid cellulose films have been prepared from the novel solvent systems by incorporation of inorganic nanoparticles in the matrices of the regenerate cellulose films. Zhou and coworkers [94] have dispersed TiO_2 nanoparticles in the cellulose solution prepared from the NaOH/urea aqueous solution by simple mechanical stirring. After coagulating the mixture solution with H_2SO_4 aqueous solution, UV-blocking and antibacterial composite films with excellent strength have been prepared. Due to the microporous structure of the wet-state regenerate cellulose

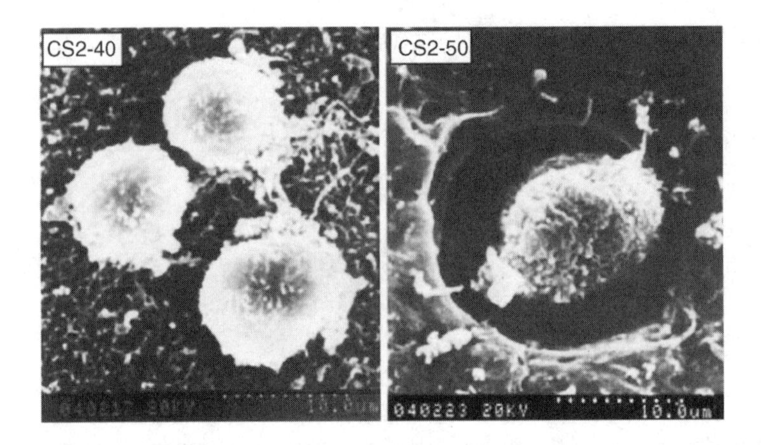

FIGURE 6.20 Vero cells cultured on the film surfaces hydrolyzed from cellulose and soy protein isolate blends. (Reproduced from Ref. [88] with permission from Elsevier © 2004.)

films prepared from the novel solvent systems, precursors of nanoparticles could be absorbed, and were then transformed in-situ into nanoscale particles [95–98]. The nanoparticles were found to be stabilized in the matrices of the regenerate cellulose films. Because of the nanoscale-distributed particles, the composite films exhibit special photoactive properties including photoluminescence emission, UV-blocking, and photocatalyzing. It is interesting to note that disk-shaped Fe_2O_3 particles with a mean diameter of ~24 nm and thickness of 2.5~3.5 nm were synthesized in-situ in the matrix of regenerate cellulose film, and then aligned to form ordered multilayer structure by the shrinkage of the hybrid film during drying [96]. The automatically aligned disk-shaped Fe_2O_3 particles resulted in anisotropic magnetic properties of the composite film (Fig. 6.21).

The porous structure of the regenerate cellulose film could also lead to the absorption of fluorescent dyes and photoluminescent pigments, so that photoluminescent materials could be fabricated (Fig. 6.22) [28]. By filling the pores of the regenerate cellulose film with a small amount of chitosan, the obtained film would

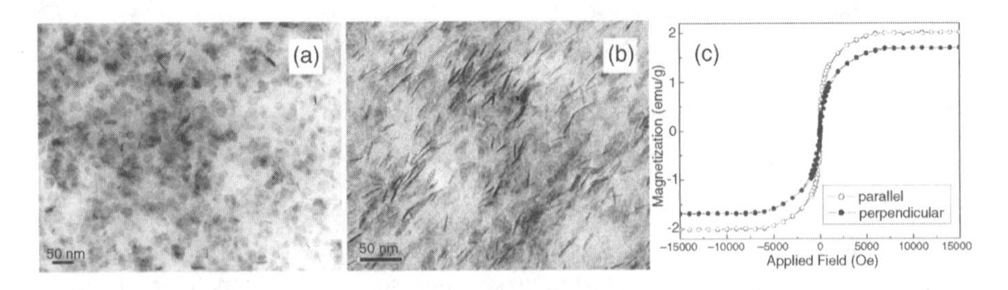

FIGURE 6.21 Alignment of iron oxide nanoparticles in regenerated cellulose film: TEM images of the slice parallel (**a**) and perpendicular (**b**) to film plane, and its anisotropic magnetic property (**c**). (Reproduced from Ref. [96] with permission from Wiley © 2006.)

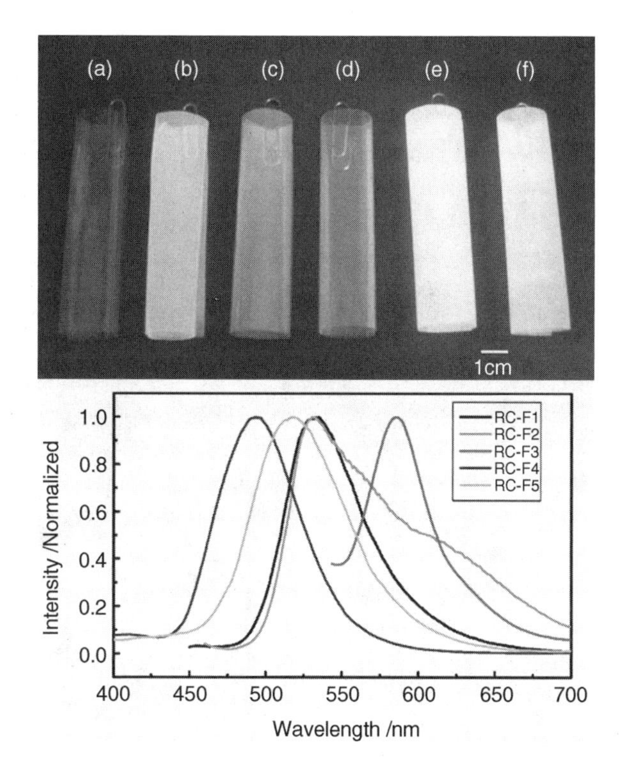

FIGURE 6.22 Fluorescence images (top) of the regenerate cellulose film (**a**) and of the photoluminescent pigments absorbed composite films (**b, RC-F1; c, RC-F2; d, RC-F3; e, RC-F4; f, RC-F5**), and their emission spectra (bottom). (Reproduced from Ref. [28] with permission from The Royal Society of Chemistry © 2009.)

show special pH sensitivity due to the different degrees of protonation of amino groups of chitosan [99]. Moreover, the pH-sensitive film retained relatively high water flux to find potential application in the field of separation technology, because the amount of the chitosan filled in the pores of the regenerate cellulose film was low and it was distributed only in the surface layer. Kuga and coworkers [100] have prepared nanocomposite films by immersing regenerate cellulose films into polystyrene (PS) and poly(methyl methacrylate) (PMMA) solutions, which penetrated into the void of cellulose gel by diffusion. By changing the polymer concentration of the immersing solution, the synthetic polymer content could be changed between 10 and 80%. The Young's modulus and tensile strength of the cellulose/PMMA composite films showed nearly linear decrease dependencies on the PMMA content. The properties of the cellulose/PS composite films, however, showed anomalous dependencies on the PS content and exhibited maxima at around 20–30% PS. Besides, the coefficients of the thermal expansion (CTE) for both kinds of composite films were low due to the restriction of the cellulose gel network, which may lead to the development of thermally stable composite films.

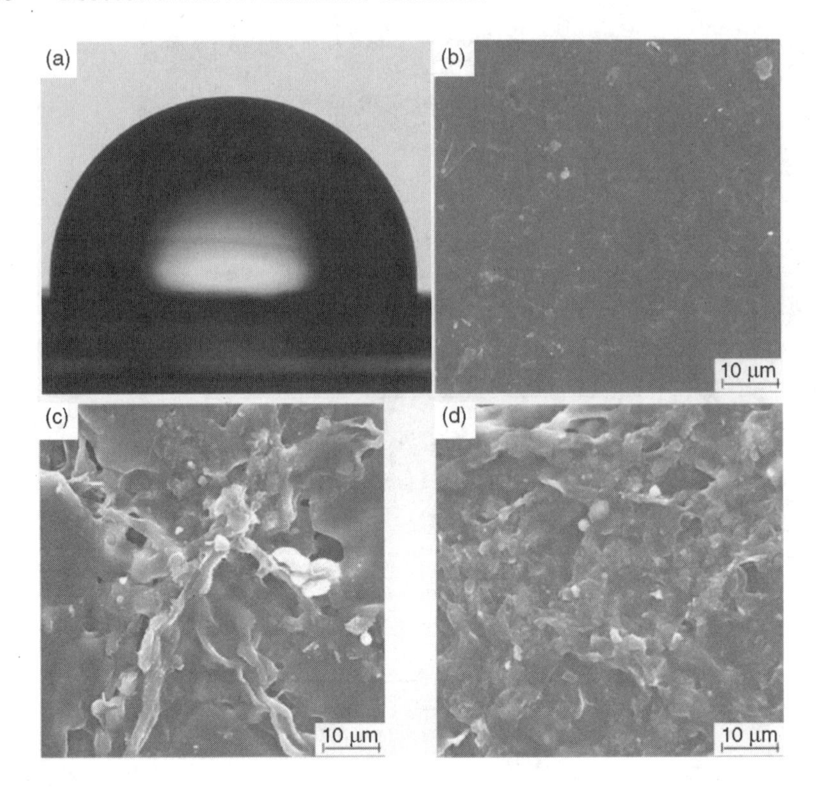

FIGURE 6.23 Photograph of a water drop on the IPN coated regenerate cellulose film surface after 30 s (**a**), and the SEM image of an IPN-coated film after decay in the soil for 7 days (**b**), 14 days (**c**), and 21 days (**d**). (Reproduced from Ref. [104] with permission from the American Chemical Society © 2006.)

The water resistivity of the regenerate cellulose film could be improved by coating with a thin layer of waterproof material such as interpenetrating polymer networks (IPNs). When the IPN or semi-IPN were prepared from biopolymers, [101–104] the coated cellulose films revealed enhanced mechanical properties and water resistivity (Fig. 6.23a), and still retained good biodegradability (Fig. 6.23b, c, and d).

6.4.3 Novel Cellulose Gels

As mentioned above, the cellulose solutions prepared from the novel solvent systems are sensitive to environmental conditions. Therefore, various gels can be prepared by changing the temperature, storage time, cellulose concentration, and the property of the coagulant.

The cellulose solution can be transformed into hydrogel directly either by heating or by cooling (Fig. 6.24), and the sol–gel transition is irreversible due to the physical cross-linking of the cellulose chains [24,78,105]. The cellulose gel could be reinforced by cellulose nanowhiskers, which have acted as physical cross-links

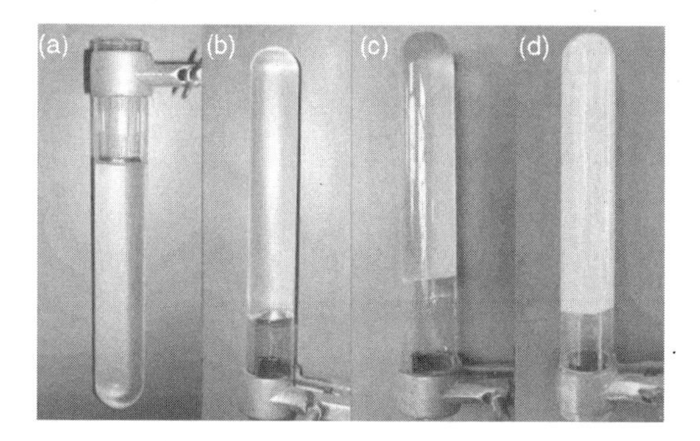

FIGURE 6.24 Photographs of cellulose solution at ambient temperature (**a**), and the gels formed at 8 °C for 4 days (**b**), at −20 °C for 1 h (**c**), and at 50 °C for 1 h (**d**). (Reproduced from Ref. [24] with permission from the American Chemical Society © 2006.)

to facilitate the cross-linking of cellulose chains during gel formation [106]. Upon cross-linking by using epichlorohydrin, [107,108] the cellulose solutions have been converted into stable hydrogels with improved mechanical strength. Biocompatible and biodegradable cellulose hydrogels have been prepared in the novel NaOH/urea solvent system by the addition of other water-soluble polymers such as poly(vinyl alcohol), [109] sodium alginate, [110] and carboxymethylcellulose (CMC) sodium [111]. Those hydrogels exhibit macroporous structures, resulting in superior water-absorbency property of the materials. The equilibrium swelling ratio of the cellulose/CMC mixture hydrogel could reach 1000% (Fig. 6.25) [111]. Alternately,

FIGURE 6.25 Photographs of cellulose/CMC mixture hydrogel: (**a**) original state, (**b**) swollen in distilled water, (**c**) vacuum dried and (d) salt balanced gel. (Reproduced from Ref. [111] with permission from Elsevier © 2010.)

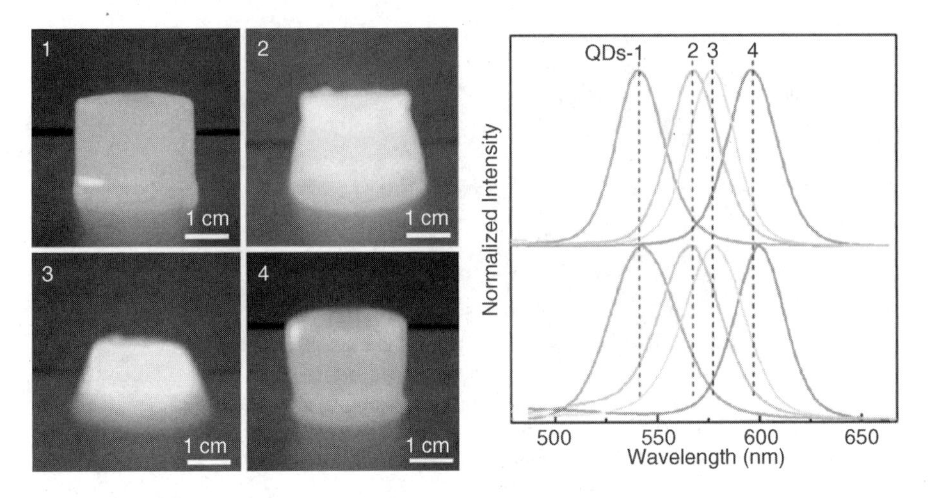

FIGURE 6.26 The appearances and photoluminescence spectra of the cellulose/QD hydrogels. (Reproduced from Ref. [114] with permission from The Royal Society of Chemistry © 2009.)

the water in the cellulose hydrogel could be exchanged for water-miscible organic solvents, followed by drying with supercritical CO_2. The aerogels prepared in this way have many potential applications such as adsorbent [112] and carrier for nanoparticles [113].

Other than pure polymer gel, hybrid gels have been prepared by mixing quantum dots (QDs) of CdSe/ZnS nanoparticles into the cellulose solution [114]. The results indicate that QDs were embedded firmly in the matrices of the cellulose hydrogels by electrostatic attraction and hydrophobic interactions. The network structure of the cellulose hydrogel played an important role in the protection of the CdSe/ZnS nanostructure to preserve the characteristics of the quantum dots. As a result, the hybrid hydrogels with size of QDs ranging from 2.8 to 3.6 nm exhibited strong photoluminescence emission and nearly pure color from green to red (Fig. 6.26).

Moreover, iron ions have also been adsorbed onto the matrices of the nanoporous cellulose microspheres, and could be synthesized in situ into nano Fe_3O_4 nanoparticles [115]. The pores of the cellulose microspheres not only provided a microchamber for the creation of inorganic nanoparticles, but also acted as a shell to prevent their aggregation. Therefore, the spherical magnetic Fe_3O_4 nanoparticles were dispersed uniformly and immobilized in the cellulose matrix to endue the hybrid microspheres with superparamagnetic property (Fig. 6.27a and b). Magnetic cellulose microspheres fabricated in this way were positioned in buffer solution to adsorb bovine serum albumin (BSA). The adsorption of BSA reached equilibrium within 30 min, and followed the Langmuir adsorption model, while the adsorption capacity varied significantly with the pH of the buffer. Moreover, the BAS adsorbed magnetic regenerate cellulose microsphere displayed pH-sensitive release (Fig. 6.27d), indicating potential application in biomaterial field such as drug carrier for targeting delivery and release.

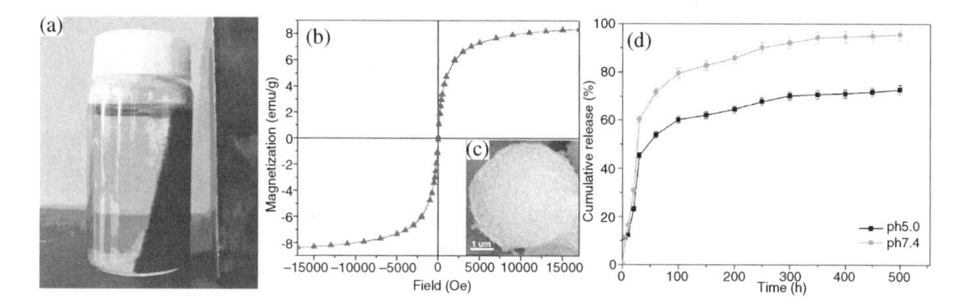

FIGURE 6.27 Photo of the magnetic regenerate cellulose microsphere in water at a magnetic field (**a**), its magnetic hysteresis loop at 25 °C (**b**), its SEM image (**c**), and the BAS release profiles at two pH conditions. (Reproduced from Ref. [115] with permission from The Royal Society of Chemistry © 2009.)

In addition, magnetic γ-Fe_2O_3 nanoparticles with a mean size of 10 nm were mixed with cellulose solution to prepare magnetic cellulose beads [116] or microspheres [117]. Similarly, the embedded γ-Fe_2O_3 nanoparticles were found to be uniformly dispersed in the cellulose substrate and to have retained their original structure and nature. Because of the magnetic property of the γ-Fe_2O_3 nanoparticles, those hybrid beads enabled their fast separation from the effluent by applying a magnetic field, indicating their application in fields such as water-pollution remediation [116]. If the magnetic cellulose microspheres were activated with epoxy chloropropane, their loading efficiency of biomacromolecules could be enhanced [117]. For example, penicillin G acylase (PGA), a biocatalyst, was immobilized successfully in magnetic cellulose microspheres as a result of the existence of the cavity and the affinity forces from –OH groups and from the Fe_2O_3 nanoparticles (Fig. 6.28) [117]. Interestingly, the immobilized PGA exhibited highly effective catalytic activity, thermal stability, and improved tolerance to pH variations than free PGA.

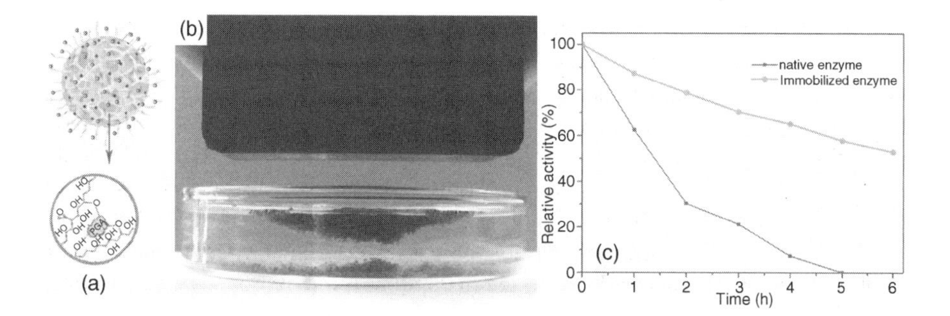

FIGURE 6.28 Scheme of enzyme immobilized in magnetic cellulose micropshere (**a**), the photograph of the microspheres attracted rapidly by a magnet (**b**), and the relative activity of free and the immobilized PGA (**c**). (Reproduced from Ref. [117] with permission from the American Chemical Society © 2010.)

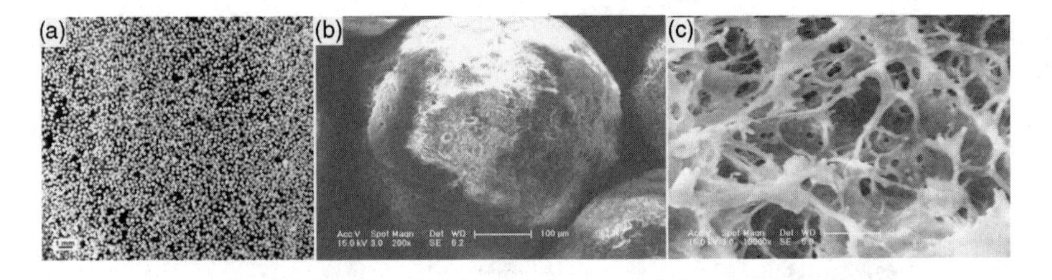

FIGURE 6.29 Photograph (**a**) and SEM images (**b** and **c**) of the porous regenerate cellulose beads. (Reproduced from Ref. [119] with permission from Elsevier © 2010.)

By using "strong" coagulants, the cellulose solutions can be regenerated to form dense gels other than the previously mentioned macroporous hydrogel. By employing microemulsion fabrication, microdroplets of cellulose solution could be obtained by adjusting the proportion of "water" to "oil" phase and the stirring speed. The microdroplets of cellulose solution were then transformed into microsphere gels with the diameter ranging from micrometer to millimeter (Fig. 6.29) [118,119]. The obtained cellulose microspheres were porous with nanoscale pores, and could be packed in glass columns to be used as preparative size-exclusion chromatography columns to satisfactorily fractionate biopolymers and synthetic polymers. If another polymer, such as chitin, with special functional groups was blended, cellulose gels prepared in this way could be used as adsorbents to remove heavy-metal ions from water [120–122]. The heavy-metal ion removal was analyzed to be based mainly on complexation adsorption by the special functional groups as well as affinity adsorption by the hydroxyl groups of the material matrix.

6.4.4 Innovative Medium for Synthesizing Cellulose Derivatives

Cellulose derivatives are pioneer compounds of cellulose chemistry, and are widely applied in coating, membrane technology, medical, biology, and food. Because of the strong intra- and intermolecular hydrogen bonds of native cellulose, industrial cellulose derivatives are generally produced through heterogeneous reactions of cellulose in solid or swollen state [1,2,123]. The heterogeneous reactions of cellulose often lead to low yield and to uneven substituent distribution, which drastically limit the applications of the obtained cellulose derivatives. By breaking the hydrogen bonds through use of specific solvents such as LiCl/DMAc [124] and DMSO/tetrabutylammonium fluoride trihydrate, [125] cellulose derivatives can be synthesized homogeneously [123,126].

The successful dissolution of cellulose in the novel solvent systems such as the NaOH/urea aqueous solution leads to the disruption of the hydrogen bonds, so that homogeneous chemical modification of cellulose becomes possible. It has been reported that various cellulose derivatives were homogeneously synthesized in those innovative media, including cellulose ethers [127–132] and cellulose polyelectrolytes [133–135]. The obtained cellulose derivatives, the preparation conditions, and their degree of substitutions (DS) are summarized in Table 6.2.

TABLE 6.2 Cellulose Derivatives Homogeneously Synthesized in the Novel Solvent Systems

	Reaction Condition			DS			
Cellulose Derivatives	Medium	Temperature (°C)	Total DS	C2	C3	C6	Ref.
Hydroxypropylcellulose (HPC)	NaOH/urea	25	0.85~1.13	0.25~0.48	0.26~0.38	0.26~0.39	[127]
Methylcellulose (MC)	NaOH/urea	25	1.48~1.69	0.63~0.66	0.42~0.45	0.40~0.59	[127]
	LiOH/urea	22	1.07~1.59				[128]
O-(2-hydroxyethyl)cellulose (HEC)	NaOH/urea	50	0.5~0.9	0.15~0.30	0.06~0.14	0.28~0.43	[129]
Cyanoethyl cellulose (CEC)	NaOH/urea	5~10	0.26~1.93	0~0.62	0~1.0	0~0.31	[130]
Carboxymethyl cellulose (CMC)	NaOH/urea	55	0.05~0.62				[131]
	LiOH/urea	55	0.1~0.65				[132]
Quaternized cellulose (QC)	NaOH/urea	25	0.20~0.63				[133]
Cellulose polyelectrolyte containing acylamino and carboxyl groups	NaOH/urea	25	0.36~0.84				[134]
Cellulose 2-hydroxypropyl-trimethylammonium chloride derivatives	NaOH/urea	RT	0.17~0.50				[135]

It is seen that mild conditions are required for the modification of cellulose in those stable media, indicating great potential to prepare cellulose derivatives with uniform microstructure. For example, hydroxypropyl cellulose (HPC) and methylcellulose (MC) with relatively high DS have been synthesized from cellulose in NaOH/urea aqueous solution at 25 °C, in satisfactory yields [127]. Heinze and coworkers [131,132] have successfully synthesized CMC in either NaOH/urea or LiOH/urea aqueous solution. They found that the total DS of the CMC could be controlled by varying the molar ratio of reagents and the reaction temperature for both reaction media. Of note, structure analysis by HPLC for the completely depolymerized CMC showed that the mole fractions of the different carboxymethylated repeating units, as well as those of unmodified glucose, follow a simple statistic pattern (Fig. 6.30), indicating distribution of the carboxymethyl groups of the anhydroglucose in the order O-6 > O-2 > O-3 position.

By reacting cellulose with 3-chloro-2-hydroxypropyl-trimethylammonium chloride (CHPTAC) in the NaOH/urea aqueous solution at room temperature, quaternized cellulose (QC) was also homogeneously synthesized [133]. As shown in Figure 6.31,

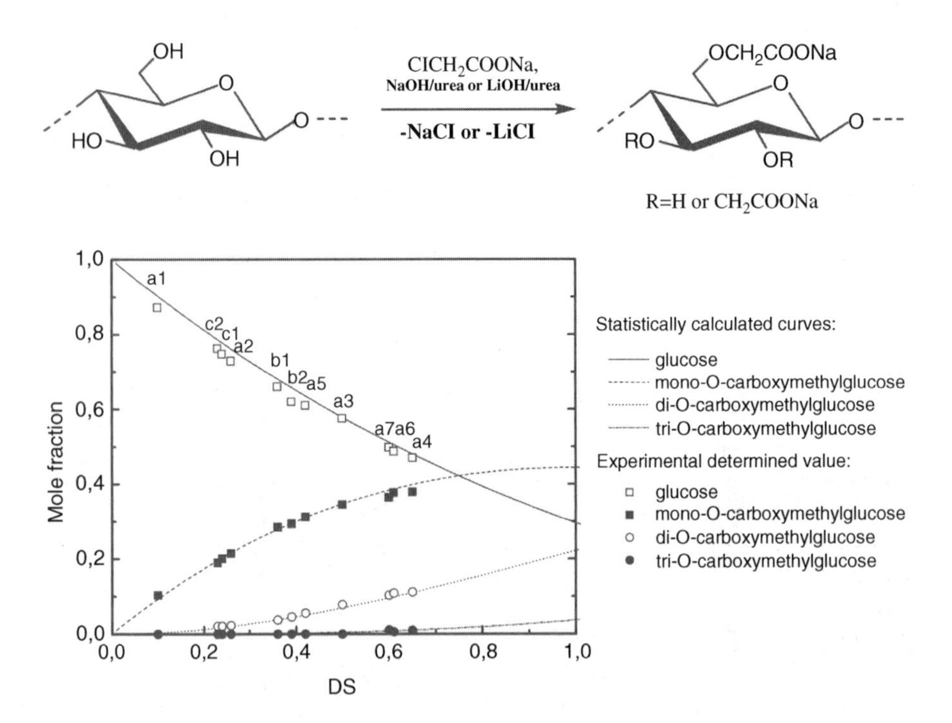

FIGURE 6.30 Reaction scheme of cellulose carboxymethylation in NaOH/urea or LiOH/urea aqueous solution (top), and the mole fractions for the different of carboxymethylated repeating units and the unmodified glucose of the CMC synthesized in LiOH/urea aqueous solution (bottom). (Reproduced from Ref. [131,132] with permission from Elsevier and Wiley © 2009, 2010.)

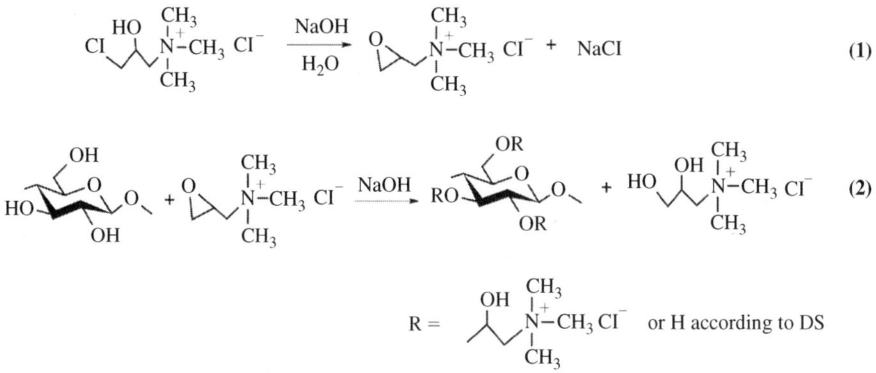

$$Cl\text{-}CH_2CH(OH)CH_2\text{-}N^{+}(CH_3)_3\ Cl^{-} \xrightarrow[H_2O]{NaOH} \text{(epoxide)}\text{-}CH_2\text{-}N^{+}(CH_3)_3\ Cl^{-} + NaCl \quad (1)$$

$$\text{Cell-OH} + \text{(epoxide)}\text{-}CH_2\text{-}N^{+}(CH_3)_3\ Cl^{-} \xrightarrow{NaOH} \text{Cell-OR} + HO\text{-}CH_2CH(OH)CH_2\text{-}N^{+}(CH_3)_3\ Cl^{-} \quad (2)$$

$$R = CH_2CH(OH)CH_2\text{-}N^{+}(CH_3)_3\ Cl^{-} \text{ or H according to DS}$$

FIGURE 6.31 Scheme of homogeneous quaternization of cellulose in NaOH/urea aqueous solution. (Reproduced from Ref. [133] with permission from the American Chemical Society © 2008.)

epoxide is produced in situ from CHPTAC under alkali condition, and QC is then formed through reaction between the cellulose sodium alkoxide and the epoxide or the CHPTAC. In addition to the main reaction of the cationization reaction of cellulose, side reaction occurs to form diols at the same time. The total DS of the obtained QCs were in the range of 0.20~0.63, and could be tuned by changing the molar ratio of the reagents and the reaction time. The QC was water soluble with typical properties of a polyelectrolyte in aqueous solution. It has been discovered that the QC exhibited high flocculation efficiency and antimicrobial activity [136]. Moreover, the QC displayed relatively low cytotoxicity, and the results of gel retardation assay indicated that QCs could condense DNA efficiently [133]. Therefore, the quaternized cellulose derivatives prepared in the NaOH/urea aqueous solution have been proposed as promising nonviral gene carriers [137]. Zhou and coworkers [137] investigated the factors affecting the gene-transfection efficiency in detail. Their results show that DNA can be bound to QC efficiently, which are affected by both molecular weight and DS of QC. The formed QC/DNA complexes exhibit effective transfection in comparison to the naked DNA. It is interesting to note that the novel QC/DNA complexes are stable in the presence of serum, and the transfection efficiency is not inhibited by the serum (Fig. 6.32), which is very important for practical applications [138,139]. Furthermore, QC synthesized in this way could be cross-linked via ionic bonding to prepare nanoparticles, whose physicochemical properties could be tuned by controlling DS of QC to be used as protein carriers [140].

If the QC with positive charges was cross-linked with negative charged CMC in water, ampholytic hydrogel could be obtained to have pH and salt-responsive properties [141]. The microstructure of the obtained ampholytic hydrogel could be adjusted by varying its composition, resulting in its tunable equilibrium swelling ratio, pH, and salt sensitivities (Fig. 6.33). It has been declared that the CMC contributed mainly to increasing the swelling due to its strong water adsorption, while the ammonium groups of the QC account for the sensitivity of the gel to environmental variations.

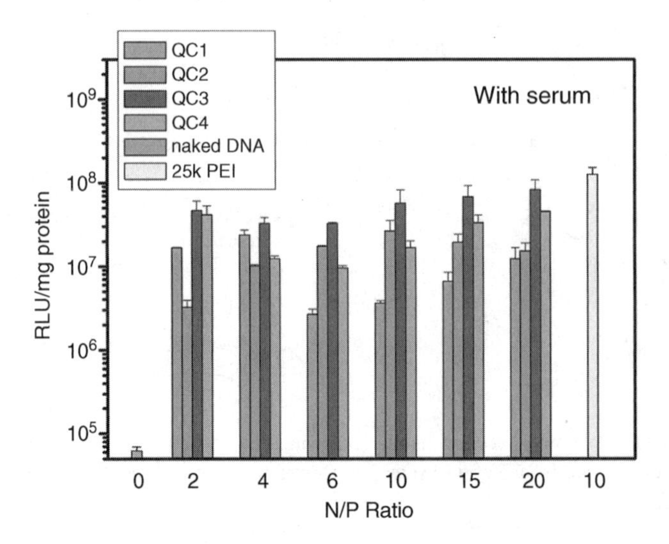

FIGURE 6.32 Luciferase expression in 293T cells transfected by QC/DNA complexes in the presence of 10% serum at different N/P ratios, and the control of the transfection efficiencies of the naked DNA and 25 kDa branched polyethylenimine (PEI). (Reproduced from Ref. [137] with permission from the American Chemical Society © 2010.)

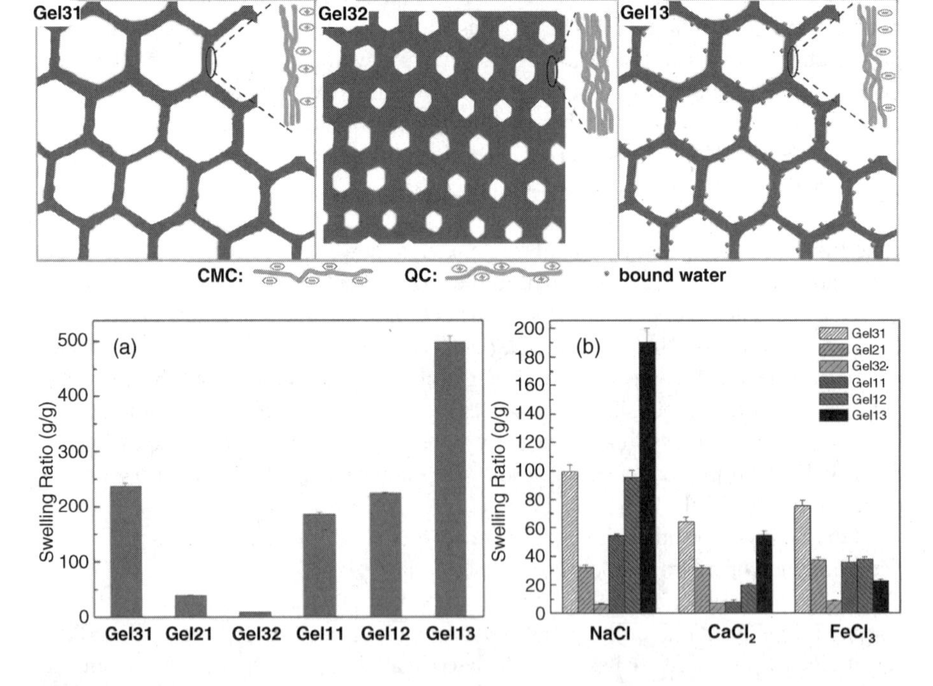

FIGURE 6.33 Schematic structures (top) and equilibrium swelling ratios (bottom) of the QC/CMC hydrogels in ultrapure water (**a**) and in 0.01 M different salt solutions (**b**). (Reproduced from Ref. [141] with permission from the American Chemical Society © 2011.)

6.5 SUMMARY AND OUTLOOK

The urgent demand for sustainable development all over the world requires more and more attention on the usage of renewable and environment-friendly raw materials [142]. Being the richest sustainable and natural polymer, cellulose is experiencing growing development in both research and application [1,2,143,144].

The invention of the dissolution of cellulose in the novel solvent systems such as the precooled NaOH/urea aqueous solution may open new routes to exploit various cellulose-based materials. The novel cellulose-dissolution method uses low-cost reagents, yields essentially nontoxic by-products, and consumes less energy than heat-treatment method to be a true "green" technology. Therefore, the discovery of dissolving cellulose in the novel solvent systems might have a vast impact on both academica and industrial processing of the oldest material. For instance, the solvent has been used for pretreatment of bamboo for bioethanol production, [145] and was used to pretreat corn starch to increase its reaction efficiency [146]. The solvent pretreatment could also lead to enhancements of enzymatic hydrolysis [147] and saccharification [148,149] of cellulose, indicating the potential in energy conversion of cellulose in the novel solvent systems. Furthermore, the NaOH/urea solvent system was found to play an important role for regenerated-cellulosic fabrics for improving their pilling behaviors [150]. Based on the dissolution of cellulose in the solvent, novel hierarchical porous materials were fabricated by replication of natural cellulosic substances and successive removal of the template cellulose components [151]. In addition, the novel solvent of NaOH/urea aqueous solution has even been utilized to dissolve another important natural polymer of chitin [152–155]. Through repeated freezing/thawing, chitin was dissolved completely in a 8 wt% NaOH/4 wt% urea aqueous solution to obtain a transparent solution, which was proved by ^{13}C-NMR results to be a nonderivation dissolution of chitin.

Moreover, the invention of the novel solvent systems has inspired extensive investigations on new aqueous solvents for cellulose. For example, NaOH/thiourea/urea aqueous solution other than single NaOH/urea or NaOH/thiourea aqueous solution was used to dissolve cellulose; [156,157] similar to urea or thiourea as acceptor of hydrogen bonding, poly(ethylene glycol) (PEG) was added into NaOH aqueous solution to dissolve cellulose, and to connect with the hydroxyl groups in cellulose and prevent the regeneration of cellulose through the inter- and intrachains association [158].

So far, the fundamental research on dissolution of cellulose in the precooled NaOH/urea aqueous solution has been extensively carried out. However, challenges remain and more detailed fundamental knowledge to fully understand the dissolving and the regenerating of cellulose in the solvent systems is required. For example, DSC measurements indicated that urea was not directly interacting with NaOH or with cellulose during dissolution, implying the requirement of a new insight into the role of urea in the system; [159] the formation of the porous microstructure of the regenerate cellulose was based on hypothesis of phase separation [69,70,78]. Furthermore, truly industrial production of regenerate cellulose fiber from the novel solvent systems has not been achieved, and neither the industrial fabrication of regenerate cellulose film nor the synthesis of cellulose derivatives has yet been started. We propose more efforts should be made in the future in order to revolutionize the traditional cellulose-processing technologies. Nevertheless, the dissolution of cellulose in the novel solvent systems removes most of the disadvantages of the traditional cellulose-processing technologies, and the progress in this area can be accessed by our own prolific creativity and effective cooperation.

ACKNOWLEDGMENTS

The authors thank Prof. L. Zhang for valuable comments and discussions. This work was financially supported by the National Basic Research Program of China (973 Program, 2010CB732203) and the Fundamental Research Funds for the Central Universities of China (No. 2010121055).

REFERENCES

1. D. Klemm, B. Heublein, H. P. Fink, and A. Bohn, *Angew. Chem. Int. Ed.* **2005**, 44, 3358.
2. J. Zhang and J. Zhang, *Acta Polym. Sin.* **2010**, 12, 1376.
3. E. Schweitzer, *J. Prakt. Chem.* **1857**, 72, 109.
4. C. F. Cross, B. T. Bevan, and C. Beadle, *Ber. Dtsch. Chem. Ges.* **1893**, 26, 1090.
5. C. F. Cross, B. T. Bevan, and C. Beadle, *Ber. Dtsch. Chem. Ges.* **1893**, 26, 2520.
6. K. Ekman, V. Eklund, J. Fors, J. I. Huttunen, J.-F. Selin and O. T. Turunen, *Cellulose Structure, Modification and Hydrolysis*, (Eds. R. A. Young and R. M. Rowell), Wiley, New York, **1986**, 131–148.
7. M. Terbojevich, A. Cosani, G. Conio, A. Ciferri, and E. Bianchi, *Macromolecules* **1985**, 18, 640.
8. K. J. Edgar, K. M. Arnold, W. W. Blount, J. E. Lawniczak, and D. W. Lowman. *Macromolecules* **1995**, 28, 4122.
9. N. Tamai, D. Tatsumi, and T. Matsumoto, *Biomacromolecules* **2004**, 5, 422.
10. H. Boerstoel, H. Maatman, J. B. Westerink, and B. M. Koenders, *Polymer* **2001**, 42, 7371.
11. S. Fischer, W. Voigt and K. Fischer, *Cellulose* **1999**, 6, 213.
12. K. Saalwächter, W. Burchard, P. Klüfers, G. Kettenbach, P. Mayer, D. Klemm, and S. Dugarmaa, *Macromolecules* **2000**, 33, 4094.
13. S. Fischer, H. Leipner, K. Thü1mmler, E. Brendler, and J. Peters, *Cellulose* **2003**, 10, 227.
14. L. Zhang and J. Zhou, Patent CN 00114486.3, **2003**.
15. J. Zhou and L. Zhang, *Polym. J.* **2000**, 32, 866.
16. L. Zhang, D. Ruan, and S. Gao, *J. Polym. Sci., Part B: Polym. Phys.* **2002**, 40, 1521.
17. H. Firgo, M. Eibl, and D. Eichinger, *Lenzinger Ber.* **1995**, 75, 47.
18. M. Yamane, M. Mori, and K. Saito and Okajima, *Polym. J.* **1996**, 28, 1039.
19. J. F. Masson and R. S. J. Manley, *Macromolecules* **1991**, 24, 5914.
20. T. Heinze and T. Liebert, *Prog. Polym. Sci.* **2001**, 26, 1689.

21. R. P. Swatloski, S. K. Spear, J. D. Holbrey, and R. D. Rogers, *J. Am. Chem. Soc.* **2002**, 124, 4974.
22. H. Zhang, J. Wu, J. Zhang and J. He, *Macromolecules.* **2005**, 38, 8272.
23. J. Cai and L. Zhang, *Macromol. Biosci.* **2005**, 5, 539.
24. J. Cai and L. Zhang, *Biomacromolecules* **2006**, 7, 183.
25. J. Cai, L. Zhang, C. Chang, G. Cheng, X. Chen, and B. Chu, *ChemPhysChem* **2007**, 8, 1572.
26. J. Cai, L. Zhang, S. Liu, Y. Liu, X. Xu, X. Chen, B. Chu, X. Guo, J. Xu, H. Cheng, C. C. Han, and S. Kuga, *Macromolecules* **2008**, 41, 9345.
27. H. Qi, C. Chang, and L. Zhang, *Cellulose* **2008**, 15, 779.
28. H. Qi, C. Chang and L. Zhang, *Green Chem.* **2009**, 11, 177.
29. R. Li, C. Chang, J. Zhou, L. Zhang, W. Gu, C. Li, S. Liu, and S. Kuga, *Ind. Eng. Chem. Res.* **2010**, 49, 11380.
30. http://www.chemcp.com
31. http://cell.sites.acs.org/
32. O. D. Bonner, J. M. Bednarek, and R. K. Arisman, *J. Am. Chem. Soc.* **1977**, 99, 2898.
33. C. Roy, T. Budtova, P. Navard, and O. Bedue, *Biomacromolecules* **2001**, 2, 687.
34. H. Qi, Q. Yang, L. Zhang, T. Liebert, and T. Heinze, *Cellulose* **2011**, 18, 237.
35. I. Nehls, W. Wagenknecht, B. Philipp, and D. Stscherbina, *Prog. Polym. Sci.* **1994**, 19, 29.
36. M. Hattori, T. Koga, Y. Shimaya, and M. Saito, *Polym. J.* **1998**, 30, 43.
37. D. Gagnaire, D. Mancier, and M. Vincendon, *J. Polym. Sci., Polym. Chem. Ed.* **1980**, 18, 13.
38. C. L. McCormick, P. A. Callais, and B. H. Hutchinson, *Macromolecules* **1985**, 18, 2394.
39. C. Cuissinat and P. Navard, *Macromol. Symp.* **2006**, 244, 19.
40. A. Lue, L. Zhang and D. Ruan, *Macromol. Chem. Phys.* **2007**, 208, 2359.
41. A. Lue and L. Zhang, *J. Phys. Chem. B* **2008**, 112, 4488.
42. A. Lue and L. Zhang, *Macromol. Biosci.* **2009**, 9, 488.
43. L. Yan, J. Chen and P. R. Bangal, *Macromol. Biosci.* **2007**, 7, 1139.
44. Q. Yang, H. Qi, A. Lue, K. Hu, G. Cheng, and L. Zhang, *Carbohydr. Polym.* **2011**, 83, 1185.
45. J. Zhou, L. Zhang, and J. Cai, *J. Polym. Sci., Part B: Polym. Phys.* **2004**, 42, 347.
46. X. Chen, C. Burger, F. Wan, J. Zhang, L. Rong, B. Hsiao, B. Chu, J. Cai, and L. Zhang, *Biomacromolecules* **2007**, 8, 1918.
47. L. Schulz, B. Seger, and W. Buchard, *Macromol. Chem. Phys.* **2000**, 201, 2008.
48. B. Seger, Ph. D. Thesis, University of Freiburg, 1996.
49. K. Fischer, *Papier* **1994**, 48, 769.
50. B. Morgenstern and T. Röder, *Papier* **1998**, 52, 713.
51. J. Cai, Y. Liu, and L. Zhang, *J. Polym. Sci., Part B: Polym. Phys.* **2006**, 44, 3093.
52. S. Zhang, F. Li, J. Yu, and Y. Hsieh, *Carbohydr. Polym.* **2010**, 81, 668.
53. J. Cai, L. Zhang, J. Zhou, H. Li, H. Chen, and H. Jin, *Macromol. Rapid Commun.* **2004**, 25, 1558.

54. X. Chen, C. Burger, D. Fang, D. Ruan, L. Zhang, B. Hsiao, and B. Chu, *Polymer*. **2006**, 47, 2839.
55. J. Cai, L. Zhang, J. Zhou, H. Qi, H. Chen, T. Kondo, X. Chen, and B. Chu, *Adv. Mater.* **2007**, 19, 821.
56. Y. Mao, L. Zhang, J. Cai, J. Zhou, and T. Kondo, *Ind. Eng. Chem. Res.* **2008**, 47, 8676.
57. D. Ruan, L. Zhang, J. Zhou, H. Jin, and H. Chen, *Macromol. Biosci.* **2004**, 4, 1105.
58. D. Ruan, L. Zhang, A. Lue, J. Zhou, H. Chen, X. Chen, B. Chu, and T. Kondo, *Macromol. Rapid Commun.* **2006**, 27, 1495.
59. N. A. El-Wakil and M. L. Hassan, *J. Appl. Polym. Sci.* **2008**, 109, 2862.
60. C. Woodings, *Regenerated Cellulose Fibres*, Woodhead Publishing Ltd, England, **2001**,
61. H. Qi, J. Cai, Zhang, Y. Nishiyama, and A. Rattaz, *Cellulose* **2008**, 15, 81.
62. S. Liu, L. Zhang, J. Zhou, and R. Wu, *J. Phys. Chem. C* **2008**, 112, 4538.
63. S. Liu, L. Zhang, J. Zhou, J. Xiang, J. Sun, and J. Guan, *Chem. Mater.* **2008**, 20, 3623.
64. H. Qi, X. Sui, J. Yuan, Y. Wei, and L. Zhang, *Macromol. Mater. Eng.* **2010**, 295, 695.
65. S. Zhang, F. Li, and J Yu, *J. Eng. Fiber. Fabr.* **2011**, 6, 31.
66. L. Zhang, D. Ruan, and J. Zhou, *Ind. Eng. Chem. Res.* **2001**, 40, 5923.
67. J. Zhou, L. Zhang, J. Cai, and H. Shu, *J. Membr. Sci.* **2002**, 210, 77.
68. L. Zhang, Y. Mao, J. Zhou, and J. Cai, *Ind. Eng. Chem. Res.* **2005**, 44, 522.
69. Y. Mao, J. Zhou, J. Cai, and L. Zhang, *J. Membr. Sci.* **2006**, 279, 246.
70. J. Cai, L. Wang, and L. Zhang, *Cellulose* **2007**, 14, 205.
71. S. Liang, L. Zhang, Y. Li, and J. Xu, *Macromol. Chem. Phys.* **2007**, 208, 594.
72. S. Liu, L. Zhang, Y. Sun, Y. Lin, X. Zhang, and Y. Nishiyama, *Macromol. Biosci.* **2009**, 9, 29.
73. S. Liu, J. Zeng, D. Tao, and L. Zhang, *Cellulose* **2010**, 17, 1159.
74. L. Yan, Y. Wang, and J. Chen, *J. Appl. Polym. Sci.* **2008**, 110, 1330.
75. M. Phisalaphong, T. Suwanmajo, and P. Sangtherapitikul, *J. Appl. Polym. Sci.* **2008**, 107, 292.
76. G. Yang, L. Zhang, C. Yamane, I. Miyamoto, M. Inamoto, and K. Okajiama, *J. Membr. Sci.* **1998**, 139, 47.
77. L. Zhang and G. Yang, *Chin. J. Appl. Chem.* **1991**, 8 (3), 17.
78. L. Weng, L. Zhang, D. Ruan, L. Shi, and J. Xu, *Langmuir* **2004**, 20, 2086.
79. H. Qi, J. Cai, L. Zhang, and S. Kuga, *Biomacromolecules* **2009**, 10, 1597.
80. N. Jia, S. Li, J. Zhu, M. Ma, F. Xu, B. Wang, and R. Sun, *Mater. Lett.* **2010**, 64, 2223.
81. D. Han, L. Yan, W. Chen, W. Li and P. R. Bangal, *Carbohydr. Polym.* **2011**, 83, 966.
82. G. Yang, L. Zhang, H. Han, and J. Zhou, *J. Appl. Polym. Sci.* **2001**, 81, 3260.
83. J. Zhou and L. Zhang, *J. Polym. Sci., Part B: Polym. Phys.* **2001**, 39, 451.
84. H. Zheng, J. Zhou, Y. Du, and L. Zhang, *J. Appl. Polym. Sci.* **2002**, 86, 1679.
85. L. Zhang, J. Guo, and Y. Du, *J. Appl. Polym. Sci.* **2002**, 86, 2025.
86. G. Yang, X. Xiong, and L. Zhang, *J. Membr. Sci.* **2002**, 201, 161.
87. Y. Chen and L. Zhang, *J. Appl. Polym. Sci.* **2004**, 94, 748.
88. Y. Chen, L. Zhang, J. Gu, and J. Liu, *J. Membr. Sci.* **2004**, 241, 393.
89. S. Liang, L. Zhang, and J. Xu, *J. Membr. Sci.* **2007**, 287, 19.

90. J. Kim, N. Wang, Y. Chen, S.-K. Lee and G.-Y. Yun, *Cellulose* **2007**, 14, 217.
91. M. Phisalaphong, T. Suwanmajo, and P. Tammarate, *J. Appl. Polym. Sci.* **2008**, 107, 3419.
92. E. V. R. Almeida, E. Frollini, A. Castellan, and V. Coma, *Carbohydr. Polym.* **2010**, 80, 655.
93. P. R. Chang, R. Jian, P. Zheng, J. Yu, and X. Ma, *Carbohydr. Polym.* **2010**, 79, 301
94. J. Zhou, S. Liu, J. Qi, and L. Zhang, *J. Appl. Polym. Sci.* **2006**, 101, 3600.
95. D. Ruan, Q. Huang, and L. Zhang, *Macromol. Mater. Eng.* **2005**, 290, 1017.
96. S. Liu, J. Zhou, L. Zhang, J. Guan, and J. Wang, *Macromol. Rapid Commun.* **2006**, 27, 2084.
97. D. Ke, S. Liu, K. Dai, J. Zhou, L. Zhang, and T. Peng, *J. Phys. Chem. C.* **2009**, 113, 16021.
98. J. Zeng, S. Liu, J. Cai, and L. Zhang, *J. Phys. Chem. C* **2010**, 114, 7806.
99. X. Xiong, J. Duan, W. Zou, X. He, and W. Zheng, *J. Membr. Sci.* **2010**, 363, 96.
100. N. Isobe, M, Sekine, S. Kimura, M. Wada, and S. Kuga, *Cellulose* **2011**, 18, 327.
101. J. Huang and L. Zhang, *J. Appl. Polym. Sci.* **2002**, 86, 1799.
102. Y. Lu and L. Zhang, *Ind. Eng. Chem. Res.* **2002**, 41, 1234.
103. Y. Lu, L. Zhang, and P. Xiao, *Polym. Degrad. Stab.* **2004**, 86, 51.
104. X. Cao, R. Deng, and L. Zhang, *Ind. Eng. Chem. Res.* **2006**, 45, 4193.
105. D. Ruan, A. Lue, and L. Zhang, *Polymer* **2008**, 49, 1027.
106. Y. Wang and L. Chen, *Carbohydr. Polym.* **2011**, 83, 1937.
107. J. Zhou, C. Chang, R. Zhang, and L. Zhang, *Macromol. Biosci.* **2007**, 7, 804.
108. C. Chang, L. Zhang, J. Zhou, L. Zhang, and J. Kennedy, *Carbohydr. Polym.* **2010**, 82, 122.
109. C. Chang, A. Lue, and L. Zhang, *Macromol. Chem. Phys.* **2008**, 209, 1266.
110. C. Chang, B. Duan, and L. Zhang, *Polymer* **2009**, 50, 5467.
111. C. Chang, B. Duan, J. Cai, and L. Zhang, *Eur. Polym. J.* **2010**, 46, 92.
112. J. Cai, S. Kimura, M. Wada, S. Kuga ,and L. Zhang, *ChemSusChem* **2008**, 1, 149.
113. J. Cai, S. Kimura, M. Wada, and S. Kuga, *Biomacromolecules* **2009**, 10, 87.
114. C. Chang, J. Peng, L. Zhang, and D. Pang, *J. Mater. Chem.* **2009**, 19, 7771.
115. X. Luo, S. Liu, J. Zhou, and L. Zhang, *J. Mater. Chem.* **2009**, 19, 3538.
116. X. Luo and L. Zhang, *J. Hazard. Mater.* **2009**, 171, 340.
117. X. Luo and L. Zhang, *Biomacromolecules* **2010**, 11, 2896.
118. X. Xiong, L. Zhang, and Y. Wang, *J. Chromatogr. A* **2005**, 1063, 71.
119. X. Luo and L. Zhang, *J. Chromatogr. A* **2010**, 1217, 5922.
120. D. Zhou, L. Zhang, J. Zhou, and S. Guo, *Water Res.* **2004**, 38, 2643.
121. D. Zhou, L. Zhang, J. Zhou, and S. Guo, *J. Appl. Polym. Sci.* **2004**, 94, 684.
122. D. Zhou, L. Zhang, and S. Guo, *Water Res.* **2005**, 39, 3755.
123. D. Klemm, B. Philipp, T. Heinze, U. Heinze, and W. Wagenknecht, *Comprehensive Cellulose Chemistry*, Wiley VCH, Weinheim, **1998**, 130–155.
124. T. R. Dawsey and C. L. McCormick, *J. Macromol. Sci., Rev. Macromol. Chem. Phys.* **1990**, 30, 405.

125. G. T. Ciacco, T. F. Liebert, E. Trollini, and T. J. Heinze, *Cellulose* **2003**, 10, 125.
126. M. Vieira, T. Liebert, and T. Heinze, *Recent Advances in Environmentally Compatible Polymers*, (Ed. J. F. Kennedy), Woodhead, Cambridge, **2001**, 53–60.
127. J. Zhou, L. Zhang, Q. Deng, and X. Wu, *J. Polym. Sci., Part A: Polym. Chem.* **2004**, 42, 5911.
128. M. C. V. Nagel, A. Koschella, K. Voiges, P. Mischnick, and T. Heinze, *Eur. Polym. J.* **2010**, 46, 1726.
129. Q. Zhou, L. Zhang, M. Li, X. Wu, and G. Cheng, *Polym. Bull.* **2005**, 53, 243.
130. J. Zhou, Q. Li, Y. Song, L. Zhang, and X. Lin, *Polym. Chem.* **2010**, 1, 1662.
131. H. Qi, T. Liebert, F. Meister, and T. Heinze, *React. Funct. Polym.* **2009**, 69, 779.
132. H. Qi, T. Liebert, F. Meister, L. Zhang, and T. Heinze, *Macromol. Symp.* **2010**, 294, 125.
133. Y. Song, Y. Sun, X. Zhang, J. Zhou, and L. Zhang, *Biomacromolecules* **2008**, 9, 2259.
134. Y. Song, J. Zhou, L. Zhang, and X. Wu, *Carbohydr. Polym.* **2008**, 73, 18.
135. L. Yan, H. Tao, and P. R. Banga, *Clean Soil Air Water* **2009**, 37, 39.
136. Y. Song, J. Zhang, W. Gan, J. Zhou, and L. Zhang, *Ind. Eng. Chem. Res.* **2010**, 49, 1242.
137. Y Song, H. Wang, X. Zeng, Y. Sun, X. Zhang, J. Zhou, and L. Zhang, *Bioconjugate Chem.* **2010**, 21, 1271.
138. G. Borchard, *Adv. Drug Delivery Rev.* **2001**, 52, 145.
139. J. H. Jeong, S. W. Kim, and T. G. Park, *Prog. Polym. Sci.* **2007**, 32, 1239.
140. Y. Song, J. Zhou, Q. Li, Y. Guo, and L. Zhang, *Macromol. Biosci.* **2009**, 9, 857.
141. C. Chang, M. He, J. Zhou, and L. Zhang, *Macromolecules* **2011**, 44, 1642.
142. A. K. Mohanfty, M. Isra, and L. T. Drzal, *J. Polym. Environ.* **2002**, 10, 19.
143. S. M. Read and T. Bacic, *Science* **2002**, 295, 59.
144. M. Jarvis, *Nature* **2003**, 426, 611.
145. M. Li, Y. Fan, F. Xu, R. Sun, and X. Zhang, *Ind. Crop. Prod.* **2010**, 32, 551.
146. F. Geng, P. R. Chang, J. Yu, and X. Ma, *Carbohydr. Polym.* **2010**, 80, 360.
147. Y. Zhao, Y. Wang, J. Zhu, A. Ragauskas, and Y. Deng, *Biotechnol. Bioeng.* **2008**, 99, 1320.
148. C. Kuo and C. Lee, *Carbohydr. Polym.* **2009**, 77, 41.
149. K. Wang, H. Y. Yang, F. Xu, and R. C. Sun, *Bioresour. Technol.* **2011**, 102, 4524.
150. A. Ehrhardt, H. Bui, H. Duelli, and T. Bechtold, *J. Appl. Polym. Sci.* **2010**, 115, 2865.
151. Y. Gu and J. Huang, *J. Mater. Chem.* **2009**, 19, 3764.
152. X. Hu, Y. Du, Y. Tang, Q. Wang, T. Feng, J. Yang, and J. F. Kennedy, *Carbohydr. Polym.* **2007**, 70, 451.
153. G. Li, Y. Du, Y. Tao, Y. Liu, S. Li, X. Hu, and J. Yang, *Carbohydr. Polym.* **2010**, 80, 970.
154. C. Chang, S. Chen, and L. Zhang, *J. Mater. Chem.* **2011**, 21, 3865.
155. X. Hu, Y. Tang, Q. Wang, Y. Li, J. Yang, Y. Du, and J. F. Kennedy, *Carbohydr. Polym.* **2011**, 83, 1128.
156. H. Jin, C. Zha, and L. Gu, *Carbohydr. Res.* **2007**, 342, 851.
157. S. Zhang, C. Fu, B. Yuan, F. Li, J. Yu, and L. Gu, *Cell. Chem. Technol.* **2008**, 42, 147.
158. L. Yan and Z. Gao, *Cellulose* **2008**, 15, 789.
159. M. Egal, T. Budtova, and P. Navard, *Cellulose* **2008**, 15, 361.

CHAPTER 7

Organosolv Biorefining Platform for Producing Chemicals, Fuels, and Materials from Lignocellulose

XUEJUN PAN

7.1 INTRODUCTION

An organosolv process refers to an operation using an organic solvent to treat lignocellulose with or without catalyst. The solvents that have been used include alcohols (methanol, ethanol, propanol, butanol, and glycol), organic acids (formic and acetic acids), phenol or cresols, acetone or other ketones, and amines (ethylenediamine, hexamethylenediamine, and methylamine). Alkalis, acids, and salts have been investigated as catalysts in organosolv processes [1–3]. Organosolv processes have been extensively investigated as pulping methods to produce fibers (pulp) from wood for papermaking [4–9] and as a biorefining platform to pretreat or fractionate lignocellulose into its major components (cellulose, lignin, and hemicellulose) for producing chemicals, liquid fuels, and materials [10–15].

Compared to traditional pulping methods (such as kraft and sulfite processes) [16] and pretreatment technologies (such as steam explosion, dilute acid, ammonia fiber expansion, etc.) [17, 18], in general, organosolv processes have the following advantages.

- Applicable to various feedstocks ranging from grasses to woods.
- Ease of solvent recovery, no need for capital-intensive chemical recovery system like that for kraft pulping.
- Able to fractionate lignocellulosic biomass into its major components for individual utilization.
- Able to generate high-purity and high-reactivity lignin for high-value coproducts development.

The Role of Green Chemistry in Biomass Processing and Conversion, First Edition.
Edited by Haibo Xie and Nicholas Gathergood.

- Cellulose fraction has excellent enzymatic digestibility, which is an ideal pretreated substrate for bioconversion.
- Good potential to convert hemicellulosic sugars to valuable chemicals.

Among the organosolv processes investigated, such as ethanol, formic acid, and acetic acid processes [7, 19, 20], the process using ethanol is the most promising one because of the simplicity of the process, the benign nature of the solvent, and the ease of the solvent recovery. In addition, (1) organosolv ethanol process has been demonstrated as a successful pulping technology for hardwood in paper industry at a precommercial scale [5]; (2) the process has been also demonstrated as an efficient biorefining platform for lignocellulose at a pilot scale [21]; (3) ethanol is an environment-friendly and inexpensive solvent; (4) the process has been extensively investigated and well understood; and (5) ethanol can be produced on site from the fermentation of biomass-derived sugars.

This chapter provides an overview of organosolv ethanol process regarding its history, status quo, and future direction. Emphasis is placed on the application of the organosolv ethanol process as a biorefining platform for converting lignocellulose to fuels, chemicals, and materials. No effort has been made to ensure comprehensive coverage of the literature [1, 3, 22, 23].

7.2 ABOUT ORGANOSOLV ETHANOL PROCESS

The organosolv ethanol process was developed by the University of Pennsylvania and the General Electric Company in the 1970s, and was later modified by the Canadian pulp and paper industry into the Alcell® pulping process for hardwoods [5, 24, 25]. The Alcell process provides an alternative to the kraft pulping process, producing comparable pulp for paper from hardwoods. The pulping process uses a blend of ethanol and water with a small amount of mineral acid to extract most of the lignin from wood chips. A pilot plant was completed in 1989 at a cost of more than 65 million dollars at Repap's Miramichi pulp and paper mill in New Brunswick, Canada. During the operation from 1989 to 1996, more than 3200 batch cooks of high-quality pulp were produced with a coproduct of more than 3700 tonnes of organosolv lignin [3, 5, 25]. In addition to the pulp for papermaking, high-quality lignin is a significant selling point of the organosolv ethanol process [26–28]. The lignin has demonstrated potential for manufacturing various industrial products such as adhesives and biodegradable polymers [29–31], as will be discussed later in this chapter. Furthermore, the process produces valuable coproducts from hemicellulose (pentoses), such as furfural and acetic acid [3, 32].

Unfortunately, with the financial collapse and consequent breakup of Repap in 1997, the demonstration plant was closed. Fortunately, the technology has since been acquired by Lignol Innovations Corporation (later Lignol Energy Corporation) of Vancouver, Canada, and is now being commercialized as a biorefinery technology with the cellulose fraction being used for ethanol production instead of pulp [21]. In

January 2008, Lignol was selected to receive an award of up to US$30 million under the US Department of Energy (DOE) "Demonstration of Integrated Biorefinery Operations for Producing Biofuels and Chemical/Materials Products" Funding Opportunity Announcement (FOA) to build a demonstration-scale cellulosic ethanol plant at approximately one-tenth of the projected scale of a first-commercial facility in Colorado. However, the project was halted in 2009 due to the recession and market volatility. Then Lignol refocused and began to modify the scale and configuration of the proposed demonstration-scale project with the goal of developing a profitable, commercial-scale project. In July 2011, Lignol announced the withdrawal from DOE funding and with the intention to pursue development of a commercial project without the assistance of an award from the DOE [33].

7.3 CHEMISTRY OF ORGANOSOLV ETHANOL PROCESS

7.3.1 Reaction of Lignin

Studies in the field of lignin chemistry concluded that the cleavage of ether linkages is primarily responsible for lignin breakdown in organosolv processes [34, 35]. Model compound studies [36] have shown that α-aryl ether linkages are more easily cleaved than β-aryl ether linkages, especially in a lignin structural unit containing a free phenolic hydroxyl group in the *para*-position. In this case, the formation of a quinonemethide intermediate is necessary (Scheme 7.1a). The hydrolysis of benzyl ether linkage is facilitated by acid catalysis. It is also possible that a nucleophilic substitution reaction occurs at the benzyl position by an S_N2 mechanism (Scheme 7.1b) to cleave α-aryl ether linkage.

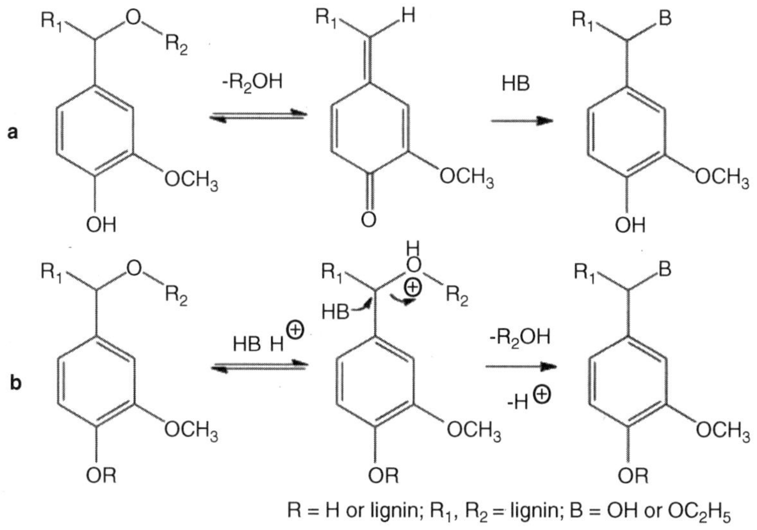

R = H or lignin; R_1, R_2 = lignin; B = OH or OC_2H_5

SCHEME 7.1 Cleavage of phenolic α-aryl ether linkage during organosolv process [34].

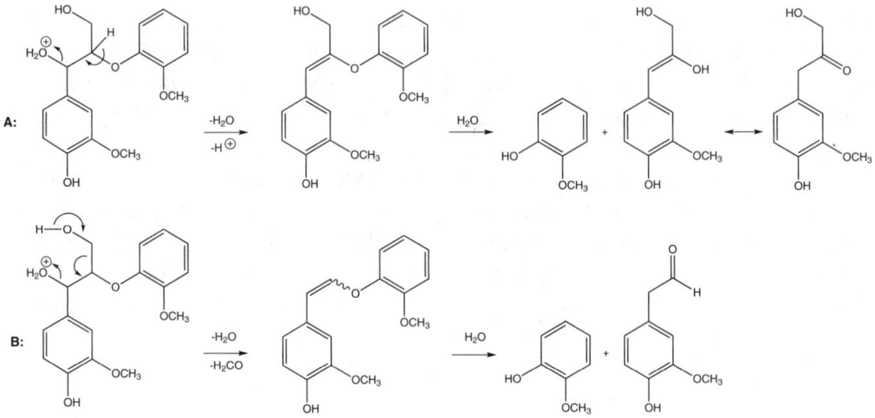

SCHEME 7.2 Cleavage of β-aryl ether linkage in lignin during organosolv process [34].

Easily hydrolysable α-aryl ether bonds are most readily broken, and β-aryl ether bonds are also broken under the conditions of organosolv pulping, in particular in more strongly acidic systems [35]. Scheme 7.2a illustrates one of the pathways for the cleavage of β-ether linkage from the study of the model compound [37]. Another proposed mechanism for the cleavage of β-ether linkage is through the elimination of γ-methylol group as formaldehyde followed by the hydrolysis of the resulting enol ether (Scheme 7.2b) [34]. In general, β-ether cleavage occurs in autocatalyzed and acid-catalyzed organosolv processes, and the extent of the cleavage is dependent on the conditions of the process. It is likely that β-ether cleavage is more significant in the processes that are more strongly acidic and in the processes where hardwood rather than softwood is the feedstock.

It has been proposed that inter- and intramolecular condensation of lignin also occurs during organosolv pulping. The direct cleavage of α-ether linkage forms a resonance-stabilized benzyl carbocation [38]. This reactive intermediate is able to undergo either inter- or intramolecular condensation. The condensation reactions were demonstrated by reactions of lignin model compounds [39], as shown in Scheme 7.3. When a veratrylglycerol-β-guaiacyl ether derivative (compound 3.1) was refluxed with propylveratrole in 85% formic acid, their condensation product (compound 3.2) formed. It is noted that the compound 3.2 is unable to undergo β-ether cleavage because of the unavailability of an α-hydroxyl to serve as a leaving group, so its formation represents an irreversible condensation process. Intermolecular condensation product (compound 3.3) was also detected. The intramolecular condensation not only makes the desirable β-ether cleavage impossible, but also protects the active benzyl carbocation from participation in the undesirable intermolecular condensation.

It has been well established that hardwoods are easier to delignify with organic solvents than softwoods [34, 40]. Besides the differences in the total

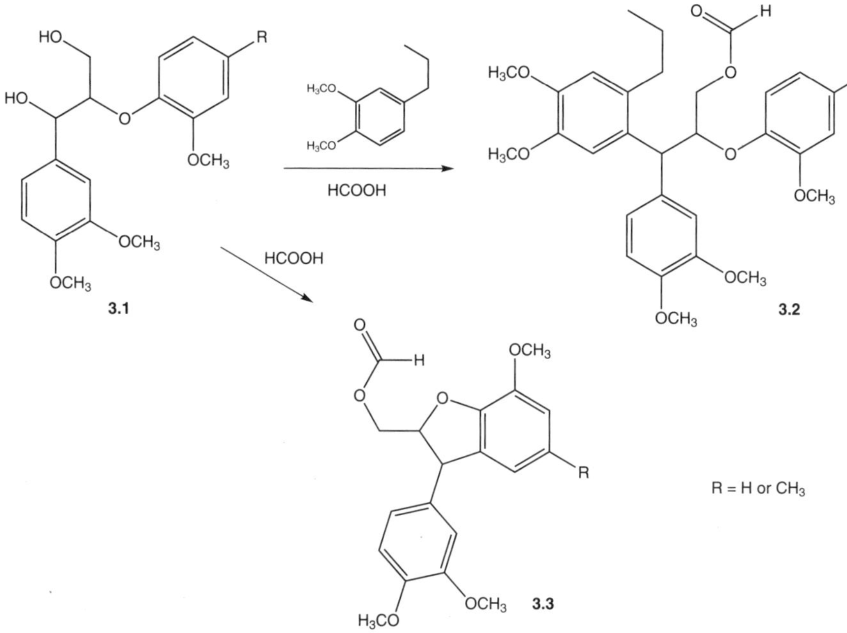

SCHEME 7.3 Inter- and intramolecular condensations of lignin model compounds [34].

lignin content, the structural differences of lignins also play an important role. For example, the higher amount of syringyl units in hardwood lignin inhibits condensation reactions [34]. Important parameters governing the course of delignification are pH, physical properties of the solvent (that govern its ability to dissolve lignin fragments), and chemical properties of the solvent (that govern its ability to participate in fragmentation reactions or inhibit recondensation of hydrolyzed lignin) [15].

7.3.2 Reaction of Cellulose and Hemicellulose

The glycosidic linkages in cellulose and hemicellulose are especially susceptible to acid-catalyzed hydrolytic attack leading to molecular degradation, the extent of which depends on the reaction conditions, such as the class of the acid, concentration, the reaction temperature, and duration. If sufficient hydrolysis occurs, then monosaccharides (sugars) will ultimately be produced; however, further acid treatment at high temperature results in the degradation/dehydration of the released monosaccharides, producing 2-furfuraldehyde (furfural, from pentoses) and 5-hydroxymethylfurfural (HMF, from hexoses). HMF tends to further degrade to levulinic acid (LA) and formic acid. In addition, furfural and HMF tend to condense under acidic condition at high temperature to form humins, which are a group of complicated and insoluble dark-brown polymers [41].

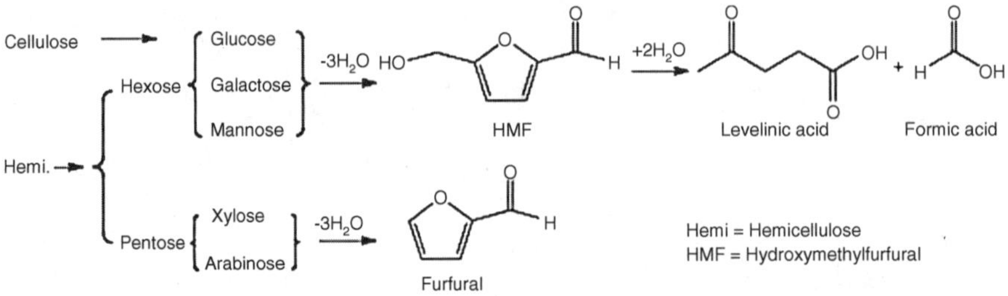

Cellulose → Hexose {Glucose, Galactose, Mannose} $\xrightarrow{-3H_2O}$ HMF $\xrightarrow{+2H_2O}$ Levelinic acid + Formic acid

Hemi. → Pentose {Xylose, Arabinose} $\xrightarrow{-3H_2O}$ Furfural

Hemi = Hemicellulose
HMF = Hydroxymethylfurfural

SCHEME 7.4 Hydrolysis and degradation reactions of cellulose and hemicellulose during acidic organosolv ethanol process.

Cellulose and hemicellulose experience similar acid-catalyzed hydrolysis and subsequent further degradation/dehydration in the autocatalyzed and acid-catalyzed organosolv ethanol processes. The hydrolysis and degradation reactions are undesirable for pulp and paper since they result in the reduction of pulp yield and strength. However, controlled prehydrolysis of cellulose is beneficial to enzymatic digestibility of cellulose since the shortened cellulose chains provide more cellulose termini for cellulases to work upon [42, 43]. Furthermore, removal of hemicellulose from biomass generates pores and increases surface area of substrate, and thereby improves the enzymatic accessibility of cellulose [44]. In some instances, sugar degradation/dehydration is desirable as well. For example, hemicellulosic sugars (both pentoses and hexoses) can be chemically converted to furfural, HMF, and LA. They are very important platform chemicals, from which liquid fuels, fertilizers, polymeric materials can be derived [41, 45]. The chemical reactions of cellulose and hemicellulose occurring during acidic organosolv ethanol process are summarized in Scheme 7.4.

7.4 ORGANOSOLV ETHANOL PROCESS AS A BIOREFINING PLATFORM

7.4.1 Description of Organosolv Ethanol Biorefining Platform

Recently, the ethanol organosolv process has been reevaluated as a pretreatment for cellulose ethanol production [10–13,21]. As shown in Figure 7.1, the organosolv ethanol process uses aqueous ethanol in the concentration range of 40–70% (v/v), at approximately 170–200°C and 300–400 psi, to extract lignin and hemicellulose from wood chips or other lignocellulosic biomass. Mineral acid such as sulfuric acid is a common catalyst for organosolv process, though alkalis and metal salts were used as catalyst as well [46–48]. After the pretreatment, spent liquor (the ethanol solution used for the pretreatment) is separated from solid cellulosic substrate by filtration. After being washed to recover ethanol and dissolved matters trapped in the fibers, the substrate is ready for bioconversion to ethanol or chemicals. Ethanol is recovered from the spent liquor by flashing evaporation. Lignin is precipitated out when the

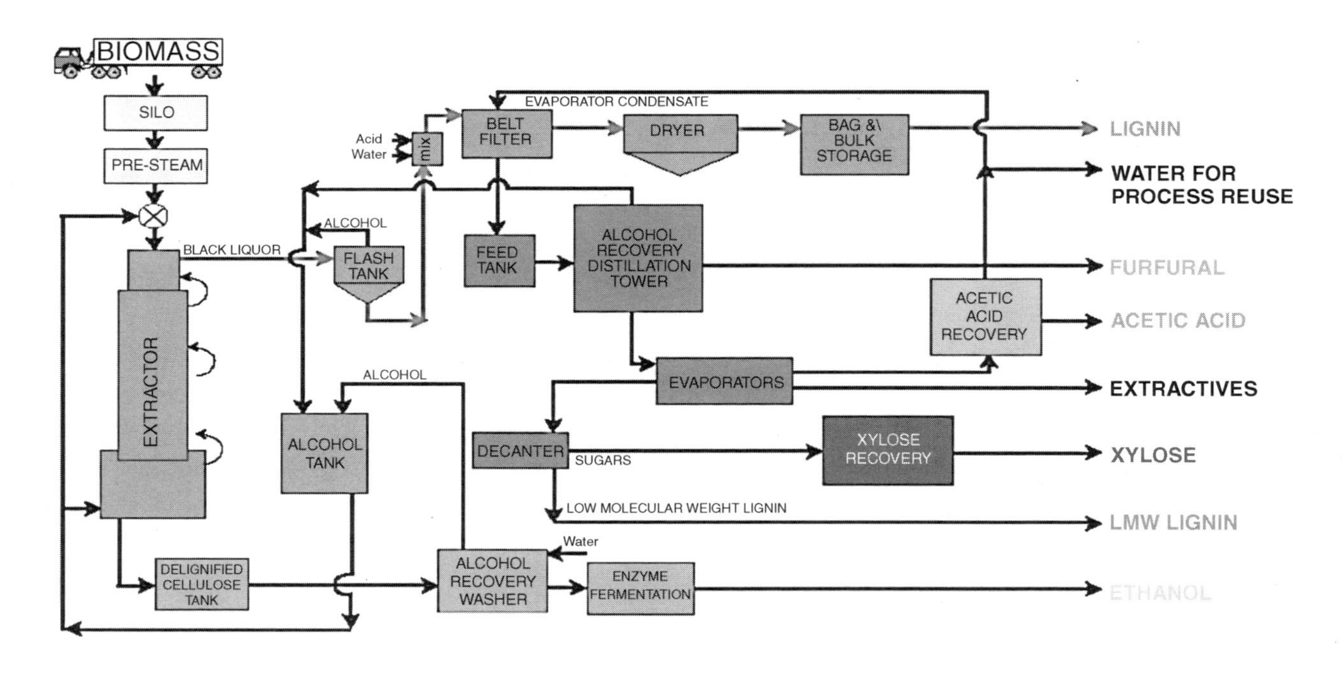

FIGURE 7.1 Commercial lignol biorefinery using organosolv ethanol process for bioconversion of wood to fuel ethanol and coproducts [11].

concentrated liquor is diluted with water. When lignin is filtered out and dried for coproducts development, ethanol left in the filtrate will be further recovered by distillation. Other products such as hemicellulose sugars, low molecular weight lignin fragments, acetic acid, furfural, HMF, and LA can be recovered from the water-soluble stream (the filtrate).

The aim of a pretreatment is to produce a cellulose-rich substrate that is susceptible to enzymatic hydrolysis for a high sugar recovery yield. As compared to a pulping process, where strength and morphology of the cellulose fiber are important factors, a pretreatment is more focused on the digestibility of the resulting cellulose substrate and overall sugar recovery. In fact, destruction of fibers and partial depolymerization of cellulose are desirable for a pretreatment, because the substrates with short fibers have a larger working surface area for enzymes, and the lower cellulose degree of polymerization (DP) favors enzymatic hydrolysability [10]. Organosolv ethanol pretreatment of mixed softwood, hybrid poplar, and lodgepole pine infested by mountain pine beetle were investigated in our previous studies [11, 12, 42, 49]. The process was optimized for the recovery of carbohydrates and lignin and the enzymatic hydrolysability of the pretreated substrate. The results indicated that the ethanol organosolv process is a unique and promising pretreatment technology for bioethanol production from lignocellulosic biomass, in particular for woody biomass. The substrates pretreated by the process had superior enzymatic digestibility over those pretreated by alternative processes. Over 90% cellulose-to-glucose conversion yield could be achieved within 24 h at a low enzyme loading, as will be discussed later in this chapter. The ethanol organosolv lignin generated had high purity, low molecular weight, narrow distribution, and more functional groups, portending potential applications in antioxidants, adhesives, polyurethane foams, and carbon fibers [11, 13]. High-value chemicals, such as furfural, HMF, and formic, acetic, and levulinic acids, were also derived from the hemicellulose fraction [11, 12]. It is worthy to mention that the ethanol organosolv process calls for directly using wood chips, thus eliminating the need for significant wood-size reduction. Size reduction, especially for wood, is energy and cost intensive and significantly compromises the process economics and energy efficiency.

7.4.2 Mass Balance of Organosolv Ethanol Pretreatment

Representative chemical composition of hardwood (hybrid poplar) and softwood (healthy and mountain beetle infested lodgepole pine) substrates pretreated with organosolv ethanol process is summarized in Table 7.1, and compared with that of unpretreated wood. It is apparent that most of hemicellulose was dissolved during the pretreatment. It is interesting that small amount of hemicellulose remained in hybrid poplar substrate, while only a trace of hemicellulose existed in softwood substrates. This might result from the differences in hemicellulose structure and their resistance against the acidic hydrolysis. The removal of hemicellulose is an important contributor to the excellent enzymatic digestibility of organosolv substrate, as more pores and greater surface area are created, which enhance the accessibility to cellulases [44]. With the dissolution of lignin, the lignin content in

TABLE 7.1 Chemical Composition of Poplar and Lodgepole Pine Before and After Organosolv Ethanol Pretreatment

	Hybrid Poplar		Lodgepole Pine		MPBI[b] Lodgepole Pine	
%[a]	Before Pretreatment	After Pretreatment	Before Pretreatment	After Pretreatment	Before Pretreatment	After Pretreatment
Lignin	21.0	11.2	27.8	14.6	26.0	10.1
Glucose	49.0	81.9	50.1	89.1	52.9	91.2
Xylose	17.9	6.4	8.4	1.0	7.6	0.5
Mannose	3.9	2.8	13.4	0.5	13.7	0.6
Galactose	0.4	ND[c]	2.9	ND	2.3	ND
Arabinose	0.3	ND	1.7	ND	1.5	ND

Source: Refs. [12, 42, 49].
[a]Percentage on dry matter.
[b]MPBI, Mountain pine beetle infested.
[c]ND, not detected. Pretreatment conditions: poplar (180°C, 60 min, 1.25% H_2SO_4, and 50% ethanol); lodgepole pine (170°C, 60 min, 1.10% H_2SO_4, and 65% ethanol).

the substrates was reduced to 10–15%, dependent on the feedstock and pretreatment conditions. Certainly, the removal of lignin is the most important reason that organosolv substrate is readily digestible. The data in Table 7.1 show that majority of cellulose was retained and enriched in the solid substrate while lignin and hemicellulose were dissolved, indicating the excellent selectivity of the organosolv ethanol process.

Mass balance data from organosolv ethanol pretreatment of three feedstocks are summarized in Table 7.2. The yield of pretreated substrate from hybrid poplar was 53%. Approximately 28% of the initial lignin in the poplar remained with the substrate after the pretreatment. Rest of the lignin was dissolved during the pretreatment and recovered as organosolv lignin, which counted for approximately 72% of the original lignin in untreated wood. Differently, the substrate yield of lodgepole pine was only 44%, which is significantly lower than that (53%) of the hybrid poplar. The lower yield of lodgepole pine substrate resulted primarily from the degradation/dissolution of cellulose and hemicellulose during the pretreatment. As shown in Table 7.2, compared to hybrid poplar, fewer cellulosic and hemicellulosic sugars retained in lodgepole pine substrate, and more sugars and sugar-degradation products were detected in water-soluble stream. When the lodgepole pine was infested by mountain pine beetle, the substrate yield further dropped to as low as 41%. The results suggested that the infestation promoted the dissolution of lignin, cellulose, and hemicellulose during organosolv pretreatment. As observed in Table 7.2, the beetle-infested pine had higher organosolv lignin yield and more sugars and sugar-degradation products in water-soluble stream.

Approximately 89% of the total glucan in untreated poplar was recovered in the solid substrate, while the value dropped to 86% and 79% in healthy and infested lodgepole pine, respectively. Significant amount of hemicellulose (19% xylose and 38% mannose) were retained in the poplar substrate, while only traces of hemicellulose

TABLE 7.2 Mass Balance of Organosolv Pretreatment of Hybrid Poplar and Lodgepole Pine

	Hybrid Poplar	Lodgepole Pine	Mountain Pine Beetle-Infested Lodgepole Pine
Substrate (g)[a]	52.7	44.3	41.3
Lignin	5.9	6.5	4.2
Glucose	43.2	39.4	37.6
Xylose	3.4	0.4	0.6
Mannose	1.5	0.2	0.6
Organosolv lignin (g)[a]	15.5	16.7	19.6
Water-soluble stream (g)[a]	17.6	23.0	24.2
Soluble lignin	5.2	3.6	4.8
Glucose	0.6	3.5	4.2
Xylose	9.4	3.7	3.2
Mannose	1.3	6.2	5.4
Galactose	0.3	2.2	1.7
Arabinose	0.2	1.0	0.8
Furfural	0.5	1.4	2.0
Hydroxymethylfurfural	0.1	1.4	2.1
Glucose yield (%)[b]	89.4	85.6	79.0
Hemicellulosic sugar yield (%)[c]	71.6	51.9	49.0
Organosolv lignin yield (%)[d]	73.8	60.1	75.4

Source: Refs. [12, 42, 49].
[a]Yield from 100 g dry unpretreated feedstock.
[b]Total glucose yield including the sugar in both substrate and water-soluble stream.
[c]Total hemicellulosic sugar (in the form of fermentable sugar) yield including the sugars in both substrate and water-soluble stream.
[d]Based on lignin in unpretreated wood. Pretreatment conditions.
Note: See Table 7.1 footnotes.

were detected in the substrates of lodgepole pine (Tables 7.1 and 7.2). Approximately 30–50% of initial hemicellulose in untreated wood were recovered in water-soluble stream in the form of monomeric sugars. Substantial amount of hemicellulosic sugars, in particular those from lodgepole pine, were further degraded into furfural and HMF. However, the total amounts of HMF and furfural detected accounted for only a small fraction of the unrecovered hemicellulosic sugars, suggesting further degradation of HMF to levulinic and formic acids or condensation of furfural and HMF to humins occurred, as discussed above.

In summary, organosolv ethanol process is able to efficiently fractionate lignocellulosic biomass into its major components for individual application. As discussed below, the cellulose-rich solid is an ideal substrate for bioconversion; the organosolv lignin is one of the best lignins in many applications; and the hemicellulosic sugars can be fermented or chemically converted to value-added products.

7.5 ENZYMATIC DIGESTIBILITY OF ORGANOSOLV ETHANOL SUBSTRATE

It is well known that organosolv-pretreated substrates have a ready enzymatic hydrolysability even at low enzyme loading [11,12,42,49–53]. Enzymatic hydrolysis of organosolv-pretreated hardwood and softwood is compared with that of steam-explosion pretreated ones in Figure 7.2. Substantial conversion of the poplar pretreated with organosolv ethanol process (PEP, containing 11.2% w/w residual lignin) was observed within 24 h, and 93% of the glucan in the substrate was hydrolyzed to glucose. When the hydrolysis was extended to 48 h, the conversion of cellulose to glucose reached 97%. The softwood substrate [11] (MSEP with 9.2% lignin) had ready digestibility as well, but was not as ready as the hardwood substrate. Though the rate of hydrolysis of the poplar substrate was slower during the first 12 h than that of the softwood substrate, the extents of conversion at 24 and 48 h were higher. The reasons of the differences in hydrolysis kinetics are unclear at this time, however, the degree of polymerization of cellulose and lignin structure are postulated as contributors [11, 12, 42].

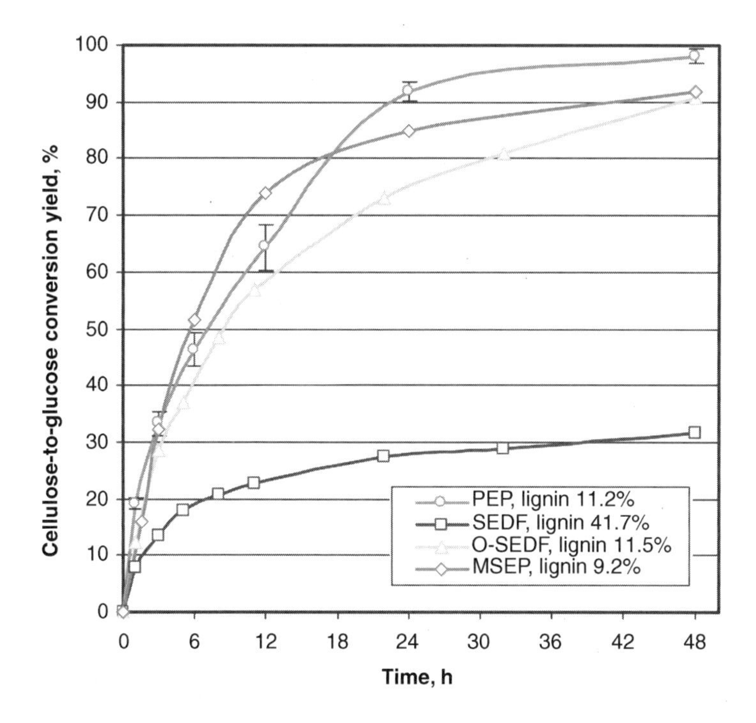

FIGURE 7.2 Enzymatic hydrolysability of pretreated woods. PEP, organosolv-pretreated poplar; MSEP, organosolv-pretreated mixed softwood; O-SEDF, oxygen-delignified steam-exploded Douglas fir; SEDF, steam-exploded Douglas fir; enzymatic hydrolysis conditions: 2% consistency (cellulose, w/v) in acetate buffer, pH 4.8, 50°C, shaker speed 150 rpm, 0.004% tetracycline as antibiotic, cellulase loading 20 FPU (21 mg total protein)/g cellulose with a supplementation of β-glucosidase 40 IU (5.7 mg total protein)/g cellulose [12].

On the other hand, SO_2-catalyzed steam explosion of softwood (Douglas fir) with 42% residual lignin showed a very poor digestibility, a conversion of cellulose to glucose of less than 30% at 24 h and 32% at 48 h. One can argue that the poor digestibility of the steam-exploded Douglas fir is due to its extremely high lignin content. However, when the lignin content of steam-exploded Douglas fir was reduced to a similar level of ~11% with that of organosolv substrate by oxygen delignification, as shown in Figure 7.2, the digestibility was substantially enhanced to 75% at 24 h and 90% at 48 h hydrolysis, but it is still not as good as the organosolv hardwood substrate (93% and 97%). This observation suggests that the excellent digestibility of organosolv substrate is not merely the result of low lignin content, and there are other contributing factors.

In another study [52], rye straw was pretreated with organosolv ethanol, hot water, and soda (NaOH) methods, and the enzymatic digestibility of pretreated rye straw substrates was compared. It was found that organosolv-pretreated rye straw was more readily digestible than hot water and soda-pretreated ones. The former needed much less enzyme for complete hydrolysis than the latter. For example, the cellulose fraction in organosolv-pretreated rye straw was completely converted into glucose in 48 h with enzyme loading lower than 2 FPU (filter paper unit)/g cellulose. Whereas hot water and soda-pretreated straw needed 25 and 15 FPU/g cellulose for a complete conversion of cellulose fraction in 48 h, respectively.

In summary, the ready digestibility of organosolv-pretreated substrate can be attributed to the following reasons. First, organosolv substrates have very low hemicellulose content, as shown in Table 7.1. The removal of hemicellulose created more pores and larger surface area, as mentioned above, facilitating the access of cellulases to cellulose. The second reason is the acid-catalyzed prehydrolysis of cellulose during the organosolv pretreatment. This lowers the molecular weight of cellulose, and the cleavage of each 1,4-glycosidic bond creates two new cellulose termini for cellulases to work upon, promoting the overall rate of cellulose hydrolysis. The differences in structure and distribution of lignin in organosolv and steam-exploded substrates might be the third factor, since the impact of lignin on cellulases is dependent not only on the content but also on structure and location [54, 55]. The residual lignin in organosolv substrate may have less impact on enzymes than steam-exploded examples.

7.6 PROPERTIES AND APPLICATIONS OF ORGANOSOLV ETHANOL LIGNIN

7.6.1 Properties of Organosolv Ethanol Lignin

Most of the lignin in biomass was extensively depolymerized and extracted during the organosolv ethanol process, and the dissolved lignin is precipitated and recovered as organosolv lignin when the pretreatment liquor is diluted with water, as described in Figure 7.1. In general, organosolv ethanol lignin has high purity, good reactivity, low molecular weight, and narrow polydispersity, being particularly suitable for a

TABLE 7.3 Chemical and Physical Properties of the Lignins Prepared from Hardwood and Softwood by Organosolv Ethanol Process

	%	
Property	Mixed Hardwood	Mixed Softwood
Moisture	2.9	1.4
Total lignin	95.5	98.3
Klason lignin	91.4	96.7
Acid-soluble lignin	4.1	1.6
Total carbohydrate	0.25	0.08
Arabinose	ND	ND
Galactose	0.03	0.02
Glucose	0.01	0.01
Xylose	0.18	ND
Mannose	0.03	0.05
Functional groups		
Phenolic hydroxyl	4.7	5.3
Aliphatic hydroxyl	4.9	5.4
Methoxyl	19.1	16.4
Molecular weight		
M_n	1340	1833
M_w	1985	2938
M_w/M_n (polydispersity)	1.48	1.60

Source: Ref. [11].
ND, not detected.

range of industrial applications, such as adhesives for wood products and polymeric composites [32, 56].

Physicochemical properties of two organosolv lignin samples from hardwood and softwood, respectively, are summarized in Table 7.3. It is apparent that the lignin samples contained almost pure lignin with only trace amounts of carbohydrate (0.08 and 0.25%, respectively), indicating high purity. In general, both samples had a low molecular weight and narrow polydispersity, compared to other types of lignin [57, 58]. Low polydispersity is important in controlling variability in lignin applications. The molecular weight (M_n and M_w) and polydispersity (M_w/M_n) of softwood lignin were slightly higher than those of hardwood lignin, which may be due to the condensation between guaiacyl structural units in softwood lignin. It is also noteworthy that unlike kraft lignin and lignosulfonate, organosolv lignin is sulfur free and has very low ash content [59, 60], which are critical for some applications (e.g., carbon fibers and lignin-based liquid fuels when being prepared by catalytic hydrodeoxygenation) [41,61].

The softwood and hardwood organosolv lignin samples were both found having a high level of phenolic hydroxyl groups, relative to the lignin prepared using other processes [62]. The softwood lignin showed slightly higher levels of phenolic and aliphatic hydroxyl groups, while the hardwood lignin has a higher level of methoxyl groups due to the presence of syringyl units. Phenolic hydroxyls are important functional groups in lignin, affecting both physical and chemical properties and reactivity of the lignin [63]. The high content of phenolic hydroxyl groups suggests the potential of the lignin in production of phenolic, epoxy, and isocyanate resins [64, 65]. Hydroxyl groups also provide opportunities for modification via esterification and etherification reactions with the goal of developing novel applications for lignin-derived compounds.

7.6.2 Polymeric Materials Based on Organosolv Ethanol Lignin

Because of high purity, good reactivity, sulfur-free, and low molecular weight, organosolv lignin has demonstrated potential uses in thermosetting resins, in polymeric composites, and in polyurethane foams [65]. For instance, organosolv lignin was used in phenolic powder resins as the binder in the manufacture of friction products. The use of 20% lignin in phenolic resin resulted in competitive performance relative to controls prepared with 100% phenolic resin. Improvements included the stability of the friction coefficient to temperature variations and improved wear behavior [66].

Oriented strand board (OSB) is one of the dominant construction-wood-panel products in North America. In its manufacture, wood strands are bonded with a binder under heat and pressure. The binders (adhesives) widely used are phenolic resins. Organosolv lignin was found to be a good replacement of phenol in the binders. When lignin was used to partially replace phenolic component (5–25%), the physical strength of the resulting boards were equal to or better than the controls. In addition to the cost benefits, the use of lignin led to an improvement of the environmental conditions in the panel-manufacturing facility, such as lower levels of dust in the area around the blenders where the resin is applied to the wood strands before board forming and pressing [67–70].

With the presence of substantial hydroxyl and carboxyl groups in organosolv lignin, it can be used to prepare rigid polyurethane foams [71, 72]. Our group has shown [73] that partial replacement of polyol with organosolv lignin could produce polyurethane foam with satisfactory cellular structure and strength, comparable with those of a control foam (no lignin). However, the compressive strength significantly decreased when the lignin addition was over 60%. Addition of chain extender such as 1,4-butanediol could significantly improve the strength of lignin-containing polyurethane foam. Figure 7.3 shows the polyurethane foam containing organosolv lignin. This type of the foam can be used as insulating and packing materials.

7.7 CARBON FIBERS FROM ORGANOSOLV LIGNIN

Carbon fiber has been widely used as a reinforcing material in composite products, in particular for the aerospace industry. A major factor is their unique properties, which in many cases are superior to metals, such as light weight, high strength, high modulus, and high stiffness [41, 48]. Recently, it has been widely recognized that if traditional metals were replaced with lower weight materials such as carbon fibers for auto parts, the fuel economy of vehicles would be improved significantly, and

FIGURE 7.3 Rigid polyurethane foam prepared from organosolv ethanol lignin.

therefore reduce the oil consumption and emissions to the environment. However, commercial carbon fibers are mostly made from oil-derived pitch (for general performance products) and polyacrylonitrile (PAN) (for high-performance products). A primary factor limiting carbon fibers from wide application in low-end products such as high-volume auto parts is the high cost of carbon fibers (over $25 per kg).

Lignin, the second largest natural polymer on earth, has been demonstrated as a potential feedstock for carbon fibers, because of its high carbon content, low price, renewability, and availability on a large scale from existing paper industry and future biorefining industry. Indeed, carbon fibers were prepared from lignin as early as in 1969 [74]. Sudo and coworkers showed that lignin could be converted into a molten viscous material with suitable properties for thermal spinning by hydrocracking, phenolation, or hydrogenolysis followed by heat treatment in vacuum [75, 76]. More recently, Sano group prepared carbon fibers from organosolv acetic acid lignin [77, 78].

Kubo and Kadla prepared carbon fibers with mechanical properties suitable for general performance products from commercial hardwood kraft lignin and organosolv ethanol lignin [61]. The lignins were first thermally pretreated to remove volatile contaminants that might disrupt fiber integrity during subsequent thermal spinning. The thermal treatment also decreased the hydroxyl content of the lignins, which condensed the lignin macromolecules and reduced intermolecular interactions. After being thermally spun to fibers, thermal stabilization was needed to change the thermoplastic character of the lignins to thermosetting, enabling the lignin-based

FIGURE 7.4 Lignin fibers spun from organosolv ethanol lignin that can be further carbonized into carbon fibers.

fibers to maintain a fiber form during subsequent carbonization. The yield of lignin-based carbon fibers ranged from 45% for the organosolv ethanol lignin to 40% for the kraft lignin. Mechanical properties (tensile strength 400–550 MPa and Young's modulus 30–60 GPa) of the lignin-based carbon fibers were equivalent or superior to those previously reported. The lignin fibers shown in Figure 7.4 were thermally spun from organosolv ethanol lignin prepared from hardwood, which could be further processed into carbon fibers through carbonization.

7.8 ANTIOXIDATION CAPACITY OF ORGANOSOLV LIGNIN

Lignin is a natural phenolic polymer and can be used as a potential antioxidant. It was reported that kraft lignin was as effective as vitamin E as an antioxidant of corn oil [79]. Lignin functions as an effective radical scavenger to prevent autoxidation and depolymerization of cellulose in pulps and papers have been reported [80, 81]. In addition, incorporation of lignin into synthetic polymer systems stabilized the

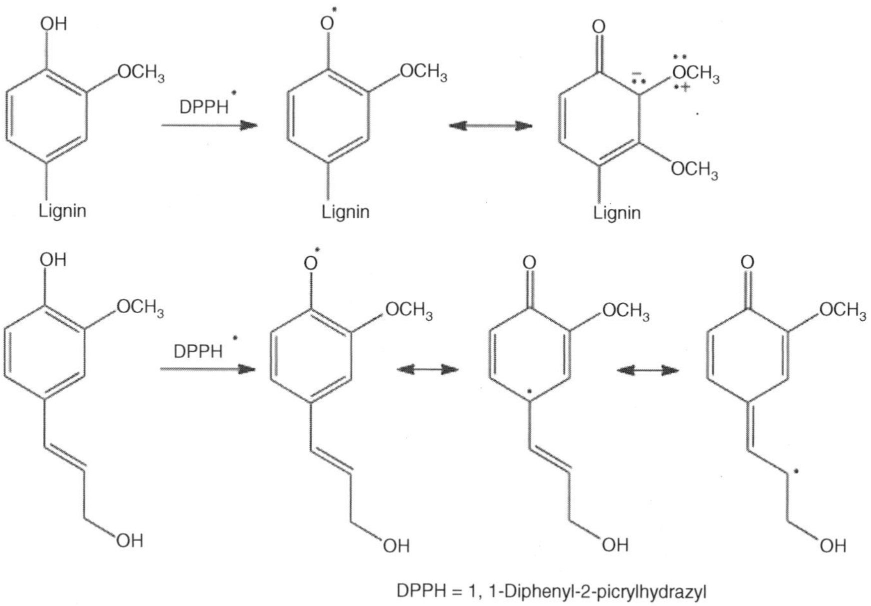

DPPH = 1, 1-Diphenyl-2-picrylhydrazyl

SCHEME 7.5 Proposed mechanisms of radical trapping and stabilization of radicals by lignin [13].

material against photo- and thermal oxidation [56–58]. As a major component in dietary fiber, lignin can inhibit the activity of enzymes related to the generation of superoxide anion radicals and obstruct the growth and viability of cancer cells [82].

Research into lignin model compounds [83–85] indicates that free phenolic hydroxyl groups are essential to antioxidant activity of lignin. Scheme 7.5 demonstrates the trapping and stabilization of radicals by lignin, proposed by Barclay et al [83]. The radical scavenging ability of phenolic compounds depends not only on the ability to form a phenoxyl radical (i.e., hydrogen atom abstraction), but also on the stability of the phenoxyl radical. In general, phenolic structures with substituents on benzene ring that can stabilize the phenoxyl radicals have higher antioxidant activity than those without. For example, *ortho* substituents (e.g., methoxyl groups) stabilize phenoxyl radicals by resonance as well as hindering them from radical propagation. Conjugated double bonds can provide additional stabilization of the phenoxyl radicals through extended delocalization. However, a conjugated carbonyl group has a negative effect on antioxidant activity.

The Radical Scavenging Index (RSI) values of organosolv ethanol lignins were investigated and compared with those of 3,5-di-*tert*-butyl-4-hydroxytoluene (BHT) and Vitamin E, which are commonly used antioxidants. It was found [13] that the ethanol lignins had much higher RSI (13–122, dependent on the extraction conditions of the lignins) than BHT (1) but significantly lower RSI than vitamin E (263). Radical scavenging index of the organosolv ethanol lignin was positively correlated to phenolic hydroxyl group content, consistent with the studies of lignin model compounds [83–85].

Conversely, aliphatic hydroxyl group content had a negative effect on antioxidant activity of the lignin. The effect of molecular weight and polydispersity of the lignins on RSI was investigated as well. The lignin preparations with low molecular weight were found to have high antioxidant activity. Meanwhile, it is likely that the antioxidant activity benefited from the narrow distribution of the lignin molecular weight. This is because the low molecular weight fraction of the lignin possessed more phenolic hydroxyl groups, the reactive center to trap radicals, than the high molecular weight fraction. As discussed in Section 7.3.1, low molecular weight fraction resulted from extensive depolymerization of lignin (i.e., cleavage of ether linkages) where cleavage of each ether linkage formed a new phenolic hydroxyl group [13].

7.9 CHEMICALS RECOVERABLE FROM ORGANOSOLV ETHANOL PROCESS

In addition to the cellulose-rich fiber fraction and organosolv lignin discussed above, other valuable chemicals are generated from organosolv ethanol process, mostly from the degradation of hemicellulose and lignin. These include monomeric and oligomeric sugars, acetic acid released from hemicellulose, low molecular weight phenolic compounds from lignin, and furfural, HMF, LA, and formic acid from sugar degradation. Furfural is a key derivative of pentoses such as xylose and has a broad spectrum of industrial applications. For example, it has been used as solvent in lubricants, coatings, adhesives, and furan resin [41]. Furfural is also a starting point for a large family of derivative chemical and polymer products [41]. HMF is a versatile biomass-derived platform compound for synthesizing a broad range of chemicals that are currently derived from petroleum. It can also be transformed to gasoline-like fuels through aldol condensation and hydrodeoxygenation [41]. Levulinic acid has been proposed as a platform chemical for producing a wide range of value-added products, such as dyestuffs, polymers, pharmaceutically active compounds, flavor substance, plasticizers, solvents, and fuel additives [45].

It was reported that furfural could be recovered at a concentration of 60–70% as a side draw in the distillation tower for recovering the pulping solvent (ethanol). The furfural was easily upgraded to 85–90% by water extraction and distillation [32]. The stillage obtained in the distillation tower after recovering ethanol contains hemicellulosic sugars, low molecular weight lignin, and acetic acid. The sugars are mostly xylose when hardwood and habergeons are feedstock, or mixture of mannose and xylose in the case of softwood. Some of the sugars are still in the form of oligomers. The sugars can be either fermented into other chemicals for example, ethanol, butanol, and single-cell protein or chemically converted into furfural (from pentoses) and HMF (from hexoses), or sugar alcohols [32]. The lignin in the stillage stream represents the small fragments of lignin that did not precipitate when recovering the organosolv lignin because of their water-solubility. This fraction of lignin has a low molecular weight (M_n 500, equivalent to DP 2$\sim$3) and low glass-transition temperature in the range of 24–60 °C. The lignin fraction is expected to be a good feedstock of phenolic resin [11, 32].

Recently, organosolv ethanol process was adapted for directly converting lignocellulosic biomass to chemicals [86]. Under the more severe conditions (higher temperature and higher acid catalyst loadings) than those used for pulping or pretreatment described above, the process was able to convert hemicellulose (xylan) and cellulose in poplar to furfural, HMF, and LA, respectively, while lignin was extracted as organosolv lignin. Yields of organosolv lignin and the sugar-derived chemicals were optimized within the ranges of temperatures 173–207 °C, ethanol concentrations 33–67% (v/v), sulfuric acid dosages 2.3–5.7% (w/w), and reaction times 15–66 min. Results indicated that temperature and sulfuric acid loading had significant effects on the yields of the lignin and chemicals. High temperature, high sulfuric acid dosage, and long reaction time enhanced the yield of organosolv lignin. Low ethanol concentration slightly promoted the lignin yield, while high ethanol concentration (over 50%, v/v) depressed delignification. In general, the conversions of hexoses to HMF/LA and pentoses to furfural were enhanced by high temperature, more sulfuric acid, and long reaction time. Ethanol concentration had less effect on the conversion of hexoses to HMF/LA than that of pentoses to furfural.

Comparative investigation of the organosolv ethanol process and conventional acid process for chemical production from biomass indicated that ethanol not only enhanced the delignification to produce the high-quality organosolv lignin (acid process generates condensed lignin), but also improved the conversion yields of pentoses to furfural and hexoses to HMF/LA. The preliminary investigation suggested that a two-step organosolv ethanol process might be a better approach for producing LA from lignocellulose. It was also observed that while pentoses were converted to furfural in high yield in the organosolv process, the overall yields of HMF and LA from hexoses were relatively low, which probably resulted from the condensation of HMF and LA to humin-like products [86].

7.10 CONCLUDING REMARKS

Organosolv ethanol process is an efficient approach to fractionate lignocellulose into its major components for individual utilization. The cellulose-rich solid is an ideal substrate for bioconversion for ethanol or other fuels and chemicals. The organosolv lignin has demonstrated many potential applications because of high purity, good reactivity, low molecular weight and narrow distribution, and unique thermal and physical properties. The hemicellulosic sugars can be fermented or chemically converted to value-added products. In addition, hemicellulose-derived chemicals, such as furfural, HMF, levulinic acid, formic acid, and acetic acid, form another stream of high-value coproducts from organosolv ethanol process.

To be successful in commercialization, the organosolv ethanol process needs improvements in the following areas. First, complete and cost-effective recovery of solvent ethanol is important to process economy. Second, it is very challenging to efficiently recover hemicellulosic sugars and covert them into high-value products. The sugars in the stillage are not suitable for direct fermentation because expensive detoxification is required to remove the fermentation inhibitors including furfural,

HMF, acetic acid, and phenolic lignin fragments. Third, though the organosolv lignin has great potential in many applications, the lignin products and market have not been developed satisfactorily. The biggest challenge of lignin products is the competition from petro-based products in price and quality. Considering the magnitude of future bio-refineries, promising lignin products must have a big market to consume the huge amount of lignin generated. In addition, since lignin is extracted from the system for developing co-products and not burned to generate heat and power for the process, organosolv ethanol process consumes more purchased energy than alternative processes.

REFERENCES

1. S. Aziz and K. Sarkanen, *Tappi J.* **1989**, 72, 169.
2. L. Paszner and H. J. Cho, *Tappi J.* **1989**, 72, 135.
3. H. L. Hergert, in *Environmentally Friendly Technologies for the Pulp and Paper Industry*, (Eds. R. A. Young and M. Akhtar), Wiley, New York, **1998**, 5–67.
4. J. H. Lora, *Tappi J.* **1992**, 75, 12.
5. E. K. Pye and J. H. Lora, *Tappi J.* **1991**, 74, 113.
6. X. J. Pan and Y. Sano, *J. Wood Sci.* **1999**, 45, 319.
7. X. J. Pan, Y. Sano and T. Ito, *Holzforschung.* **1999**, 53, 49.
8. J. Baeza, A. M. Fernandez, J. Freer, A. Pedreros, E. Schmidt and N. Duran, *Appl. Biochem. Biotechnol.* **1991**, 31, 273.
9. B. Saake, S. Lummitsch, R. Mormanee, R. Lehnen and H. H. Nimz, *Papier.* **1995**, 49, V1.
10. X. Pan, D. Xie, K. Y. Kang, S. L. Yoon and J. N. Saddler, *Appl. Biochem. Biotechnol.* **2007**, 137, 367.
11. X. J. Pan, C. Arato, N. Gilkes, D. Gregg, W. Mabee, K. Pye, Z. Z. Xiao, X. Zhang and J. Saddler, *Biotechnol. Bioeng.* **2005**, 90, 473.
12. X. J. Pan, N. Gilkes, J. Kadla, K. Pye, S. Saka, D. Gregg, K. Ehara, D. Xie, D. Lam and J. Saddler, *Biotechnol. Bioeng.* **2006**, 94, 851.
13. X. J. Pan, J. F. Kadla, K. Ehara, N. Gilkes and J. N. Saddler, *J. Agric. Food Chem.* **2006**, 54, 5806.
14. N. Brosse, R. El Hage, P. Sannigrahi and A. Ragauskas, *Cellulose Chem. Technol.* **2010**, 44, 71.
15. J. H. Lora and S. Aziz, *Tappi J.* **1985**, 68, 94.
16. C. J. Biermann, *Handbook of Pulping and Papermaking*, 2nd ed., Academic Press, San Diego, **1996**, 55–100.
17. P. Stroeve, P. Kumar, D. M. Barrett and M. J. Delwiche, *Ind. Eng. Chem. Res.* **2009**, 48, 3713.
18. C. E. Wyman, B. E. Dale, R. T. Elander, M. Holtzapple, M. R. Ladisch and Y. Y. Lee, *Bioresour. Technol.* **2005**, 96, 1959.
19. A. Rodriguez and L. Jimenez, *Afinidad.* **2008**, 65, 188.
20. A. Leponiemi, *Appita J.* **2008**, 61, 234.
21. C. Arato, E. K. Pye and G. Gjennestad, *Appl. Biochem. Biotechnol.* **2005**, 121, 871.
22. E. Muurinen, *Organosolv Pulping*, Oulu University Library, Oulu, **2000**, 16–123.
23. F. Lopez, A. Alfaro, L. Jimenez and A. Rodriguez, *Afinidad.* **2006**, 63, 174.

24. P. N. Williamson, *Sven. Papperstidn.* **1988**, 91, 21.
25. P. Stockburger, *Tappi J.* **1993**, 76, 71.
26. J. H. Lora, G. C. Goyal and M. Raskin in *Proceedings of the 7th International Symposium on Wood and Pulping Chemistry*, Beijing, **1993**, 1, 327–336.
27. M. M. Hepditch and R. W. Thring, *Can. J. Chem. Eng.* **1997**, 75, 1108.
28. Y. Liu, S. Carriero, K. Pye and D. S. Argyropoulos, in *Lignin: Historical, Biological, and Materials Perspectives*, (Eds. W. G. Glasser, R. A. Northey and T. P. Schultz), ACS Publications, Washington, DC, **1999**, 447–464.
29. D. G. B. Boocock and J. J. Balatinecz, *Abstr. Papers Am. Chem. Soc.* **1992**, 203, 106-Cell.
30. R. W. Thring, M. N. Vanderlaan and S. L. Griffin, *Biomass Bioenerg.* **1997**, 13, 125.
31. S. Kubo and J. F. Kadla, *Macromolecules.* **2004**, 37, 6904.
32. J. H. Lora, A. W. Creamer, L. C. F. Wu and G. C. Goyal in *Proceedings of the 6th International Symposium on Wood and Pulping Chemistry, Appita*, **1991**, 431–438.
33. Lignol, can be found under http://www.lignol.ca/news/News-2011/Lignol_DOE_NR_July_15_2011_FINAL.pdf, **2011**.
34. T. J. Mcdonough, *Tappi J.* **1993**, 76, 186.
35. E. Sjostrom, in *Wood Chemistry: Fundamentals and Applications*, (Eds. E. Sjostrom), Academic Press, San Diego, **1993**, 70–90.
36. A. Sakakiba, H. Takeyama and N. Morohosh, *Holzforschung.* **1966**, 20, 45.
37. E. Adler, J. M. Pepper and E. Eriksoo, *Ind. Eng. Chem.* **1957**, 49, 1391.
38. A. H. Lohrasebi and L. Paszner, *Tappi J.* **2001**, 84, 69.
39. R. M. Ede and G. Brunow, in *Proceedings of TAPPI International Symposium on Wood and Pulping Chemistry*, TAPPI Press, Atlanta, **1989**, 139–143.
40. K. V. Sarkanen, and H. L. Hergert, in *Lignins: Occurrence, Formation, Structure, and Reactions*, (Eds. K.V. Sarkanen and C. H. Ludwig), Wiley-Interscience, New York, **1971**, 95.
41. B. Kamm, M. Kamm, M. Schmidt, T. Hirth and M. Schulze, in *Biorefineries – Industrial Processes and Products*, (Eds. B. Kamm, P.R. Gruber and M. Kamm), Willey-VCH, Weinheim, **2006**, 97–150.
42. X. J. Pan, D. Xie, R. W. Yu and J. N. Saddler, *Biotechnol. Bioeng.* **2008**, 101, 39.
43. J. N. Saddler, S. D. Mansfield and C. Mooney, *Biotechnol. Prog.* **1999**, 15, 804.
44. C. E. Wyman and B. Yang, *Biotechnol. Bioeng.* **2004**, 86, 88.
45. D. J. Hayes, S. Fitzpatric, M. H. B. Hayes and J. R. H. Ross, in *Biorefineries – Industrial Processes and Products*, (Eds. B. Kamm, P.R. Gruber and M. Kamm), Willey-VCH, Weinheim, **2006**, 139–164.
46. L. Paszner and N. C. Behera, *Holzforschung.* **1989**, 43, 159.
47. D. Yawalata and L. Paszner, *Holzforschung.* **2006**, 60, 239.
48. H. T. Sahin, *J. Chem. Technol. Biotechnol.* **2003**, 78, 1267.
49. X. J. Pan, D. Xie, R. W. Yu, D. Lam and J. N. Saddler, *Ind. Eng. Chem. Res.* **2007**, 46, 2609.
50. J. N. Saddler, L. F. Del Rio and R. P. Chandra, *Biotechnol. Bioeng.* **2011**, 108, 1549.
51. V. Arantes and J. N. Saddler, *Biotechnol. Biofuels.* **2011**, 4, 3.
52. T. Ingram, K. Wormeyer, J. C. I. Lima, V. Bockemuhl, G. Antranikian, G. Brunner and I. Smirnova, *Bioresour. Technol.* **2011**, 102, 5221.
53. A. J. Ragauskas, B. B. Hallac, M. Ray and R. J. Murphy, *Biotechnol. Bioeng.* **2010**, 107, 795.

54. X. J. Pan, D. Xie, N. Gilkes, D. J. Gregg and J. N. Saddler, *Appl. Biochem. Biotechnol.* **2005**, 121, 1069.
55. X. J. Pan, *J. Biobased Mater. Bioenergy.* **2008**, 2, 25.
56. A. W. Creamer, B. A. Blackner and J. H. Lora, in *Proceedings of the 9th International Symposium on Wood and Pulping Chemistry*, Montreal, **1997**, 21, 1–4.
57. B. L. Browning, *The Chemistry of Wood*, Interscience Publishers, New York, London, **1963**, 271.
58. F. Pla, in *Methods in Lignin Chemistry*, (Eds. S.Y. Lin and C. W. Dence), Springer-Verlag, Berlin, **1992**, 498–507.
59. G. Petty, *Paper Technol.* **1989**, 30, 10.
60. E. K. Pye and J. H. Lora, *Tappi J.* **1991**, 74, 113.
61. J. F. Kadla, S. Kubo, R. A. Venditti, R. D. Gilbert, A. L. Compere and W. Griffith, *Carbon.* **2002**, 40, 2913.
62. Y.-Z. Lai, in *Methods in Lignin Chemistry*, (Eds. S.Y. Lin and C. W. Dence), Springer-Verlag, Berlin, **1992**, 423–434.
63. E. Adler, *Wood Sci. Technol.* **1977**, 11, 169.
64. M. Olivares, J. A. Guzman, A. Natho and A. Saavedra, *Wood Sci. Technol.* **1988**, 22, 157.
65. J. H. Lora and W. G. Glasser, *J. Polym. Environ.* **2002**, 10, 39.
66. N. J. Nehez, Canadian Patent Application 2,242,554, 1997.
67. P. C. Muller and W. G. Glasser, *J. Adhes.* **1984**, 17, 157.
68. P. C. Muller, S. S. Kelley and W. G. Glasser, *J. Adhes.* **1984**, 17, 185.
69. T. Sellers, J. H. Lora and M. Okuma, *Mokuzai Gakkaishi.* **1994**, 40, 1073.
70. T. Sellers, J. H. Lora and M. Okuma, *Mokuzai Gakkaishi.* **1994**, 40, 1079.
71. C. Bonini, M. D'Auria, L. Ernanuele, R. Ferri, R. Pucciariello and A. R. Sabia, *J. Appl. Polym. Sci.* **2005**, 98, 1451.
72. H. Nadji, C. Bruzzese, M. N. Belgacem, A. Benaboura and A. Gandini, *Macromol. Mater. Eng.* **2005**, 290, 1009.
73. X. J. Pan, Unpublished data, **2011**.
74. S. Otani, Y. Fukuoka, and B. Igarashi, US Patent. 3,461,082, 1969.
75. K. Sudo, and K. Shimizu, *J. Appl. Polym. Sci.* **1992**, 44, 127.
76. K. Sudo, K. Shimizu, N. Nakashima and A. Yokoyama, *J. Appl. Polym. Sci.* **1993**, 48, 1485.
77. Y. Uraki, S. Kubo, N. Nigo, Y. Sano and T. Sasaya, *Holzforschung.* **1995**, 49, 343.
78. S. Kubo, Y. Uraki and Y. Sano, *Carbon.* **1998**, 36, 1119.
79. G. L. Catignani and M. E. Carter, *J. Food Sci.* **1982**, 47, 1745.
80. J. A. Schmidt, C. S. Rye and N. Gurnagul, *Polym. Degrad. Stab.* **1995**, 49, 291.
81. C. Pouteau, P. Dole, B. Cathala, L. Averous and N. Boquillon, *Polym. Degrad. Stab.* **2003**, 81, 9.
82. F. J. Lu, L. H. Chu and R. J. Gau, *Nutri. Cancer-Int. J.* **1998**, 30, 31.
83. L. R. C. Barclay, F. Xi and J. Q. Norris, *J. Wood Chem. Technol.* **1997**, 17, 73.
84. T. Dizhbite, G. Telysheva, V. Jurkjane and U. Viesturs, *Bioresour. Technol.* **2004**, 95, 309.
85. M. Ogata, M. Hoshi, K. Shimotohno, S. Urano and T. Endo, *J. Am. Oil Chem. Soc.* **1997**, 74, 557.
86. D. E. Kim and X. J. Pan, *Ind. Eng. Chem. Res.* **2010**, 49, 12156.

CHAPTER 8

Pyrolysis Oils from Biomass and Their Upgrading

QIRONG FU, HAIBO XIE, and DIMITRIS S. ARGYROPOULOS

8.1 INTRODUCTION

The ongoing energy crisis and the accompanying environmental concerns, have caused the development of intense interest toward alternative fuels based on sources other than petroleum. Bio-oil is a renewable liquid fuel, having negligible contents of sulfur, nitrogen, and ash, and is widely recognized as one of the most promising renewable fuels that may one day replace fossil fuels. Fast pyrolysis of biomass technologies for the production of bio-oil have been developed extensively in recent years. The conversion of solid biomass into bio-oil using fast pyrolysis is carried out by the rapid (a few seconds) raising of temperature to around 450–550 °C under atmospheric pressure and anaerobic conditions. The resulting products are short-chain molecules, which should be rapidly quenched to liquids. However, such biomass-based bio-oils are of high oxygen content, high viscosity, thermal instability, corrosiveness, and chemical complexity. These characteristics create many obstacles to the applications of bio-oils, precluding it from being used directly as a liquid fuel [1]. Consequently, bio-oils need to be upgraded to improve its fuel properties. Current upgrading techniques are hydrogenation, catalytic cracking, steam reforming, emulsification, converting into stable oxygenated compounds, extracting chemicals from the bio-oils. In this chapter, the preparation and properties of bio-oils will be reviewed with emphasis on advanced upgrading techniques.

8.2 BIO-OIL PREPARATION

Pyrolysis is the thermal decomposition of a material in the absence of oxygen. For biomass, the product of pyrolysis is a mixture of solids (char), liquids (bio-oil),

The Role of Green Chemistry in Biomass Processing and Conversion, First Edition.
Edited by Haibo Xie and Nicholas Gathergood.

TABLE 8.1 Range of Typical Fast-Pyrolysis Conditions

Temperature (°C)	450–550
Gas residence time (s)	0.5–2
Particle size (mm)	0.2–2
Moisture (wt%)	2–12
Cellulose (wt%)	45–55
Ash (wt%)	0.5–3
Yields (wt%)	
Organic liquid	60–75
Water	10–15
Char	10–15
Gas	10–20

and gas (methane, carbon monoxide, and carbon dioxide). The pyrolysis processes are generally divided into "slow pyrolysis" and "fast pyrolysis." Just as its name implies, slow pyrolysis is operated by heating biomass to ~500 °C at a slow heating rate. The vapor residence time varies from 5 min to 30 min and the main product is charcoal [2]. On the contrary, fast pyrolysis is carried out by heating biomass feedstock at a rapid heating rate to around 450–550 °C, followed by the rapid quenching of the product vapors to liquids. The aim of fast pyrolysis is to maximize the yields of bio-oil.

Currently, fast pyrolysis is the only feasible technology for the production of bio-oils at an industrial scale. To maximize the yields of bio-oil in a fast-pyrolysis process, various pyrolysis parameters such as temperature, heating rate, vapor residence time, feedstock properties, particle size, and moisture content, need to be optimized. Typical fast-pyrolysis conditions are shown in Table 8.1 [3].

Westerhof et al. [4] showed that the conventional view on pyrolysis taking place between 400 and 550 °C with the aim to maximize oil yields must be reconsidered for temperatures below 400 °C. Such temperatures have been shown to produce bio-oils of better quality for certain applications. Another factor determining the efficiency of such processes is the ash content of the feedstock biomass with dominant effects on the yield and composition of bio-oils [3]. In general, the yields of char and gas increase significantly for higher ash contents, while the yields of bio-oil decrease.

Present fast-pyrolysis reactors include bubbling fluidized-bed, circulating fluidized-bed, ablative, rotating cone, auger, and vacuum reactors. The major features of the first four reactors are listed in Table 8.2 [2,5] and their schematic illustration is shown in Figure 8.1 [6]. The essential characteristics of a fast-pyrolysis reactor to maximize the yield of bio-oil are a very rapid heating rate, a reaction temperature of around 500 °C, and a rapid quenching of the produced vapors [3]. With the development of fast-pyrolysis technologies, the primary method of heat transfer varies from solid–solid to gas–solid, and there is also a corresponding change of dominant mode of heat transfer from conduction to convection [5].

TABLE 8.2 Summary of Characteristics of Some Common Pyrolysis Systems

Reactor Type	Ablative	Bubbling Fluidized Bed	Circulating Fluidized Bed	Rotating Cone
Carrier gas	No	Yes	Yes	No
Heating method	• Reactor wall/disc	• Heated recycle gas • Hot inert gas	• Particle gasification • Fire tubes	• Gasification of char to heat sand
Primary heat-transfer method	• Solid–solid	• Solid–solid • Gas–solid	• Solid–solid • Gas–solid	• Solid–solid • Gas–solid
Modes of heat-transfer (suggested)	–	• 95% conduction • 4% convection • 1% radiation	• 80% conduction • 19% convection • 1% radiation	• 95% conduction • 9% convection • 1% radiation
Main features	• Accepts large size feedstock • Very high mechanical char abrasion from biomass • Compact design • Heat supply problematical • Particulate transport gas not always required	• High heat-transfer rates • Heat supply to fluidizing gas or to bed directly • Limited char abrasion • Very good solids mixing • Particle size limit <2 mm in smallest dimension • Simple reactor configuration • Residence time of solids and vapors controlled by the fluidizing gas flow rate	• High heat-transfer rate • High char abrasion from biomass and char erosion leading to high char in product • Char/solid heat carrier separation required • Solid recycle required • Increased complexity of system • Maximum particle size up to 6 mm • Possible liquid cracking by hot solids • Possible catalytic activity from hot char • Greater reactor wear possible	• Centrifugal force moves heated sand and biomass • Small particle sizes needed

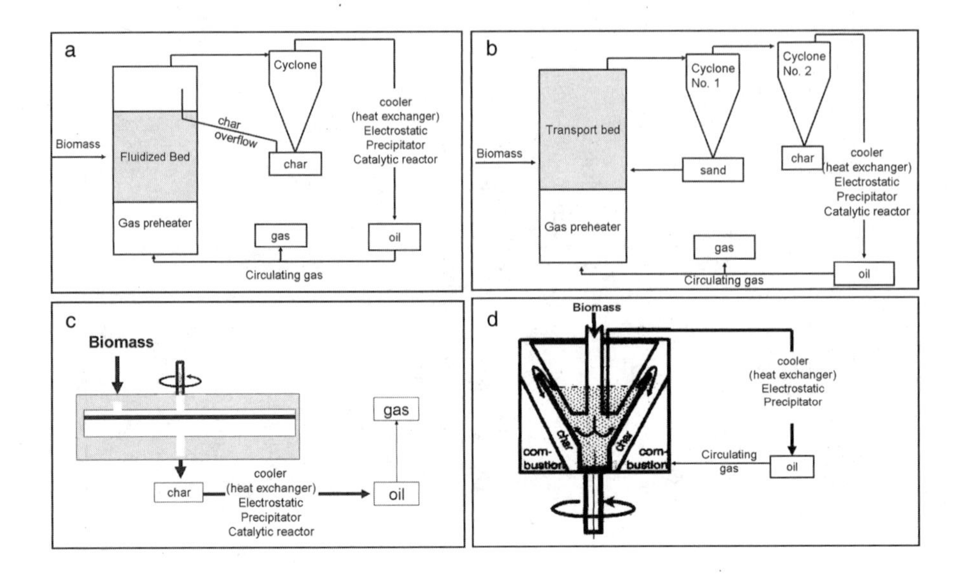

FIGURE 8.1 Schematic illustration of the reactor types for the fast pyrolysis of biomass: (a) bubbling fluidized bed; (b) circulating fluidized bed; (c) ablative pyrolysis; (d) rotating cone reactor. (Reproduced from Ref. [6] with permission from Elsevier © 1999.)

8.3 BIO-OIL

8.3.1 Composition and Physicochemical Properties

There are big differences between bio-oil and petroleum-derived fuel, which can be shown in Table 8.3 [7].

8.3.1.1 Water The water content of bio-oils ranges between 15% and 30% depending on the feedstock and the operating conditions of the pyrolysis process. The presence of water has significant effects on the oil properties since it lowers its heating value and flame temperature. On a positive note, however, water reduces the viscosity of the oil, improving its fluidity leading to uniform combustion characteristics.

8.3.1.2 Oxygen Bio-oil contains 35–40% oxygen as an integral part of the more than 300 chemical compounds present within. The distribution of these compounds can be altered by various biomass feedstocks and pyrolysis process conditions. The high oxygen content in bio-oil represents the biggest difference between bio-oils and hydrocarbon fuels. Due to the high oxygen content, bio-oils have about 50% lower energy density than conventional fuel oils [7] and are immiscible with hydrocarbon fuels. Moreover, bio-oils contain abundant reactive oxygen-containing functional

TABLE 8.3 A Comparison of Various Typical Properties of Wood Pyrolysis Bio-Oil and of Heavy Fuel Oil

Physical Property	Bio-Oil	Heavy Fuel Oil
Moisture content, wt%	15–30	0.1
pH	2.5	–
Specific gravity	1.2	0.94
Elemental composition, wt%		
C	54–58	85
H	5.5–7.0	11
O	35–40	1.0
N	0–0.2	0.3
Ash	0–0.2	0.1
HHV (higher heating value), MJ kg^{-1}	16–19	40
Viscosity (at 50°C), cP	40–100	180
Solids, wt%	0.2–1	1
Distillation residue, wt%	Up to 50	1

groups such as carbonyl, carboxyl, methoxyl, and hydroxyl groups. Thus, bio-oils are acidic and unstable.

8.3.1.3 Viscosity The viscosity of bio-oils can vary widely from 35 cP to 1000 cP at 40 °C depending on the biomass source and operating conditions. The major factors affecting viscosity are temperature and water content. The temperature dependence of the viscosity becomes more pronounced when the viscosity of the oil increases. Minor factors that may affect oil viscosity were their acidity, particulates content, and micro-/nanostructure [8]. The viscosity of bio-oils can be reduced by the addition of a polar solvent such as methanol.

8.3.1.4 Acidity The pH value of fast-pyrolysis bio-oils is low ranging between 2 and 3. Such acidity values are mainly due to the presence of large amounts of volatile acids (60–70%), with acetic and formic acids being the main constituents. Other groups of compounds present in fast-pyrolysis bio-oils, that also affect their acidity, include phenolics, fatty and resin acids, and hydroxy acids (e.g. glycolic acid) [9]. The acids in biomass fast-pyrolysis oils are mainly derived from the degradation of hemicelluloses in wood. The high acidity of bio-oils can lead to severe corrosion of the storage containers and transportation lines such as carbon steel and aluminum [10]. For these reasons bio-oils should be stored in acid-resistant vessels (e.g. stainless steel or polyolefin lined).

8.3.1.5 Heating Value Due to its high oxygen content and the presence of a significant amount of water, bio-oils are of lower heating value (16–19 MJ kg^{-1}), compared to their fossil fuel counterparts (42–44 MJ kg^{-1}).

8.3.2 Compositions of Bio-Oil

Bio-oil is a complicated mixture of highly oxygenated organic compounds including aldehydes, carboxylic acids, phenols, sugars, and aliphatic and aromatic hydrocarbons. A typical example of compositions of bio-oil is listed in Table 8.4 [11].

Understanding the composition of bio-oils is extremely valuable if one is to evaluate the bio-oils' stabilities, properties, and toxicity [12]. For example, raw bio-oils may contain some highly reactive oxygenated organic compounds which make pyrolysis oils unstable. During storage, chemical reactions can occur between these compounds to form larger molecules, resulting in increased viscosities over time [7].

Mullen and Boateng [12] studied the chemical composition of bio-oils produced by fast pyrolysis of switchgrass and alfalfa stems. It was found that more nitrogen-containing compounds were found in the alfalfa stem derived bio-oils with correspondingly higher nitrogen content of the alfalfa stems biomass versus the switchgrass. Nitrogen-containing compounds found in the alfalfa stems bio-oils are 2,2,6,6-tetramethylpiperidone, benzylnitrile, pyridinol, indole, and methylindole. But for these nitrogen-containing compounds, only benzylnitrile was found in switchgrass.

TABLE 8.4 Yields of Bio-Oil Compounds from the Pyrolysis of Southern Pine Wood

Compound	Yield (wt%)
Hydroxyacetaldehyde	3.07
Acetic acid	1.87
Hydroxyacetone	1.36
2-Furaldehyde	0.34
Furfuryl alcohol	0.37
Furan-(5H)-2-one	1.10
Phenol	0.04
Guaiacol	0.41
o-Cresol	0.05
p-Cresol	0.07
Levoglucosenone	0.19
4-Methyl guaiacol	0.65
2,4-Dimethyl phenol	0.13
4-Ethyl-guaiacol	0.12
Eugenol	0.22
5-(Hydroxy-methyl)-furaldehyde	0.99
Catechol	0.62
Isoeugenol	0.51
Vanillin	0.35
Acetoguaiacone	0.23
Guaiacyl acetone	0.45
Levoglucosan	4.86

A large amount of the water-soluble compounds is mainly derived from the decomposition of cellulose and hemicellulose [13]. Switchgrass biomass contains higher levels of cellulose and hemicelluloses than the alfalfa stems explaining why higher concentration of the water solubles can be determined for switchgrass-derived bio-oil.

8.4 UPGRADING OF BIO-OILS

While bio-oil production has achieved commercial success, overall poor characteristics limits the application as transportation fuels. The high oxygen and water contents of crude bio-oil are two key issues that make bio-oils unstable with high viscosity, thermal instability, corrosive characteristics, and chemical complexity. Therefore, it is imperative to develop efficient techniques to upgrade bio-oils. Nowadays, bio-oil upgrading techniques include hydrogenation, catalytic cracking, steam reforming, emulsification, converting them into stable oxygenated compounds, extracting chemicals from the bio-oils. In the following sections, such modern refining treatments are described in detail.

8.4.1 Hydrogenation

Hydrogenation is considered to be amongst the most effective method for bio-oil upgrading [13]. Typical catalysts used in the conventional hydrogenation process are metal sulfide catalysts such as cobalt or nickel doped molybdenum sulfides. However, these catalysts are sensitive to water and are readily poisoned by high concentrations of oxygen-containing compounds [14]. In addition, there are some drawbacks in the hydrogenation process, such as target products contaminated by sulfur, coke accumulation, and water-induced catalyst deactivation. Therefore, new catalysts are under development with the goal of increasing their catalytic selectivity and product yields.

In the conversion of aqueous phenolic bio-oil components to alkanes, Zhao et al. [15] developed a new and efficient (nearly 100% cycloalkane and methanol yields) catalytic route based on bifunctional catalysts combining Pd/C-catalyzed hydrogenation with H_3PO_4-catalyzed hydrolysis/dehydration. The final alkane products were easily separated from the aqueous phase. Moreover, Zhao et al. [14] developed a new green route based on low-cost RANEY® Ni catalysts and an environmentally friendly Brønsted solid acid, that is, Nafion/SiO_2. This process was claimed to convert the aqueous phenolic monomers (phenols, guaiacols and syringols) within bio-oil to hydrocarbons and methanol. The new catalyst combination opens the possibility for the application in hydrodeoxygenation and hydrogenation of lignin-derived bio-products.

Due to the presence of many unstable compounds, bio-oils are of poor stability with aldehydes being the most reactive functional groups present in them. Consequently, it is essential to convert aldehydes to more stable compounds. A homogeneous catalyst based on ruthenium, ($RuCl_2(PPh_3)_3$) was prepared for the hydrogenation of bio-oils, with a very significant catalytic performance on aldehydes and ketone hydrogenation in a single-phase system [16]. In this work, it was found that most of the aldehydes in a bio-oil fraction extracted with ethyl acetate could be converted to the corresponding alcohols under mild conditions (70 °C, 3.3 MPa H_2). Thus the stability of bio-oils and as such the fuel quality could be improved.

Ionic liquids themselves are known to exhibit catalytic effect and be used as solvents and as catalysts for chemical transformations. Moreover, chemical reactions can be performed under mild conditions in ionic liquids. Yan et al. [17] reported the hydrodeoxygenation of lignin-derived phenols into alkanes by using metal nanoparticle catalysts combined with ionic liquids with Brønsted acidity, under mild conditions. This bifunctional system was used to transform a variety of phenolic compounds in both [Bmim][BF_4] and [Bmim][Tf_2N]. Only the Rh-nanoparticle containing system was able to convert the branched phenols into alkanes in high yields. Compared to previous systems with metal sulfite or with mineral acid-/supported metal catalysts in water, this system was found to be an efficient and less energy-demanding process to upgrade lignin derivatives.

Another novel strategy used to improve the properties of bio-oils involves hydrotreating the raw bio-oil under mild conditions reducing carboxylic acid compounds to alcohols which could subsequently be esterified with unconverted acids within the bio-oil [18,19]. The traditional severe conditions required for a hydrotreatment of bio-oil (high temperature 300–400 °C and high hydrogen pressure 10–20 MPa) could thus be avoided. For example, bio-oil was upgraded as it emerged from the fast/vacuum pyrolysis of biomass over MoNi/γ-Al_2O_3 catalysts. The resulting GC-MS spectrometric analyses showed that both hydrotreatment and esterification had occurred over the 0.06MoNi/γ-Al_2O_3(873) catalyst during the upgrading process. Furthermore, the data showed that the reduced Mo-10Ni/γ-Al_2O_3 catalyst had the highest activity with 33.2% acetic acid conversion and the ester compounds in the upgraded bio-oil was found to be increased threefold.

Murata et al. [20] have reported that the pretreatment of cellulose in 1-hexanol at 623 K, followed by hydrocracking, catalyzed by Pt/H-ZSM-5(23) at 673 K, yielded up to 89% of C2–C9 alkanes with only 6% of CH_4/CO_x. The combination of alcohol pretreatment and Pt/H-ZSM-5-catalyzed hydrocracking was effective for the reaction. The findings suggested that the alcohol treatment could lead to lowering of the molecular weight of the cellulosic material, producing oxygenated intermediates such as monosaccharides and disaccharides, which could be further converted to C2–C9 alkane products by successive hydrocracking and condensation.

Pyrolytic lignins affect the bio-oil properties such as high viscosity, high reactivity, and low stability, which is difficult for bio-oil upgrading due to their nonvolatility and thermal instabilities. Tang et al. [21] converted pyrolytic lignins to stable liquid compounds through hydrocracking at 260 °C in supercritical ethanol under a hydrogen atmosphere by the use of Ru/ZrO_2/SBA-15 or Ru/SO_4^{2-}/ZrO_2/SBA-15 catalyst. The accumulated data demonstrated that under supercritical ethanol conditions, $Ru/ZrO_2/SBA$-15 and $Ru/SO_4^{2-}/ZrO_2/SBA$-15 were effective catalysts converting pyrolytic lignins to stable monomers such as phenols, guaiacols, anisoles, esters, light ketones (with the C_5–C_6 ring), alcohols, long-chain alkanes (C_{13}–C_{25}).

A potential way of valorizing bio-oils as a fuel is their cohydrotreatment with petroleum fractions. To meet environmental fuel standards, there is a need to study the simultaneous hydrodeoxygenation and hydrodesulfurization reactions before considering such a cotreatment. Pinheiro et al. [22] investigated the impact of oxygenated compounds from lignocellulosic biomass pyrolysis oils on gas oil hydrotreatment. They found that 2-propanol, cyclopentanone, anisole, and guaiacol were not inhibitors of catalytic performances under such operating conditions. On the contrary, propanoic acid and ethyl decanoate had an inhibiting effect on hydrodesulfurization, hydrodenitrogenation, and aromatic ring hydrogenation reactions.

8.4.2 Catalytic Cracking

Bio-oil vapors can be upgraded via catalytic deconstruction to hydrocarbons with their oxygen being removed as H_2O, CO_2, or CO in the absence of added hydrogen. Zeolites have been found to be promising catalysts for such upgrading. French and Czernik [23] have evaluated the catalytic performance of a set of commercial and laboratory-synthesized catalysts for upgrading of bio-oils via this route. In this effort, ZSM-5 type catalysts performed the best while larger-pore zeolites presented less deoxygenation activity. The highest yield of hydrocarbons (approximately 16 wt%, including 3.5 wt% of toluene) was obtained over nickel, cobalt, iron, and gallium-substituted ZSM-5 materials.

The presence of oxygenated compounds in the pyrolysis products results in a high level of acidity, which can lead to corrosion on internal combustion engines. Quirino et al. [24] studied the influence of alumina catalysts doped with tin and zinc oxides in the soybean oil pyrolysis reaction. It was observed that the presence of alumina catalysts doped with tin and zinc oxides during the pyrolysis can decrease undesirable carboxylic acid content up to 30%. Higher deoxygenating activities were achieved over solid $(SnO)_1(ZnO)_1(Al_2O_3)_8$ catalyst. Pyrolysis of vegetable oils is an acceptable process to convert vegetable oils into gasoline and diesel fuel. A. Demirbas [25] had documented that a gasoline like material can be obtained from sunflower oil via a pyrolysis process in the presence of Al_2O_3 catalyst treated with sodium hydroxide. The highest yield of gasoline was found to be 53.8% based on sunflower oil in the presence of 5% of catalyst.

8.4.3 Steam Reforming

Catalytic steam reforming of bio-oil offers a feasible option to produce hydrogen sustainably. Many literature accounts provide new developments in the area of hydrogen production via steam reforming of bio-oil in recent years. But due to the complexity of bio-oil and carbon deposition on the catalyst surface during the reaction process, currently research studies mainly focus on the steam reforming of model compounds in bio-oil and reforming catalysts [26].

Xu et al. [27] studied hydrogen production via catalytic steam reforming of bio-oil in a fluidized-bed reactor and selected nickel-based catalyst (Ni/MgO) as the reforming catalyst. It was found that the carbon deposition was not the main reason for catalyst deactivation. In fact, the fresh catalyst deactivation can be explained by the NiO grain sintered on the support surface.

Li et al. [28] applied the Ni/Mg/Al catalysts to the steam reforming of tar from pyrolysis of biomass. The optimized Ni/Mg/Al catalyst with a composition of Ni/Mg/Al=9/66/25 was found to exhibit high activity with high resistance to coke deposition, in particular, to coke formed by the disproportionation of CO which is an important product of the steam-reforming process.

Kan et al. [29] developed an efficient method for the production of hydrogen from the crude bio-oil via an integrated gasification-electrochemical catalytic reforming (G-ECR) process using a NiCuZnAl catalyst. The accumulated data showed that a maximum hydrogen yield of 81.4% and carbon conversion of 87.6% were obtained via the integrated G-ECR process. Compared to direct reforming of crude bio-oil, the deactivation of the catalyst was significantly suppressed by using the integrated gasification-reforming method. It was thus concluded that the integrated G-ECR process could be a potentially useful route to produce hydrogen from crude bio-oil.

8.4.4 Emulsification

One of the methods used to upgrade bio-oil to transportation fuels is to combine the bio-oil with other fuel sources by forming an emulsion. This results in a liquid fuel of a low viscosity, high calorific value, and high cetane number. It was observed that the emulsion of bio-oil with diesel fuel, at a suitable volume ratio, can lead to more stable emulsions compared to the original bio-oil [30, 31]. A stable bio-crude oil/diesel oil emulsion can be seen from Figure 8.2 [30]. The viscosity of emulsified bio-oil was substantially lower than the viscosity of bio-oil itself and the corrosivity of the emulsified fuels was also found to be reduced [32].

Crossley et al. [33] reported a family of solid catalysts that can simultaneously stabilize water–oil emulsions and catalyze reactions at the liquid/liquid interface. By depositing palladium onto carbon nanotube–inorganic oxide hybrid nanoparticles, a biphasic hydrodeoxygenation and condensation catalytic reaction occurred. To illustrate the application of the catalytic nanohybrids in emulsions, a hydrodeoxygenation reaction was examined at the water/oil interface with vanillin as a test substrate and Pd-containing nanohybrid as the catalyst. During the reaction, different products were obtained depending on the reaction temperature and the degrees of hydrogenation, hydrogenolysis, and decarbonylation reactions. At 100 °C, the primary product was found to be vanillin alcohol that remained within the aqueous phase. As reaction time progressed, the vanillin alcohol was consumed by hydrogenolysis to form *p*-creosol, which was found to migrate to the organic phase upon formation, preventing further conversion. At 250 °C, the dominant reaction is the decarbonylation of the aldehyde group, leading primarily to guaiacol

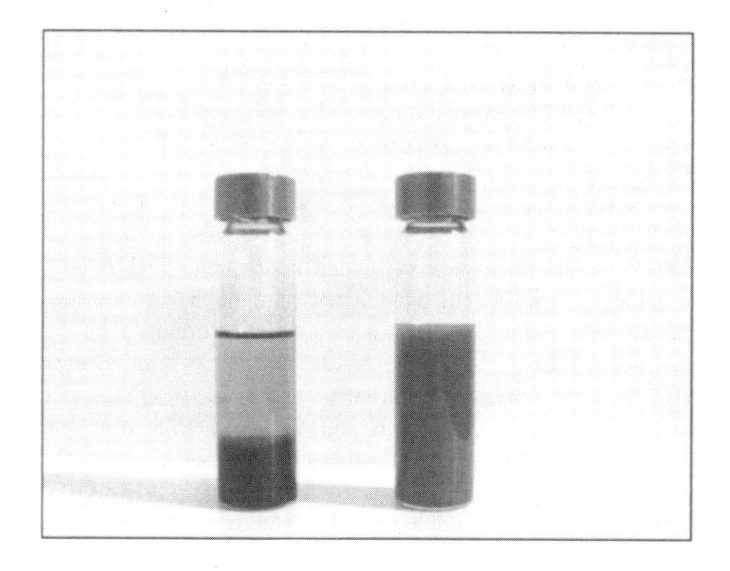

FIGURE 8.2 Bio-crude oil/diesel oil mixture (left) and emulsion (right). (Reproduced from Ref. [30] with permission from Elsevier © 2002.)

which was also found to migrate to the organic phase. The carbon chains migrate to the organic phase after growing long enough to get the desirable products, whereas the shorter chains remained in the aqueous phase for further growth. This data illustrates that the concept of simultaneous reactions and separation of the intermediate products is possible since sequential reactions can be conducted within a single reactor.

Garcia-Perez et al. [34] reported on the fuel properties of fast-pyrolysis oil/bio-diesel blends. Commercial biodiesels are of lower oxidation stability and comparatively poor cold flow properties. When pyrolysis oil was blended with bio-diesel, the bio-diesel could extract selectively some of the fractions of the bio-oil, particularly those high in phenolic compounds. This could partially use bio-oils as additives for transportation fuels, while the oxidation stability of the bio-diesel could also be improved since phenolic compounds are known to be excellent antioxidants. However, other fuel properties such as solid residue and the acid number were found to deteriorate. This is because the solubilization of lignin-derived oligomers within the bio-diesel resulted in an increase of the solid residue. Consequently, solid residues need to be carefully monitored. Moreover, the solubility of bio-oil in bio-diesel was also found to be improved on addition of ethyl acetate.

8.4.5 Converting into Stable Oxygenated Compounds

Most bio-oil upgrading methods are based on deoxygenation of the crude bio-oils to reduce its oxygen content. Such processes are known to increase the upgrading costs while consuming large amounts of hydrogen. A useful approach is to convert

SCHEME 8.1

chemically unstable and corrosive oxygenated components (acids, phenols, aldehydes) into stable and flammable oxygenated compounds (esters, alcohols, and ketones). These stable oxygenated compounds can be developed into oxygenated fuels and can also be added into petroleum fuel to raise its combustion efficiency.

This upgrading approach can be classified based on two reaction categories: esterification and ketonization reactions. As far as esterification reactions are concerned, Tang et al. [35] designed hydrogenation–esterification of aldehydes and acids (Scheme 8.1) over a 5%Pt/HZSM-5 and a 5%$Pt/Al_2(SiO_3)_3$ at 150 °C and 1.5 MPa of H_2 pressure. Acetaldehyde and butyl aldehyde were reduced *in situ* to ethanol and butanol, respectively, then found to react with acetic acid forming ethyl acetate and butyl acetate.

Peng et al. [36] upgraded the pyrolysis bio-oil from rice husks in sub- and supercritical ethanol using HZSM-5 as the catalyst. The data showed that a supercritical upgrading process was superior to a subcritical upgrading process. Acidic HZSM-5 was found to promote esterification reactions converting acids into a wide range of esters in supercritical ethanol. During supercritical upgrading, stronger acidic HZSM-5 (low Si/Al ratio) can more effectively facilitate the cracking of heavy components of crude bio-oil. Similar results were also obtained when the pyrolysis bio-oil was upgraded in supercritical ethanol using aluminum silicate as the catalyst [37]. Acidic aluminum silicate can facilitate the esterification to convert most carboxylic acids contained within the crude bio-oil into esters.

Upgrading bio-oil by catalytic esterification over solid acid ($40SiO_2/TiO_2$–SO_4^{2-}) and solid base ($30K_2CO_3/Al_2O_3$–NaOH) catalysts can lower the bio-oil's dynamic viscosity, enhance fluidity, and improve stability over time. The solid acid catalyst was found to achieve higher catalytic activity upon esterification than the solid base catalyst [38].

Xiong et al. [39] synthesized a dicationic ionic liquid $[C_8(mim)_2][HSO_4]_2$ (Fig. 8.3) and used it as the catalyst to upgrade bio-oil through the esterification reaction of organic acids and ethanol at room temperature. It was found that no coke and deactivation of the catalyst were observed. The yield of upgraded oil reached 49%, and its properties were significantly improved with higher heating

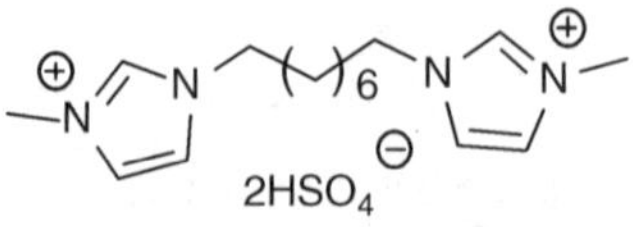

FIGURE 8.3 Structure of ionic liquid $[C_8(mim)_2][HSO_4]_2$.

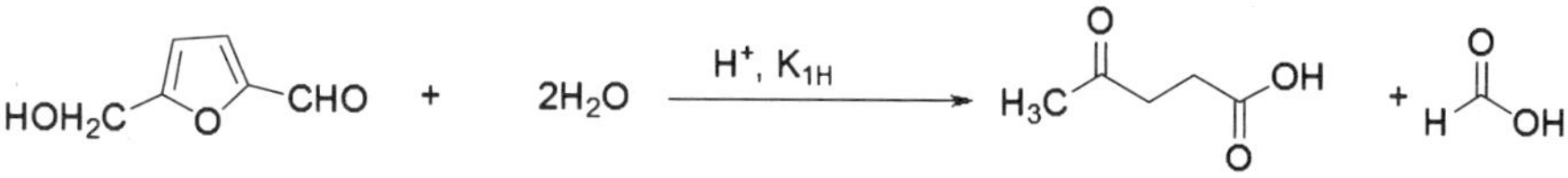

HOH_2C–(furan)–CHO + $2H_2O$ $\xrightarrow{H^+, K_{1H}}$ H_3C–CO–CH_2CH_2–COOH + HCOOH

SCHEME 8.2

value of 24.6 MJ kg^{-1}, an increase of pH value to 5.1, and a decrease of moisture content to 8.2 wt%. The data showed that organic acids could be successfully converted into esters and that the dicationic ionic liquid can facilitate the esterification to upgrade bio-oil.

Wang et al. [40] upgraded bio-oil by catalytic esterification over 732- and NKC-9-type ion-exchange resins. It was shown that the acid number of bio-oil was significantly reduced by 88.54 and 85.95% after bio-oil was upgraded over 732 and NKC-9 resins respectively, which represented that organic acids were converted to neutral esters. The heating values increased significantly, while the moisture contents and the densities decreased. Specifically, the viscosity was lowered from 81.27 mm^2 s^{-1} to 2.45 mm^2 s^{-1} (at 40 °C) when bio-oil was upgraded over NKC-9-type ion-exchange resin. It was observed that the stability of upgraded bio-oil was improved.

With respect to the ketonization reaction, Deng et al. [41] proposed a novel method to upgrade the acid-rich phase of bio-oil via ketonic condensation over weak base CeO_2 catalysts. Most acetic acid was effectively transformed to acetone in model reactions:

$$2\text{RCOOH} \rightarrow \text{RCOR} + \text{H}_2\text{O} + \text{CO}_2.$$

During this reaction, furfural is known to exhibit significant deactivation. Consequently it is recommended to remove or decompose furans before upgrading the feed from bio-oil by hydrothermal treatment [42], which is presented in Scheme 8.2.

Gärtner et al. [43] showed that ceria-zirconia was an effective catalyst for the upgrading of acids in bio-oils to produce larger ketones through ketonization reactions. At the same time, it was also found that esterification was an important side reaction that could compete with ketonization when alcohols were present in the hydrophobic mixture. Since esterification cannot be avoided over ceria-zirconia, it was suggested that esters could be converted to the desired ketones over the same catalyst through the direct ketonization, without the need to add water to the feed for the hydrolysis of the ester.

8.4.6 Chemicals Extracted From Bio-Oils

With the development of new pyrolysis technologies many studies have been carried out to increase the yields of the target products in bio-oils through specific pretreatments of biomass or catalytic pyrolysis of biomass. The pyrolytic treatment of biomass in the presence of various metal oxides is one such example. Six nanostructured metal oxides (MgO, CaO, TiO_2, Fe_2O_3, NiO, and ZnO) were used as catalysts to upgrade biomass fast-pyrolysis vapors aimed at maximizing the formation of various valuable chemicals [44]. CaO was found to significantly decrease the yields of phenols and anhydrosugars, and eliminate the acids, while also increasing the formation of cyclopentanones, hydrocarbons, and several lighter products such as acetaldehyde, acetone, 2-butanone, and methanol. ZnO was also a mild catalyst which only slightly altered the pyrolytic product composition. The remaining four catalysts (MgO, TiO_2, Fe_2O_3, NiO) all decreased the yield of linear aldehydes dramatically while increasing the yields of ketones and cyclopentanones. With the exception of NiO, they also decreased the anhydrosugars content. Furthermore, Fe_2O_3 was found to promote the production of various hydrocarbons.

In general, the fast pyrolysis of cellulose generates low yields of furan compounds. Lu et al. [45] studied the catalytic pyrolysis of cellulose with three sulfated metal oxides (SO_4^{2-}/TiO_2, SO_4^{2-}/ZrO_2, and SO_4^{2-}/SnO_2) in order to obtain high yields of light furan compounds. The oligomers were cracked into monomeric compounds over these catalysts through catalytic cracking of the pyrolysis vapors. The final primary pyrolytic products (such as levoglucosan and hydroxyacetaldehyde) were found to be decreased or completely eliminated while the yields of three light furan compounds (5-methyl furfural, furfural, and furan) increased greatly. The catalysts presented different selectivities on the targeted products with the formation of 5-methyl furfural favored by SO_4^{2-}/SnO_2, furfural favored by SO_4^{2-}/TiO_2, and furan favored by SO_4^{2-}/ZrO_2, respectively.

Lu et al. [46] found that Pd/SBA-15 catalysts could remarkably promote the formation of monomeric phenolic compounds when biomass fast-pyrolysis vapors were catalytically cracked over these catalysts. The Pd/SBA-15 catalysts presented cracking capabilities to convert the lignin-derived oligomers to monomeric phenolic compounds and further convert them to phenols. The removal of carbonyl group and unsaturated C–C bond from the phenolic compounds indicated that Pd/SBA-15 catalysts presented decarbonylation activity and might have some hydrotreating capability.

Lin et al. [47] examined the direct deoxygenation effect of CaO on bio-oil during biomass pyrolysis in a fluidized-bed reactor. It was shown that at a CaO/white pine mass ratio of 5, the oxygen content of the organic components in the bio-oil was reduced by 21%. With increasing mass ratio, the oxygen-rich compounds in the bio-oil, such as laevoglucose, formic acid, acetic acid, and D-allose, decreased dramatically, which could reduce the total oxygen content of the bio-oil.

CCA (chromated copper arsenate) treated wood originally impregnated with such metals for the preservation purposes was also examined as a source of chemicals under low-temperature pyrolysis conditions, as an alternate method for its disposal [11]. The work showed that the presence of chromated copper arsenate within the structure of wood had a significant effect on the yields of the main carbohydrate-degradation products under mild pyrolytic conditions. More specifically, the yield of levoglucosan from treated wood was found to increase while the yields of hydroxyacetaldehyde and hydroxyacetone were seen to decrease.

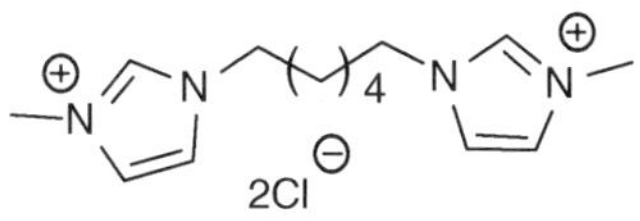

FIGURE 8.4 Structure of ionic liquid $[C_6(mim)_2]Cl_2$.

In order to obtain high-yield commodity chemicals from pyrolysis oil, Vispute et al. [48] used an integrated catalytic approach that combines low-temperature hydroprocessing of the bio-oils over a Ru-based catalyst and at higher temperature over a Pt-based catalyst, followed by a zeolite-conversion step. This combination of the hydrogenation steps with a zeolite-conversion step can reduce the overall hydrogen requirements as compared to hydrogen used for a complete deoxygenation of pyrolysis oil. The intrinsic hydrogen content of the pyrolysis oil increased through the hydroprocessing reaction. Polyols and alcohols could be produced through the hydroprocessing reaction. The zeolite catalyst then converted these hydrogenated products into light olefins and aromatic hydrocarbons with a yield much higher than that produced with the pure pyrolysis oil. Thus the combination of the hydrogenation steps with a zeolite-conversion step significantly raised the yields of olefins and aromatics. The direct zeolite upgrading of the water-soluble fraction of a pinewood bio-oil could offer 26.7% carbon yield of olefins and aromatics. Low-temperature hydrogenation before zeolite upgrading raised the yield to 51.8%, whereas the high-temperature hydrogenation resulted in higher yield of olefins and aromatics to 61.3%.

Ionic liquids have also been used in the field of biomass pyrolysis. More recently, Sheldrake and Schleck [49] have reported that dicationic molten salts ionic liquids were used as solvents for the controlled pyrolysis of cellulose to anhydrosugars. It was demonstrated that the use of dicationic $[C_6(mim)_2]Cl_2$ (Fig. 8.4) for the pyrolysis of cellulose gave levoglucosenone as the dominant anhydrosugar product at 180 °C. A variety of other special catalytic systems have also been reported to effectively favor the production of various chemicals [44], including the formation of levoglucosenone by using H_3PO_4 [50], 1-hydroxy-3,6-dioxabicyclo[3.2.1]octan-2-one by nano aluminium titanate [51], acetol by NaOH or Na_2CO_3 [52], furfural by $MgCl_2$ [53] or $Fe_2(SO_4)_3$ [54].

8.4.7 Other Bio-Oil Upgrading Methods

High pressure thermal treatment (HPTT) [55] is a new process developed by BTG and University of Twente with the potential to economically reduce the oxygen and water content of oil obtained by fast pyrolysis. During the HPTT process, pyrolysis oil undergoes a phase separation at temperatures of 200–350 °C with a residence time of several minutes (1.5–3.5 min) at 200 bar, yielding a gas phase, an aqueous phase, and an oil phase. The results showed that the oil obtained had lower oxygen (reduced from 40 to 23 wt%) and water content, and higher energy density (wet HHV ranging from 21.8 to 28.4 $MJ \cdot kg^{-1}$). However, the adverse formation of high molecular weight components occurred during HPTT of pyrolysis oil, which was probably due to polymerization of the sugars present within the pyrolysis oil. Miscibility tests showed that HPTT oil was immiscible with a conventional heavy refinery stream. It was recommended that further upgrading of the HPTT oil by hydrodeoxygenation was an option that could reduce the H_2 consumption during hydrodeoxygenation as compared to direct hydrodeoxygenation of pyrolysis oil.

Vispute and Huber [56] reported a new approach for the conversion of bio-oils via aqueous phase processing (APP). During the process, hydrogen, alkanes (ranging from C1 to C6) and polyols (ethylene glycol, 1,2-propanediol, and 1,4-butanediol) can be produced from the aqueous fraction of wood-derived pyrolysis oils. The pyrolysis oil was first phase separated into aqueous and nonaqueous fractions by mixing with distilled water. Then the aqueous fraction was subjected to a low-temperature hydrogenation with Ru/C catalyst at 125–175 °C at 68.9 bar in order to thermally convert all unstable compounds to thermally stable compounds prior to APP. After the hydrogenation step, the polyols can be separated out if desired. In the ensuing steps, hydrogen was produced with high selectivity of 60% from the water-soluble part of bio-oil by aqueous-phase reforming. This is a feasible way to produce hydrogen from bio-oil. Alkanes can be produced from the water-soluble bio-oils by aqueous-phase dehydration/hydrogenation over a bifunctional catalyst (Pt/Al_2O_3–SiO_2). The results showed that an alkane selectivity of 77% was obtained with hydrogen being co-fed to the reactor. Alternatively, an alkane selectivity of 45% was achieved when hydrogen was generated *in situ* from bio-oil. It can be seen that the advantage of this approach is that the aqueous phase of the bio-oil is processed differently from the organic phase, which can allow us to design catalysts that are well-suited for conversion of both the aqueous and organic phases, achieving higher overall yields for conversion of bio-oils into liquid fuels and chemicals.

8.5 CONCLUSIONS

Bio-oil has the potential to replace petroleum oil and also offers a source of valuable chemicals. The science and engineering pertaining to bio-oil upgrading has seen great progress in recent years with numerous challenges and limitations investigated, specially when considering large-scale application of bio-oil as a fuel. More specifically; the following issues need to be considered.

- Reducing the cost of bio-oil as compared to petroleum oil.
- Availability of raw material sources, handling and transportation issues.
- Addressing catalysts deactivation and coke-deposition issues.
- Effective procedures need to be developed when combining two or more bio-oil upgrading techniques.
- Reactors need to be designed so as to meet the product requirements.
- Environmental health and safety issues need to be given greater priority.
- Standards need to be set up for the quality monitoring and testing of bio-oils.

REFERENCES

1. E. Furimsky, *Appl. Catal. A.* **2000**, 199, *147.*
2. D. Mohan, C. U. Pittman and P. H. Steele, *Energ. Fuel* **2006**, 20, *848.*
3. R. H. Venderbosch and W. Prins, *Biofuels Bioprod. Bioref.* **2010**, 4, *178.*
4. R. J. M. Westerhof, D. W. F. Brilman, W. P. M. V. Swaaij and S. R. A. Kersten, *Ind. Eng. Chem. Res.* **2010**, 49, *1160.*
5. A. V. Bridgwater, D. Meier and D. Radlein, *Org. Geochem.* **1999**, 30, *1479.*
6. D. Meier and O. Faix, *Bioresource Technol.* **1999**, 68, *71.*
7. S. Czernik and A. V. Bridgwater, *Energ. Fuel* **2004**, 18, *590.*
8. M. W. Nolte and M. W. Liberatore, *Energ. Fuel* **2010**, 24, *6601.*
9. A. Oasmaa, D. C. Elliott and J. Korhonen, *Energ. Fuel* **2010**, 24, *6548.*
10. E. J. Soltes and J.-C. K. Lin, *Progress in Biomass Conversion*, (Eds. D. A. Tillman, E. C. Jahn), Academic Press: New York, **1984**.
11. Q. Fu, D. S. Argyropoulos, D. C. Tilotta and L. A. Lucia, *Ind. Eng. Chem. Res.* **2007**, 46, *5258.*
12. C. A. Mullen and A. A. Boateng, *Energ. Fuel* **2008**, 22, *2104.*
13. G. W. Huber, S. Iborra and A. Corma, *Chem. Rev.* **2006**, 106, *4044.*
14. C. Zhao, Y. Kou, A. A. Lemonidou, X. Li and J. A. Lercher, *Chem. Commun.* **2010**, 46, *412.*
15. C. Zhao, Y. Kou, A. A. Lemonidou, X. Li and J. A. Lercher, *Angew. Chem. Int. Ed.* **2009**, 48, *3987.*
16. F. Huang, W. Li, Q. Lu and X. Zhu, *Chem. Eng. Technol.* **2010**, 33, *2082.*
17. N. Yan, Y. Yuan, R. Dykeman, Y. Kou and P. J. Dyson, *Angew. Chem. Int. Ed.* **2010**, 49, *5549.*
18. Y. Xu, T. Wang, L. Ma, Q. Zhang and L. Wang, *Biomass Bioenerg.* **2009**, 33, *1030.*
19. Y. Xu, T. Wang, L. Ma, Q. Zhang and W. Liang, *Appl. Energ.* **2010**, 87, *2886.*
20. K. Murata, Y. Liu, M. Inaba and I. Takahara, *Catal. Lett.* **2010**, 140, *8.*
21. Z. Tang, Y. Zhang and Q. Guo, *Ind. Eng. Chem. Res.* **2010**, 49, *2040.*
22. A. Pinheiro, D. Hudebine, N. Dupassieux and C. Geantet, *Energ. Fuel* **2009**, 23, *1007.*
23. R. French and S. Czernik, *Fuel Process. Technol.* **2010**, 91, *25.*
24. R. L. Quirino, A. P. Tavares, A. C. Peres, J. C. Rubim and P. A. Z. Suarez, *J. Am. Oil Chem. Soc.* **2009**, 86, *167.*
25. A. Demirbas, *Energ. Source* **2009**, 31, *671.*
26. M. Balat, *Energ. Source* **2011**, 33, *674.*
27. Q. Xu, P. Lan, B. Zhang, Z. Ren and Y. Yan, *Energ. Fuel* **2010**, 24, *6456.*
28. D. Li, L. Wang, M. Koike, Y. Nakagawa and K. Tomishige, *Appl. Catal. B: Environ.* **2011**, 102, *528.*
29. T. Kan, J. Xiong, X. Li, T. Ye, L. Yuan, Y. Torimoto, M. Yamamoto and Q. Li, *Int. J. Hydrogen Energ.* **2010**, 35, *518.*
30. D. Chiaramonti, M. Bonini, E. Fratini, G. Tondi, K. Gartner, A. V. Bridgwater, H. P. Grimm, I. Soldaini, A. Webster and P. Baglioni, *Biomass Bioenerg.* **2003**, 25, *85.*
31. X. Jiang and N. Ellis, *Energ. Fuel* **2010**, 24, *1358.*

32. M. Ikura, M. Stanciulescu and E. Hogan, *Biomass Bioenerg*. **2003**, 24, *221*.
33. S. Crossley, J. Faria, M. Shen and D. E. Resasco, *Science* **2010**, 327, *68*.
34. M. Garcia-Perez, J. Shen, X. Wang and C. Li, *Fuel Process. Technol.* **2010**, 91, *296*.
35. Y. Tang, W. Yu, L. Mo, H. Lou and X. Zheng, *Energ. Fuel* **2008**, 22, *3484*.
36. J. Peng, P. Chen, H. Lou and X. Zheng, *Bioresource Technol.* **2009**, 100, *3415*.
37. J. Peng, P. Chen, H. Lou and X. Zheng, *Energ. Fuel* **2008**, 22, *3489*.
38. Q. Zhang, J. Chang, T. Wang and Y. Xu, *Energ. Fuel* **2006**, 20, *2717*.
39. W. Xiong, M. Zhu, L. Deng, Y. Fu and Q. Guo, *Energ. Fuel* **2009**, 23, *2278*.
40. J. Wang, J. Chang and J. Fan, *Energ. Fuel* **2010**, 24, *3251*.
41. L. Deng, Y. Fu and Q. Guo, *Energ. Fuel* **2009**, 23, *564*.
42. B. Girisuta, L. P. B. M. Janssen and H. J. Heeres, *Green Chem.* **2006**, 8, *701*.
43. C. A. Gärtner, J. C. Serrano-Ruiz, D. J. Braden and J. A. Dumesic, *ChemSusChem*. **2009**, 2, *1121*.
44. Q. Lu, Z. Zhang, C. Dong and X. Zhu, *Energies*. **2010**, 3, *1805*.
45. Q. Lu, W. Xiong, W. Li, Q. Guo and X. Zhu, *Bioresource Technol.* **2009**, 100, *4871*.
46. Q. Lu, Z. Tang, Y. Zhang and X. Zhu, *Ind. Eng. Chem. Res.* **2010**, 49, *2573*.
47. Y. Lin, C. Zhang, M. Zhang and J. Zhang, *Energ. Fuel* **2010**, 24, *5686*.
48. T. P. Vispute, H. Zhang, A. Sanna, R. Xiao and G. W. Huber, *Science* **2010**, 330, *1222*.
49. G. N. Sheldrake and D. Schleck, *Green Chem.* **2007**, 9, *1044*.
50. G. Dobele, T. Dizhbite, G. Rossinskaja, G. Telysheva, D. Mier, S. Radtke and O. Faix, *J. Anal. Appl. Pyrol.* **2003**, 68–69, *197*.
51. D. Fabbri, C. Torri and I. Mancini, *Green Chem.* **2007**, 9, *1374*.
52. M. Chen, J. Wang, M. Zhang, M. Chen, X. Zhu, F. Min and Z. Tan, *J. Anal. Appl. Pyrol.* **2008**, 82, *145*.
53. Y. Wan, P. Chen, B. Zhang, C. Yang, Y. Liu, X. Lin and R. Ruan, *J. Anal. Appl. Pyrol.* **2009**, 86, *161*.
54. M. Chen, J. Wang, M. Zhang, M. Chen, X. Zhu, F. Min and Z. Tan, *J. Anal. Appl. Pyrol.* **2008**, 82, *145*.
55. F.de Miguel Mercader, M. J. Groeneveld, S. R. A. Kersten, R. H. Venderbosch and J. A. Hogendoorn, *Fuel* **2010**, 89, *2829*.
56. T. P. Vispute and G. W. Huber, *Green Chem.* **2009**, 11, *1433*.

CHAPTER 9

Microwave Technology for Lignocellulosic Biorefinery

TAKASHI WATANABE and TOMOHIKO MITANI

9.1 INTRODUCTION

Excessive use of fossil resources causes global warming and depletes readily accessible crude oil. The conversion of lignocellulosic biomass into biofuels and chemicals is becoming vital to the solution of these problems. Lignocellulosic feedstocks, however, have a complex composite structure, and their efficient utilization requires separation of their main polymeric components: cellulose, hemicelluloses, and lignin. So far, various pretreatment methods have been developed to increase the accessibility of cellulolytic and hemicellulolytic enzymes to the cell-wall polysaccharides. These include steam explosion, ammonia fibre/freeze explosion (AFEX), ammonia recycle percolation (ARP), hydrothermolysis, milling, microwave irradiation, electron beam irradiation and pretreatments with alkalis, acids, supercritical carbon dioxide, lime, white rot fungi, peroxide, and organic solvents (organosolvolysis) [1]. Among them microwave irradiation is attractive owing to its high efficiency of internal-energy transfer depending on a dielectric loss factor (ε''), selectivity for reactions, and short reaction time. These factors are advantageous for scaling up of the process. So far, microwave irradiation has been applied to the pretreatment of lignocellulosics by hydrothermolysis with and without acids or alkalis [2–17]. More recently, microwave pretreatments have been applied to organosovloysis with different types of catalysts [18–21]. Microwave-assisted pretreatments with hydrogen peroxide and metal complexes in aqueous media have also been studied [22]. For industrial applications, scaling up of microwave irradiators is essential, though some industrial microwave systems have been achieved for rubber preheating, ceramics processing, and food processing [23].

The Role of Green Chemistry in Biomass Processing and Conversion, First Edition.
Edited by Haibo Xie and Nicholas Gathergood.

In this chapter, we describe the principle of microwave heating, development of continuous bench-scale microwave irradiators and microwave-assisted reactions for the pretreatments of lignocellulosic biomass. The pretreated biomass can be converted to bioethanol and chemicals that are essential outputs for the biorefinery.

9.2 PRINCIPLES AND FEATURES OF MICROWAVE HEATING

Microwave heating is a method of material heating by electromagnetic waves [24]. Various definitions of the microwave are provided in terms of wavelength and frequency. Although the microwave wavelength (frequency) range is generally defined between 1 m (300 MHz) and 1 mm (300 GHz), well-known microwave heating systems are allocated outside the microwave range. The ISM (Industrial, Science, and Medical) frequency band of 2.45 GHz, which is allocated for commercial microwave ovens, is most commonly used for microwave heating. The ISM bands of 915 MHz and 5.8 GHz are also used for some microwave heating applications, for example, the 915 MHz ISM band is allocated for industrial microwave heating in some countries.

The microwave heating consists of three types of electromagnetic heating: dielectric heating, magnetic heating, and Joule heating. The dielectric heating arises from the time delay between the applied electric field to dielectric materials and the orientation polarization response of the materials to the electric field. The magnetic heating is generated by the hysteresis property of ferromagnetic materials. The Joule heating is caused by two types of current flow in conductive materials; the current driven by the electric field directly and the eddy current induced by the magnetic field. Since most microwave heating systems including biomass pretreatment systems are based on the dielectric heating, only the dielectric heating is described hereafter.

A property of dielectric materials is defined by permittivity. In general, the permittivity ε is normalized by the permittivity of free space $\varepsilon_0 = 8.85 \times 10^{-12}$ F·m^{-1}, and the relative permittivity $\varepsilon_r = \varepsilon/\varepsilon_0$ is expressed as a complex number $\varepsilon_r = \varepsilon_r' - j\varepsilon_r''$, where j is the imaginary unit. The real part ε_r' is related to the ability to store electric energy or the wavelength shortening in the material, and the imaginary part ε_r'' is related to the absorption and consumption (generally converted to heat) of electric energy in the material. ε_r is dependent on the electromagnetic wave frequency, and most dielectric materials have a peak of ε_r'' in the microwave range, due to their orientation-polarization response.

Features of microwave heating in comparison with external heating include shortening of heating time, improvement of heating efficiency, and capability of selective heating. One can experience these features by comparing food on a ceramic plate in a microwave oven and in a traditional oven, as shown in Figure 9.1. One can touch and pick up the plate with one's hands, even after the food is heated by microwaves, because ceramic is less susceptible to the microwave than the food at a low temperature. However, it is extremely difficult (indeed hazardous!) to touch the plate directly when heating the food in the traditional oven, because the plate and even

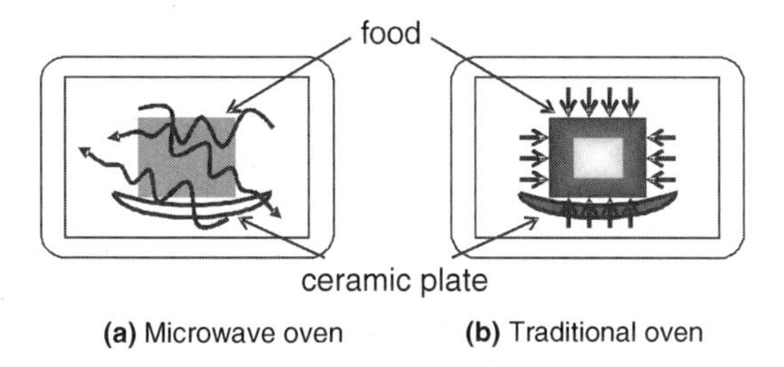

FIGURE 9.1 Comparison of material heating between by a microwave oven and a traditional oven.

the atmosphere are simultaneously heated with the food. In addition, the food can be heated rapidly by the microwave; heated gradually in the traditional oven, under the same energy consumption. These experiences prove the microwave can heat particular materials quickly, effectively, and selectively. In principle, the dielectric heating relies heavily on the permittivity. High ε_r'' materials for instance, water and ethylene glycol are heated easily by microwaves; whereas low ε_r'' materials for instance, quartz and crystal, are not. Therefore, temperature difference between materials can be observed by the microwave heating. In terms of shortening of heating time, numerous examples of rapid chemical synthesis have been also reported [24].

9.3 ABSORBED MICROWAVE POWER AND PENETRATION DEPTH

The absorbed microwave power and penetration depth are used as indices of microwave heating properties. The absorbed microwave power P is expressed as the following equation:

$$P = 2\pi f \varepsilon_0 \varepsilon_r'' |E|^2 = 2\pi f \varepsilon_0 \varepsilon_r' \tan\delta |E|^2,$$

where f is the microwave frequency, E the microwave electric field, and $\tan\delta = \varepsilon_r''/\varepsilon_r'$ the loss tangent, representing the microwave energy dissipated per unit wavelength. The penetration depth d of a material is defined as the depth to which the electromagnetic wave power is attenuated by $1/e$ from the surface, where e is the base of natural logarithm, and expressed as the following equation:

$$d = \frac{c}{2\pi f} \frac{1}{\sqrt{2\varepsilon_r'\left(\sqrt{\tan\delta^2 + 1} - 1\right)}},$$

where c is the speed of light.

Hence, the microwave heating properties are characterized by the relative permittivity and the loss tangent. Since the penetration depth at a microwave frequency is much deeper than at far-infrared, microwave heating can directly

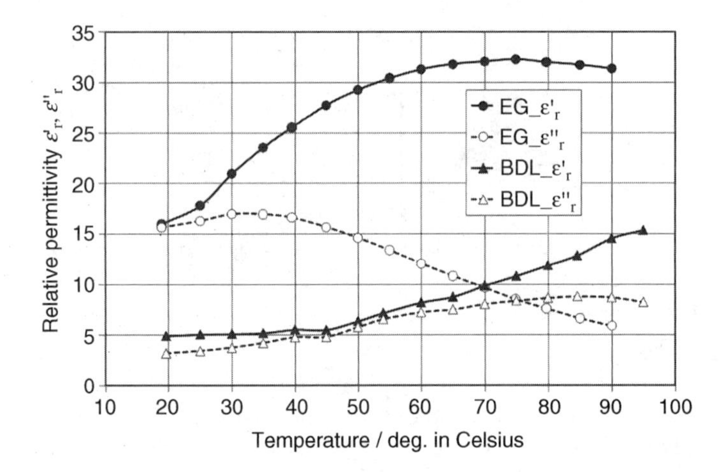

FIGURE 9.2 Temperature–permittivity characteristics of ethylene glycol (EG) and butanediol (BDL).

heat the material inside compared with radiation heating. This feature contributes not only to the shortening of heating time, but also the temperature uniformity between the material inside and the surface.

Permittivity is dependent on temperature as well as frequency. As measurement examples, measured temperature–permittivity characteristics of ethylene glycol and butanediol are shown in Figure 9.2. Solid lines and dashed lines show the real part (ε_r') and imaginary part (ε_r'') of the relative permittivity, respectively. As one can see, temperature drastically affects the material permittivity. It is therefore essential to take the permittivity characteristics into consideration when designing a highly efficient microwave heating device; otherwise a large amount of microwave energy will be reflected at the material boundary and return to the microwave device.

9.4 MICROWAVE IRRADIATION SYSTEM FOR WOODY BIOMASS PRETREATMENT

Woody biomass is a recalcitrant material to decompose for enzymatic saccharification compared with herbaceous plants such as sugarcane and corn, because cellulose and hemicelluloses in plant cell walls are coated with lignin, which makes the access of enzymes difficult. Development of an efficient pretreatment process prior to the enzymatic saccharification is therefore essential for profitable bioethanol production from the woody biomass. In order to unbind the lignin coating effectively and enhance enzymatic susceptibility, microwave irradiation is expected as an efficient and energy-cost-saving pretreatment method. Effectiveness of microwave irradiation to the woody biomass was reported and a continuous-flow-type microwave

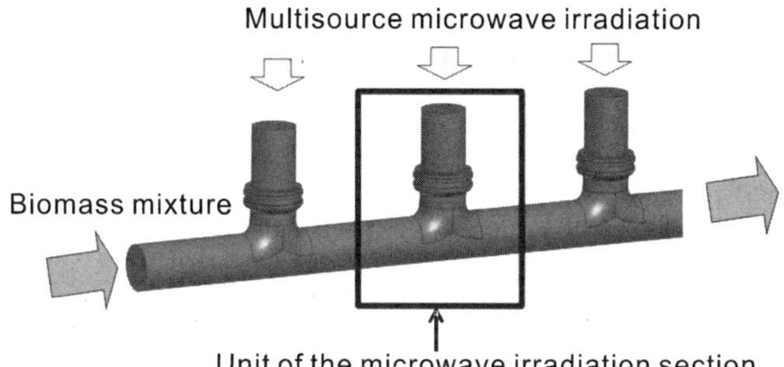

FIGURE 9.3 Schematic of the newly developed continuous-flow-type microwave pretreatment system [20].

irradiation system was developed in the 1980s [6]. In this section, a state-of-the-art microwave pretreatment system for woody biomass [20] is mainly described.

Figure 9.3 shows a schematic of a newly developed continuous-flow-type microwave pretreatment system. Slurry consisting of woody biomass, catalysts, and solvents such as water, ethylene glycol, glycerol, and ethanol flows through a metal pipe, and the mixture is irradiated with 2.45 GHz microwave at T-junction metal pipe sections. A unit of the microwave irradiation section, shown in the black square in Figure 9.3, is removable and the number of the units is changeable, depending on the volume of the slurry, flow rate of the mixture, microwave output, and pretreatment time. Thus, the new system is a multisource microwave irradiator applicable to a wide range of pretreatment system ranging from a small-scale on-site production to a large-scale production using huge amount of plantation biomass.

Figure 9.4 shows a photograph of a prototype of a continuous-flow-type microwave pretreatment system. There are three units present at the microwave irradiation section. A 5 kW microwave generator is attached to each unit, and one can control the microwave power independently at each irradiation section. The biomass mixture temperature is monitored at several points and the temperature data is in a feedback

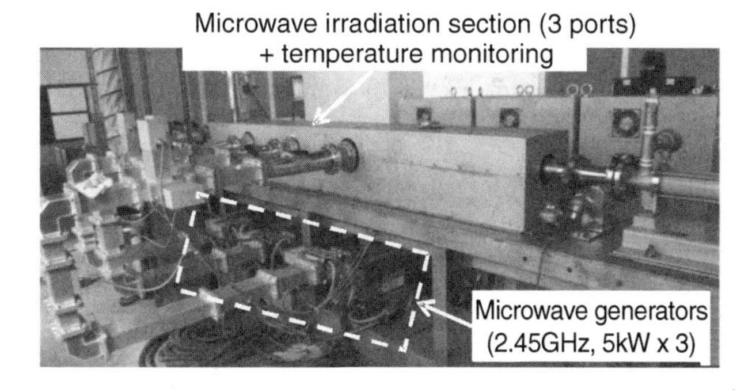

FIGURE 9.4 Photograph of a prototype of a continuous-flow-type microwave pretreatment system.

loop to the microwave power control to enable real-time monitoring and control of the reaction conditions. From experimental results, the organic solvent-based mixture with 10 wt% woody biomass could be pretreated under 170 °C and 30 min microwave irradiation. The pretreated woody biomass provided over 90% of the saccharide yield after enzyme saccharification.

9.4.1 Microwave-Assisted Reactions for Lignocellulosic Biorefinery

Microwave irradiation has been applied to the pretreatments of lignocellulosics for enzymatic saccharification and fermentation. In particular, hydrothermolysis with or without acids or alkalis has been extensively applied to the microwave-assisted pretreatments of various lignocellulosic biomass, such as Japanese red pine (*Pinus densiflora*) [5,6,14,15], Sakhalin fir (*Abies sachalinensis*) [14,15], bald cypress (*Taxodium distichum*) [14,15], Yezo spruce (*Picea jezoensis*) [14,15], Japanese cypress (*Chamaecyparis obtusa*) [14,15], Japanese larch (*Larix leptolepis*) [14,15], Japanese cedar (*Cryptomeria japonica*) [14,15,18,19], loblolly pine (*Pinus taeda*) [14,15], slash pine (*Pinus elliottii*) [14,15], beech (*Fagus crenata*) [5,6,22], Japanese white birch (*Betula platyphylla*) [12], bamboo [5], bagasse [2-4], rice straw [2,4,10–12], rice hulls [2], wheat straw [17], switchgrass [9,13], and coastal bermudagrass [9]. Barks of softwoods, Japanese red pine, Yezo spruce, Japanese cypress and Japanese larch and Japanese cedar have also been pretreated with microwave hydrothermolysis for enzymatic saccharification to give 35.5–73.5% sugar yields [16]. Sequential pretreatments with acid, alkali, and alkali–H_2O_2 by microwave irradiation were used to accelerate enzymatic saccharification of rice straw [8].

Recently, microwave-assisted organosolvolysis and oxidation with metal complexes and H_2O_2 have been applied to the pretreatments of woody biomass [18–22]. In general, softwood is more resistant than hardwood to conventional hydrothermolysis. In particular, Japanese cedar wood, which comprises around 60% of the plantation forest in Japan, is highly resistant to the conventional hydrothermolysis and organosolvolysis with ethanol (ethanolysis). Therefore, Japanese cedar wood is a good model for the development of an effective pretreatment system applicable to recalcitrant softwood species. Watanabe and coworkers developed two different microwave-assisted pretreatment systems (i) ammonium molybdate/H_2O_2 reactions in aqueous media [22] and (ii) glycerolysis with organic or inorganic acids [19], which can be applied to a wide range of lignocellulosic biomass including the recalcitrant softwood.

9.4.2 Microwave-Assisted Reactions with Ammonium Molybdate and H_2O_2 for Enzymatic Saccharification of Lignocellulosics

Thermal and nonthermal microwave effects have been discussed to explain unusual observations in the microwave chemistry. Although the microwave effects have not been fully understood, it is known that a number of chemical reactions exhibited enhanced reactivity and selectivity on irradiation with microwave in the presence of a microwave sensitizer. However, to our knowledge, no direct evidence has been

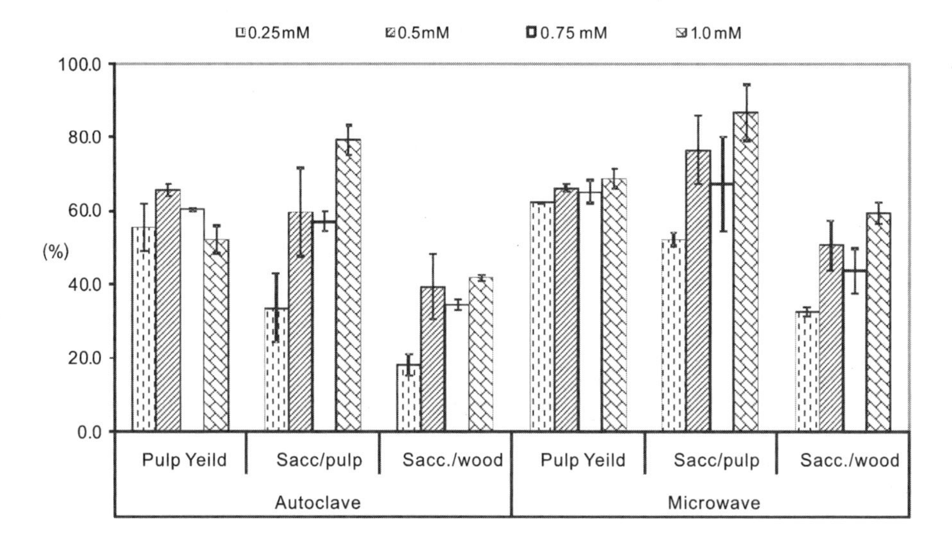

FIGURE 9.5 Comparison of external heating (autoclave) and 2.45 GHz microwave-assisted pretreatment of beech wood at different ammonium molybdate concentrations [22]. Pretreatments by microwave and external heating were carried out with 0.88 M H_2O_2 at 140 °C for 30 min. (Reproduced from Ref. [22] with permission from Elsevier © 2011.)

reported for the microwave effects on pretreatments of biomass for enzymatic saccharification and fermentation. Therefore, Watanabe and coworkers compared effects of microwave irradiation and external heating on pretreatments of woody biomass under the same reaction temperature, time, and catalyst concentration using ammonium molybdate and H_2O_2 [22]. Ammonium molybdate reacts with H_2O_2 to produce permolybdate that in turn oxidizes alkyl aromatic compounds [25]. The metal catalyst and peroxidized intermediate are expected to act as microwave sensitizer. Figure 9.5 shows pulp yield, saccharification ratio, and sugar yield from Japanese beech wood after pretreatments by external heating or microwave irradiation with ammonium molybdate and H_2O_2. The microwave process produced a higher pulp yield and saccharification ratio [22]. At a catalyst concentration of 1 mM, the pulp yield and saccharification ratio with microwave irradiation were 68.7% and 86.7%, whereas those with the autoclave were 52.2% and 83.4%, respectively. The maximum sugar yields for the autoclave and microwave processes were 43.5% and 59.5%, respectively (Fig. 9.5). Thus, the sugar yield with the microwave process was approximately 16% higher than that with external heating. The differences are ascribed to the higher pulp yield, higher susceptibility of the pulp with cellulase mainly due to the lower lignin content of the microwave-treated pulp. At a catalyst concentration of 1.0%, the lignin contents with microwave and autoclave heating were 3.9% and 5.1%, respectively. Thus, lignin degradation and enzymatic saccharification were promoted more by microwave irradiation than by external heating in ammonium molybdate and H_2O_2 system. This phenomenon can be explained by microwave sensitization of molybdate ions and peroxymolybdate.

9.4.3 Microwave-Assisted Glycerolysis of Recalcitrant Softwood

Organosolvolysis is a potential delignification process for the production of second-generation bioethanol from woody biomass which may deliver high-yield recoveries of pulp and coproducts (e.g., fermentable sugars, lignin, and extractives). However, there are several drawbacks, including the risks associated with high-pressure operations, use of volatile flammable solvents and energy loss from evaporation, of organosolv pretreatment with low-boiling-point solvents that reduce feasibility for small- and medium-scale bioethanol plants. An organosolv pulping process with high-boiling-point solvents has been reported for paper production [26] and is potentially applicable to a wide range of biofuel plants.

Glycerol, a high-boiling-point organic solvent is the main by-product of the biodiesel industry. Since soaring petroleum prices have caused the biodiesel industry to flourish, glycerol production now exceeds demand. A major research effort is directed towards novel economic applications of glycerol. However, few studies have been carried out on the alternative use of glycerol for pretreatment of lignocellulosic biomass. Demirba in 1998 and Kücük in 2005 showed delignification in aqueous glycerol or alkaline glycerol organosolv pulping [27,28]. High-temperature reactions result in increased delignification, but increased loss of cellulose was also observed. Sun and Chen [29] reported in 2007 that autocatalytic solvolysis with glycerol at 240 °C for 4 h was effective for pretreating wheat straw. We focused on microwave-assisted organosolvolysis with glycerol because microwave irradiation of the glycerol system enhances the energy balance for heating due to a larger dielectric loss factor (ε''). Compared with conventional heating, microwave irradiation requires much less energy input for heating when solvents with higher ε'' are irradiated. In addition to this effect, it should be noted that a mixture of solid and highly viscose solvent requires longer time for heat transfer. Due to the high viscosity of glycerol and porous structure of wood, microwave heating is advantageous for the rapid heat transfer. This attracts the use of microwave organosolvolysis with aqueous glycerol for the conversion of woody biomass, a porous organic material with extractives and a trace amount of transition metals.

Our group examined microwave-assisted pretreatment of recalcitrant softwood in aqueous glycerol containing a series of organic and inorganic acids with different pK_a values [19]. The pulp obtained by organosolvolysis with 0.1% hydrochloric acid (p$K_a = 6$) at 180 °C for 6 min gave the highest sugar yield, 53.1%, based on the weight of original biomass. The pretreatment efficiency correlated linearly with the pK_a of the acids, with the exception of malonic and phosphoric acids (Fig. 9.6). Organosolvolysis with 1.0% phosphoric acid (p$K_a = 2.15$) gave a saccharification yield (50.6%) higher than that expected from its pK_a, while the catalytic effect of malonic acid (p$K_a = 2.83$) was negligible. Extensive exposure of crystalline and noncrystalline cellulose by the microwave-assisted glycerolysis with strong inorganic acids was demonstrated by using fluorescent-labeled recombinant carbohydrate-binding modules (CBMs) [19]. Because of the low concentration of the acid catalysts and availability of glycerol as a by-product from biodiesel and fatty-acid production, organosolvolysis in glycerol is an attractive process for pretreatment of recalcitrant softwood.

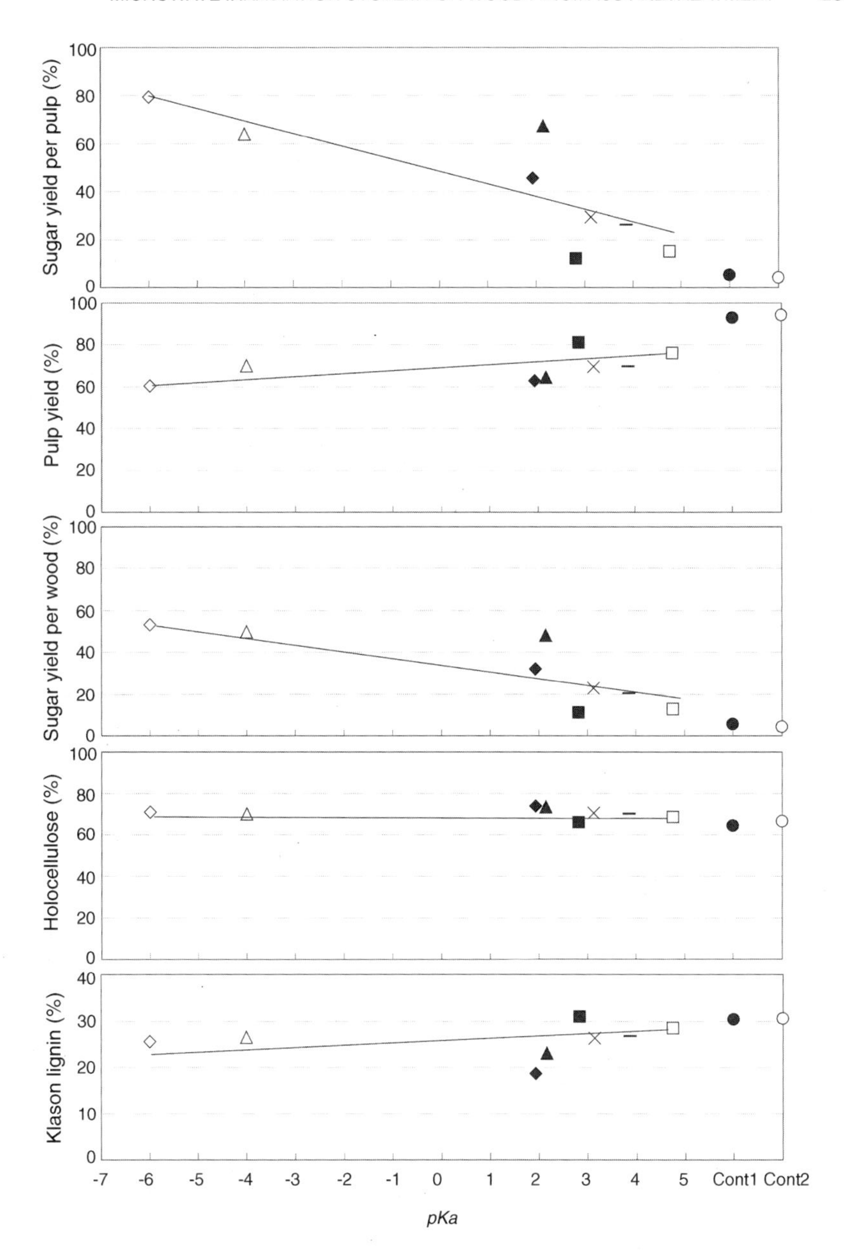

FIGURE 9.6 Saccharification yield per pulp, pulp yield, sugar yield per wood, holocellulose, and Klason lignin content of microwave-assisted pretreatment in aqueous glycerol containing acids with various pK_a [19]. HCl, ◇; H_2SO_4, △; maleic acid, ◆; H_3PO_4, ▲; malonic acid, ■; citric acid, ×; lactic acid, –; acetic acid, □; Control 1, ●; Control 2, ○. All the reactions except for HCl and H_2SO_4 were carried out at 200 °C for 12 min. Reactions with HCl and H_2SO_4 were performed at 180 °C for 6 min due to their high acidic effects in organosolvolysis. Two corresponding control experiments without acids (Control 1:200 °C for 12 min; Control 2: 180 °C for 6 min) were carried out. (Reproduced from Ref. [19] with permission from the American Chemical Society © 2010.)

9.5 CONCLUSIONS AND FUTURE PROSPECTS

Microwave reactions are advantageous over external heating in terms of rapid energy transfer by internal heating. Reactions with high heating efficiency can be achieved if the microwave irradiator is designed to minimize unfavorable reflections and the biomass samples are irradiated in solvents with high ε_r'' value, such as ethylene glycol, glycerol, and water. The reaction rate can also be accelerated by the use of catalysts with high microwave sensitizer effects. We have demonstrated that microwave irradiation with ammonium molybdate and H_2O_2 increased enzymatic saccharification yields than those of external heating at the same reaction temperature and catalyst concentration. For industrial applications, development of cost-effective microwave irradiators with high-energy efficiency is essential. To design highly efficient irradiators and reaction systems, the measurement of permittivity characteristics of the reaction mixture is of uttermost importance. The permittivity depends on the slurry samples containing solvents and biomass, temperature and frequency of microwave. Recently, we have developed a continuous-flow-type microwave irradiator by applying three-dimensional electromagnetic simulation, with feedback experiments to maximize the efficiency of microwave adsorption to the biomass.

Future prospects of microwave technology in biorefinery, depend on the development of low-cost irradiators. Due to the limited penetration depth of microwaves, it is important to develop continuous-flow-type microwave irradiators and rapid microwave reactions that enable high flow rate of the sample feeding.

In conclusion, microwave technology has great potential in the lignocellulosic biorefinery. Applications of microwave irradiation to liquefaction and gasification of biomass by pyrolytic reactions should also be studied, in addition to the pretreatments for enzymatic saccharification and fermentation.

REFERENCES

1. A. T. W. M. Hendriks and G. Zeeman, *Biores. Technol.* **2009**, 100, 10.
2. J.-I. Azuma, F. Tanaka and T. Koshijima, *J. Ferment. Technol.* **1984**, 62, 377.
3. H. Ooshima, K. Aso and Y. Harano, *Biotechnol. Lett.* **1984**, 6, 289.
4. P. Kitchaiya, P. Intanakul and M. Krairish, *J. Wood Chem. Technol.* **2003**, 23, 217.
5. J.-I. Azuma, F. Tanaka and T. Koshijima, *Mokuzai Gakkaishi.* **1984**, 30, 501.
6. K. Magara, S. Ueki, J.-I. Azuma and T. Koshijima, *Mokuzai Gakkaishi.* **1988**, 34, 462.
7. H. Ma, W. W. Liu, X. Chen, Y. J. Wu and Z. L. Yu, *Biores. Technol.* **2009**, 100, 1279.
8. S. Zhu, Y. Wu, C. Wang, F. Yu, S. Jin, Y. Ding, R. Chi, J. Liao and Y. Zhang, *Biosyst. Eng.* **2006**, 93, 279.
9. D. R. Keshwani and J. J. Cheng, *Biotechnol. Prog.* **2010**, 6, 644.
10. S. Zhu, Y. Wu, Z. Yu, J. Liao and Y. Zhang, *Process Biochem.* **2005**, 40, 3082.
11. S. Zhu, Y. Wu, Z. Yu, X. Zhang, C. Wang, F. Yu, S. Jin, Y. Zhao, S. Tu and Y. Xue, *Biosyst. Eng.* **2005**, 92, 229.

12. S. Zhu, Y. Wu, Z. Yu, X. Zhang, H. Li and M. Gao, *Biores. Technol.* **2006**, 97, 1964.
13. Z. Hu and Z. Wen, *Biochem. Eng. J.* **2008**, 38, 369.
14. J.-I. Azuma, J. Higashino and T. Koshijima, *Wood Res.* **1984**, 70, 17.
15. J.-I. Azuma, J. Higashino and T. Koshijima, *Wood Res.* **1985**, 71, 13.
16. J.-I. Azuma, J. Higashino and T. Koshijima, *Mokuzai Gakkaishi.* **1986**, 32, 351.
17. S. Zhu, Y. Wu, Z. Yu, Q. Chen, G. Wu, F. Yu, C. Wang and S. i Jin, *Biosyst. Eng.* **2006**, 94, 437.
18. Y. Baba, T. Tanabe, N. Shirai, T. Watanabe, Y. Honda and T. Watanabe, *Biomass Bioenerg.* **2011**, 35, 320.
19. J. Liu, R. Takada, S. Karita, T. Watanabe, Y. Honda and T. Watanabe, *Biores. Technol.* **2010**, 101, 9355.
20. K. Yano, T. Mitani, N. Shinohara, M. Oyadomari, M. Daidai and T. Watanabe, Asia-Pacific Microwave Conference Proceedings 53, Yokohama, Japan, Dec. 2010. **2010**.
21. T. Mitani, H. Suzuki, M. Oyadomari, N. Shinohara, T. Watanabe, T. Tsumiya and H. Sego, Renewable Energy 2010 Proceedings 53, Yokohama, Japan, No. O-Bm-6-1, Jul. **2010**.
22. P. Verma, T. Watanabe, Y. Honda and T. Watanabe, *Biores. Technol.* **2011**, 102, 3941.
23. J. M. Osepchuk, *IEEE Transactions on Microwave Theory and Techniques* **2002**, 50, 975.
24. H. M. Kingston and S. J. Haswell, *Microwave-Enhanced Chemistry: Fundamentals, Sample Preparation, and Applications*, American Chemical Society Publication, Washington, DC, **1997**.
25. M. P. Chaudhari and S. B. Sawant, *Chem. Eng. J.* **2005**, 106, 111.
26. T. Kishimoto and Y. Sano, J. *Wood Chem. Technol.* **2003**, 23, 279.
27. A. Demirba, *Biores. Technol.* **1998**, 63, 179.
28. M. Kucuk, *Energy Sources* **2005**, 27, 1245.
29. F. Sun and H. Chen, *J. Chem. Technol. Biotechnol.* **2007**, 82, 1039.

CHAPTER 10

Biorefinery with Microbes

CUIMIN HU and ZONGBAO K. ZHAO

10.1 BRIEF INTRODUCTION OF BIOTRANSFORMATION

Microorganisms can be categorized into heterotrophs and autotrophs depending on the energy source used for growth. Heterotrophic microorganisms are responsible for recycling the huge masses of organic matter synthesized by plants. In this respect, it is not surprising to explore microbial transformation of organic substances. These known conversion processes, brewing, aerobic digestion, fermentation, and composting, involve a variety of microorganisms. Industrial biotransformation converts organic substrates (glucose, starch, proteins, lipids, etc.) to a variety of useful products at mesophilic temperatures (typically 20–40 °C) within a limited pH range (typically pH 3–8). Although there are copious examples where only a single step is involved in the transformation [1], the current chapter will focus on those involving multiple reactions, and thus, the microorganisms typically being cultivated. In general, the culture medium contains organic substances as carbon and energy sources as well as minerals, and is sterilized (typically by autoclaving at 120 °C, ~0.1 MPa) to ensure the presence of only one type of microorganism. Large quantities of a range of products, such as vinegar, beer, wine, citric acid, glutamic acid, fuel ethanol, and vitamin C, are being produced using controllable bioprocesses with microorganisms either isolated from the environment or genetically engineered [1].

The processes of biotransformation use microorganisms as the equivalents of chemical catalysts. When carbohydrates are used as the carbon and energy sources, there are multiple metabolic steps as well as transportation events, leading to the formation of the products. Take the conversion of glucose into ethanol by yeast as an example; there are at least 12 biochemical steps to convert one glucose molecule into two ethanol molecules. The theoretical yield of ethanol production is 0.511 g ethanol per gram of glucose. At industrial operation facilities, yields of 0.475 g ethanol per gram of glucose has been achieved. This transforms as a mass yield of 93% of the

The Role of Green Chemistry in Biomass Processing and Conversion, First Edition.
Edited by Haibo Xie and Nicholas Gathergood.

theoretical. A little loss is due largely to the production of yeast cell mass and side products such as glycerol and higher alcohols [2]. In this respect, there have been no compatible chemical processes that can make products using so many steps with such a high selectivity and efficiency.

10.2 THE GREENNESS OF MICROBIAL TRANSFORMATION

Green Chemistry is defined as the "design of chemical products and processes to reduce or eliminate the use and generation of hazardous substances." It is characterized by careful planning of chemical synthesis and molecular design to reduce adverse consequences. To achieve these goals, the Twelve Principles of Green Chemistry are widely adopted as "design rules" to help chemists [3]. In order to establish a sustainable production of biofuels and biochemicals, the integration of green chemistry, along with the use of low environmental impact technologies, is mandatory. The overall goal of green chemistry combined with a biorefinery is the production of genuinely green and sustainable chemical products with a minimum production of waste [4].

Although the Twelve Principles of Green Chemistry were originally proposed to design synthetic procedures, they have also been applied to evaluate the greenness of individual process/product. As microbial transformation is, in fact, the conversion of a substrate into a product using the living cells as the "catalyst," we can assess the greenness of microbial transformation according to these principles. Of the twelve principles of green chemistry, the 4th, 8th, 9th, and 11th principles are irrelevant to microbial transformation discussed here, because these principles are largely associated with conventional chemical processes. However, the 1st, 2nd, 3rd, 5th, 6th, 7th, 10th, and 12th principles are closely related to microbial transformation, since (i) microbial transformations are dedicated to producing less wastes and making full use of raw materials (1st and 2nd principles), (ii) the process is performed in aqueous environment at ambient temperature and pressure (5th and 6th principles), (iii) the substrate is usually renewable and biodegradable (7th principle), (iv) the product is usually nontoxic and biodegradable (3rd and 10th principles), and (v) the potential for accidents such as fires and explosions is minimal (12th principle). Next, we will try using examples to demonstrate why and how green chemistry ideas are applied in microbial transformation of renewable materials.

10.3 BIOCONVERSION OF BIOMASS INTO BIOFUELS

10.3.1 Biomass as the Feedstock

Biomass, when considered as a renewable energy source, refers to organic matter that has stored energy through the process of photosynthesis. More specifically, it includes agricultural wastes such as straw, corn stalks, forest residues, the manure of farm animals and dedicated industrial materials from plants, including miscanthus, switchgrass, corn, poplar, willow, and sugarcane. Biomass is produced in a large quantity, with an estimated 140 billion metric tons per year. Although wheat, corn, rice, and soybean are all biomass by definition, in this chapter, we focus on biomass, also called as lignocellulose, which is mainly composed of cellulose, hemicellulose, and lignin. Therefore, we are using biomass and lignocellulose interchangeably. The main components of plant biomass are cellulose (40–50%, dry weight), hemicellulose (20–40%), lignin (20–25%), and extractives [5]. For example, the sugarcane bagasse contains cellulose (40–50%), hemicellulose (20–30%), lignin (20–25%), and ash (1.5–3.0%) [6].

Coal, oil, and gas are called fossil fuels because they have been formed from the organic remains of prehistoric plants and animals. Therefore, fossil fuels are limited resources. Growing worldwide concerns regarding environmental consequences of heavy dependence on fossil fuels have stimulated much effort to use biomass for the production of fuels and chemicals. Biomass is one of the most plentiful and well-utilized sources of renewable energy in the world. It is estimated that by 2025, up to 30% of raw materials for the chemical industry will be produced from renewable sources. However, it remains challenging to effectively utilize biomass, especially for lignocellulosic materials.

Generally, the methods available for biomass conversion can be divided into two main categories: thermo/chemical and biological conversion routes. Both routes are being intensively pursued. However, the biological routes are generally operated under mild conditions and with higher product selectivity. In this chapter, we will focus on the biological routes. In terms of end products, there are gaseous products such as hydrogen and biogas (methane) and liquid products such as ethanol, butanol, and related products. To further limit our efforts, we will only try to cover biological production of liquid biofuel here. An overview of the bioconversion of lignocellulosic biomass into fuels and chemicals is shown in Figure 10.1.

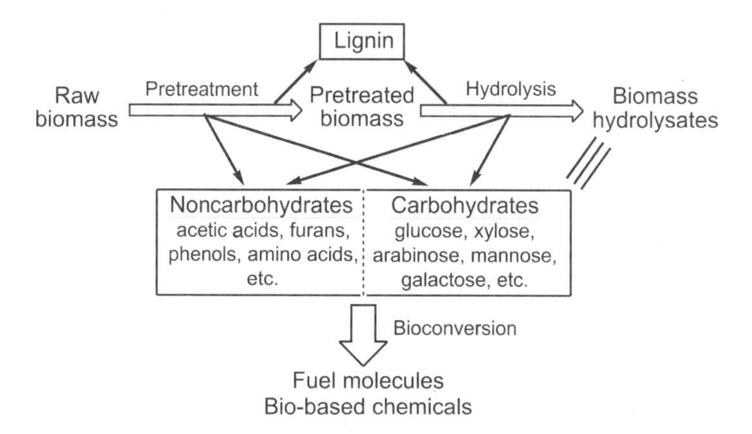

FIGURE 10.1 An overview of the production of fuels and chemicals using lignocellulosic biomass as the feedstock and bioconversion as the key step.

To make the biological process practical and efficient, carbohydrates must be liberated from the key components of biomass, cellulose and hemicelluloses, and also be separated from the nonfermentable lignin component. Usually, this is done in a two-step process involving the pretreatment followed by the hydrolysis step (Fig. 10.1). In practice, biomass is firstly pretreated to reduce the biomass volume and the cellulose crystallinity and increase the cellulose porosity. Pretreatment methods can be classified as mechanical, chemical, enzymatic, and combinations thereof. Several other methods are available for pretreatment, for example, hot-water treatment, steam/peroxide explosion, dilute acid, dilute acid and elevated temperature, ammonia explosion, alkaline pretreatment, organic solvent, sulfur dioxide, etc [6]. Depending on the technology, either hemicellulose or lignin can be targeted for degradation. Pretreatment by dilute acid is one of the most extensively applied ones so far, but it has major drawbacks such as inorganic side-streams, high capital costs, small particle sizes, and high temperature and pressure. Currently, the biomass pretreatment technologies are energy intensive, and result in accumulation of inhibitors that are toxic to microorganisms [7]. The development of advanced pretreatment technologies is one of the main issues in biorefinery.

Following the pretreatment, the material is subjected to hydrolysis. There are three types of hydrolysis processes—acid, enzymatic, and hydrothermal (involving the use of hot water or supercritical methods). It should be emphasized that partial hydrolysis can occur during the pretreatment step, particularly for hemicellulose. For the dilute-acid pretreatment process, hemicellulose is hydrolyzed to sugars under milder conditions. In case of concentrated-acid hydrolysis, decrystallization of cellulose is followed by hydrolysis of sugars with dilute acid. Enzymatic hydrolysis has been developed rapidly, leading to high yields of monomeric sugars. In practice, these enzymes are complex mixture of cellulase, hemicellulase, glucosidase, etc [8]. These enzymes are of natural origin and biodegradable, albeit they remain relatively low in specific activities, and thus, costly. Generally speaking, carbohydrates presented in the final hydrolysate include glucose and xylose, followed by arabinose, galactose, and mannose [9].

As shown in Figure 10.1, it is obvious that various noncarbohydrate compounds were present in the stream, biomass hydrolysate. These compounds include furan derivatives, weak carboxylic acids, and phenolic compounds. The concentrations of these chemicals varied substantially depending on the source of the feedstock as well as the pretreatment method [10]. Table 10.1 shows the main by-products formed during the pretreatment of corn stover by some representative methods. These compounds may inhibit cell growth and product formation, thus reducing the bioavailability of the hydrolyzed feedstocks, and therefore the productivity of the process. Although various approaches have been adopted to mitigate this problem [11], real-world biomass hydrolysate are by no means free of inhibitors. It should be pointed out that even though those noncarbohydrate compounds may be inert to some microorganisms, they may have detrimental effects on the downstream processes, such as waste-water treatment. Therefore, it is important to control the production of these compounds in terms of process efficiency and greenness. Because of the complex feature of the biomass hydrolysate, it is pivotal to take

TABLE 10.1 Concentrations of Main Degradation Products Observed in Corn Stover Hydrolysates

		Analytes (mg L^{-1})					
Pretreatment Methods	Chemical Environments	Acetic Acid	Formic Acid	Furfural	HMF	Vanillin	4-Hydroxy-benzaldehyde
0.7% H_2SO_4	0.7% (w/w) aqueous sulfuric acid	170	120	220	44	4.0	3.6
Liquid hot water	Deionized water	34	55	8	2.3	2.6	2.7
Wet oxidation	Water saturated with O_2 (g) at 174 psi	58	79	6.5	2.8	6.7	4.4
NH_3	0.1% (w/w) aqueous NH_4OH	180	250	0.40	0.89	2.6	1.5
Lime	Water containing 0.1 g $Ca(OH)_2$ per g biomass	120	43	1.5	2.3	3.6	2.3

Source: Ref. [10a].
Note: All pretreatments were carried out using 0.8 g dry biomass at 180°C, for 8 min at a solids concentration of 10 g L^{-1}.

green chemistry principles more seriously to ensure the bioconversion *per se* and its downstream processes being done with less wastes, higher efficiency, and better economics.

The next common step is to convert sugars in the hydrolysate by microorganisms, such as *Saccharomyces cerevisiae, Escherichia coli, Zymomonas mobilis, Clostridium* sp., *Rhodosporidium toruloides, Lipomyces starkeyi, Lactobacillus* sp., *Actinobacillus* sp., and others [12]. Two common features of all such processes are: (1) the presence of molecules as a result of biomass pretreatment that may be inhibitory to the fermentation process for the production of the desirable molecule(s) and (2) the inhibitory nature of the desirable product (such as ethanol or butanol, other alcohols, hydrocarbons, succinate, butyrate, and other carboxylic acids) or bioproducts. Accumulation of products during a fermentation, whether desirable or not, can be toxic [13], inhibiting cell growth or resulting in death. Similarly, toxic contaminants in the biomass hydrolysate can inhibit cell growth and product formation [14]. Inhibition from such chemicals typically limits product titers, affects fermentation performance and operational options (continuous vs. batch or fed-batch), and profoundly impacts process economics.

There are alternative processes to the aforementioned two-step process in terms of biomass utilization. Simultaneous saccharification and fermentation (SSF) combines enzymatic hydrolysis and microbial transformation, such that sugars produced *in situ* are converted into a product, such as ethanol. A few microbial species such as *Neurospora*, *Monilia*, *Paecilomyces*, and *Fusarium* have been reported [15] to possess the ability to ferment cellulose directly to bioproducts by SSF. Consequently, SSF results in rapid throughput, which in turn, improves economic viability.

However, SSF requires the production of inexpensive and even more robust cellulases with higher activity and longer lifetime for breakdown of cellulose microfibrils to sugars. Because *in situ* production of cellulase remains challenging technically and economically, cellulases from other microbes should be added to complete this process. Another process called consolidated bioprocessing (CBP), which features cellulase production, cellulose hydrolysis, and fermentation in one step, is an alternative approach with outstanding potential [16]. The CBP microbes can realize the direct conversion of biomass to bioproducts.

However, there are incredible driving forces for innovation in terms of biomass pretreatment and hydrolysis, and progress is expected to be rapid. In the following discussion, we assume that biomass hydrolysates are available, and try to focus on the opportunities and technologies that can make liquid biofuels and commodity chemicals using microorganisms. It should be noted that we did not intend to cover the state-of–the-art of microbial production of biofuels and chemicals, because most of those excellent results have been achieved using starch-based materials, and more specifically, glucose, as the carbon source. Rather, to stick to the core value of green chemistry, we emphasized on the transformations using biomass hydrolysate, waste materials, and other nonconventional materials as the carbon source. As described above, these real-world feedstocks have major differences compared with starch-based materials, and therefore, require distinct strains, processes, and technologies.

10.3.2 Bioethanol

Conversion of sugar coming from crops to ethanol is a well-established technology, but has certain limits. These crops have high value for food applications and their sugar yield per hectare is low compared with the most prevalent forms of sugar, lignocellulosic biomass [15]. The potential of lignocellulosic biomass is therefore considerably larger than that of the crops.

The general steps including pretreatment of biomass, saccharification, fermentation of sugars, and distillation to recover ethanol are shown in Figure 10.2. As bioethanol is the major target of biofuel research and commercial production, some leading biomass pretreatment and hydrolysis technologies are developed to match the processes for ethanol fermentation. Biomass hydrolysates are complex mixtures of hexose and pentose sugars together with other compounds, some of which can act as fermentation inhibitors. The optimal microorganisms for the conversion of biomass hydrolysate should combine the following properties [17]: efficient fermentation of both, hexose and pentose sugars, preferentially simultaneously, high tolerance to

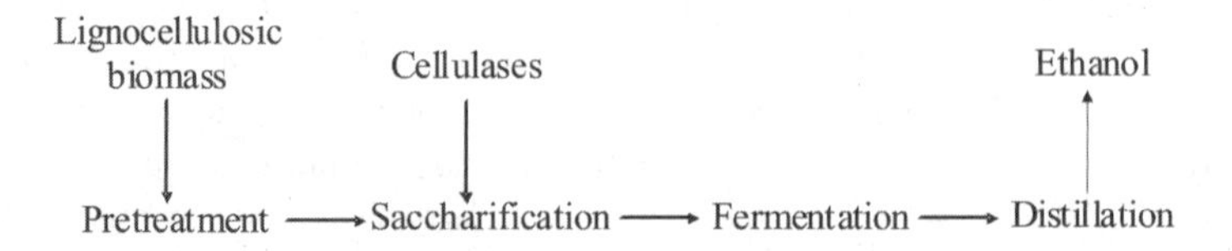

FIGURE 10.2 The general steps for ethanol production from lignocellulosic biomass.

fermentation inhibitors and fermentation products, resistance against microbial contamination, high productivities and yields. However, such microorganisms remain to be identified or engineered.

S. cerevisiae is one of the most effective bioethanol-producing yeasts. It has many advantages owing to its high bioethanol production from hexoses and high tolerance to ethanol. However, natural *S. cerevisiae* is unable to ferment pentoses into ethanol, which significantly reduced its usefulness in direct fermentation of lignocellulosic feedstocks. The commercial viability for bioethanol from biomass would call for developing microorganisms to effectively ferment hexoses and pentoses. Currently, there are intensive researches on genetic engineering of *S. cerevisiae* for better pentose-metabolizing capacity [18] as well as other phenotypic features [19]. Sugarcane bagasse hydrolysate was fermented with the recombinant xylose-utilizing *S. cerevisiae* TMB 3001, and ethanol yield was $0.18\,g\,g^{-1}$ dry bagasse when the hydrolysates were detoxified by overliming [20]. The ethanol yield from consumed sugars reached 93% of the theoretical value when using corn stover hydrolysate by engineered strain *S. cerevisiae* 424A(LNH-ST) [21]. A derivative of *S. cerevisiae* 424A(LNH-ST) was further developed for efficient anaerobic fermentation of L-arabinose [18b]. The new strain was constructed by overexpression of two genes from fungal L-arabinose utilization pathways, and the new 424A(LNH-ST) strain exhibited the production of ethanol from L-arabinose in more than 40% yield. Ethanol production in 72.5% yield was achieved from five-sugar mixtures containing glucose, galactose, mannose, xylose, and arabinose. Pilot-scale fermentation of aqueous-ammonia-steeped switchgrass was carried out in a 50-L reactor using SSF strategy. A maximum ethanol yield of 73% was achieved by *S. cerevisiae* D_5A (ATCC 200062) at 48 h after inoculation. Other progresses were shown in Table 10.2. These results were promising and similar to laboratory-scale experiments [22].

A few species among the bacteria (thermophilic anaerobic bacteria), yeast, and fungi can convert pentoses into ethanol with a satisfactory yield and productivity. *Pichia stipitis* is one of the few natural xylose-fermenting yeasts and thus is a promising candidate for ethanol production. An enhanced inhibitor-tolerant strain of *P. stipitis* was identified, which afforded an ethanol yield of $0.45\,g\,g^{-1}$ total sugar, equivalent to 87% of the theoretical yield, when cultivated in rice straw hydrolysate at pH 5.0 [29]. *E. coli* has the ability to utilize hexoses as well as pentoses for biofuel production from lignocellulosic biomass. With glucose phosphotransferase mutation

TABLE 10.2 Bioethanol Production by *S. cerevisiae* Using Various Raw Materials

S. cerevisiae Strains	Raw Materials	Ethanol Yield from Material	Ref.
GIM-2	Paper sludge	$190\,g\,kg^{-1}$	[23]
F12	Wheat straw	$243.6\,g\,kg^{-1}$	[25]
KNU5377	Waste paper	$263.5\,g\,kg^{-1}$	[26]
Y5	Lodgepole pine	$211.8\,g\,kg^{-1}$	[27]
424A(LNH-ST)	Corn stover	$191.5\,g\,kg^{-1}$	[28]
KCCM50549	Jerusalem artichoke	$201.1\,g\,kg^{-1}$	[24]

to eliminate carbon catabolite repression, *E. coli* metabolized pentoses simultaneously with glucose rather than sequentially during mixed-sugar fermentation with ethanol yields of 87–94% of the theoretical [30]. The performance of CBP organism *C. phytofermentans* ATCC 700394 on AFEX-treated corn stover was analyzed, and it was found that 76% of glucan and 88.6% of xylan were hydrolyzed in 10 days, and that ethanol titer was 2.8 $g\,L^{-1}$ with a process yield of more than 65% [31].

10.3.3 Biobutanol

Butanol is also considered as an attractive biofuel, as it exhibits superior chemical properties in terms of energy content, volatility, and corrosiveness. Currently almost all butanol is produced through chemical process depending on fossil resources. The most common chemical route is the Oxo process, which involves the hydrocarbonylation of propylene in the presence of an appropriate catalyst. The resulting butyraldehydes are hydrogenated to the corresponding butyl alcohols, and further distillation is required to recover butanol and its isomers. The chemical process for butanol production uses nonrenewable, toxic, and explosive feedstocks, which is not green. The background information and challenges of biobutanol production have been included in an excellent review published recently [32]. Historically, the process of biobutanol production is known as acetone–butanol–ethanol (ABE) fermentation and has already intensively been used in times of insufficient oil supply. Several species of *Clostridium* bacteria are capable of making butanol and other solvents using different feedstocks. A typical ABE fermentation using *C. acetobutylicum* yields acetone, butanol, and ethanol in the ratio of 3:6:1 [33]. Corn, molasses, and whey permeate from the cheese industry are the traditional raw materials for butanol production [34].

As discussed in the case of bioethanol production, feedstock for future biobutanol is also expected to shift to lignocellulosic biomass. Similarly, lignocellulosic materials are processed by pretreatment and hydrolysis prior to fermentation. *C. beijerinckii* P260 has been used for butanol fermentation using biomass hydrolysate. Specifically, when wheat straw hydrolysate with a total sugar concentration of 60 $g\,L^{-1}$ was used in batch fermentation, and total solvent, product yield, and productivity reached 25.0 $g\,L^{-1}$, 0.42 $g\,g^{-1}$, and 0.28 $g\,L^{-1}\,h^{-1}$, respectively [34]. When barley straw hydrolysate was used, total solvent yield and productivity were 0.33 $g\,g^{-1}$ and 0.10 $g\,L^{-1}\,h^{-1}$, respectively. However, when the hydrolysate was detoxified with lime to reduce the potential inhibitors, results were a total solvent yield of 0.43 $g\,g^{-1}$ and productivity of 0.39 $g\,L^{-1}\,h^{-1}$ [35]. These results demonstrate that *C. beijerinckii* P260 has excellent capacity to convert lignocellulosic biomass-derived sugars to butanol.

10.3.4 Biodiesel and Related Products

Biodiesel is traditionally produced using vegetable oil as the feedstock. However, it remains a major challenge to secure the feedstock for biodiesel production because vegetable oil supply is arable land-dependent and resource competitive with food and

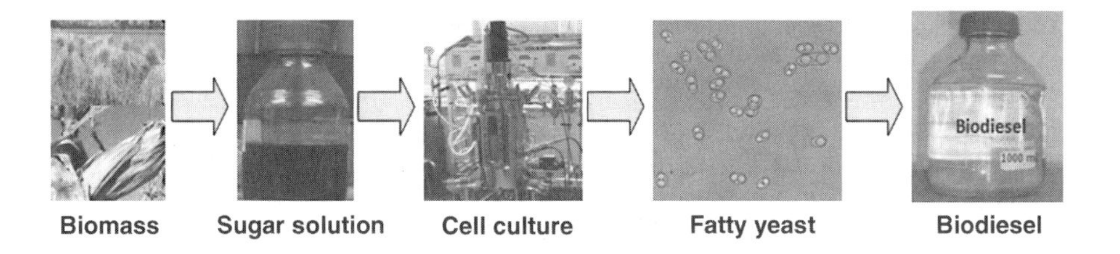

FIGURE 10.3 Schemes of biodiesel production from biomass.

oleochemical industry. Production of biodiesel starting from lignocellulosic biomass has been proposed recently [36]. The route is similar with the one for bioethanol production, except that, microbial lipid is accumulated intracellularly after fermentation which will be used for biodiesel production through transesterification (Fig. 10.3). The key to realize such transformation is to use oleaginous microorganisms that convert carbohydrates into microbial lipid with triacylglycerol as a major component. Generally, lipid is extracted from yeast cells for transesterification. Conversion of fatty yeast directly into biodiesel has also been documented [37]. The technology to convert carbohydrates into lipids and biodiesel by oleaginous microorganisms is unique because there is no chemical counterpart for such transformation. Thermochemical processes provide possibilities to convert biomass into diesel [38], but not in the forms of fatty acid derivatives as is done by biological routes.

Oleaginous microorganisms, mainly oleaginous yeasts, could accumulate intracellular lipid up to 70% of cell mass when glucose was used as a carbon source [39]. *Rhodosporidium toruloides*, *Trichosporon cutaneum*, *Cryptococcus curvatus*, and *Lipomyces starkeyi* are some representative microbial lipid producers with high efficiency [40]. Unlike most *S. cerevisiae* species used in ethanol fermentation, oleaginous yeasts are usually able to assimilate pentoses [41]. For example, *L. starkeyi* AS 2.1560 was shown to coferment glucose and xylose (2:1 wt wt^{-1}), and cellular lipid content reached 61.5% under optimized culture conditions [40c]. Recently, our laboratory also identified a couple of oleaginous yeast strains that have better ability for xylose utilization. When cultivated on a glucose/xylose mixture (2:1 wt wt^{-1}) in a 3-L stirred tank bioreactor, the yeast *T. cutaneum* AS 2.571 assimilated glucose and xylose simultaneously, and lipid content and lipid coefficient were 59% and 0.18 g g^{-1} sugar, respectively [42].

Studies on microbial lipid production from lignocellulosic biomass have made several progresses. First, common inhibitors presented in biomass hydrolysate were tested to identify robust lipid producers [43]. Among several inhibitors, furfural was found to be the most severe inhibitor to *R. toruloides* [43a]. *R. toruloides* growth was inhibited by 45% in the presence of 1 mM furfural in the culture medium. However, *R. toruloides* exhibited noticeable resistance to other inhibitors, including 5-hydroxymethylfurfural, vanillin, syringaldehyde, *p*-hydroxybenzaldehyde, and acetate. More significantly, this strain had nearly identical fermentation performance in a culture with a mixture of six typical inhibitors at their respective concentrations

TABLE 10.3 Microbial Lipid Production from Natural Raw Materials

Strains	Raw Materials	Lipid Content	Lipid Yield	Ref.
T. cutaneum B3	Cassava starch	42.8%	0.20 g g^{-1} sugar	[48]
R. mucilaginosa TJY15a	Cassava starch	45.9%	0.25 g g^{-1} sugar	[49]
R. mucilaginosa TJY15a	Hydrolysate of Jerusalem artichoke	48.6%	0.23 g g^{-1} sugar	[50]
R. toruloides Y4	Extract of Jerusalem artichoke	43.0%	0.26 g g^{-1} sugar	[51]
Microsphaeropsis sp.	Steam-exploded wheat straw mixed with wheat bran	10.2% of dry fermented mass	0.074 g g^{-1} dry substrate	[52]
R. toruloides Y4	Hydrolysate of Jerusalem artichoke	39.5%	0.21 g g^{-1} sugar	[53]

usually found in biomass hydrolysate as in the absence of any inhibitors. A number of oleaginous yeasts have been screened for their resistance to hydrolysis by-products [43b]. It was found that *T. cutaneum* 2.1374 showed an excellent tolerance to lignocellulose degradation compounds. In another study, the inhibition effects of acetic acid and furfural on *R. glutinis* were analyzed. The net specific growth rates indicated that the presence of these two inhibitors in artificial acid hydrolysate inhibited the growth of *R. glutinis* on glucose, but not on xylose [43c].

Real biomass hydrolysate has also been used in lipid production. *R. glutinis* was identified for the phenotype of xylose-assimilation and high lipid content. When cultivated on *Populus uramevicana* leaves hydrolysate [44], *R. glutinis* accumulated lipid to 28.6%; however, lipid content was improved to 34.2% when the hydrolysates were detoxified [45]. Similarly, *T. fermentans* was able to accumulate lipid up to 40.1% when detoxified rice straw hydrolysates were used as carbon and energy source [46]. More experiments have also been conducted on other raw materials, which are summarized in Table 10.3. The theoretical lipid yield from glucose was reported about 0.32 g g^{-1} sugar [47], and the results from cell mass materials below are comparable. A lower lipid yield loss is due largely to the production of yeast cell mass. All the data reported indicated the feasibility of microbial lipid production from biomass feedstocks.

10.4 BIOCONVERSION OF BIOMASS INTO COMMODITY CHEMICALS

With the decreasing oil reserves and the rising concerns on climate change, it is urgent to develop new, petrochemical-independent technologies to meet the future demand for chemicals. Microbial conversion of renewable feedstocks has huge potential to develop novel technologies for chemical production, as many

TABLE 10.4 Representative Commodity Chemicals Production from Natural Raw Materials

Commodity Chemicals	Strains	Raw Materials	Product Titer ($g\,L^{-1}$)	Yield ($g\,g^{-1}$ Sugar)	Ref.
Lactic acid	*Lactobacillus brevis* S3F4	Corncob	39.2	0.69	[55]
	L. brevis ATCC 14869	Rice straw	30.5	0.92	[56]
	Bacillus sp. strain XZL9	Corncob molasses	74.7	0.50	[57]
Succinic acid	*Actinobacillus succinogenes* BE-1	Cotton stalk	15.8	1.23	[58]
	A. succinogenes NJ113	Corn fiber	35.4	0.73	[59]
	A. succinogenes CGMCC1593	Corn stover	47.4	0.72	[60]

petroleum-derived chemicals can be directly or functionally substituted with chemicals from renewable feedstocks [54]. In this part, we intend to use lactic acid and succinic acid as examples to highlight recent progresses in the production of commodity chemicals from lignocellulosic biomass. Representative data are summarized in Table 10.4.

Lactic acid is an important platform chemical, and is also widely applied in food, pharmaceutical, cosmetics, and plastics industries [61]. Although lactic acid can be manufactured chemically, a major disadvantage of chemical synthesis is that the product is a racemic mixture, unless a stereoselective transformation is included. Because most applications require enantiomerically pure lactic acid, microbial production of lactic acid is preferred. Currently, commercial production of lactic acid is largely dependent on microbial fermentation of sucrose or starch-derived glucose. However, use of lignocellulosic biomass has some additional advantages [62].

A lactic-acid producing strain *L. brevis* S3F4 was isolated from sour cabbage. This strain could utilize various sugars present in biomass hydrolysate, and also showed strong resistance to potential inhibitors. After fermentation in corncob dilute acid hydrolysate with an initial total sugar concentration of 56.9 $g\,L^{-1}$, lactic acid concentration of 39.2 $g\,L^{-1}$ was achieved [55]. Rice straw utilization was also explored for lactic acid production by *L. brevis* [56]. When the sequential hydrolysis and fermentation strategy was applied, glucose, xylose, and arabinose were utilized simultaneously and lactic acid concentration of 30.5 $g\,L^{-1}$ was achieved after 24 h. Corncob molasses containing a high content of xylose has been used for L-lactic acid production by a xylose-utilizing *Bacillus sp.* strain XZL9. Lactic acid concentration reached 74.7 $g\,L^{-1}$ from the molasses with an initial total sugar concentration of 91.4 $g\,L^{-1}$ in fed-batch fermentation [57].

When aqueous ammonia pretreated corn stover was used as the feedstock under SSF condition, *L. pentosus* produced lactic acid of 75.0 $g\,L^{-1}$ [63]. Lime-treated wheat straw was hydrolyzed and fermented simultaneously to lactic acid by an enzyme preparation and *Bacillus coagulans* DSM 2314 [64]. Decrease in pH because of lactic acid formation was partially adjusted by automatic addition of the alkaline substrate. A fermentation efficiency of 81% and a chiral L-(+)-lactic acid purity of 97.2% were obtained. In total, 711 g lactic acid was produced out of 2706 g lime-treated straw, representing 43% of the overall theoretical yield. Of the lime added during the pretreatment of straw, 61% was used for the neutralization of lactic acid. Thus, the alkaline pretreatment of lignocellulosic biomass and the pH control in fermentation was successfully combined, resulting in a significant saving of lime consumption and avoiding the necessity to recycle lime. This work offered an excellent example of applying green chemistry principles in biorefinery. A strain of *B. coagulans* was isolated from composted dairy manure that was tolerant to a number of aldehyde inhibitors. Furfural, 5-hydroxymethylfurfural, vanillin, and *p*-hydroxybenzaldehyde at 2.5 g L^{-1}, 2.5 g L^{-1}, 2.5 g L^{-1}, and 1.2 g L^{-1}, respectively, had little effects on the performance by this yeast. In an SSF process, the strain converted a dilute-acid hydrolysate of corn fiber of 100 g L^{-1} to lactic acid of 39 g L^{-1} in 72 h at 50 °C [65].

Although lactic acid bacteria have good capability of utilizing sugars from lignocellulosic biomass, cultivation of these microorganisms requires complex media compositions and careful pH control. While *S. cerevisiae* does not naturally produce lactic acid in large quantities, its robustness, pH tolerance, and simple nutrient requirements make it an excellent candidate organic acid producer. Metabolically engineered *S. cerevisiae* has been examined for lactic-acid production and a high yield was achieved from glucose [66]. Producing lactic acid by such a strain will be carried out using lignocellulosic materials in the near future.

Succinic acid is a C_4-dicarboxylic acid widely used in chemical, food, and pharmaceutical industries. Currently, succinic acid is largely manufactured chemically by hydrogenation of maleic anhydride followed by a hydration reaction. However, it has been recognized as one of the most important platform chemicals that can be produced via a biochemical route [67]. In terms of cheap raw materials used for succinic acid fermentation, wood hydrolysate [68], corn straw hydrolysate [60,69], crop stalk hydrolysate [58], and corn fiber hydrolysate [59] were reported. Batch and continuous cultures of *Mannheimia succiniciproducens* MBEL55E have been carried out with NaOH-treated wood hydrolysate for the production of succinic acid. *M. succiniciproducens* MBEL55E utilized xylose as well as glucose in the wood hydrolysate based medium as a carbon source for the succinic acid production. In batch cultures, the final succinic acid concentration of 11.7 g L^{-1} was obtained from the pretreated wood hydrolysate based medium, resulting in a succinic acid yield of 56% and a succinic acid productivity of 1.17 g L^{-1} h^{-1}, while the corresponding continuous cultures gave the succinic acid yield and productivity of 55% and 3.19 g L^{-1} h^{-1}, respectively [68a]. *A. succinogenes* BE-1 produced 15.8 g L^{-1} succinic acid with the yield of 1.23 g g^{-1} using crop stalk hydrolysate [58] as the feedstock. Both glucose and xylose were utilized at the same time, while cellobiose was not consumed until glucose and xylose were completely consumed. When *A. succinogenes* NJ113 was cultivated on dilute acid hydrolysate of corn fiber in a 7.5-L fermentor for 36 h, the concentration of succinic acid reached 35.4 $g L^{-1}$ with a yield of 0.73 $g g^{-1}$ [59]. *A. succinogenes* CGMCC1593 was found as a good succinic acid producer using renewable materials. It utilizes both glucose and xylose in the straw hydrolysate. In batch fermentation, 45.5 $g L^{-1}$ succinic acid concentration and 80.7% yield were attained after 48 h incubation with 58 $g L^{-1}$ of initial sugar from corn straw hydrolysate in a 5-L stirred bioreactor [69a]. When SSF technique was applied using corn stover as the raw material, *A. succinogenes* CGMCC1593 generated succinic acid with product concentration and yield of 47.4 $g L^{-1}$ and 0.72 $g g^{-1}$ total sugar, respectively, after incubation for 48 h at 38 °C [60].

The recombinant *E. coli* strains with homologous or cyanobacterial *ppc* overexpression and *ldhA*, *pflB*, *ptsG* mutations were constructed for succinic acid overproduction. When a modeled corn stalk hydrolysate containing 30 $g L^{-1}$ glucose, 10 $g L^{-1}$ xylose, and 2.5 $g L^{-1}$ arabinose was applied with the prominent strain *E. coli* SD121, succinic acid yield was 0.77 $g g^{-1}$ total sugar [69b]. Fermentation of corn stalk hydrolysate with SD121 produced a final succinic acid concentration of 36.5 $g L^{-1}$ with a higher yield of 0.83 $g g^{-1}$ total sugar. In the two-stage fermentation process in the bioreactor, initial aerobic growth facilitated the subsequent anaerobic succinic acid production with a final concentration of 57.8 $g L^{-1}$, and a yield of 0.87 $g g^{-1}$ total sugar. Thus, metabolically engineered *E. coli* showed a great potential usage of renewable biomass as a feedstock for an economical succinic acid production [69b].

10.5 OPPORTUNITIES AND CHALLENGES

The use of biomass as raw materials for biofuels and commodity biochemicals is encouraged by the reduction of fossil CO_2 emission, the need for a secure energy supply, and a revitalization of rural areas. Accordingly, the research is on to develop new technologies and create novel processes, products, and capabilities to ensure that the growth is sustainable from economic, environmental, and social perspectives. Biomass energy and material recovery is maximized if a biorefinery approach is considered, where many technological processes are jointly applied to different kinds of biomass feedstock for producing a wide range of products. A lot of biorefinery pathways, from feedstock to products, can then be envisioned according to the different types of feedstock, conversion technologies, and products [70].

As discussed above, biomass is a recalcitrant material that needs to be treated to release carbohydrates. It should be kept in mind that, regardless of the procedures used for carbohydrates release, the resulting materials are complex feedstocks (Fig. 10.1). Two features are obvious: (1) the presence of hexoses (glucose, mannose, galactose, etc.) and pentoses (xylose and arabinose) with various ratios depending on the upstream processes, (2) the presence of inhibitory components, such as acetate, furfural, HMF, and phenolic compounds. Up to now, microorganisms that can assimilate hexoses and pentoses simultaneously are rare [42], not to mention producing products with high efficiency. Although different microorganisms may have excellent resistance to some of these inhibitory components, robust microorganisms with improved resistance to a broader spectrum of inhibitors remain to be identified. Moreover, resistance is not enough. Inhibitory components should preferentially be assimilated and converted into the targeted product, because it will reduce the costs for product separation and waste-water treatment. All those are requirements of green chemistry. Consequently, advanced microorganisms should be engineered. Recently, synthetic biology has come into the era [71]. Specific pathways may be assembled to metabolize one specific inhibitory component to produce a dedicated microbial strain. In this way, it should be possible to engineer superior host with the outstanding capacity to metabolize diversified components of biomass origin.

The performances of inhibitors and the inhibition mechanisms have been investigated in-depth which can be used as guidance for further application. Furfural and HMF are the key toxic furan derivatives in acid-pretreated biomass hydrolysate. These toxins caused a decrease in microbe growth, product yield, and productivity [14,72]. Adaptation method has been used to obtain tolerant strains in bioethanol production [73]. One research showed that, NADH- or NADPH-dependent oxidoreductases played major roles in the resistance mechanisms. Another study reached similar conclusion, where numerous enzymes possessing aldehyde reductase activities contributed to the detoxification of the aldehyde inhibitors [74]. When furfural and HMF were reduced to furanmethanol and furan-2,5-dimethanol, respectively, cellular toxicity was significantly reduced [75]. Detailed metabolic information for biological degradation of inhibitors remains to be elucidated. One example is the established degradation pathways of furfural and HMF in *Cupriavidus basilensis* HMF14 [76]. Both aldehydes were converted to 2-furoic acid, and ultimately to 2-oxoglutaric acid, which could enter the tricarboxylic acid cycle for further utilization. There results pave the way for mechanism-based *in situ* removal of these inhibitors biologically which is a key step toward improving the efficiency of utilization of lignocellulosic feedstocks.

Inhibition of cell growth and product formation by the inhibitory compounds in the biomass hydrolysate can be reduced by treatment with enzymes such as the lignolytic enzymes [77], for example, laccase and microorganisms such as *Trichoderma reesei*, *Trametes versicolor*, *Pseudomonas putida* Fu1, *Candida guilliermondii*, and *Ureibacillus thermosphaericus* [19]. Microbial and enzymatic detoxifications of lignocellulosic hydrolysate are mild and more specific. Adaptation to the biomass hydrolysate is an alternative approach to detoxification. Increases in fermentation rate and product yield by adapted microorganisms to lignocellulosic hydrolysate have been reported in some studies. Another approach to alleviate the inhibition problem is to introduce genes encoding enzymes for resistance against specific inhibitors and altering cofactor balance into the hosts. For example, cloning of the laccase gene followed by heterologous expression in yeasts was shown to provide higher enzyme yields and permit production of laccases with desired properties for detoxification of lignocellulose hydrolysate [78].

In principle, biomass can be converted into a range of valuable fuels, chemicals, materials, and products. Thus, the concept of biorefinery has been developed to pursue multiple products much like petrochemical plants do. In the petroleum industry, feedstock is available in dense and very large quantity, and fractionation largely by distillation is possible. Therefore, large investment on facilities for multiple products is possible, because profits are secured upon long-time, large-scale operation of these facilities. However, biomass is a dispersed feedstock, and fluctuates seasonally. It is hard to sustainably gather and process more than a certain amount in one place [79]. Although the key point of refinery, that is, maximal usage of a feedstock for multiple products, holds true when we are trying to use biomass, care should be taken to limit final products, because there will be inadequate material to distribute among them, while keeping the investment profit [80]. Another issue is the flexibility in feed, for example, the possibility to use multiple types of raw material. The possibility to use different ligno-cellulosic sources (agriculture and forest residues, municipal wastes, herbaceous energy crops) is in high demand.

The use of biomass as raw materials for biofuels and commodity biochemicals is encouraged by the reduction of fossil CO_2 emission, the need for a secure energy supply, and a revitalization of rural areas. Accordingly, the research is to develop new technologies and create novel processes, products, and capabilities to ensure that the growth is sustainable from economic, environmental, and social perspectives. Biomass energy and material recovery is maximized if a biorefinery approach is considered, where many technological processes are jointly applied to different kinds of biomass feedstock for producing a wide range of products. A lot of biorefinery pathways, from feedstock to products, can then be envisioned according to the different types of feedstock, conversion technologies, and products [70].

10.6 OUTLOOK

Biotransformations are normally operated under relatively mild conditions. When considering the conversion of complex materials such as biomass hydrolysate, the biological processes are robust and attractive because it is possible to achieve higher substrate utilization and product selectivity. Therefore, greener processes are expected. However, the pretreatment technology for biomass could be expensive and inefficient. When developing biological processes for future biorefinery, it is important that the methods and techniques used minimize impact on the environment and the final products are truly green and sustainable. Biomass has to be considered as an alternative source of energy that is abundant in a wide-scale yet nondisruptive manner, since it is capable of being implemented at all levels of society. Biorefinery with microorganisms are expected to have significant contributions in many aspects of sustainable development. However, as discussed above, the challenges are tremendous in terms of establishing biomass-based, commercially viable bioprocesses for the production of fuels and commodity chemicals. A number of aspects should be paid attention to in further efforts.

- Robust microorganisms: Develop or identify superior microbes that have a wide substrate spectrum and work properly under diversified environments including the presence of a variety of inhibitory components.

- Higher product selectivity and productivity: Both product selectivity and productivity are very important for a large-scale microbial production.
- Innovative product recovery technology: Product recovery is a very important part of biorefinery. It can be costly and depreciate the greenness of biotransformation. Readers are suggested to check recent reviews in this area [81].

REFERENCES

1. A. C. Murphy, *Nat. Prod. Rep.* **2011**, 28, 1406–1425.
2. L. R. Lynd, *Annu. Rev. Energ. Env.* **1996**, 21, 403–465.
3. P. Anastas and N. Eghbali, *Chem. Soc. Rev.* **2010**, 39, 301–312.
4. J. H. Clark, F. E. I. Deswarte and T. J. Farmer, *Biofuel. Bioprod. Bior.* **2009**, 3, 72–90.
5. P. Kumar, D. M. Barrett, M. J. Delwiche and P. Stroeve, *Ind. Eng. Chem. Res.* **2009**, 48, 3713–3729.
6. (a) A. M. J. Kootstra, H. H. Beeftink, E. L. Scott and J. P. M. Sanders, *Biochem. Eng. J.* **2009**, 46, 126–131; (b) T. Marzialetti, M. B. V. Olarte, C. Sievers, T. J. C. Hoskins, P. K. Agrawal and C. W. Jones, *Ind. Eng. Chem. Res.* **2008**, 47, 7131–7140; (c) C. Wyman, B. Dale, R. Elander, M. Holtzapple, M. Ladisch and Y. Lee, *Bioresour. Technol.* **2005**, 96, 1959–1966; (d) R. Zhao, Z. Zhang, R. Zhang, M. Li, Z. Lei, M. Utsumi and N. Sugiura, *Bioresour. Technol.* **2010**, 101, 990–994.
7. (a) S. Sakai, Y. Tsuchida, S. Okino, O. Ichihashi and H. Kawaguchi, *Appl. Environ. Microbiol.* **2007**, 73, 2349–2353; (b) M. G. D. Gutierrez-Padilla and M. N. Karim, *J. Am. Sci.* **2005**, 1, 24–27; (c) J. Zaldivar, A. Martinez and L. O. Ingram, *Biotechnol. Bioeng.* **1999**, 65, 24–33.
8. T. Shimokawa, M. Ishida, S. Yoshida and M. Nojiri, *Bioresour. Technol.* **2009**, 100, 6651–6654.
9. (a) J. Lee, *J. Biotechnol.* **1997**, 56, 1–24; (b) X. P. Ye, L. Liu, D. Hayes, A. Womac, K. L. Hong and S. Sokhansanj, *Bioresour. Technol.* **2008**, 99, 7323–7332.
10. (a) B. Du, L. N. Sharma, C. Becker, S. F. Chen, R. A. Mowery, G. P. van Walsum and C. K. Chambliss, *Biotechnol. Bioeng.* **2010**, 107, 430–440; (b) H. B. Klinke, A. B. Thomsen and B. K. Ahring, *Appl. Microbiol. Biotechnol.* **2004**, 66, 10–26.
11. P. T. Pienkos and M. Zhang, *Cellulose*. **2009**, 16, 743–762.
12. M. A. Rude and A. Schirmer, *Curr. Opin. Microbiol.* **2009**, 12, 274–281.
13. L. P. Yomano, S. W. York and L. O. Ingram, *J. Ind. Microbiol. Biot.* **1998**, 20, 132–138.
14. J. R. M. Almeida, M. Bertilsson, M. F. Gorwa-Grauslund, S. Gorsich and G. Liden, *Appl. Microbiol. Biotechnol.* **2009**, 82, 625–638.
15. Y. Lin and S. Tanaka, *Appl. Microbiol. Biotechnol.* **2006**, 69, 627–642.
16. (a) L. R. Lynd, P. J. Weimer, W. H. van Zyl and I. S. Pretorius, *Microbiol. Mol. Biol. Rev.* **2002**, 66, 739–739; (b) L. R. Lynd, W. H. Zyl, J. E. McBride and M. Laser, *Curr. Opin. Biotechnol.* **2005**, 16, 577–583.
17. (a) C. R. Fischer, D. Klein-Marcuschamer and G. Stephanopoulos, *Metab. Eng.* **2008**, 10, 295–304; (b) K. Rumbold, H. J. J. van Buijsen, K. M. Overkamp, J. W. van Groenestijn, P. J. Punt, and M. J. van der Werf, *Microb. Cell. Fact.* **2009**, 8, 64.

18. (a) O. Bengtsson, B. Hahn-Hagerdal and M. F. Gorwa-Grauslund, *Biotechnol. Biofuels.* **2009**, *DOI: 10.1186/1754-6834-2-9*; (b) A. K. Bera, M. Sedlak, A. Khan and N. W. Y. Ho, *Appl. Microbiol. Biotechnol.* **2010**, 87, 1803–1811.
19. W. Parawira and M. Tekere, *Crit. Rev. Biotechnol.* **2011**, 31, 20–31.
20. C. Martin, M. Galbe, C. F. Wahlbom, B. Hahn-Hagerdal and L. J. Jonsson, *Enzyme. Microb. Tech.* **2002**, 31, 274–282.
21. M. Sedlak and N. W. Y. Ho, *Appl. Biochem. Biotech.* **2004**, 113, 403–416.
22. A. Isci, J. N. Himmelsbach, J. Strohl, A. L. Pometto, 3rd, D. R. Raman and R. P. Anex, *Appl. Biochem. Biotech.* **2009**, 157, 453–462.
23. L. C. Peng and Y. C. Chen, *Biomass. Bioenerg.* **2011**, 35, 1600–1606.
24. S. H. Lim, J. M. Ryu, H. Lee, J. H. Jeon, D. E. Sok and E. S. Choi, *Bioresour. Technol.* **2011**, 102, 2109–2111.
25. E. Tomas-Pejo, M. Ballesteros, J. M. Oliva and L. Olsson, *J. Ind. Microbiol. Biot.* **2010**, 37, 1211–1220.
26. I. Park, I. Kim, K. Kang, H. Sohn, I. Rhee, I. Jin and H. Jang, *Process. Biochem.* **2010**, 45, 487–492.
27. S. Tian, X. L. Luo, X. S. Yang and J. Y. Zhu, *Bioresour. Technol.* **2010**, 101, 8678–8685.
28. M. W. Lau and B. E. Dale, *Proc. Natl. Acad. Sci. USA.* **2009**, 106, 1368–1373.
29. C. F. Huang, T. H. Lin, G. L. Guo and W. S. Hwang, *Bioresour. Technol.* **2009**, 100, 3914–3920.
30. N. N. Nichols, B. S. Dien and R. J. Bothast, *Appl. Microbiol. Biotechnol.* **2001**, 56, 120–125.
31. M. Jin, V. Balan, C. Gunawan and B. E. Dale, *Biotechnol. Bioeng.* **2011**, 108, 1290–1297.
32. V. Garcia, J. Pakkila, H. Ojamo, E. Muurinen and R. L. Keiski, *Renew. Sust. Energ. Rev.* **2011**, 15, 964–980.
33. Y. N. Zheng, L. Z. Li, M. Xian, Y. J. Ma, J. M. Yang, X. Xu and D. Z. He, *J. Ind. Microbiol. Biot.* **2009**, 36, 1127–1138.
34. N. Qureshi, B. C. Saha and M. A. Cotta, *Bioprocess. Biosyst. Eng.* **2007**, 30, 419–427.
35. N. Qureshi, B. C. Saha, B. Dien, R. E. Hector and M. A. Cotta, *Biomass. Bioenerg.* **2010**, 34, 559–565.
36. Z. B. Zhao, *China. Biotechnol.* **2005**, 25(2), 8–11.
37. B. Liu and Z. K. Zhao, *J. Chem. Technol. Biot.* **2007**, 82, 775–780.
38. C. M. Hu, S. G. Wu, Q. Wang, G. J. Jin, H. W. Shen and Z. K. Zhao, *Biotechnol. Biofuels.* **2011**, 4, 25.
39. Y. H. Li, B. Liu, Z. K. Zhao and F. W. Bai, *Chin. J. Biotechnol.* **2006**, 22(4), 650–656.
40. (a) Y. H. Li, Z. K. Zhao and F. W. Bai, *Enzyme. Microb. Tech.* **2007**, 41, 312–317; (b) S. G. Wu, C. M. Hu, X. Zhao and Z. B. K. Zhao, *Eur. J. Lipid. Sci. Tech.* **2010**, 112, 727–733; (c) X. Zhao, X. L. Kong, Y. Y. Hua, B. Feng and Z. B. K. Zhao, *Eur. J. Lipid. Sci. Tech.* **2008**, 110, 405–412.
41. Y. H. Li, B. Liu, Y. Sun, Z. B. K. Zhao and F. W. Bai, *China. Biotechnol.* **2005**, 25(12), 39–44.
42. S. Zinoviev, F. Müller-Langer, P. Das, N. Bertero, P. Fornasiero, M. Kaltschmitt, G. Centi and S. Miertus, *ChemSusChem.* **2010**, 3, 1106–1133.
43. (a) C. M. Hu, X. Zhao, J. Zhao, S. G. Wu and Z. B. K. Zhao, *Bioresour. Technol.* **2009**, 100, 4843–4847; (b) X. Chen, Z. H. Li, X. X. Zhang, F. X. Hu, D. D. Y. Ryu and J. Bao,

Appl. Biochem. Biotech. **2009**, 159, 591–604; (c) G. C. Zhang, W. T. French, R. Hernandez, E. Alley and M. Paraschivescu, *Biomass. Bioenerg.* **2011**, 35, 734–740.

44. C. C. Dai, J. Tao, F. Xie, Y. J. Dai and M. Zhao, *Afr. J. Biotechnol.* **2007**, 6, 2130–2134.
45. J. Tao, C. C. Dai, Q. Y. Yang, X. Y. Guan and W. L. Shao, *Int. J. Green. Energy.* **2010**, 7, 387–396.
46. C. Huang, M. H. Zong, H. Wu and Q. P. Liu, *Bioresour. Technol.* **2009**, 100, 4535–4538.
47. C. Ratledge, in *Single Cell Oil*, (Ed. R. S. Moreton), Longman, London, **1988**, pp. 33–70.
48. J. Y. Yuan, Z. Z. Ai, Z. B. Zhang, R. M. Yan, Q. G. Zeng and D. Zhu, *Chin. J. Biotech.* **2011**, 27(3), 453–460.
49. M. Li, G. L. Liu, Z. Chi and Z. M. Chi, *Biomass. Bioenerg.* **2010**, 34, 101–107.
50. C. H. Zhao, T. Zhang, M. Li and Z. M. Chi, *Process. Biochem.* **2010**, 45, 1121–1126.
51. X. Zhao, S. G. Wu, C. M. Hu, Q. Wang, Y. Y. Hua and Z. B. K. Zhao, *J. Ind. Microbiol. Biot.* **2010**, 37, 581–585.
52. X. W. Peng and H. Z. Chen, *Bioresour. Technol.* **2008**, 99, 3885–3889.
53. Y. Y. Hua, X. Zhao, J. Zhao, S. F. Zhang and Z. B. K. Zhao, *China. Biotechnol.* **2007**, 27 (10), 59–63.
54. K. T. Shanmugam and L. O. Ingram, *J. Mol. Microb. Biotech.* **2008**, 15, 8–15.
55. W. Guo, W. Jia, Y. Li and S. Chen, *Appl. Biochem. Biotech.* **2010**, 161, 124–136.
56. J. H. Kim, D. E. Block, S. P. Shoemaker and D. A. Mills, *Appl. Microbiol. Biotechnol.* **2010**, 86, 1375–1385.
57. L. M. Wang, B. Zhao, B. Liu, B. Yu, C. Q. Ma, F. Su, D. L. Hua, Q. G. Li, Y. H. Ma and P. Xu, *Bioresour. Technol.* **2010**, 101, 7908–7915.
58. Q. Li, M. Yang, D. Wang, W. Li, Y. Wu, Y. Zhang, J. Xing and Z. Su, *Bioresour. Technol.* **2010**, 101, 3292–3294.
59. K. Q. Chen, M. Jiang, P. Wei, J. M. Yao and H. Wu, *Appl. Biochem. Biotech.* **2010**, 160, 477–485.
60. P. Zheng, L. Fang, Y. Xu, J. J. Dong, Y. Ni and Z. H. Sun, *Bioresour. Technol.* **2010**, 101, 7889–7894.
61. G. Reddy, M. Altaf, B. J. Naveena, M. Venkateshwar and E. V. Kumar, *Biotechnol. Adv.* **2008**, 26, 22–34.
62. M. Neureiter, H. Danner, L. Madzingaidzo, H. Miyafuji, C. Thornasser, J. Bvochora, S. Bamusi and R. Braun, *Chem. Biochem. Eng. Q.* **2004**, 18, 55–63.
63. Y. M. Zhu, Y. Y. Lee and R. T. Elander, *Appl. Biochem. Biotech.* **2007**, 137, 721–738.
64. R. H. W. Maas, R. R. Bakker, M. L. A. Jansen, D. Visser, E. De Jong, G. Eggink and R. A. Weusthuis, *Appl. Microbiol. Biotechnol.* **2008**, 78, 751–758.
65. K. M. Bischoff, S. Q. Liu, S. R. Hughes and J. O. Rich, *Biotechnol. Lett.* **2010**, 32, 823–828.
66. D. A. Abbott, E. Suir, A. J. A. van Maris and J. T. Pronk, *Appl. Environ. Microbiol.* **2008**, 74, 5759–5768.
67. (a) S. K. C. Lin, C. Du, A. Koutinas, R. Wang and C. Webb, *Biochem. Eng. J.* **2008**, 41, 128–135; (b) K. Q. Chen, J. Li, J. F. Ma, M. Jiang, P. Wei, Z. M. Liu and H. J. Ying, *Bioresour. Technol.* **2011**, 102, 1704–1708.
68. (a) D. Y. Kim, S. C. Yim, P. C. Lee, W. G. Lee, S. Y. Lee and H. N. Chang, *Enzyme. Microb. Tech.* **2004**, 35, 648–653; (b) D. B. Hodge, C. Andersson, K. A. Berglund and U. Rova, *Enzyme. Microb. Tech.* **2009**, 44, 309–316.

69. (a) P. Zheng, J. J. Dong, Z. H. Sun, Y. Ni and L. Fang, *Bioresour. Technol.* **2009**, 100, 2425–2429; (b) D. Wang, Q. A. Li, M. H. Yang, Y. J. Zhang, Z. G. Su and J. M. Xing, *Process. Biochem.* **2011**, 46, 365–371.
70. S. Octave and D. Thomas, *Biochimie.* **2009**, 91, 659–664.
71. J. W. Lee, T. Y. Kim, Y. S. Jang, S. Choi and S. Y. Lee, *Trends Biotechnol.* **2011**, 29, 370–378.
72. E. N. Miller, L. R. Jarboe, P. C. Turner, P. Pharkya, L. P. Yomano, S. W. York, D. Nunn, K. T. Shanmugam and L. O. Ingram, *Appl. Environ. Microbiol.* **2009**, 75, 6132–6141.
73. D. Heer and U. Sauer, *Microb. Biotechnol.* **2008**, 1, 497–506.
74. Z. L. Liu, *Appl. Microbiol. Biotechnol.* **2011**, 90, 809–825.
75. (a) Z. L. Liu, M. G. Ma and M. Z. Song, *Mol. Genet. Genomics.* **2009**, 282, 233–244; (b) N. N. Nichols, L. N. Sharma, R. A. Mowery, C. K. Chambliss, G. P. van Walsum, B. S. Dien and L. B. Iten, *Enzyme. Microb. Tech.* **2008**, 42, 624–630.
76. F. Koopman, N. Wierckx, J. H. De Winde and H. J. Ruijssenaars, *Proc. Natl. Acad. Sci.* **2010**, 107, 4919–4924.
77. L. J. Jonsson, E. Palmqvist, N. O. Nilvebrant and B. Hahn-Hagerdal, *Appl. Microbiol. Biotechnol.* **1998**, 49, 691–697.
78. W. T. Huang, R. Tai, R. S. Hseu and C. T. Huang, *Process. Biochem.* **2011**, 46, 1469–1474.
79. D. Simon, W. E. Tyner and F. Jacquet, *Bioenerg. Res.* **2010**, 3, 183–193.
80. J. D. Stephen, W. E. Mabee and J. N. Saddler, *Biofuels Bioprod. Bioref.* **2010**, 4, 503–518.
81. (a) A. Oudshoorn, C. Van den Berg, C. P. M. Roelands, A. J. J. Straathof and L. A. M. Van der Wielen, *Process. Biochem.* **2010**, 45, 1605–1615; (b) D. J. Roush and Y. F. Lu, *Biotechnol. Prog.* **2008**, 24, 488–495.

CHAPTER 11

Heterogeneous Catalysts for Biomass Conversion

AIQIN WANG, CHANGZHI LI, MINGYUAN ZHENG, and TAO ZHANG

11.1 BRIEF INTRODUCTION OF GREENNESS OF HETEROGENEOUS CATALYSIS

With the ever-increasing concerns about the depletion of fossil resources and the accompanying environmental issues, development of new alternative resources is becoming a hot topic. Biomass, composed mainly of carbohydrates, lignin, and glycerides, is being considered as a renewable and alternative source to the traditional fossil (petroleum, coal, and natural gas) for producing fuels and chemicals [1]. However, how to convert biomass in an economical and environmentally benign way is still a grand challenge. For example, it has been realized that lignocellulose, the most abundant source of biomass on the earth, is a promising feedstock for producing fuels and chemicals; and more importantly, lignocellulose can not be digested by human beings and therefore its large-scale consumption will not impose a negative impact on food supplies. However, the complex and very strong structure of lignocellulose makes it rather difficult to depolymerize under milder conditions. Actually, strong liquid acids or bases are usually employed to depolymerize lignocellulose, which causes severe environmental issues. To solve this problem, development of green approaches, for example, heterogeneous catalysis is expected.

Catalysis approaches can be classified into three categories: homogeneous, heterogeneous, and enzymatic catalysis [2]. All the three methods have their advantages and limitations. Homogeneous catalysis offers a high reaction rate but the recovery of catalysts is often difficult. Similarly, enzymatic catalysis is featured with high activity and selectivity, but still has the limitation of nonrecyclability. On the other hand, solid, heterogeneous catalysts have the advantage of ease of recovery and recycling, and consequently, it is often considered as a kind of green catalysts. Nevertheless, heterogeneous catalysts generally possess low activities, as a

The Role of Green Chemistry in Biomass Processing and Conversion, First Edition.
Edited by Haibo Xie and Nicholas Gathergood.

result, harsh reaction conditions are required, which often causes concurrence of a number of undesired reactions. Therefore, development of highly active, selective, and robust catalysts is the main target for the design of heterogeneous catalysts.

11.2 DESIGN AND SELECTION OF HETEROGENEOUS CATALYSTS

To design a heterogeneous catalyst, four aspects must be considered: activity, selectivity, stability, and accessibility. The requirement for activity and selectivity is well known since they are common for both homogeneous and heterogeneous catalysts. The stability of a heterogeneous catalyst is particularly important for biomass conversion because most of the reactions involved in biomass conversion proceed under hydrothermal conditions which require excellent hydrothermal stability of the catalyst. The accessibility of a heterogeneous catalyst is also a crucial factor determining the activity and selectivity in biomass conversion. For example, the transformation of lignocellulose involves depolymerization to water-soluble oligosaccharides, which will further convert into fuels and chemicals under the catalysis of a solid catalyst. To facilitate the reaction, the active sites on the catalyst surface must be accessible to the oligosaccharides as well as intermediates. For this purpose, the design of the catalyst should allow most of the active sites exposed on the outer surface and the support structure preferably be meso- or macroporous. It is widely reported that meso- or macroporous materials offer significant advantages over microporous ones with respect to the transportation of large molecules [3].

In addition to the above fundamental properties, industrial applications require that a catalyst be easily regenerated, reproducible, while being mechanically and thermally stable. All these characteristics are critically dependent on the physical and chemical properties of the catalysts, which in turn are closely related with many parameters inherent to the method used to prepare the catalyst.

In the past ten years, great advances have been made in the synthesis of catalyst materials, from support materials with tunable chemical compositions, pore structures and morphologies, to catalytically active metal particles with well-controlled particle size and morphology [4]. On the other hand, development of advanced characterization techniques such as Sub-Ångström-resolution aberration-corrected scanning transmission electron microscopy (STEM) and various in-situ spectroscopies, together with the great progress in theoretical calculations, has made it possible to understand the heterogenous catalysis reaction mechanism at a molecular level. All these advances lay the basis for the design of a heterogenous catalyst towards the efficient conversion of biomass. Nevertheless, biomass is a complex substance, and its transformation often involves a series of different, consecutive or parallel reactions, such as hydrolysis, dehydration, C—C cleavage, C—O hydrogenolysis, hydrogenation, C—C coupling (aldol condensation), isomerization, and selective oxidation. Therefore, it is very difficult to give a general guide for the design of solid catalysts for biomass conversion although some desired properties of solid catalysts have been addressed in a recent review by

Rinaldi and Schüth [3]. Moreover, the chemical transformation of lignocellulosic materials using heterogeneous catalysts is in its infancy, and much effort is being devoted to the process rather than catalyst optimization. Based on the importance and application of heterogeneous catalysts in biomass conversion, in this chapter we focus on two kinds of solid catalysts. First, solid acid/base catalysts and second supported metal catalysts. In practice, the two kinds of catalysts are often combined together to constitute a bifunctional or multifunctional catalysts in biomass conversion. In the following section, we first discuss how to design or select an appropriate catalyst for a given reaction. Then, we give an overview of recent advances on new heterogeneous catalytic systems in biomass conversion. This includes production of 5-hydroxymethylfurfural and polyols from cellulose and biodiesel synthesis through transesterification of triglycerides or esterification of free fatty acids with short chain alcohols.

11.2.1 Solid Acid and Base Catalysts

Solid acids are widely involved in biomass transformation, such as hydrolysis of cellulose to water-soluble oligosaccharides and glucose [5], isomerization of glucose to fructose [6], transesterification of glycerides to produce biodiesel [7], and other reactions [3]. Since the use of mineral and some Lewis acids (e.g., $AlCl_3$) produces a large quantity of waste, the search for a suitable replacement by recyclable, less corrosive solid acids has attracted a great deal of attention. A wide variety of solid acid catalysts is available, including acidic clays, zeolites, silica-occluded heteropoly acids, sulfonated carbons and resins, as well as sulfated and phosphated mesoporous metal oxides. When designing a solid acid catalyst, the strength of acidity, the number of acid sites, the specific surface area, and the accessibility of the acid sites are all important factors to be considered. Moreover, when the reaction proceeds for a long time under hydrothermal conditions (e.g., hydrolysis of cellulose), good durability of the catalysts under such harsh reaction conditions is also required. A typical application of solid acids in biomass conversion is the hydrolysis of cellulose [8]. The hydrolysis of cellulose to glucose and sugar oligomers with sulfuric acid has been investigated widely and applied in a large scale [9]. However, the use of sulfuric acid inevitably causes significant problems in the disposal of waste acid, specialist equipment which can be resistant to corrosion is also needed. Solid acid catalysts are readily separated from the reactants, are easy to reuse, and cause less or even no corrosion problem. Developing efficient solid acids for cellulose conversion is therefore one attractive way to meet the demand of the principles of green chemistry [10], and has gained much attention in recent years.

Table 11.1 gives a summary of the recent work on solid acid-catalyzed hydrolysis of cellulose. It can be seen that solid acids are generally much less effective than the liquid acids in the hydrolysis of cellulose. To achieve a high conversion of cellulose, the amount of solid acids used is rather high, and even greater than the amount of cellulose in some cases. Since the breakage of β-1,4-glycosidic bonds are difficult, the hydrolysis of cellulose usually proceeds under

TABLE 11.1 Hydrolysis of Cellulose on Solid Acid Catalysts

Solid Acids	SA ($m^2 g^{-1}$)	Acid Density ($mmol\ g^{-1}$)	Reaction Conditions	Yields (%)	Ref.
Sulfonated mesoporous silicas	868	1.11	0.2 g starch, 0.1 g catalyst, 10 g water, 130 °C, 6 h	39 (glucose) 18 (maltose)	[8a]
Acid-treated kaolinite	nd	nd	1 g microcrystalline cellulose, 1 g catalyst, ball-milling for 3 h	84 (soluble oligosaccharides)	[8b]
Fe_3O_4-SBA-SO_3H	464	1.09	1 g amorphous cellulose, 1.5 g catalyst, 15 mL water, 150 °C, 3 h	50 (glucose)	[8c]
Fe_3O_4-SBA-SO_3H	464	1.09	1.5 g microcrystalline cellulose, 1.5 g catalyst, 15 mL water, 150 °C, 12 h	42 (glucose)	[8c]
AC-SO_3H	806	1.63	0.045 g ball-milled cellulose, 0.05 g catalyst, 5 mL water, 150 °C, 24 h	40.5 (glucose)	[8d]
Sulfated ZrO_2	52	1.60	0.045 g ball-milled cellulose, 0.05 g catalyst, 5 mL water, 150 °C, 24 h	10–15 (glucose)	[8d]
Amberlyst-15	15	1.8	0.045 g ball-milled cellulose, 0.05 g catalyst, 5 mL water, 150 °C, 24 h	25 (glucose)	[8d]
H-ZSM-5 (45)	124	0.3	0.045 g ball-milled cellulose, 0.05 g catalyst, 5 mL water, 150 °C, 24 h	10 (glucose)	[8d]
AC-SO_3H-250	762	2.23	0.27 g ball-milled cellulose, 0.3 g catalyst, 27 mL water, 150 °C, 24 h, water	62.6 (glucose)	[8e]

CMK-3-SO_3H-250 (mesoporous carbon)	412	2.39	0.27 g ball-milled cellulose, 0.3 g catalyst, 27 mL water, 150 °C, 24 h, water	74.5 (glucose)	[8e]
Amberlyst 15DRY (acidic ion-exchange resin)	36	nd	5 g microcrystalline cellulose, 1.0 g catalyst, 100 g C_4mimCl 100 °C, 5 h	>90 (cellooligomers)	[8f,8g]
Amorphous carbon bearing SO_3H, COOH, and OH groups	2	SO_3H: 1.9 COOH: 0.4 Phenolic OH: 2.0	0.025 g microcrystalline cellulose, 0.3 g catalyst, 0.7 g water, 100 °C, 3 h	4 (glucose) 64 (others)	[8h]
Layered $HNbMoO_6$	5	1.9	0.1 g microcrystalline cellulose, 0.2 g catalyst, 5 mL water, 130 °C, 12 h	8.5 (glucose and cellobiose)	[8i]
Sulfonated bamboo carbon bearing SO_3H, COOH, and OH groups	nd	SO_3H: 1.99 COOH: 0.13 Phenolic OH: 3.43	0.2 g microcrystalline cellulose, 0.1 g catalyst, 1.5 mL water, 90 °C, 1 h, microwave irradiation	19.8 (glucose) 8 (cellooligosaccharides)	[8j]
Sulfonated SiO_2/carbon	nd	0.37	0.05 g ball-milled cellulose, 0.05 g catalyst, 5 mL water, 150 °C, 24 h	50 (glucose)	[8k]

nd = not determined.

hydrothermal conditions. In this case, the durability of the catalysts is an important factor to be considered. Zeolites, which possess good hydrothermal stability and have been widely used in hydrocarbon conversion in petrochemical processes, are much less effective than the sulfonic acid-functionalized materials in the hydrolysis of cellulose. This is due to two factors, first the effective acidity of zeolites in aqueous solution is very low and second most of the acid sites located in the micropores of zeolites (<1 nm) are inaccessible to the relatively large molecules such as polysaccharides. Among the sulfonated materials, sulfonated zirconia, as well as sulfonated resins (Amberlyst 15), exhibited remarkable activities in the first run for the hydrolysis of cellulose. However, they lost the high activities in the following runs due to elution of SO_4^{2-} ions in the solution. Therefore, their behavior is similar to homogeneous catalysts during the reaction [8d]. In contrast, sulfonated carbons exhibited relatively high activities and selectivities and could be recycled without any significant decrease in performance. Sulfonated carbon materials, in particular, offer great advantages such as high surface areas, excellent hydrothermal stabilities, tunable surface hydrophiliciy/hydrophobicity, large pore size from mesopore to macropore, and wide availability from diverse raw biomass materials [8j], which bestow them with great potentials as a class of green catalysts in the hydrolysis of cellulose. To understand the outstanding hydrolytic catalysis of sulfonated carbon material, Hara and coworkers made a detailed study of the mechanism of biomass hydrolysis [8h]. They suggested that the phenolic OH groups in carbon material have a strong affinity for β-1,4 glucan, and the SO_3H groups bonded to the carbon function as effective active sites for both breaking the hydrogen bonds and hydrolyzing the β-1,4-glycosidic bonds in the adsorbed long-chain water-soluble β-1,4-glucan compounds. Both the sulfonation temperature and the nature of the carbon materials impose a significant impact on the performances of the resulting sulfonated carbons in the hydrolysis of cellulose. Zhang and coworkers [8e] found that the acid density on the surface of sulfonated carbons increased with the sulfonation temperature, while the density of sulfonic groups ($-SO_3H$), as well as the specific surface areas, reached the maximum value at the sulfonation temperature of 250 °C. The efficiencies of sulfonated carbons in cellulose hydrolysis are closely correlated with the density of sulfonic groups, the 250 °C-sulfonated carbons gave rise to the highest cellulose conversion and glucose yield. It is interesting to note that when sulfonated ordered mesoporous carbon (CMK-3) was used as the solid acid catalyst, the cellulose conversion attained 94.4% and the glucose yield was as high as 74.5% [8e], which is the highest value reported so far for the solid acid-catalyzed hydrolysis of cellulose, even comparable to the efficiency of liquid H_2SO_4 [9]. The superior performance of the sulfonated CMK-3 to that of sulfonated active carbon should be attributed to its high density of accessible acid sites located in the mesopores. A recent study indicated that even nonsulfonated CMK-3 could promote the hydrolysis of cellulose to oligosaccharide [11]. A recent interesting approach to catalyst formulation was reported by Fu and coworkers [8c] who incorporated magnetic Fe_3O_4 nanoparticles in the structure of sulfonated silica. With this magnetic solid acid as the catalyst for hydrolysis of cellulose, the maximum yield of glucose from microcrystalline cellulose was 26% and the catalyst was easily separated from the reaction mixture by applying a magnetic force. With the advancement of novel materials [12] and new preparation methods [13], development of a more active, selective, and robust solid acid catalyst can be expected in future.

Compared to solid acids, solid base catalysts have been less developed in the past years. However, solid base catalysts are finding more and more applications in various reactions involved in biomass conversion, such as transesterification of glycerides with methanol to synthesize biodiesel, C—C forming (aldol), and C—C breaking (retro-aldol) reactions [14]. The solid-base-catalyzed biodiesel synthesis will be addressed in Section 11.3.3. Here we give a typical example of solid-base-catalyzed C—C forming reaction involved in the biomass conversion, in which the cross-condensation of furfural or HMF with acetone to form C_8–C_{15} large molecules that can be further converted to liquid alkanes by hydrogenation. This process was firstly proposed by Dumesic and coworkers [15]. Since furfural or HMF can be obtained from C5 or C6 sugars that are the main components of hemicellulose and cellulose, this process is being considered as a promising approach for the sustainable production of liquid fuels from biomass.

The sequential reactions shown in Figure 11.1 clearly demonstrate that the activity and selectivity of the aldol condensation of furfural (or HMF) with acetone determine the overall distribution of the finally produced hydrocarbons. It is therefore very important to design an active, selective, and robust catalyst in aqueous media. Initially, Dumesic and coworkers used liquid base (NaOH) to catalyze the aldol-condensation reaction [14]. However, the use of liquid base catalysts produces significant amount of waste water which causes severe environmental impact. Consequently, they employed solid base catalysts in their subsequent work [15a]. It is known that various solid base catalysts are effective for the aldol reaction, such as alkali [15b] and alkaline earth oxides [16], anion-exchange resins [17], organic base-functionalized mesoporous silica [18], and hydrotalcites [19]. However, since the aldol reaction of furfural (or HMF) with acetone proceeds in aqueous media, the resistance to water poisoning and to water dissolution will be the paramount requirement in choosing a solid base catalyst. In their early study, Dumesic et al. used a mixed Mg-Al-oxide derived from hydrotalcite to catalyze this reaction [15a]. However, this catalyst could not be recycled because approximately 80% of its initial activity was lost after the second run, and this deactivation continued in the subsequent runs. The authors attributed the deactivation to the

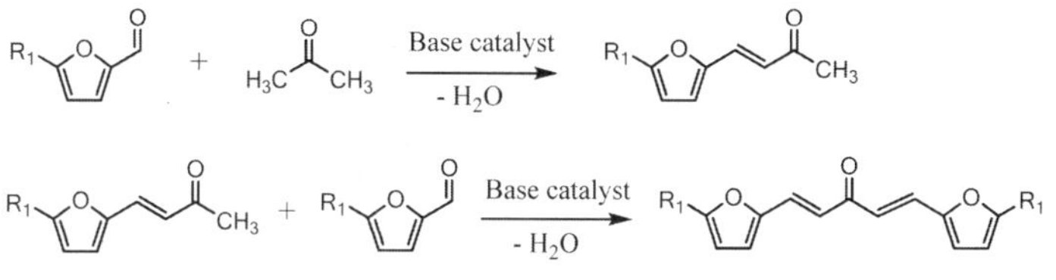

FIGURE 11.1 Single and double aldol condensation of fural derivative (for furfural, $R_1 = H$) and acetone.

reversible structural changes upon interaction with water in an aqueous reaction medium. In the following studies, they found that mixed oxide $MgO–ZrO_2$ gave much better recycling ability thanks to its stable structure during hydrothermal reaction for extended periods. In 2011, Sádaba et al. [20] reported that the active component in $MgO–ZrO_2$ catalyst was Mg^{2+}, and the leaching of Mg^{2+} during the reaction was still inevitable although the leaching extent was greatly alleviated in comparison with Mg-Al-oxide. Therefore, the challenge in the design of solid base catalysts for biomass conversion occurring in aqueous phase would be hydrothermal stability of the catalyst materials in addition to activity and selectivity, which is distinct from those reactions taking place in organic phase or gas phase.

Actually, there has been an intense research effort devoted to improvement of resistance of solid bases to poisoning by water and CO_2. For example, Jong and coworkers [21] precipitated hydrotalcites on the carbon nanofibers to obtain hydrotalcites platelets with a lateral size of $\sim$20 nm. Such a small lateral size of crystallites led to a great increase in the number of edge sites which have been thought as the actual active sites in base catalysis [22]. After activation by heat treatment and subsequent rehydration, the carbon nanofibers supported hydrotalcites exhibited a high number of accessible Brønsted base sites and therefore gave rise to high specific activity and good stability in the liquid phase self-condensation of acetone. Similarly, Duan and coworkers synthesized alumina-supported hydrotalcites by *in situ* growth method, and the resulting material exhibited excellent recyclability during five repetitive runs for liquid-phase self-condensation of acetone [23]. In 2009, Matsuhashi reported that by decomposition of $Al(OCH(CH_3)_2)_3$ over the surface of Mg $(OH)_2$, the MgO surface was covered with a layer of Al_2O_3 and the resulting composite exhibited high activity for acetone-aldol reaction and good resistance to water dissolution [24].

11.2.2 Supported Metal Catalysts

Supported metal catalysts have been widely applied to industrially important reactions, including hydrogenation/dehydrogenation, oxidation, C—C bond formation, etc., and all these reactions have been involved in the biomass conversion. Generally, a supported metal catalyst is composed of highly dispersed metal nanoparticles on a high-surface-area support. The size and shape of metal particles, the chemical nature and textural properties of the support, and the metal–support interactions all have strong influences on the catalytic performances of the catalysts. Supported catalysts can be prepared by a variety of methods including decomposition, impregnation, precipitation, coprecipitation, adsorption, or ion exchange. Selection of an appropriate preparation method is critically dependent on the nature of active components as well as the characteristics of the supports.

For biomass conversion, most of the supported metal catalysts are involved in hydrogenation/dehydrogenation reaction steps. For that purpose, the platinum group metals, including Pt, Ru, Rh, Pd, Os, and Ir are often selected as the active components because they usually possess better antisintering and antileaching properties under harsh hydrothermal reaction conditions in comparison with base metals such as Ni and Cu. Silica, alumina, and carbon are preferred supports due to their large surface areas that can well disperse the active metals. Specifically, porous carbon materials possess greater hydrothermal stabilities than the other supports, which are particularly important for biomass-involved reactions in aqueous solution.

When bifunctional or multifunctional catalysts are required, the design of a supported metal catalyst must consider both the hydrogenation/dehydrogenation capability of the metal component and the acid/base property of the support. For example, Dumesic and coworkers found that bifunctional Pd/MgO–ZrO_2 catalyst was effective and robust for the aqueous phase aldol-condensation and hydrogenation of furfural (or HMF) to large water-soluble intermediates [15b]. In this catalyst, MgO–ZrO_2 acts both as a basic catalyst for the aldol-condensation and as a support for dispersing the metallic Pd. Nevertheless, when lignocellulosic biomass (e.g., corncobs) was used directly as the feed stock, Pd/WO_3–ZrO_2 proved to be a more effective catalyst, even superior to a mixture of WO_3–ZrO_2 and Pd/MgO–ZrO_2 [25]. Pd/WO_3–ZrO_2 is a multifunctional catalyst in which WO_3–ZrO_2 plays dual roles in hydrolysis/dehydration of lignocellulose and aldol reaction of furfural with acetone.

In cases where dual function can not be accomplished over a single catalyst, a mixture of two different catalysts has to be employed. For example, Lercher and coworkers [26] used Raney Ni and Nafion/SiO_2 catalysts for converting phenolic components in bio-oil to hydrocarbons and methanol with nearly 100% yields. Since crude bio-oil contains approximately 30% water and a large number of unstable oxygen-containing compounds, the catalyst for bio-oil upgrading must be water-tolerant and highly active under milder reaction conditions. A combination of Raney Ni and Nafion/SiO_2 proved to be efficient for aqueous-phase hydrodeoxygenation of phenolic monomers in bio-oil where Raney Ni functioned as hydrogenation catalyst and Nafion/SiO_2 acts as the Brønsted solid acid for hydrolysis and dehydration.

Although the platinum group metals are the most popular choice as hydrogenation/dehydrogenation active components in biomass conversion reactions, the high cost of these precious metals has limited their application. In this context, searching for a less expensive but efficient catalyst has become a major impetus. For example, tin-promoted Raney nickel has been proved to be a good substitute of Pt/Al_2O_3 for hydrogen production from biomass [27]. The addition of Sn to Ni decreased the rate of methane formation from C—O breaking while maintaining the high activity for C—C bond cleavage, which resulted in the high selectivity of hydrogen production. Zhang and coworkers [28] reported that tungsten carbide (W_2C) could replace Pt for polyols production from cellulose. Of note, supported W_2C catalyst exhibited a uniquely high selectivity towards ethylene glycol, and this selectivity was further enhanced greatly by the addition of the second metal component nickel. As demonstrated by the above Sn–Ni and Ni–W_2C, supported bimetallic catalysts usually offer a better activity, selectivity, or stability in comparison with monometallic catalysts, and the synergy between the two metals can be explained by electronic or geometric effect.

As described previously, we gave a brief discussion in this section on how to design a solid catalyst for biomass conversion. Taking into account the basic properties of biomass, such as macromolecular and robust structure, rich oxygen-containing functional groups, and high boiling point and low thermal stability (difficult to vaporize), the biomass conversion in many cases proceeds in aqueous solution, which requires the catalyst be hydrothermally stable in addition to activity and selectivity considerations. Moreover, the accessibility of the solid catalyst should be maximized, for example, via employment of meso- or macroporous support, in designing a solid catalyst. However, we must emphasize that the biomass conversion involves many different types of reactions, and the design of a solid catalyst critically depends on the understanding of the reaction mechanism of a given reaction. In the following section, we will give an overview of the recent advances on new heterogenous catalyst systems, including production of HMF from biomass, production of polyols from biomass, as well as biodiesel synthesis, to further demonstrate the special requirements for a solid catalyst in converting the biomass.

11.3 OVERVIEW OF RECENT ADVANCES ON NEW HETEROGENEOUS CATALYTIC SYSTEMS

11.3.1 Heterogeneously Catalyzed Production of HMF from Biomass

11.3.1.1 Introduction The five-membered ring compound, 5-hydroxymethylfurfural (5-HMF or simply called as HMF), is one of the leading bio-based platform compounds [43]. HMF can be produced from hexose through dehydration reaction by the loss of three water molecules. Various substrates can be used for the production of HMF: hexoses, oligosaccharides, polysaccharides including cellulose [29], and raw biomass such as inulin [30] (see Fig. 11.2). HMF can be further transformed into levulinic acid and 2,5-disubstituted furan derivatives [31], such as 2,5-furandicarboxylic acid (FDCA), which can be used as a replacement for terephthalic acid in the production of polyesters (e.g., PET and PBT) [56].

Moreover, through hydrogenation, HMF can be converted into biofuel molecules 2,5-dimethylfuran and other liquid alkanes that can be used, for example, in diesel engines. Therefore, production of HMF will be of great significance for the transformation of biomass into bio-based chemicals and biofuels.

Commonly, the synthesis of HMF is more efficient and more selective from ketohexoses (e.g., fructose) rather than from aldohexoses (e.g., glucose), mainly because the structure of aldohexose is very stable, it enolyses in a very low degree while the enolization to enols (see in Fig. 11.3, enols 1 & 2) is a determining factor of the HMF formation from hexoses. Moreover, aldohexose can condense to form

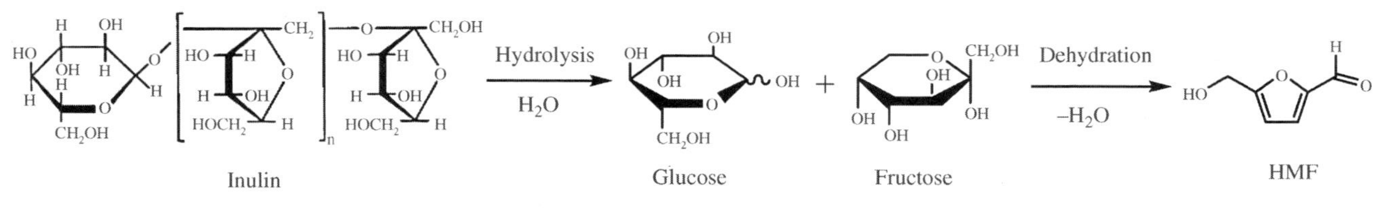

FIGURE 11.2 Reaction path from inulin to HMF.

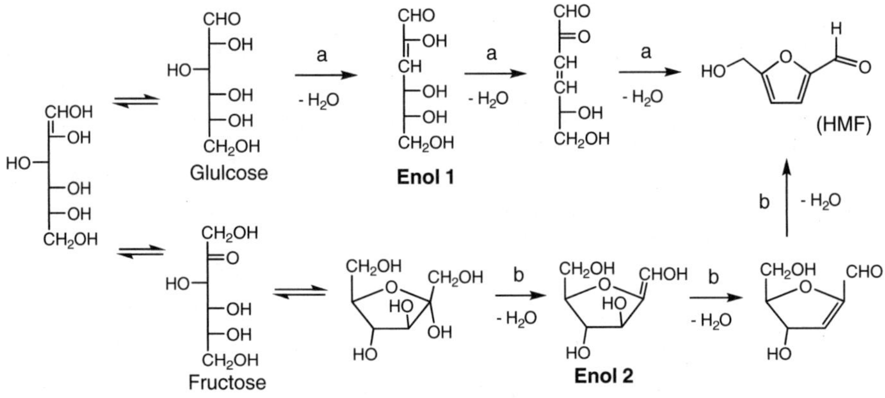

FIGURE 11.3 Possible mechanisms for the dehydration reaction of hexose. (**a**) open-chain route. (**b**) cyclic route.

oligosaccharides bearing reducing groups, which may react with intermediates or with HMF itself.

Dehydration of hexose to produce HMF is predominantly catalyzed by acidic catalyst. Independent of the catalyst selected, the dehydration of hexoses goes through two possible pathways, namely, the open-chain (Fig. 11.3, path a) and the cyclic fructofuransyl intermediate (path b) pathway [32]. Although the reaction mechanism is in debate, it is clear that the chemistry of the formation of HMF is very complex and includes a series of side-reactions, such as isomerization, dehydration, fragmentation, and condensation, which influence strongly the selectivity of HMF.

Since the 19th century, studies on HMF production have been reported in various solvent systems. For example, in water, organic solvents, biphasic systems, ionic liquids, and near- or supercritical fluids, using a variety of acidic catalysts, such as mineral acids (HCl, H_2SO_4), heterogeneous acids (zeolites, ion-exchange resins, etc.), and salts ($ZnCl_2$, $LaCl_3$) [33]. Because heterogeneous catalysts are less corrosive, easier to handle and separate, and reusable, they are considered to have greater potential than homogeneous catalysts in industrial applications. Based on different solvent systems, recent advances on heterogeneous catalyzed HMF production are summarized here.

11.3.1.2 Heterogeneous Catalysis in Aqueous Media In fructose dehydration, water is an excellent solvent for fructose as well as HMF, and has been frequently used in the reaction. Among the various heterogeneous catalysts, research concerning the application of ion-exchange resins for the synthesis of HMF are the most prevalent. In the past 30 years, many researchers investigated ion-exchange resins such as Levatit® SPC-108, Amberlite-15, and Diaion® PK-228 for the production of HMF [34]. Highly acidic cation-exchanged resins, such as examples functionalized with sulfonic acid groups, are effective catalysts, providing the acidity of mineral acids meanwhile possessing the advantages of heterogeneous catalysts.

Niobium catalysts [35] such as niobic acid ($Nb_2O_5 \cdot nH_2O$) and niobium phosphate ($NbOPO_4$) have been reported to exhibit an intermediate selectivity of about 70% for the production of HMF at about 70% conversion for different substrates, such as fructose, sucrose, and inulin. With a $VOPO_4$-based heterogeneous catalyst, HMF was produced from fructose after reaction at 80 °C for 2 h under nitrogen atmosphere, with a selectivity higher than 80% at the fructose conversion of 50% [36].

Layered zirconium- and titanium-hydrogenphosphates in the *a* and *r* structural arrangements, their corresponding *a*- and *r*-layered pyrophosphates as well as cubic pyrophosphates were used as acid catalysts in the dehydration of fructose and inulin to HMF [37]. All the catalysts presented promising performances (TN > 6 mmol of HMF/g of catalyst × h and selectivity to HMF > 98%). The comparison reactions suggested that both Brønsted and Lewis acid sites are involved in the catalytic process, whereas the Lewis acid sites are more active in generating HMF.

Particularly interesting, TiO_2 and ZrO_2 in subcritical water were found to be active for direct conversion of glucose, cellulose, and lignocellulose with high HMF yields and less by-products formation [38]. The anatase-TiO_2 catalyst showed both basic and acidic properties. The basic properties of the catalyst were thought to catalyze the isomerization of glucose to fructose, whereas the acidic sites catalyzed the dehydration of fructose [38]. The direct conversion of glucose to HMF can be enhanced up to fivefold compared to the hydrothermal dehydration, by employing an anatase-TiO_2 at 200 °C [39]. Importantly, it was also found that the sulfur-doping content (for SO_4–ZrO_2) and the calcination temperature strongly affected the catalyst reactivity. The main disadvantage of using subcritical water is the high temperature and pressure required which leads to a high cost of the specialist equipment.

11.3.1.3 Heterogeneous Catalysis in Modified Aqueous Media and Two-Phase Systems As an abundant and nonhazardous solvent, water is the preferred choice when exploring green and sustainable chemistry. However, the dehydration of hexose to yield HMF in aqueous media is significantly hampered by a series of competitive side-reactions, such as rehydration, isomerization, and condensation of hexoses, reaction intermediates, and HMF, which influence strongly the efficiency of the process.

In order to limit the side reactions and improve the yield of HMF, polar phase modifiers that are miscible with water such as acetone [40], dimethylsulfoxide (DMSO) [41], dimethylformamide (DMF), and methyl isobutyl ketone (MIBK) [42], are employed in the dehydration of hexose. A further modification of the aqueous phase system is the introduction of a second immiscible phase to create a two-phase reaction system. The organic phase extracts HMF from the aqueous phase upon its formation and consequently reduces the occurrence of rehydration and polymeric by-products. This effective catalytic system was first developed by Dumesic's group [43]. They obtained HMF from D-fructose in high yields (>80%) at relatively high fructose concentrations (10–50 wt%). In acetone–water biphasic media, Qi et al. [40] achieved HMF yield of 73% at 94% conversion of fructose with ion-exchanged resin. In the modified aqueous media or two-phase systems, the heterogeneous catalysts applied include vanadyl phosphate [36], ion-exchange resins [40,43], aerosil [44], and zeolites in their protonic forms such as H-Y, H-mordenite, H-beta, and H-ZSM5 [42]. Hexoses conversion and selectivity to HMF were found to depend on acidic and structural properties of the catalysts used, as well as on the micropore versus mesopore volume distributions [42]. In the two-phase systems, however, most extracting solvents used show poor partitioning of HMF into the organic phase, which decreases the possibility in industrial application. The addition of a salt (e.g., NaCl) to the aqueous phase may improve the partitioning of HMF into the extracting phase by means of the "salting-out effect".

11.3.1.4 Heterogeneous Catalysis in Non-Aqueous Organic Solvents

Before ionic liquids were employed in this area, the best results for the dehydration of hexoses to HMF have been achieved in high-boiling organic solvents. The anhydrous circumstance suppresses unwanted side reactions such as the rehydration of HMF to levulinic acid and formic acid, and generates high yield of HMF.

The first excellent yield of HMF was obtained by Nakamura and Morikawa [45] using a strongly acidic ion-exchanged resin as the catalyst in DMSO at 80 °C. These conditions gave HMF yield of 90% from fructose. On increasing the reaction temperature to 150 °C, a yield of 92% was reached in 2 h even without a catalyst [46]. By monitoring the reaction with ^{1}H- and ^{13}C-NMR spectra, Amarasekara [47] proposed a DMSO-catalyzed mechanism, in which a key intermediate was identified as (4*R*,5*R*)-4-hydroxy-5-hydroxymethyl-4,5-dihydrofuran-2-carbaldehyde (Fig. 11.4).

In nonaqueous solvent mixture such as acetone–DMSO, heterogeneous catalysts such as ion-exchanged resins [34] and SO_4^{2-}/ZrO_2 [38a] also show much better catalytic activities than those exhibited in water. The excellent result was explained by the structural similarities between acetone and DMSO, which would promote the furanoid form of fructose and hence favor the formation of HMF [48]. By simply removing water through reduced pressure distillation, enhanced production of HMF from fructose with various solid acid catalysts (heteropoly acids, zeolites, and acidic resins) were realized [49].

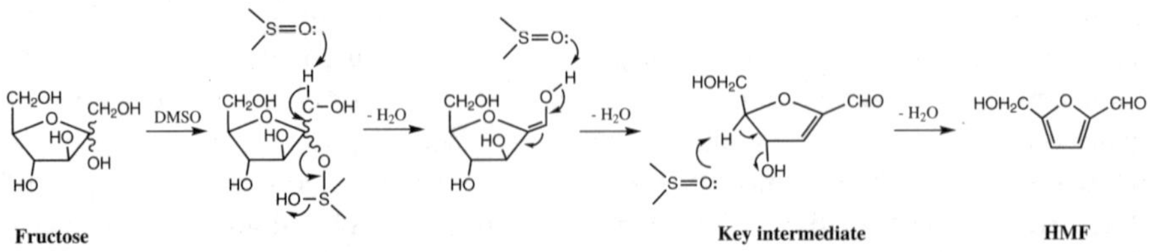

FIGURE 11.4 Proposed mechanism for the dehydration of fructose in DMSO at 150 °C.

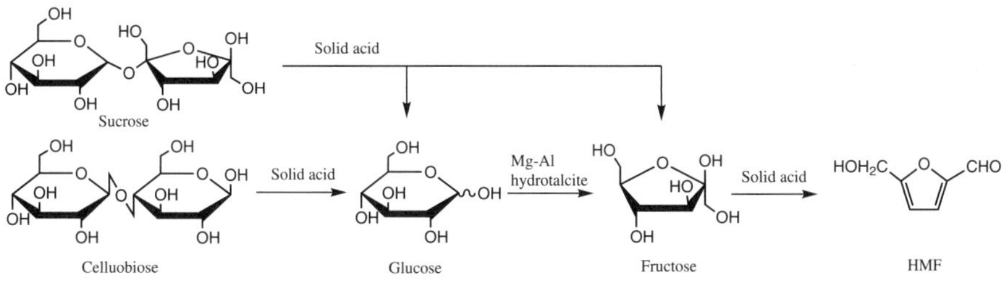

FIGURE 11.5 Schemate of one-pot synthesis of HMF from mono- and disaccharides using solid acid and Mg–Al hydrotalcite.

It is interesting to point out the strategy of adopting solid acid–base pair in a simple one-pot reaction for the production of HMF from disaccharides (sucrose and cellobiose) [50]. Figure 11.5 clearly shows the role of acid–base in the process. This process includes (1) hydrolysis of disaccharides and dehydration of fructose over solid acid, for example, Amberlyst-15; and (2) isomerization of glucose into fructose over solid base Mg–Al hydrotalcite. Such an approach has never been reported for liquid acid–base pair due to their neutralization. With a similar idea, Hu's group [51] synthesized SO_4^{2-}/ZrO_2 and $SO_4^{2-}/ZrO_2–Al_2O_3$ by impregnation of $Zr(OH)_4$ and $Zr(OH)_4–Al(OH)_3$ with ethylene dichloride solution of chlorosulfonic acid. CO_2– and NH_3– TPD showed that both base sites and acid sites, which were responsible for the glucose–fructose isomerization and fructose dehydration, respectively, existed on the catalysts. The synthesized catalysts were used in the catalytic conversion of glucose to HMF, the optimized yield of 47.6% was obtained within 4 h at 130 °C in DMSO.

Although remarkable HMF yield could be obtained in nonaqueous organic solvents with heterogeneous catalysts, none of the above examples are, at this stage, suitable for production on a large-scale. High-boiling aprotic solvents such as DMSO, DMF, and 1-methyl-2-pyrrolidinone (NMP) are all miscible with water as well as many other common organic solvents. This makes separation of the desired products very difficult. Therefore, it is essential to address better separation strategy for the production of HMF in nonaqueous organic solvents.

11.3.1.5 Ionic Liquids The use of ionic liquids has emerged in recent years as a feasible alternative to organic solvents [52]. Their unique properties such as strong solubility for carbohydrates, pure ionic circumstance, and tunable anion–cation composition make them particularly suitable solvents for catalytic biomass conversion [29–30]. As some examples, Zhang [53] and coworkers proved recently that ILs that comprise halide anions, such as C_4mimBr, C_4PyCl, C_2mimCl, AmimCl, and C_6mimCl, are excellent solvents for HMF production. Some representative ionic liquids used for production of HMF are listed in Figure 11.6 [53,54]. Heterogeneous catalytic transformation of biomass into HMF in ILs will be summarized in detail in Chapter 3.

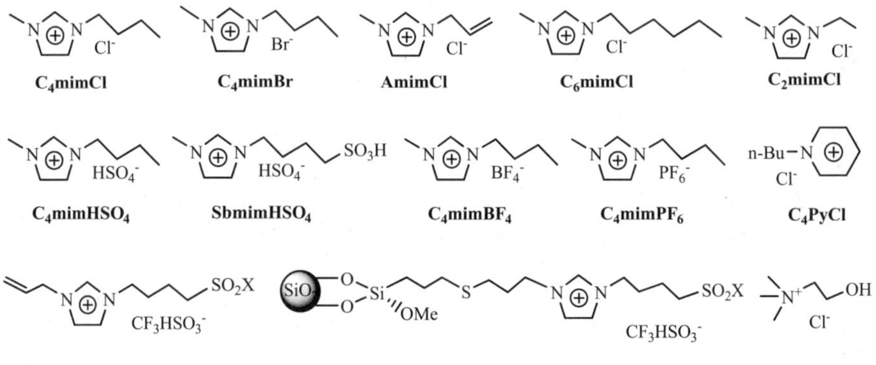

C_4mimCl C_4mimBr AmimCl C_6mimCl C_2mimCl

$C_4mimHSO_4$ $SbmimHSO_4$ C_4mimBF_4 C_4mimPF_6 C_4PyCl

FIGURE 11.6 Ionic liquids used in the production of HMF.

As one of the top bio-based intermediates, HMF is a versatile molecule that can be converted to many valuable chemicals, for example, pharmaceuticals [55], antifungals [56], and polymer precursors [57]. For this reason, heterogeneous catalytic transformation of HMF into chemicals becomes a hot research area recently. This part of work has been reviewed comprehensively by Afonso [58] and Corma [59].

11.3.1.6 Challenges and Outlook Dehydration of carbohydrates leads to the formation of HMF, which is an essential intermediate for the production of chemicals and liquid fuel. In the past decade, some advances have been achieved with heterogeneous catalysts in various solvent systems. Despite the above-discussed progresses, the substrate used in most of their catalytic systems is fructose, the conversion of aldohexoses, for example, glucose, or its polymer cellulose, even the raw lignocelluose material is still a huge challenge. For heterogeneous catalyzed HMF production, both the substrates and HMF contain active hydroxyl and carbonyl groups, which readily react to form various by-products, making the selectivity very low.

From the application point of view, research and development focuses on low toxic catalyst with high activity for the conversion of cheap and renewable biomass are more attractive. Before one technology is moved toward practical application, the problems of product isolation and purification, as well as catalyst activity and recycling should be well settled.

11.3.2 Cellulose Conversion to Polyols with Heterogenous Catalysts

11.3.2.1 Direct Conversion of Cellulose to Hexitols Since cellulose is composed of glucan units, it is possible to couple the hydrolysis of cellulose and the subsequent hydrogenation of glucose in one pot to form hexitols. In this way, the reaction process is simplified; meanwhile, the product yields might be improved due to the better stability of hexitols than glucose under reaction conditions.

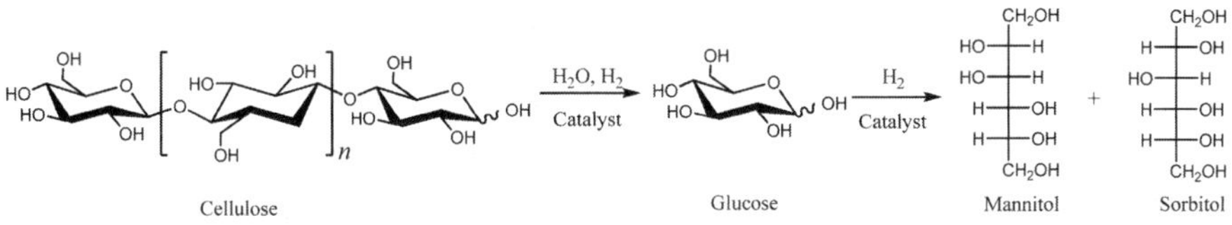

FIGURE 11.7 Catalytic conversion of cellulose into hexitols.

Heterogeneous catalysts bearing both acid sites and hydrogenating sites are suitable candidates to meet the requirements of this reaction. Although this kind of bifunctional catalysts have been applied commercially in petroleum refinery, such as zeolites supported noble-metal catalyst, good hydrothermal stability of the bifunctional catalysts in an aqueous solution is required for the long period of operation for the cellulose conversion.

A highlight of cellulose conversion to hexitols with heterogenous catalysts was first disclosed by Fukuoka et al. [60] They studied a series of hydrogenating metals loaded on different supports for cellulose conversion. The reaction was performed in hydrothermal condition at 190 °C and 6 MPa H_2 for 24 h. A Pt/Al_2O_3 catalyst showed the best catalytic performance, giving sorbitol and mannitol at yields of 25 mol% and 6 mol%, respectively. The solid catalyst was ready to be recycled by filtration after the reaction and was reused for at least three times with stable performance. Among various hydrogenating metal catalysts, Pt and Ru catalysts were more effective than Rh, Ir, Ni, and Pd catalysts. The acidic properties of the supports were important for the cellulose conversion. As shown in Figure 11.7, cellulose first undergoes hydrolysis to produce glucose, followed with hydrogenation to form sorbitol and mannitol. The metals on support played dual roles: providing active sites for the hydrogenation and generating H^+ sites by dissociation of H_2 and spillover of H atoms to the support. A similar work by Kou et al. demonstrated that the (β-1,4) glycosidic bond in cellobiose could be selectively broken to form hexitols in one step over soluble Ru nano clusters in an aqueous solution [61].

In this reaction, the hydrolysis of cellulose is the rate-determining step for cellulose conversion to hexitols [62]. Thus, the efforts to improve the efficiency of cellulose hydrolysis over acid sites will eventually benefit all reaction pathways to form hexitols. In order to enhance the yield of hexitols from cellulose, the following three factors can be considered:

(1) The reaction temperature. At high temperatures, protonic acid will be formed *in situ* from hot water and favor the cellulose hydrolysis. Liu [63] et al. performed the cellulose conversion at 245 °C under catalysis of Ru/AC (AC = active carbon), and got remarkably enhanced efficiency compared to that at 190 °C by Fukuoka et al. The hexitols yield reached 39.3 wt% with 85.5% conversion of microcrystalline cellulose after 30 min reaction.

(2) The support acidity. Carbon materials are preferred supports for cellulose conversion in hot water due to their good hydrothermal stability and tunable surface properties. The acidified carbon supports of the catalysts can provide acid sites for the cellulose hydrolysis. When carbon nanotubes (CNT) were pretreated with concentrated nitric acid, acid groups were extensively generated on the surface of CNT [62b, 64]. After deposition of Ru nanoparticles, the resultant Ru/CNT catalyst showed notably higher activity than that of Ru supported on CNT without nitric acid pretreatment. According to Hara et al., carbon materials with a large number of hydroxyl groups and phenolic groups exhibited good activity for cellulose hydrolysis [8h]. Therefore, it is reasonable to speculate that the acidified materials will impose positive effect on the cellulose conversion to hexitols when they are used as the catalyst supports.

(3) Additional acid sites on catalysts. Besides the acid sites of catalyst supports, introducing additional acid sites onto the catalysts is a good option. Zhang's group studied Ni_2P/AC catalysts for cellulose conversion [65]. Nickel phosphide catalysts possess bifunctional sites, one is the hydrogenation sites given by metallic nickel, and the other is the acid sites provided by phosphates residue on the support. With this bifunctional Ni_2P/AC catalyst, cellulose was completely converted with 48.4 mol% yield of sorbitol under optimized conditions, that is, 225 °C, 6 MPa H_2, and 90 min reaction time. Furthermore, the ratio of acid sites to hydrogenation sites can be tuned by the ratio of nickel to phosphorus, and the best performance was obtained at the acid sites/hydrogenation sites of 1/6. On the other hand, although the nickel phosphide catalyst presents even better activity than the noble-metal catalysts, the leaching of P is serious under hydrothermal reaction conditions, which significantly limits its recyclability. Therefore, using the catalyst in nonaqueous solution might be a future direction towards the application of nickel phosphide in cellulose conversion.

In addition, since the cellulose is a macromolecule substrate, the compatibility between bulky cellulose and solid catalyst should be considered in the cellulose conversion. Employing CNT as a support with opened structure can increase the accessibility of the reactant to the active sites [66]. The Ni particles fixed on the tip of carbon nanofibers (Ni/CNF) exhibited a good activity for the conversion of cellulose into hexitols. The yields of sorbitol and mannitol were 30% and 5%, respectively, with 87% cellulose conversion after reaction at 210 °C for 24 h, which is the best result among the nickel catalysts reported to date.

To further improve the efficiency of cellulose conversion, some researchers combined homogenous catalysts, such as heteropoly acids, with solid hydrogenation catalysts [67]. With such mixed catalysts, the cellulose conversion reaction could be conducted at relatively milder conditions and shorter time. However, the recyclability of the heteropoly acids is still a challenging task.

Besides water, ILs are also a class of green solvents. In the past few years, ILs have attracted much attention for cellulose conversion due to their good dissolving ability for cellulose, acceptable thermal stability, and low vapor pressure. There is also the possibility that the dissolved cellulose in IL can be transformed into valuable chemicals under milder conditions. Initially, it was reported that cellulose could not be readily degraded in ionic liquid C_4mimCl over Rh/AC and Pt/AC catalysts although it was dissolved in C_4mimCl [68]. However, the same group performed the reaction with a combination of a heterogeneous catalyst Rh/AC and a homogeneous catalyst $HRuCl(CO)(PPh_3)_3$, giving a yield of 74% sorbitol after reaction at 150 °C for 48 h. The synergistic effect of heterogeneous catalysts and homogeneous catalysts was attributed to the remarkable improvement of H_2 solubility in IL in the presence of the homogeneous catalyst, which provided an active hydrogen source for the hydrogenation reaction over the heterogeneous catalysts.

Zhu [69] et al., reported that ILs not only dissolve cellulose, but also stabilize nanoparticles in small sizes and high surface areas, which contribute to the high performance of catalysts. By using a boronic acid binding agent and Ru nanoparticles, 98% cellulose was converted with 89% yield of sorbitol and 3% yield of mannitol after 5 h reaction at 80 °C in C_4mimCl.

11.3.2.2 Direct Conversion of Cellulose to Ethylene Glycol

Tungsten Carbide Catalysts In the past several decades, transition-metal carbides, nitrides, and phosphides have been investigated extensively in reactions involving hydrogen transfer which typically occur on noble-metal catalysts [70], such as hydrazine decomposition [71], ammonia synthesis or decomposition [72], hydrodesulferization [73], hydrodenitrogenatin [74], and electrochemistry [75].

Noble-metal catalysts show promising performance in cellulose conversion to hexitols under hydrothermal and hydrogen atmosphere conditions. However, the prohibitively high cost of precious metals would be an obstacle in the large-scale commercial applications. It is therefore highly desirable to develop less expensive but more efficient catalysts for the cellulose conversion.

Zhang and coworkers for the first time employed tungsten carbide in the catalytic conversion of cellulose under hydrothermal conditions [28]. Surprisingly, the tungsten carbide catalysts exhibited very good activity for cellulose conversion, even better than that of noble-metal catalysts. After a typical set of reaction conditions (30 min, 245 °C, and 6 MPa H_2), the cellulose conversion reached 98% over a W_2C/AC catalyst, in contrast to cellulose conversions less than 80% over noble metal catalysts such as Ru/AC [63] and Pt/Al_2O_3 [60]. Of particular interest was that the product selectivity was different from that obtained on noble metals. With W_2C/AC catalyst, the main product was ethylene glycol (EG, 27% yield) rather than the hexitols (2% yield). The high selectivity of the formation of ethylene glycol, a C_2 small molecule, demonstrates that the tungsten carbide catalysts possess unique properties different from the noble-metal catalysts.

As reported in the literature [76], a second metal modification to the tungsten carbide can result in remarkable synergy effects in many reactions. Also, nickel addition promotes the tungsten carbide formation at lower temperatures [77]. To further improve the EG yield in cellulose conversion, Ni–W_2C/AC catalysts were prepared and tested in Zhang's work [28]. On a 2% Ni–30% W_2C/AC catalyst, EG yield was increased to as high as 61%, and the cellulose conversion reached 100%. The Ni–W_2C/AC catalysts showed moderate reusability, producing EG at a yield of higher than 40% in the third run. Liquid product analyses showed that leaching of either nickel or tungsten was negligible. The reduction of the EG yield was attributed to the partial oxidation of the W_2C phase by the reaction medium during recycling. Zhang's work is the first report that EG can be directly produced from cellulose in such a high yield, opening a new route for biomass conversion as well as reducing dependence on petroleum resources.

Catalyst supports usually impose significant impacts on both the activity and selectivity of catalysts due to the different dispersion and accessibility of the active sites on the supports. The conventional active carbon (AC) support often leads to low dispersion and poor accessibility of the active sites owing to the microporous structure, in spite of a large surface area. In contrast, mesoporous carbon supports present remarkable advantages over microporous examples in respect of accessibility of the active sites and the diffusion of molecules. Accordingly, in a subsequent work on cellulose conversion, Zhang's group employed a mesoporous carbon (MC) with three-dimensional (3D) interconnected mesopores as a support for the tungsten-carbide catalysts [78]. The MC was prepared by a nanocasting approach using commercial silica as the hard template and sucrose as a carbon source. Compared with CMK-3 with 1D-ordered mesochannels or AC with microporous structure, the MC-supported tungsten carbide catalyst (WCx/MC) exhibited a much higher selectivity towards EG; the EG yield reported up to 72.9%, which is even higher than that on the Ni–W_2C/AC (61.0% EG yield) reported previously [28]. The excellent performance of the WCx/MC was ascribed to a high dispersion of the tungsten carbide, which was identified with TEM and CO chemisorption experiments. The addition of small amount of nickel to the W_2C/MC further increased the EG yield to 75%. Moreover, the WCx/MC showed a much better resistance to deactivation. After four repeated runs, the EG yield over the WCx/MC only decreased by ca. 15%, in contrast to the 40% EG yield loss in the fourth run over the conventional WCx/AC catalysts. After the fourth run, the partially deactivated WCx/MC catalyst was regenerated in H_2 at 550 °C for 2 h, the EG yield (65.9%) almost restored to the original level, again suggesting the deactivation was mainly caused by the oxidation of tungsten carbide.

Tungsten-Based Bimetallic Catalysts Following the discovery of cellulose conversion to ethylene glycol (EG) over tungsten carbide catalysts, Zhang's group further developed a series of tungsten-based bimetallic catalysts for this reaction, including the supported Ni–W, Pd–W, Pt–W, Ru–W, and Ir–W catalysts [79]. Over the bimetallic catalysts of M(8,9,10)-W, high EG yields were obtained in the range of 50–76% with 100% conversion of cellulose. Particularly on a 5Ni-15W/SBA-15 catalyst, the EG yield was as high as 76.1%. In contrast, the monometallic W/AC catalyst, or the monometallic M(8,9,10) catalysts including Ni/AC, Pd/AC, Pt/AC, Ru/AC, and Ir/AC, merely gave EG yields less than 10%.

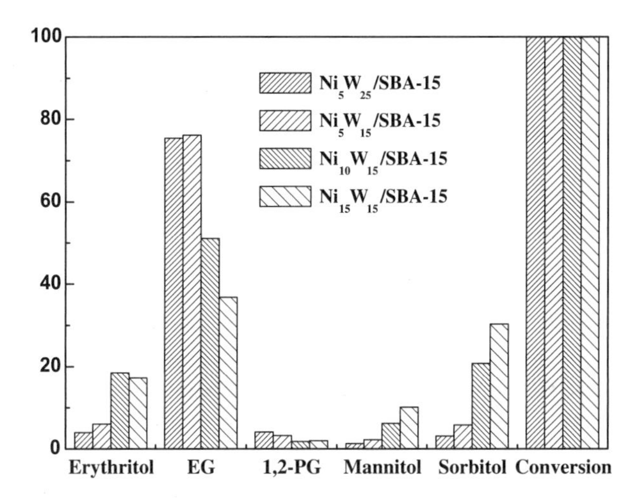

FIGURE 11.8 Influence of Ni–W metal ratio on polyols production in cellulose conversion.

Evidently, over M(8,9,10)-W bimetallic catalysts, synergistic effect was present between tungsten and the hydrogenating metals. A control experiment showed that a notable amount of EG (43.0% yield) was produced upon a Ni/AC and a W/AC catalyst being mechanically mixed in the reaction, in contrast to the EG yields less than 10% over the individual catalysts. Accordingly, the synergistic effect between Ni and W does not necessarily require intimate contact between the two metallic components.

Tuning the weight ratios of Ni to W on the Ni-W/SBA-15 catalysts led to the change of polyol products distribution. As shown in Figure 11.8, when increasing Ni/W ratios, the yields of hexitols and C_4 sugar alcohol were significantly increased at the expense of the EG yield. The CO chemisorption measurements showed that more active hydrogenation sites are present on the Ni-W/SBA-15 catalysts at higher Ni/W ratios. Thus, it is speculated that there are several reaction routes taking place and competing with the EG-formation reaction. As shown in Figure 11.9, at least three types of reactions are involved, that is, hydrolysis (R1), hydrogenation (R2), and cracking (R3).

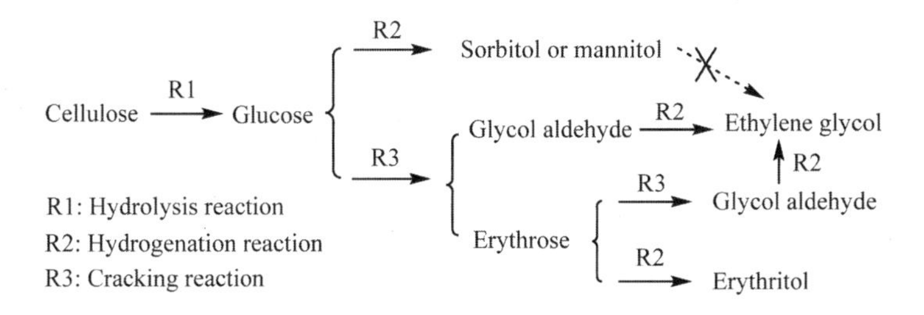

FIGURE 11.9 Possible mechanism for catalytic conversion of cellulose into ethylene glycol over tungsten-based catalysts.

The cracking reaction (such as retro-aldol condensation reaction [80]) happens most likely over the W sites, as suggested by the complete conversion of cellulose and the formation of a large amount of unsaturated compounds over W/AC. Meanwhile, the hydrogenation reaction takes place on the Ni and noble-metal active sites. In the case of bimetallic catalysts, glucose or sugar oligomers derived from cellulose hydrolysis undergo degradation (R2) over the W active sites to form C_4 species (erythrose) and C_2 species (glycol aldehyde), competing or followed with subsequent hydrogenation (R3) over the metallic M(8,9,10) active sites to form EG and other polyols. An excess of hydrogenating sites will make the hydrogenation reaction overwhelm the cracking reactions, resulting in an increase in the yields of hexitols and erythritol but at the expense of EG formation, and vice versa. Thus, there is an optimal regime for the Ni/W ratio, at which hydrogenation and C–C cracking reactions reach a good balance to produce a high EG yield. In addition, the hexitols (mannitol and sorbitol) were found to be stable and did not form EG when they were used as feedstock under current reaction conditions.

For the tungsten carbide catalysts mentioned above [28,78], they comprise multifunctional catalytic sites, including hydrogenating and cracking sites, for the cellulose conversion. As a result, they exhibit unique selectivity for EG production. The ratio of the two types of active sites was optimized in the case of addition of a small amount of nickel to W_2C catalysts, thus leading to a large enhancement of the EG yield over the Ni–W_2C catalysts.

Based on the proposed mechanism, one can effectively tune the competition between hydrogenation and C—C cracking capabilities, and accordingly tune the selectivity towards ethylene glycol or other polyols.

11.3.2.3 Summary and Perspectives Research on biomass conversion to polyols with heterogenous catalysts has made great progress in the last five years. Several groups have developed catalytic processes with green solvents and coupled one-pot process to obtain hexitols and ethylene glycols with good yields. These investigations give valuable reference to the exploration of sustainable ways for chemicals and energy production.

On the other hand, many problems still need to be resolved during the development of green chemistry. The efficiency of cellulose conversion still has great room for improvement. The reaction conditions are generally harsh and therefore need moderation by developing more active catalysts. When the reactions are conducted under hydrothermal conditions, the stability of catalysts is of great concern and must endure long periods of operation and repetitious use. Catalyst materials undergo decomposition, oxidation, or dissolving in hot water, leading to the loss of active sites which is still a major issue. Development of novel catalytic materials or application of nonaqueous solvents might be suitable ways to resolve the problems.

Furthermore, cellulose degradation to polyols, particularly to small polyols such as ethylene glycol and propylene glycol, is a very complex process. Many fundamental reactions are involved, including hydrolysis, hydrogenation, degradation, and so on. These reactions are competing or cooperating to produce various products.

However, the knowledge on reaction mechanism and reaction kinetics is very limited. The progress on this aspect will eventually provide guidance for designing better catalysts and optimizing reaction processes, producing a high yield of the target products.

Finally, cellulose usually coexists with hemicellulose and lignin components in the raw biomass materials. Separating cellulose from biomass to a high purity is a high-cost process. In view of practical application, the influence of multicomponents on the cellulose conversion should be considered and investigated systematically in the future.

11.3.3 Biodiesel Production Over Heterogeneous Catalysts

11.3.3.1 Introduction Biodiesel is a nonpetroleum-based liquid fuel that consists of alkyl esters derived from either the transesterification of triglycerides (TGs) or the esterification of free fatty acids (FFAs) with short-chain alcohols [81]. Compared to fossil fuels, biodiesel is renewable, biodegradable, nontoxic, and has lower emission of CO, SO_x, and unburned hydrocarbons [82], and thus has attracted continual attention in the past decade as a promising energy choice [83].

The common way to produce biodiesel is by transesterification [84], which refers to a catalyzed chemical reaction involving an alcohol, and a TG source such as vegetable oils, to yield fatty acid alkyl esters (i.e., biodiesel) and glycerol. Currently, most of the commercial processes employ a homogeneous base [85], for example, alkaline metal alkoxides, hydroxides, or sodium or potassium carbonates as catalysts due to their high catalytic activity, wide availability, and low cost. In this method, however, the use of a soluble base suffers from several limitations [86] in cost-effective production of biodiesel. First, the amount of FFAs in TG source must not exceed 0.5 wt% to avoid the formation of soap between FFAs and the base catalyst. Second, both the alcohol and the catalyst must be essentially anhydrous; otherwise, the presence of water in the feedstock promotes hydrolysis of the alkyl esters to FFAs and, consequently, soap formation [87].

To overcome the problems associated with the use of liquid bases, great efforts have been made towards the development of heterogeneous catalytic process. Heterogeneous catalysts are less corrosive, easier to handle and separate, reusable, and do not generate neutralization wastes. The use of heterogeneous catalysts in transesterification reactions prevents the undesirable saponification, allows process simplification, and thus lowers the processing costs.

11.3.3.2 Overview of the Reaction Mechanisms For a complete transesterification reaction of a TG molecule, three consecutive reactions are required. One mole of TG molecule consumes three moles of alcohol and produces three moles of alkyl esters (biodiesel) with one mole of side product glycerol. All types of glyceride species, including triglyceride, diglyceride, and monoglyceride, participate in the reaction (Fig. 11.10).

Solid acids and bases are the most studied heterogeneous catalysts. According to the nature of active sites, they are also classified as Brønsted or Lewis catalysts. In

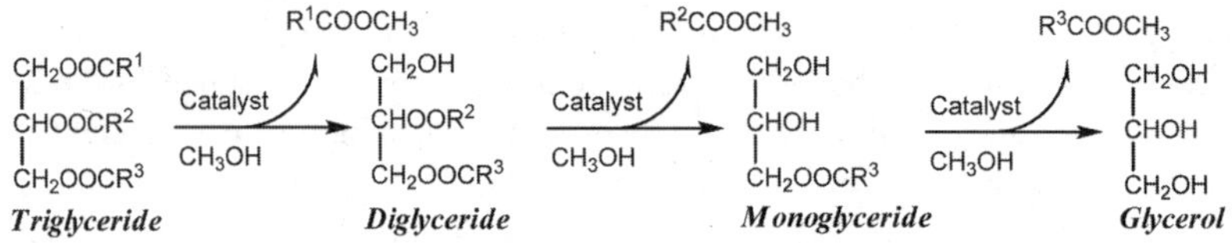

$$\begin{array}{c} CH_2OOCR^1 \\ | \\ CHOOCR^2 \\ | \\ CH_2OOCR^3 \end{array} \xrightarrow[CH_3OH]{Catalyst} R^1COOCH_3 + \begin{array}{c} CH_2OH \\ | \\ CHOOR^2 \\ | \\ CH_2OOCR^3 \end{array} \xrightarrow[CH_3OH]{Catalyst} R^2COOCH_3 + \begin{array}{c} CH_2OH \\ | \\ CHOH \\ | \\ CH_2OOCR^3 \end{array} \xrightarrow[CH_3OH]{Catalyst} R^3COOCH_3 + \begin{array}{c} CH_2OH \\ | \\ CHOH \\ | \\ CH_2OH \end{array}$$

Triglyceride *Diglyceride* *Monoglyceride* *Glycerol*

FIGURE 11.10 Transesterification reactions of glycerides with methanol.

many cases, both types of active sites are present in one heterogeneous catalyst and it is not easy to evaluate the contribution of the two types of sites in the reaction.

Studies [86,88] have shown that for base catalysis, no matter what base (heterogeneous or homogeneous, Brønsted or Lewis) used, the active species are alkoxide groups (RO^-) which follow similar mechanism [86] for alkyl ester production although the formation of RO^- may vary [86,89].

For heterogeneous acid catalysis, the key step in the transesterification is the protonation (Brønsted acid)/coordination (Lewis acid) of the ester group in TG molecule, this in turn increases the electrophilicity of the group, making it more susceptible to nucleophilic attack by an alcohol [90]. It is different from the mechanism of base catalysis, that is, the formation of a more electrophilic species (acid catalysis) versus that of a stronger nucleophile (base catalysis), which is ultimately responsible for the observed differences in activity.

The esterification of carboxylic acids is another important reaction catalyzed by solid acid for the production of biodiesel. The mechanism [91] is similar to the transesterification pathway. Generally, Brønsted acid sites are proposed to be active mainly in esterification reactions while Lewis acid sites are more active in transesterification reactions [92].

11.3.3.3 Solid Base Catalysis for Biodiesel Production Similar to their homogeneous counterparts, solid base catalysts are more active than solid acid catalysts. The development of solid base catalysts for biodiesel production has been widely described in the literature [88b, 88c, 93] including alkaline oxides, rare earth metal oxides, nanometer solid base catalysts, alkali metal exchanged zeolites, anionic resins, and clay minerals such as hydrotalcites. Some of these catalysts have shown a good catalytic performance even under reaction conditions similar to those used for the homogeneous catalysts [88b, 93b, 93d].

Supported Alkaline Metal and Alkaline Earth Metal Catalysts Alkali metals are the most common source of super basicity, while alkaline earth metals show moderate basicity but less solubility in methanol in comparison to supported alkaline. The loading of alkaline and alkaline earth metals on supports, in their various compounds such as hydroxides or carbonates, leads to solid base catalysts for biodiesel production. The supports for these catalysts can be Al_2O_3, ZnO, SiO_2, and CeO_2 [92, 93b, 94].

An example of a commercialized super base catalyst is $Na/NaOH/Al_2O_3$ [95], it shows almost the same activity as homogeneous NaOH catalyst under optimized reaction conditions for the transesterification of soybean oil with methanol. The basicity is believed to be associated with the Lewis base concept according to the O 1s XPS results. Furthermore, many papers report alumina and other supports loaded with alkaline metal, for example, K_2CO_3, KF, $LiNO_3$, NaOH, and Na_2SiO_3 as catalysts for the production of biodiesel [88b, 93b, 94]. The transformation of TGs catalyzed with calcined KF/γ-Al_2O_3 [93b], for example, afforded biodiesel yield over 99% under the optimized conditions. XRD spectra showed that K_3AlF_6 which was formed at 550 °C calcination was active for the transesterification reaction. Xiu [88b] et al. examined calcined sodium silicate as a solid base catalyst for the transesterification of soybean oil, biodiesel yield of almost 100% was obtained under the optimized conditions. This catalyst displays excellent tolerant ability to water and FFAs, and could be reused for at least five cycles without loss of activity. The tolerance of calcined sodium silicate to water and FFAs is related to its special crystal and porous structure.

For alkaline earth metal catalysts, Ca- and Mg-derived bases are selected more often for biodiesel production as they are inexpensive, less toxic, and hold moderate basicity. The origin of basic sites in alkali earth oxides has been debated, and it is generally believed that they are generated by the presence of M^{2+}–O_2^- ion pairs in different coordination environments [96]. Commonly, there is a clear correlation between base strength and activity [97]. Lithium incorporation was proposed to increase the base strength of alkaline earth metal oxide, and thus enhanced catalytic activity. Other types of solid oxide catalysts, such as $(Al_2O_3)_4(SnO)$ [98], $(Al_2O_3)_4(ZnO)$ [99], have also been reported for the transesterification reactions, producing biodiesel with yield close to 100% and purity higher than 99%, which has been commercialized in 2006.

Basic Zeolites Among the numerous heterogeneous catalysts for transesterification, zeolites modified by alkali cation ion exchange have emerged as interesting solid bases [100]. The base strength of the alkali ion-exchanged zeolites increases with increasing electropositive nature of the exchanged cation. Ion exchange can also affect the water tolerant behavior of the basic zeolite system.

The zeolites NaX and titanosilicate structure-10 (ETS-10), a microporous inorganic lithium containing zeolite, have been shown to be excellent solid base catalysts for transesterification. In particular, the ETS-10 catalyst gave better conversions of triglycerides than the zeolite-X type catalysts. The higher performance of ETS-10 was attributed to its unique large pore structure, novel chemical composition, and strong basic character [101]. The basicity of zeolites, including ETS-10, NaX, and several Y-type zeolites, could be further enhanced by ion exchange with higher electropositive metals such as K and Cs [100a, 100b].

Hydrotalcites Hydrotalcites (HT) are a class of anionic and basic clays known as layered double hydroxides with brucite-like ($Mg(OH)_2$) hydroxide layers containing octahedrally coordinated M^{2+} and M^{3+} cations. The structures of pure hydrotalcites were reported to have Al content (x) in the range of $0.25 < x < 0.44$ [102]. Variations in the Al content can modify the basic property and the pore structure of the material. Al-rich hydrotalcites tend to exhibit higher basicity, and have larger pores than Al_2O_3 or MgO, which are responsible for their high activities in transesterification reaction [103].

Zeng et al. reported the transesterification of rapeseed oil with methanol using Mg–Al hydrotalcite (molar ratio of Mg to Al was 3.0) to afford methanol ester yield of about 90.5% [104]. In another work, Corma [105] et al. reported the calcined Li–Al hydrotalcites are more active in glycerolysis of fatty acid methyl esters than the Mg–Al material (or MgO) due to their higher Lewis basicity. With this catalyst, near-quantitative conversion of the soybean oil was achieved at low catalyst loading (2–3 wt %) in a short time ($\sim$2 h) [106].

Other Heterogeneous Basic Catalysts The solid basic resin is an alternative catalyst to be used to produce fatty acid ethyl esters. Shibasaki-Kitakawa [107] et al. investigated the transesterification of triolein with ethanol using various commercial anion-exchange resins, such as Diaion PA308, PA306s, and HPA25, and they achieved the best yield over 80%. The transesterification reaction of acid oil using solid resin Dowex monosphere 550 A as the catalyst yielded a final conversion over 90% albeit the reaction time was much longer than the conventional process employing a homogeneous basic catalyst [93a].

Catalysts derived from renewable biont materials, such as shrimp shell, crab shell [108], oyster shell [109] and egg shell [110], which are generally considered as waste, have been employed for conversion of oils to FAME. The major components of these biont shells are chitin, protein, and $CaCO_3$. By supporting KF or simple calcination of the biont shells, these materials proved active for biodiesel synthesis [108–110]. However, further work is required to limit the amount of Ca leaching in order to improve the catalyst life and tolerance to water and FFA in oil feedstock.

11.3.3.4 Solid Acid Catalysis for Biodiesel Production Even though heterogeneous base catalysts are highly active for biodiesel synthesis and the resistance to FFAS is increased, they still cannot tolerate acidic oils with FFA content >3.5%. It is therefore a big obstacle for base catalysts to be applied in conversion of waste oil with high FFAs content. In this respect, acid catalysts prove to be advantageous. Since acid catalysts show a much higher tolerance to FFAs and can simultaneously catalyze both esterification and transesterification, it presents great potential in processing low-cost, low-quality feedstocks, and thereby lowering overall production costs.

Zeolite Zeolites can be synthesized with extensive variation of textural properties, such as crystal structures, pore sizes, framework Si/Al ratios, and proton-exchange levels. According to the substrate's size and polarity, zeolites with proper characteristics could be designed and synthesized to make them active for both the transesterification and esterification reactions [111]. The acidity can also be adjusted to meet the requirement of the reaction [112]. Too low acidity may cause the reaction to proceed at a low rate; too high acidity may cause deactivation by coking or producing undesirable by-products.

Multifunctionalization of mesoporous silica with both organosulfonic acid and hydrophobic organic groups was shown to be effective in esterifying the free fatty acid [113]. By incorporating certain species such as heteropoly acids into the pore structure of zeolites, they can exhibit hydrophobic characteristics without compromising their functionalized acidic sites, which is crucial for increasing the performance of the catalysts [114]. By introducing aluminum, zirconium, titanium, or tin into the silica matrix can also significantly improve the acidity of the mesoporous silica. However, metal-doped materials behave more like weak acids and can only be used for reactions that do not require strong acidity [114]. In general, the catalytic activity of zeolites increases with increasing of acid site strength as well as surface hydrophobicity. Other properties, such as pore size, dimensionality of the channel system (related to the diffusion of reagents and products), and aluminum content of the zeolite framework can also strongly affect the catalytic activity.

Heteropoly Acids In the search for water-tolerant acid catalysts, heteropoly acids (HPAs) appear to be an attractive candidate for the rapid preparation of biodiesel from vegetable oils [115]. Most of these systems have acidity in the range of super acids with the possibility of tailoring the porous architecture as well as solubility in water.

As an example, pure HPAs such as $H_3PW_{12}O_{40}$, $H_4SiW_{12}O_{40}$, $H_3PMo_{12}O_{40}$, and $H_4SiMo_{12}O_{40}$, exhibit higher activities than liquid H_2SO_4 for the transesterification of TGs [116], but they are soluble in alcohol and thus are difficult to separate from water or alcohol. Interestingly, by doping with large monovalent ions, such as Cs^+, NH_4^+, or Ag^+, the salts of HPA became insoluble in water and organic solvents. Meanwhile, the activities remained almost the same as the parent acids. The best yield higher than 96% was achieved for the conversion of yellow horn oil by using a low amount (1% w/w of oil) of $Cs_{2.5}H_{0.5}PW_{12}O_{40}$ at a low molar ratio of methanol/oil (12 : 1) in a short reaction time (10 min) at 60 °C. This procedure is environmentally benign and economical since the catalysts are tolerant to the free fatty acid and moisture, and are easy to recycle. The high catalytic performance of HPAs as well as high resistance to leaching has also been demonstrated by the recent work of Chai [117] and Alsalme [118].

Carbon-Based Solid Acid Catalysts Carbon-based solid acids can be readily prepared by incomplete carbonization of sulfopolycyclic aromatic compounds in concentrated H_2SO_4 or sulfonation of incompletely carbonized natural organic matter [8h, 119], such as sugar and cellulosic materials. These materials usually contain SO_3H, COOH, and phenolic OH groups and have received a great deal of research attention in biomass conversion as stable and efficient catalysts [120].

It was reported that a carbon-based solid acid catalyst, which was prepared by the sulfonation of carbonized vegetable oil asphalt and petroleum asphalt [121], was employed to simultaneously catalyze esterification and transesterification of a waste vegetable oil to synthesis biodiesel. The maximum conversion of TG and FFAs reached 80.5 wt% and 94.8 wt%, respectively, after 4.5 h at 220 °C, when using a 16.8 molar ratio of methanol to oil and 0.2 wt% of catalyst to oil. The high catalytic activity and stability of this catalyst was related to its high acid site density (—OH, Brønsted acid sites), hydrophobicity that prevented the hydration of —OH species, hydrophilic functional groups (—SO_3H) that gave improved accessibility of methanol to TG and FFAs, and large pores that provided more acid sites for the reactants. It is also suggested that electron-withdrawing COOH groups increase the electron density between the carbon and sulfur atoms, which may result in the stability of the material under such harsh conditions [122].

Other Solid Acid Catalysts Acidic resins [90b, 91] such as Nafion® SAC-13 (silica-supported perfluorinated-based ion-exchange resin) and Nafion® NR50 (unsupported Nafion) are another kind of catalysts for conversion of low-quality feedstocks with high FFAs. For Amberlyst-15, mild reaction conditions are necessary to avoid catalyst degradation [123]. Sulfated zirconia (SO_4/ZrO_2) [124], various tin complexes [125], and WO_3/ZrO_2 catalysts [126] have also been reported as promising catalysts for esterification due to their high acid strength. However, the conventional SO_4/ZrO_2 suffers from sulfate leaching [124b]. SO_4/ZrO_2 prepared by using chlorosulfonic acid as precursor could avoid this problem [124a].

11.3.3.5 Summary and Outlook As described in the present review, the use of heterogeneous catalysts for biodiesel production is an emerging field of research which has been gradually growing during the last two decades. Heterogeneous catalysts would result in simpler, cheaper separation processes, a reduced water effluent load, as well as capital and energy costs. There would be fewer inputs and less waste, as no soap by-product would be formed. Furthermore, the catalyst would not have to be continuously added and would be easier to reuse.

With relation to heterogeneous base catalysts, they are commonly more effective than acid catalysts, and some of these catalysts even show a good catalytic performance similar to those of the homogeneous base catalysts. Despite these advantages, the heterogeneous base catalysts cannot tolerate acidic oils with FFA content higher than 3.5%. Therefore, strict feedstock limitations still persist as the main issue with this process. In future, for the commercial introduction of these catalysts important progress on the continuous processes using packed bed flow reactors, as well as enhancing their resistance to high FFA concentration is demanded.

On the other hand, heterogeneous acid catalysts can tolerate high-FFA feedstocks, simultaneously carrying out esterification and transesterification. However, only few studies focus on solid acid catalysts for biodiesel synthesis due to pessimistic expectations about catalytic activity and harsh reaction conditions. In general, the use of solid acid catalysts to produce biodiesel requires a better understanding of the factors that govern their reactivity. It seems that an ideal solid acid catalyst should contain an interconnected system of large pores, a high concentration of strong acid sites that are stable at high temperatures, and a hydrophobic surface.

11.4 OPPORTUNITIES AND CHALLENGES

The use of carbon-neutral and renewable biomass, especially the second generation of biomass (lignocellulosic materials) for production of fuels and chemicals represents an important direction towards reducing the dependence on fossil resource and alleviating the climate changes caused by CO_2 emission. Chemical transformation of biomass by using a heterogeneous catalyst is a desirable green approach. Although great progress has been made in recent years, the heterogeneously catalyzed biomass conversion still faces significant challenges in the following aspects:

(i) *Activity*: The activity of a heterogeneous catalyst is yet to be enhanced greatly. To date, most of the heterogeneous catalysts are much less active than the homogeneous counterparts in biomass conversion (e.g., solid acids vs. liquid acids in cellulose hydrolysis and solid base vs. liquid base in biodiesel synthesis), mainly due to the poor accessibility of the active sites on the heterogeneous catalysts. Therefore, the design of a highly active heterogeneous catalyst should be guided towards maximization of the accessibility of the active sites, for example, by using nanometer sized particles, by employing mesoporous/macroporous materials as the supports, or by using a subcritical/supercritical fluid as the solvent. The great advances in material synthesis are believed to contribute to the development of novel active catalysts for biomass conversion.

(ii) *Selectivity*: The selectivity of a heterogeneous catalyst may not be inferior to its homogenous counterpart. However, due to the presence of a wide range of functional groups in biomass and the low thermal stability of these groups, the conversion of biomass to a target product is most often associated with a variety of side reactions, especially under harsh reaction conditions. This would result in a low to medium selectivity to the desired product, which leads to great difficulty and additional cost in the separation and purification. Moreover, the production of some unwanted chemicals from biomass may poison the catalyst, which in turn decreases the activity and life of the catalyst. Therefore, selectivity control is a central task for biomass conversion. In order to achieve a high selectivity, a one-pot process may be a good choice because this combines a series of consecutive reactions in one pot thus minimizing the unwanted decomposition of unstable intermediate (e.g., for one-pot conversion of cellulose to hexitols). In addition to the process optimization, the design of a highly selective heterogeneous catalyst is also important. To this end, the influence of catalyst properties, such as the particle size of active components, the acidity/basicity and the textural properties of the support, and the metal–support interactions on the selectivity to the formation of the target product needs to be investigated in detail. More importantly, the reaction mechanism understanding and the kinetic studies, for example, the adsorption of reactants on the catalyst surface, the desorption of products from the catalyst surface, the reactivities of all the involved compounds under the reaction conditions, would contribute much to the design of a highly selective catalyst. Unfortunately, the progress in this fundamental research area is very limited.

(iii) *Stability*: The stability of a heterogeneous catalyst is particularly important for biomass conversion. There are many factors affecting the stabilities of heterogeneous catalysts, such as the sintering of small metal particles, the oxidation of active metallic sites, and the leaching of active species in the solution. Among them, the most significant factor responsible for the deactivation of the catalyst is the leaching out of active species. This is because most of the biomass conversion reactions proceed in aqueous phase, and the leaching of the active components will be more significant in comparison with those in gas phase. In addition, the biomass as well as its derived compounds usually contains plenty of unsaturated groups which will interact strongly with the metal catalysts to form complex, again leading to leaching of active metals. The leaching of catalytically active species into the solution will inevitably cause the decrease or even total deactivation of a heterogeneous catalyst. If a heterogeneous catalyst can not be recycled due to the leaching of active components, it may not be a genuine heterogeneous catalyst; instead, the species dissolved in the liquid phase might function as the catalyst. In this case, one must carefully differentiate heterogeneous from homogeneous catalysis by some control experiments, for example, via hot filtration. To date, although great efforts have been devoted to improve the stability of heterogeneous catalysts, for example, by pretreatment under harsh conditions, by doping a more antileaching component, etc., the stability of the catalyst under reaction conditions is still a big challenge for its practical application in biomass conversion.

11.5 OUTLOOK

Presently, the biomass conversion is becoming a hot topic in the field of catalysis. As the first generation of biomass, the production of bio-ethanol from sugar or starch and biodiesel synthesis by transesterification of vegetable oils has been commercialized. In contrast, the catalytic conversion of second generation of biomass, that is, lignocellulose, is still in its infancy. In recent years, there have been intensive studies on the catalytic conversion of lignocellulose, among them have appeared some encouraging achievements. For example, the important platform chemical HMF has been produced directly from cellulose in IL at a yield as high as 60% by using $CrCl_3$ and microwave irradiation [29], and the important petroleum-derived chemical ethylene glycol has been produced directly from cellulose at a yield up to 74.4% by using a Ni–W_2C/MC catalyst [78]. These achievements have demonstrated the great potential of catalysis in the conversion of lignocellulose to chemicals with a high selectivity. However, as we have addressed above, the heterogeneously catalyzed conversion of biomass still faces significant challenges; the activity, selectivity, in particular stability, needs to be improved further. Based on these challenges, great efforts should be devoted to the following aspects in future.

Fundamental research on the catalytic conversion of biomass on solid catalyst, including understanding the mechanism, modeling the reaction kinetics, systematically investigating the factors governing the activity and selectivity, and clarifying the deactivation reason.

Development of catalyst, include porosity control, acid/base property, metal–support interaction, metal dispersion, hydrothermal stability, and cost. It must be stressed that the development of an effective catalyst not only depends on the understanding of the reaction mechanism, but also benefits from the advances in material science.

Process design and large-scale production, including reactor configuration, fluid type, operation parameters. When doing this, one must consider the pretreatment of raw lignocellulose and the impact of lignin and hemicellulose on the conversion efficiency of cellulose and product selectivity.

REFERENCES

1. D. L. Klass, *Biomass for Renewable Energy, Fuels, and Chemicals*, Academic Press, San Diego, **1998**.
2. G. Ertl, H. Knözinger, F. Schüth and J. Weitkamp, *Handbook of Heterogeneous Catalysis*, Wiley-VCH, Weinheim, **2008**.
3. R. Rinaldi and F. Schüth, *Energy Environ. Sci.* **2009**, 2, 610.
4. (a) M. R. Buchmeiser, *Chem. Rev.* **2009**, 109, 303; b) Z. Wang, G. Chen and K. Ding, *Chem. Rev.* **2009**, 109, 322.
5. R. Rinaldi and F. Schüth, *ChemSusChem* **2009**, 2, 1096.
6. M. Moliner, Y. Román-Leshkov and M. E. Davis, *Proc. Natl. Acad. Sci.* **2010**, 107, 6164.
7. J. A. Melero, J. Iglesias and G. Morales, *Green Chem.* **2009**, 11, 1285.
8. For some recent examples, see: (a) P. L. Dhepe, M. Ohashi, S. Inagaki, M. Ichikawa and A. Fukuoka, *Catal. Lett.* **2005**, 102, 163; (b) S. M. Hick, C. Griebel, D. T. Restrepo, J. H. Truitt, E. J. Buker, C. Bylda and R. G. Blair, *Green Chem.* **2010**, 12, 468; (c) D. M. Lai, L. Deng, J. Li, B. Liao, Q. X. Guo and Y. Fu, *ChemSusChem* **2011**, 4, 55; (d) A. Onda, T. Ochi and K. Yanagisawa, *Green Chem.* **2008**, 10, 1033; (e) J. Pang, A. Wang, M. Zheng and T. Zhang, *Chem. Commun.* **2010**, 46, 6935; (f) R. Rinaldi, N. Meine, J. v. Stein, R. Palkovits and F. Schüth, *ChemSusChem* **2010**, 3, 266; (g) R. Rinaldi, R. Palkovits and F. Schüth, *Angew. Chem. Int. Ed.* **2008**, 47, 8047; (h) S. Suganuma, K. Nakajima, M. Kitano, D. Yamaguchi, H. Kato, S. Hayashi and M. Hara, *J. Am. Chem. Soc.* **2008**, 130, 12787; (i) A. Takagaki, C. Tagusagawa and K. Domen, *Chem. Commun.* **2008**, 5363; (j) M.-M. Titirici and M. Antonietti, *Chem. Soc. Rev.* **2010**, 39, 103; (k) S. V. d. Vyver, L. Peng, J. Geboers, H. Schepers, F. d. Clippel, C. J. Gommes, B. Goderis, P. A. Jacobs and B. F. Sels, *Green Chem.* **2010**, 12, 1560.
9. M.von Sivers and G. Zacchi, *Bioresour. Technol.* **1995**, 51, 43.
10. P. Anastas and N. Eghbali, *Chem. Soc. Rev.* **2010**, 39, 301.
11. H. Kobayashi, T. Komanoya, K. Hara and A. Fukuoka, *ChemSusChem*, **2010**, 3, 440.
12. (a) R. J. White, V. Budarin, R. Luque, J. H. Clark and D. J. Macquarrie, *Chem. Soc. Rev.* **2009**, 38, 3401; (b) D. Jagadeesan and M. Eswaramoorthy, *Chem. Asian J.* **2010**, 5, 232.

13. S. Kubo, R. Demir-Cakan, L. Zhao, R. J. White and M.-M. Titirici, *ChemSusChem* **2010**, 3, 188.
14. (a) Z. Helwani, M. R. Othman, N. Aziz, J. Kim and W. J. N. Fernando, *Appl. Catal. A* **2009**, 363, 1; (b) J. N. Chheda and J. A. Dumesic, *Catal. Today* **2007**, 123, 59.
15. (a) G. W. Huber, J. N. Chheda, C. J. Barrett and J. A. Dumesic, *Science* **2005**, 308, 1146; (b) C. J. Barrett, J. N. Chheda, G. W. Huber and J. A. Dumesic, *Appl. Catal. B* **2006**, 66, 111.
16. Z. Zhang, Y. W. Dong and G. W. Wang, *Chem. Lett.* **2003**, 32, 966.
17. J. I. D. Cosimo, V. K. Diez and C. R. Apesteguia, *Appl. Catal. A* **1996**, 137, 149.
18. V. Serra-Holm, T. Salmi, J. Multamaki, J. Reinik, P. Maki-Arvela, R. Sjoholm and L. P. Lindfors, *Appl. Catal. A* **2000**, 198, 207.
19. (a) J. Yu, S. Y. Shiau and A. Ko, *Catal. Lett.* **2001**, 77, 165; (b) J.C.A.A. Roelofs, D. J. Lensveld, A. J. van Dillen and K. P.de Jong, *J. Catal.* **2001**, 203, 184.
20. I. Sádaba, M. Ojeda, R. Mariscal, J. L. G. Fierro and M. L. Granados, *Appl. Catal. B* **2011**, 101, 638.
21. F. Winter, V. Koot, A. J.van Dillen, J. W. Geus and K. P.de Jong, *J. Catal.* **2005**, 236, 91.
22. J. C. A. A. Roelofs, D. J. Lensveld, A. J.van Dillen and K. P.de Jong, *J. Catal.* **2001**, 203, 184.
23. Z. Lü, F. Zhang, X. Lei, L. Yang, S. Xu and X. Duan, *Chem. Eng. Sci.* **2008**, 63, 4055.
24. H. Matsuhashi, *Top. Catal.* **2009**, 52, 828.
25. W. Dedsuksophon, K. Faungnawakij, V. Champreda and N. Laosiripojana, *Bioresour. Technol.* **2011**, 102, 2040.
26. C. Zhao, Y. Kou, A. A. Lemonidou, X. Li and J. A. Lercher, *Chem. Commun.* **2010**, 46, 412.
27. G. W. Huber, J. W. Shabaker and J. A. Dumesic, *Science* **2003**, 300, 2075.
28. N. Ji, T. Zhang, M. Zheng, A. Wang, H. Wang, X. Wang and J. Chen, *Angew. Chem. Int. Ed.* **2008**, 47, 8510.
29. C. Li, Z. Zhang and Z. K. Zhao, *Tetrahedron Lett.* **2009**, 50, 5403.
30. J. B. Binder and R. T. Raines, *J. Am. Chem. Soc.* **2009**, 131, 1979.
31. L. D. Schmidt and P. J. Dauenhauer, *Nature* **2007**, 447, 914.
32. B. M. F. Kuster, *Starch* **1990**, 42, 314.
33. B. H. Zheng, Z. J. Fang, J. Cheng and Y. H. Jiang, *Z. Baturforsch. B* **2010**, 65, 168.
34. (a) X. Qi, M. Watanabe, T. M. Aida and R. L. Smith, Jr., *Ind. Eng. Chem. Res.* **2008**, 47, 9234; (b) F. Du, X. H. Qi, Y. Z. Xu, and Y. Y. Zhuang, *Chem. J. Chin. Univ.-Chin.* **2010**, 31, 548.
35. C. Carlini, M. Giuttari, A. M. Raspolli Galletti, G. Sbrana, T. Armaroli and G. Busca, *Appl. Catal. A* **1999**, 183, 295.
36. C. Carlini, P. Patrono, A. M. R. Galletti, G. Sbrana and V. Zima, *Appl. Catal. A* **2005**, 289, 197.
37. F. Benvenuti, C. Carlini, P. Patrono, A. M. Raspolli Galletti, G. Sbrana, M. A. Massucci and P. Galli, *Appl. Catal. A* **2000**, 193, 147.
38. (a) X. H. Qi, M. Watanabe, T. M. Aida and R. L. Smith, *Catal. Commun.* **2009**, 10, 1771; (b) A. Chareonlimkun, V. Champreda, A. Shotipruk and N. Laosiripojana, *Fuel* **2010**, 89, 2873; (c) M. Watanabe, Y. Aizawa, T. Iida, T. M. Aida, C. Levy, K. Sue and H. Inomata, *Carbohydr. Res.* **2005**, 340, 1925.

39. M. Watanabe, Y. Aizawa, T. Iida, R. Nishimura and H. Inomata, *Appl. Catal. A* **2005**, 295, 150.
40. X. Qi, M. Watanabe, T. M. Aida and R. L. Smith, Jr., *Green Chem.* **2008**, 10, 799.
41. J. N. Chheda, Y. Román-Leshkov and J. A. Dumesic, *Green Chem.* **2007**, 9, 342.
42. C. Moreau, R. Durand, S. Razigade, J. Duhamet, P. Faugeras, P. Rivalier, P. Ros and G. Avignon, *Appl. Catal. A* **1996**, 145, 211.
43. Y. Román-Leshkov, J. N. Chheda and J. A. Dumesic, *Science* **2006**, 312, 1933.
44. A. J. Crisci, M. H. Tucker, J. A. Dumesic and S. L. Scott, *Top. Catal.* **2010**, 53, 1185.
45. Y. Nakamura and S. Morikawa, *Bull. Chem. Soc. Jpn.* **1980**, 53, 3705.
46. R. M. Musau and R. M. Munavu, *Biomass* **1987**, 13, 67.
47. A. S. Amarasekara, L. D. Williams and C. C. Ebede, *Carbohydr. Res.* **2008**, 343, 3021.
48. M. Bicker, J. Hirth and H. Vogel, *Green Chem.* **2003**, 5, 280.
49. K. Shimizu, R. Uozumi and A. Satsuma, *Catal. Commun.* **2009**, 10, 1849.
50. (a) A. Takagaki, M. Ohara, S. Nishimura and K. Ebitani, *Chem. Commun.* **2009**, 41, 6276. (b) M. Ohara, A. Takagaki, S. Nishimura and K. Ebitani, *Appl. Catal. A* **2010**, 383, 149.
51. H. P. Yan, Y. Yang, D. M. Tong, X. Xiang and C. W. Hu, *Catal. Commun.* **2009**, 10, 1558.
52. H. Olivier-Bourbigou, L. Magna and D. Morvan, *Appl. Catal. A* **2010**, 373, 1.
53. (a) C. Li, Z. K. Zhao, A. Wang, M. Zheng and T. Zhang, *Carbohydr. Res.* **2010**, 345, 1845; (b) C. Li, Z. K. Zhao, H. Cai, A. Wang and T. Zhang, *Biomass Bioenerg.* **2011**, 35, 2013.
54. (a) H. B. Zhao, J. E. Holladay, H. Brown and Z. C. Zhang, *Science* **2007**, 316, 1597; (b) J. B. Binder, A. V. Cefali, J. J. Blank and R. T. Raines, *Energy Environ. Sci.* **2010**, 3, 765; (c) L. Lai and Y. Zhang, *ChemSusChem* **2010**, 3, 1257.
55. A. Gandini and M. N. Belgacem, *Prog. Polym. Sci.* **1997**, 22, 1203.
56. C. Moreau, M. N. Belgacem and A. Gandini, *Top. Catal.* **2004**, 27, 11.
57. A. Sanborn and P. D. Bloom, World Pat. WO 2006063287, **2006**.
58. A. A. Rosatella, S. P. Simeonov, R. F. M. Frade and C. A. M. Afonso, *Green Chem.* **2011**, 13, 754.
59. M. J. Climent, A. Corma and S. Iborra, *Green Chem.* **2011**, 13, 520.
60. A. Fukuoka and P. Dhepe, *Angew. Chem. Int. Ed.* **2006**, 45, 5161.
61. N. Yan, C. Zhao, C. Luo, P. Dyson, H. Liu and Y. Kou, *J. Am. Chem. Soc.* **2006**, 128, 8714.
62. (a) H. Kobayashi, Y. Ito, T. Komanoya, Y. Hosaka, P. Dhepe, K. Kasai and K. Hara, A. Fukuoka, *Green Chem.* **2011**, 13, 326; (b) W. Deng, M. Liu, X. Tan, Q. Zhang and Y. Wang, *J. Catal.* **2010**, 271, 22.
63. C. Luo, S. Wang and H. Liu, *Angew. Chem. Int. Ed.* **2007**, 46, 7636.
64. W. Deng, X. Tan, W. Fang, Q. Zhang and Y. Wang, *Catal. Lett.* **2009**, 133, 167.
65. L. Ding, A. Wang, M. Zheng and T. Zhang, *ChemSusChem* **2010**, 3, 818.
66. S. Vyver, J. Geboers, M. Dusselier, H. Schepers, T. Vosch, L. Zhang, G. Tendeloo, P. Jacobs and B. Sels, *ChemSusChem* **2010**, 3, 698.
67. (a) R. Palkovits, K. Tajvidi, J. Procelewska, R. Rinaldia and A. Ruppert, *Green Chem.* **2010**, 12, 972; (b) J. Geboers, S. Vyver, K. Carpentier, K. Blochouse, P. Jacobs and B.

Sels, *Chem. Commun.* **2010**, 46, 3577; (c) R. Palkovits, K. Tajvidi, A. Ruppert and J. Procelewska, *Chem. Commun.* **2010**, 47, 576.

68. I. Ignatyev, C. Doorslaer, P. Mertens, K. Binnemans and D. Vos, *ChemSusChem* **2010**, 3, 91.
69. Y. Zhu, Z. Kong, L. Stubbs, H. Lin, S. Shen, E. Anslyn and J. Maguire, *ChemSusChem* **2010**, 3, 67.
70. (a) R. Levy and M. Boudart, *Science* **1973**, 181, 547; (b) J. Chen, *Chem. Rev.* **1996**, 96, 1477.
71. For some examples, see: (a) X. Chen, T. Zhang, P. Ying, M. Zheng, W. Wu, L. Xia, T. Li, X. Wang and C. Li, *Chem. Commun.* **2002**, 3, 288. (b) X. Chen, T. Zhang, M. Zheng, L. Xia, T. Li, W. Wu, X. Wang and C. Li, *Ind. Eng. Chem. Res.* **2004**, 43, 6040; (c) J. Sun, M. Zheng, X. Wang, A. Wang, R. Cheng, T. Li and T. Zhang, *Catal. Lett.* **2008**, 123, 150; (d) H. Wang, A. Wang, X. Wang and T. Zhang, *Chem. Commun.* **2008**, 22, 2565.
72. C. Jacobsen, *Chem. Commun.* **2000**, 12, 1057.
73. K. McCrea, J. Logan, T. Tarbuck, J. Heiser and M. Bussell, *J. Catal.* **1997**, 171, 255.
74. E. Furimsky, *Appl. Catal. A* **2003**, 240, 1.
75. E. Weigert, A. Stottlemyer, M. Zellner and J. Chen, *J. Phys. Chem. C* **2007**, 111, 14617.
76. (a) J. Kitchin, J. Norskov, M. Barteau and J. Chen, *J. Chem. Phys.* **2004**, 120, 10240; (b) J. Norskov and P. Liu, *Phys. Chem. Chem. Phys.* **2001**, 3, 3814; (c) J. Chen, C. Menning and M. Zellner, *Surf. Sci. Rep.* **2008**, 63, 201; (d) M. Gauthier, S. Padovani, E. Lundgren, V. Bus, G. Kresse, J. Redinger and P. Varga, *Phys. Rev. Lett.* **2001**, 87, 036103.
77. C. Liang, F. Tian, Z. Li, Z. Feng, Z. Wei and C. Li, *Chem. Mater.* **2003**, 15, 4846.
78. Y. Zhang, A. Wang and T. Zhang, *Chem Commun.* **2010**, 46, 2010, 862.
79. M. Zheng, A. Wang, N. Ji, J. Pang, X. Wang and T. Zhang, *ChemSusChem* **2010**, 3, 63.
80. (a) M. Sasaki, K. Goto, K. Tajima, T. Adschiric and K. Arai, *Green Chem.* **2002**, 4, 285; (b) M. Sasaki, M. Furukawa, K. Minami, T. Adschiri and K. Arai, *Ind. Eng. Chem. Res.* **2002**, 41, 6642.
81. M. Balat, *Energy Sources Part A* **2009**, 31, 1300.
82. (a) I. M. Atadashi, M. K. Aroua and A. A. Aziz, *Renew. Sust. Energ. Rev.* **2010**, 14, 1999; (b) E. P. Feofilova, Y. E. Sergeeva and A. A. Ivashechkin, *Appl. Biochem. Microbiol.* **2010**, 46, 369.
83. (a) L. Singaram, *Therm. Sci.* **2009**, 13, 185; (b) L. Fjerbaek, K. V. Christensen and B. Norddahl, *Biotechnol. Bioeng.* **2009**, 102, 1298; (c) E. M. Shahid and Y. Jamal, *Renew. Sust. Energ. Rev.* **2008**, 12, 2484.
84. D. Y. C. Leung, X. Wu and M. K. H. Leung, *Appl. Energy* **2010**, 87, 1083.
85. (a) J. C. Ye, S. Tu and Y. Sha, *Bioresour. Technol.* **2010**, 101, 7368; (b) J. M. Dias, M. C. M. Alvim-Ferraz and M. F. Almeida, *Fuel* **2008**, 87, 3572.
86. E. Lotero, Y. Liu, D. E. Lopez, K. Suwannakarn, D. A. Bruce, J. James and G. Goodwin, *Ind. Eng. Chem. Res.* **2005**, 44, 5353.
87. A. Banerjee and R. Chakraborty, *Resour. Conserv. Recycl.* **2009**, 53, 490.
88. (a) D.W. Lee, Y.M. Park and K.Y. Lee, *Catal. Surv. Asia* **2009**, 13, 63; (b) F. Guo, Z.-G. Peng, J.-Y. Dai and Z.-L. Xiu, *Fuel Process. Technol.* **2010**, 91, 322; (c) A. P. Vyas, N. Subrahmanyam and P. A. Patel, *Fuel* **2009**, 88, 625.

89. (a) E. Lotero, J. G. Goodwin, Jr., D. A. Bruce, K. Suwannakarn, Y. Liu and D. E. Lopez, *Catalysis* **2006**, 19, 41; (b) A. Kawashima, K. Matsubara and K. Honda, *Bioresour. Technol.* **2009**, 100, 696.

90. (a) Y. Liu, E. Lotero and J. G. Goodwin, Jr., *J. Catal.* **2006**, 243, 221; (b) D. E. López, J. G. Goodwin, Jr. and D. A. Bruce, *J. Catal.* **2007**, 245, 381.

91. R. Tesser, L. Casale, D. Verde, M. D. Serio and E. Santacesaria, *Chem. Eng. J.* **2010**, 157, 539.

92. M. D. Serio, R. Tesser, L. Pengmei and E. Santacesaria, *Energy Fuels* **2008**, 22, 207.

93. (a) J. M. Marchetti and A. F. Errazu, *Biomass Bioenerg.* **2010**, 34, 272; (b) G. Teng, L. Gao, G. Xiao and H. Liu, *Energy Fuels* **2009**, 23, 4630; (c) X. Liu, X. Piao, Y. Wang and S. Zhu, *J. Phys. Chem. A* **2010**, 114, 3750; (d) C. Liu, P. Lv, Z. Yuan, F. Yan and W. Luo, *Renew. Energy* **2010**, 35, 1531.

94. (a) X. H. Liu, X. Y. Xiong, C. M. Liu, D. Y. Liu, A. J. Wu, Q. L. Hu and C. L. Liu, *J. Am. Oil Chem. Soc.* **2010**, 87, 817; (b) I. Lukic, J. Krstic, D. Jovanovic and D. Skala, *Bioresour. Technol.* **2009**, 100, 4690.

95. H. J. Kim, B. S. Kang, M. J. Kim, Y. M. Park, D. K. Kim, J. S. Lee and K. Y. Lee, *Catal. Today* **2004**, 93–95, 315.

96. A. A. Davydov, M. L. Shepotko and A. A. Budneva, *Catal. Today* **1995**, 43, 225.

97. C. S. MacLeod, A. P. Harvey, A. F. Lee and K. Wilson, *Chem. Eng. J.* **2008**, 135, 63.

98. W. Xie and X. Huang, *Catal. Lett.* **2005**, 107, 53.

99. L. Bournay, D. Casanave, B. Delfort, G. Hillion and J. A. Chodorge, *Catal. Today* **2005**, 106, 190.

100. (a) A. Brito, M. E. Borges and N. Otero, *Energy Fuels* **2007**, 21, 3280; (b) W. Xie, X. Huang and H. Li, *Bioresour. Technol.* **2007**, 98, 936; (c) T. Augustine, N. S. Chong, J. M. Childress, B. M. Armstrong and K. Hill, *Abstr. Am. Chem. Soc.* **2006**, 231; (d) G. J. Suppes, M. A. Dasari, E. J. Doskocil, P. J. Mankidy and M. J. Goff, *Appl. Catal. A* **2004**, 257, 213.

101. A. Philippou and M. W. Anderson, *J. Catal.* **2000**, 189, 395.

102. G. Carja, R. Nakamura, T. Aida and H. Niiyama, *Micropor. Mesopor. Mater.* **2001**, 47, 275.

103. M. D. Serio, M. Ledda, M. Cozzolino, G. Minutillo, R. Tesser and E. Santacesaria, *Ind. Eng. Chem. Res.* **2006**, 45, 3009.

104. H.-Y. Zeng, Z. Feng, X. Deng and Y.-Q. Li, *Fuel* **2008**, 87, 3071.

105. A. Corma, S. B. A. Hamid, S. Iborra and A. Velty, *J. Catal.* **2005**, 234, 340.

106. J. L. Shumaker, C. Crofcheck, S. A. Tackett, E. Santillan-Jimenez and M. Crocker, *Catal. Lett.* **2007**, 115, 56.

107. N. Shibasaki-Kitakawa, H. Honda, H. Kuribayashi, T. Toda, T. Fukumura and T. Yonemoto, *Bioresour. Technol.* **2007**, 98, 416.

108. J. Xie, X. Zheng, A. Dong, Z. Xiao and J. Zhang, *Green Chem.* **2008**, 11, 355.

109. N. Nakatani, H. Takamori, K. Takeda and H. Sakugawa, *Bioresour. Technol.* **2009**, 100, 1510.

110. Z. Wei, C. Xu and B. Li, *Bioresour. Technol.* **2009**, 100, 2883.

111. Q. Shu, B. L. Yang, H. Yuan, S. Qing and G. L. Zhu, *Catal. Commun.* **2007**, 8, 2159.

112. A. Corma and H. Garcia, *Catal. Today* **1997**, 38, 257.

113. I. K. Mbaraka, D. R. Radu, V. S. Y. Lin and B. H. Shanks, *J. Catal.* **2003**, 219, 329.

114. J. Perez-Pariente, I. Diaz, F. Mohino and E. Sastre, *Appl. Catal., A* **2003**, 254, 173.

115. N. Katada, T. Hatanaka, M. Ota, K. Yamada, K. Okumura and M. Niwa, *Appl. Catal. A* **2009**, 362, 164.

116. S. Zhang, Y.-G. Zu, Y.-J. Fu, M. Luo, D.-Y. Zhang and T. Efferth, *Bioresour. Technol.* **2010**, 101, 931.

117. F. Chai, F. Cao, F. Zhai, Y. Chen, X. Wang and Z. M. Su, *Adv. Synth. Catal.* **2007**, 349, 1057.

118. A. Alsalme, E. F. Kozhevnikova and I. V. Kozhevnikov, *Appl. Catal. A* **2008**, 349, 170.

119. A. M. Toda, Takagaki, M. Okamura, J. N. Kondo, S. Hayashi, K. Domen and M. Hara, *Nature* **2005**, 438, 178.

120. (a) A. M. Dehkhoda, A. H. West and N. Ellis, *Appl. Catal. A* **2010**, 382, 197; (b) M. Hara, *Top Catal.* **2010**, 53, 805.

121. (a) Q. Shu, J. Gao, Z. Nawaz, Y. Liao, D. Wang and J. Wang, *Appl. Energy* **2010**, 87, 2589; (b) Q. Shu, Z. Nawaz, J. Gao, Y. Liao, Q. Zhang, D. Wang and J. Wang, *Bioresour. Technol.* **2010**, 101, 5374.

122. (a) M. Kitano, D. Yamaguchi, S. Suganuma, K. Nakajima, H. Kato, S. Hayashi and M. Hara, *Langmuir* **2009**, 25, 5068; (b) X. Mo, D. E. López, K. Suwannakarn, Y. Liu, E. Lotero, J. G. Goodwin, Jr. and C. Lu, *J. Catal.* **2008**, 254, 332.

123. G. Vicente, A. Coteron, M. Martinez and J. Aracil, *Ind. Crops Prod.* **1998**, 8, 29.

124. (a) G. D. Yadav and A. D. Murkute, *J. Catal.* **2004**, 224, 218; (b) F. Omota, A. C. Dimian and A. Bliek, *Chem. Eng. Sci.* **2003**, 58, 3175.

125. S. Einloft, T. O. Magalhães, A. Donato, J. Dullius and R. Ligabue, *Energy Fuels* **2008**, 22, 671.

126. (a) D. E. López, K. Suwannakarn, G. J. Goodwin, Jr. and D. A. Bruce, *Ind. Eng. Chem. Res.* **2008**, 47, 2221; (b) C. V. McNeff, L. C. McNeff, B. Yan, D. T. Nowlan, M. Rasmussen, A. E. Gyberg, B. J. Krohn, R. L. Fedie and T. R. Hoye, *Appl. Catal., A* **2008**, 343, 39.

CHAPTER 12

Catalytic Conversion of Glycerol

JIE XU, WEIQIANG YU, HONG MA, FENG WANG, FANG LU, MUKUND GHAVRE, and NICHOLAS GATHERGOOD

12.1 INTRODUCTION

Glycerol is a simple polyol compound, 1,2,3-propane triol, known for centuries as an important chemical [1]. As fossil resources are insufficient to meet the increasing demand for fuels and chemicals, glycerol has been prioritized as one of the 12 biomass-derived building blocks selected from a list of more than 300 candidates based on the petrochemical model of building blocks, chemical data, known market data, and properties (Fig. 12.1) [2]. Due to the three hydroxyl groups, glycerol is a potential starting material for various high-value fine chemicals, and can also be a model for other biomass-derived higher polyols.

Commercial production of glycerol was realized many years ago, and can be derived from natural glycerol, synthetic glycerol, and nowadays the main source is from biodiesel (Fig. 12.2). Natural glycerol is named from the production of soaps, fatty acids, and fatty esters [3], and is one of the by-products from these processes. Synthetic glycerol is produced by using propylene as the starting material.

Biodiesel-derived glycerol is produced by transesterification reaction between fats or oils and methanol. In general, the fatty acid methyl esters are referred as "biodiesel." Biodiesel is considered as the fuel with the greatest potential today due to the renewable and clean properties. Glycerol is one by-product of the transesterification, which accounts for up to 10% of the total products. Due to the continuously increasing amount of biodiesel production, availability of glycerol is increasing correspondingly and is expected to increase in the foreseeable future. This phenomenon has influenced the glycerol market remarkably, and therefore greatly impacted the traditional methods of glycerol production [1,4,5].

The low price and wide availability recommend glycerol as a potential feedstock for the production of valuable chemicals. Application of glycerol needs to be explored across the entire chemical industry, not limited to the traditional uses in

The Role of Green Chemistry in Biomass Processing and Conversion, First Edition.
Edited by Haibo Xie and Nicholas Gathergood.

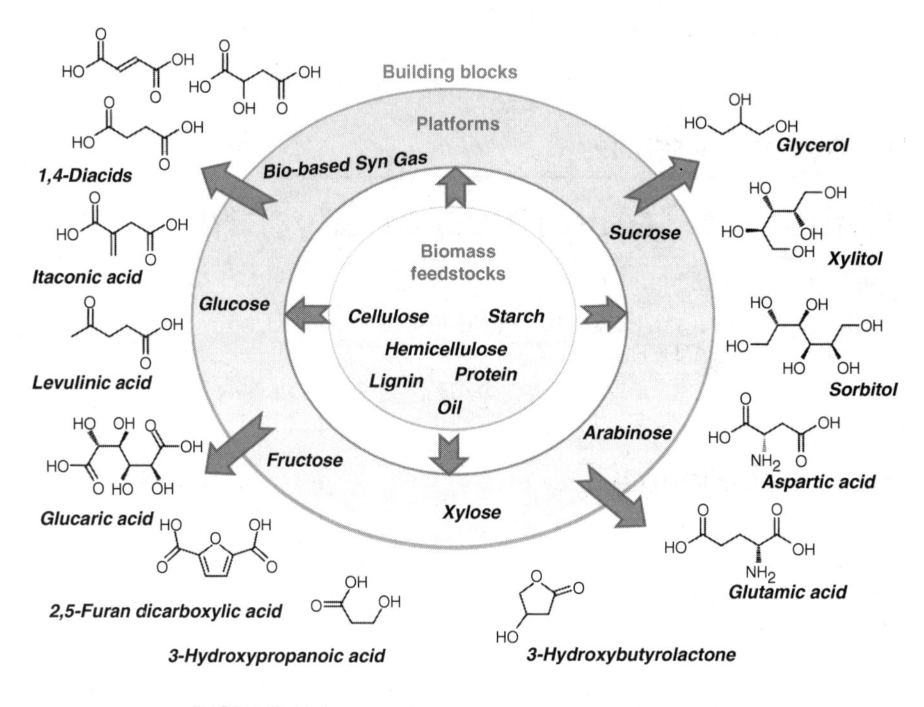

FIGURE 12.1 Top 12 biomass-derived building blocks.

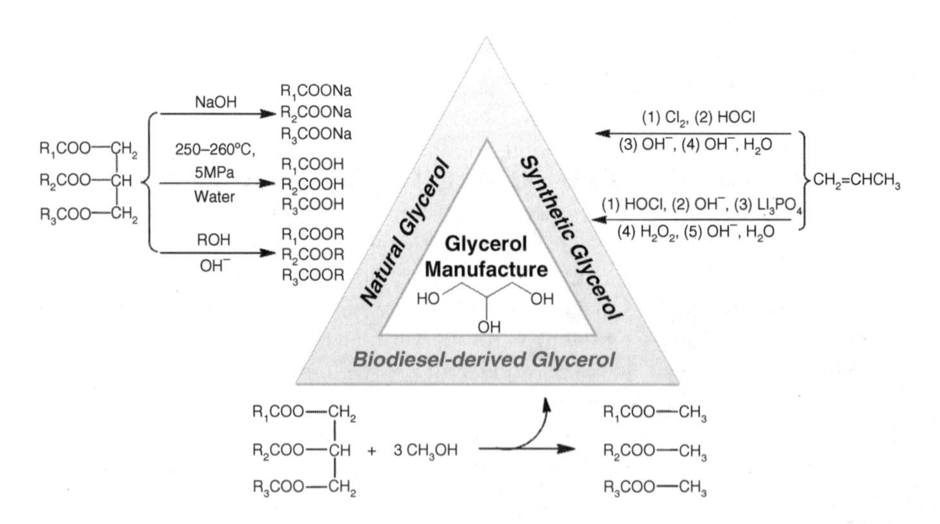

FIGURE 12.2 Glycerol manufacture.

pharmaceuticals, cosmetics, or as a wetting agent in tobacco [4]. Therefore, new synthetic methodologies need to be developed to realize glycerol-derived value-added products, which can also advance sustainable development of biodiesel manufacture. Although glycerol contains only one functional group, an alcohol (albeit primary and secondary), is a reactive center and can be transformed into many different classes of products. Recently, there have been a plethora of studies reporting glycerol conversion through different methods, such as hydrogenolysis, oxidation, dehydration, reforming, etherification, and carboxylation (Fig. 12.3). Through these conversions, valuable chemicals and fuels are obtained, including H_2, diols, acids, esters, or ethers. This chapter will highlight and discuss the recent representative progress for the utilization of glycerol.

12.2 CATALYTIC HYDROGENOLYSIS OF GLYCEROL INTO DIOLS

The catalytic hydrogenolysis route of glycerol is an economical and environmentally friendly process, which occurs via the breaking of a C–C bond and/or C–OH bond with simultaneous addition of hydrogen. Three diols, ethylene glycol (EG), 1,2-propylene glycol (1,2-PG), or 1,3-propylene glycol (1,3-PG) can be produced from the catalytic hydrogenolysis route of glycerol (Fig. 12.4). This is an atom-economic process as all the C and O atoms in the diol products are from the initial feedstock. In particular, the by-product from hydrogenolysis of glycerol to 1,2-PG and 1,3-PG is only water. The development of high atom economy reactions is a fundamental principle of green chemistry.

These important diols are widely used for agricultural adjuvants, plastics, liquid fuels, pharmaceuticals, as well as liquid detergents (such as emulsifiers, surface active agents, dehumidifying agents, antifreeze, lubricants, and solvents). 1,2-PG is used extensively for the synthesis of lactic acid and polyesters. Lactic acid is the key intermediate for the production of biodegradable polymers (such as polylactic acid) and has become a bulk chemical in great demand in recent years. A process of converting 1,2-PG to lactic acid via a very efficient and environment-friendly route was reported by Xu et al. in 2010 [6]. 1,3-PG is an excellent feedstock for producing polyester with high performance, especially when copolymerized with terephthalic acid to produce polytrimethylene terephthalate (PTT). Both DuPont and Shell Corporation have developed this polyester which is commercially available as SORONA® and CORTERRA®, respectively. Such a polyester has special properties including chemical resistance, light stability, elastic recovery, and dyeability [1]. Ethylene glycol is also an important raw material for the synthesis of polyester fibers and resins [7], such as polyethylene terephthalate (PET) and polyethylene glycol naphthalate (PEN).

Currently, the production of these diols is based on petroleum feedstocks. 1,2-PG and EG are produced from the hydration of propylene oxide or ethylene oxide. 1,2-PG can also be prepared by the hydrolysis of 1,2-dichloropropane. There are currently two typical routes for producing 1,3-PG. Shell company investigated

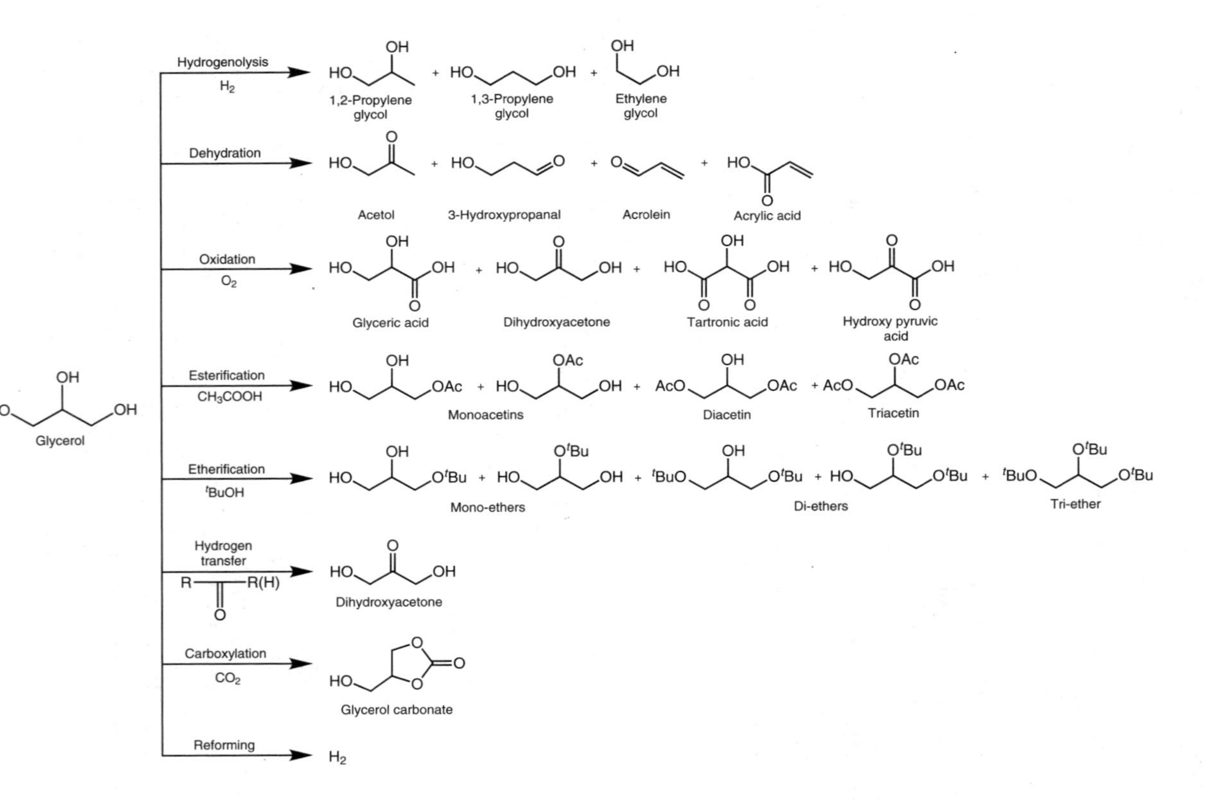

FIGURE 12.3 Routes for catalytic conversion of glycerol.

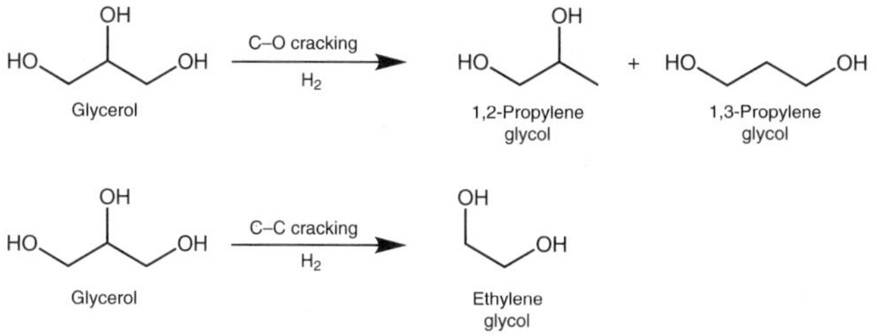

FIGURE 12.4 The diols produced in glycerol hydrogenolysis.

the carbonylation of ethylene oxide using syngas, to produce the 3-hydroxyl-propionaldehyde intermediate, followed by hydrogenation. Alternatively, Degussa-DuPont used acrolein as the feedstock, and developed a process including hydration and hydrogenation steps. However, both the current production methods still have many limitations. First, the increasing cost of petroleum-based feedstock restricts the development of these diols and raises issues with the commercial viability of these raw materials from fossil fuels; second, negative effects on the environment are needed to be reduced. Finally, some routes require complex special equipments and have high production cost [1].

The study of glycerol hydrogenolysis to 1,2-PG, 1,3-PG, and EG has attracted considerable interest from researchers and many catalysts for glycerol conversion have been developed. Different reaction pathways are proposed depending on the specific conditions and catalyst applied.

12.2.1 The Production of 1,2-Propylene Glycol

Ruthenium-based catalysts are widely studied noble-metal catalysts in the hydrogenolysis of glycerol to 1,2-PG, with an acid-catalyzed mechanism proposed. Tomishige's group extensively studied Ru/C catalyst for the hydrogenolysis reaction [8] and found that the addition of Amberlyst resin enhanced the catalyst performance. They proposed that the acidic Amberlyst resin promotes the dehydration of glycerol. Final products are formed via hydrogenation of the dehydrated intermediate over the metal catalyst. Balaraju et al. [9] reported that solid acid could also greatly enhance the hydrogenolysis performance of Ru/C catalysts, especially in the case of Nb_2O_5. They established a similar process as put forward by Tomishige. These reports proved that addition of solid acid promoted the catalytic performance, including conversion and selectivity. They proposed that the reaction proceeds via dehydration of glycerol to acetol over solid acid then hydrogenation to 1,2-PG on metal catalyst (Fig. 12.5).

The Ru/C catalyst has a strong ability of cracking C–C bonds, which can lead to undesirable formation of EG. Therefore, efforts have been made to reduce the selectivity of EG. Bolado et al. [10] explored the possibility of forcing the dehydration step, without hydrogen over Ru/C catalyst, with the promotion of cationic exchange resin. Under those conditions, glycerol hydrogenolysis was

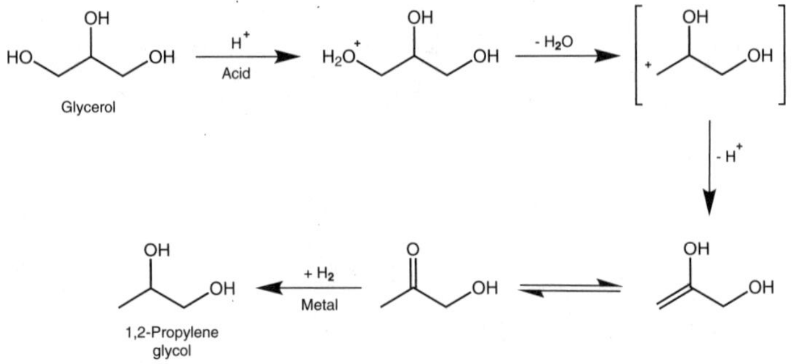

FIGURE 12.5 Reaction mechanism for the glycerol hydrogenolysis over acid catalysts.

conducted in two stages: (1) dehydration under nitrogen and (2) reduction under hydrogen. This two-stage process leads to a doubling of the selectivity ratio of 1,2-PG/EG compared to one-step hydrogenolysis. Amberlyst 70 as cationic-exchange resin greatly improved the dehydration of glycerol.

In 2008, bi-functional metal-acid catalysts were attempted to be used for the hydrogenolysis of glycerol to 1,2-PG. More recently, Alhanash et al. [11] used Ru supported on a heteropoly salt of cesium phosphorus tungsten (CsPW) as the bifunctional catalyst for glycerol hydrogenolysis, by which they achieved 21% glycerol conversion and 96% selectivity of 1,2-PG at 150 °C after 10 h. Gandarias's group [12] prepared the bifunctional catalyst of Pt supported on amorphous silica–alumina (Pt/ASA). They studied the role of acid and metal sites of Pt/ASA and concluded that ASA acid sites are responsible for glycerol dehydration to acetol while Pt-metal sites catalyze acetol hydrogenation to 1,2-PG. In addition, they propose that Pt sites also promote glycerol dehydration due to the higher yield of acetol over Pt/ASA than ASA.

The alkaline effect on the glycerol hydrogenolysis has also been studied by researchers. Maris's group [13] tested the activity and selectivity of Ru/C and Pt/C in glycerol hydrogenolysis by the addition of NaOH or CaO. Results showed that the addition of NaOH or CaO enhanced the rate of glycerol hydrogenolysis. A significant enhancement was observed for Pt/C than Ru/C. The by-products in the hydrogenolysis of glycerol to 1,2-PG were EG and lactic acid. A mechanism proposed accounting for the product and by-products formed is shown in Figure 12.6 [13]. Glycerol is first dehydrogenated to glyceraldehyde on the metal catalyst with the promotion of base. 1,2-PG was obtained by base-catalyzed dehydration of glyceraldehyde and subsequent hydrogenation. Yuan et al. [14] reported an efficient solid-base supported catalyst for glycerol hydrogenolysis to 1,2-PG in a base-free aqueous solution. The hydrotalcite supported Pt catalyst gave excellent selectivity and activity. Selectivity of 1,2-PG was up to 93% with 92.1% of glycerol conversion. Hydrotalcite was used as an alkaline oxide which plays the same role as NaOH or CaO in previous studies.

Non-noble metal catalysts have attracted great interest in recent years. Ni is one of the commonly selected non-noble metals for catalysis studies. Xu's group [15] in 2010 reported two kinds of Ni-based bifunctional catalysts, Ni/AC catalyst

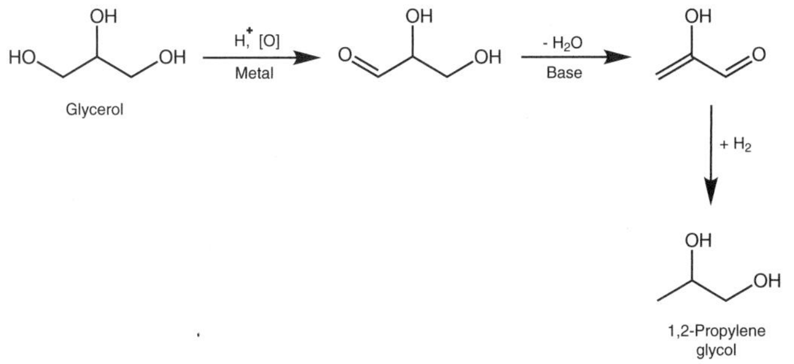

FIGURE 12.6 Reaction pathway for glycerol hydrogenolysis on Ru- and Pt-based catalysts in the existence of base. (Reproduced from Ref. [13] with permission from The Royal Society of Chemistry © 2007.)

(AC = activated carbon) and Ni/NaX catalyst (X = zeolite). Ni/AC catalysts were prepared by a novel procedure via impregnation, carbothermal reduction, and KBH_4-treatment. Through this procedure, acidity (and activity) of the catalyst was improved compared to other preparation methods. Consequently, conversion of glycerol to 1,2-PG was very efficient and catalytic performance was increased remarkably as well. Further study revealed that catalytic activity was greatly enhanced after adding cerium to the Ni/AC catalyst. The glycerol conversion increased to 90.4% from 43.3% in 6 h at 200 °C with 5 MPa of H_2. Different zeolites, Al_2O_3 and SiO_2 were used as the supports to prepare the Ni-based catalyst and their performance assessed (Table 12.1). Excellent results were obtained for glycerol hydrogenolysis by Ni/NaX catalyst. Selectivity of 1,2-PG was up to 80.4% with 86.6% glycerol conversion and catalyst performance was attributed to the higher acidity of NaX. The function of these two catalysts for glycerol hydrogenolysis is also attributed to the acid-catalyzed mechanism, both of which have potential advantages in the future due to their non-noble properties.

TABLE 12.1 Results of Glycerol Hydrogenolysis over Ni-Based Catalysts Supported on Different Materials

		Selectivity (%)		
Catalyst	Conversion (%)	1,2-PG	EG	Others[a]
Ni/NaMOR	14.0	56.7	13.3	30.0
Ni/NaZSM-5	47.8	9.4	13.0	77.6
Ni/NaA	65.3	46.8	14.1	39.1
Ni/NaX	94.5	72.1	11.1	16.8
Ni/SiO_2	56.9	44.4	8.8	46.8
$Ni/\gamma\text{-}Al_2O_3$	97.1	44.2	7.5	48.3

Source: Ref. [15].

[a]Others: ethanol, *n*-propanol, acetol, CO, CO_2, and CH_4.

TABLE 12.2 The Catalytic Performance of Cu-Based Catalysts in the Glycerol Hydrogenolysis to 1,2-PG

Catalysts	Conversion (%)	1,2-PG Selectivity (%)	Ref.
Cu–Cr	54.8	85	[16]
Cu/SiO_2	19	98	[17]
Cu/Al_2O_3	49.6	96.8	[18]
$Cu–Ag/Al_2O_3$	27	96	[19]
Cu–ZnO	19	100	[20]
Cu–ZnO	22.5	83.6	[21]
Cu–ZnO	34	93	[22]
Cu–ZnO	46	90	[23]
Cu–ZnO	ca.25	93.9	[24]

Cu-based catalysts are another series of non-noble metal catalysts, which have shown superior ability for selective cleavage of C–O bonds than C–C bonds. This property is important for improving the selectivity of 1,2-PG via glycerol hydrogenolysis. Many catalysts have been reported with Cu as the active metal (Cu–Cr [16], Cu/SiO_2 [17], Cu/Al_2O_3 [18], $CuAg/Al_2O_3$ [19], and Cu–Zn [20–24]) for the hydrogenolysis of glycerol to 1,2-PG and their catalytic performances are summarized in Table 12.2. These Cu-based catalysts display excellent selectivity to 1,2-PG. Function of these catalysts has been studied by Huang et al. [17] who prepared the Cu/SiO_2 catalyst by precipitation-gel method. They found that Cu^0 and (inadequately) oxidized Cu^+ coexisted and were dispersed on the surface of the catalyst leading to a highly active catalyst. Cu^0 was the active site, which generated acidic sites *in situ* to catalyze the dehydration of glycerol to acetol, followed by hydrogenation to give 1,2-PG over the copper surface. Cu^+ species were inactive in reaction but can inhibit the sintering of active phases. In the study of the function of the Cu–Zn catalyst, Wang et al. [24] reported that highly strained Cu particles were formed on interaction with ZnO. A good linear correlation between the microstrain of Cu particles and their TOFs was exhibited. They propose that PG formation over Cu–ZnO was via dehydrogenation to glyceraldehyde, dehydration of glyceraldehyde to pyruvaldehyde, and then hydrogenation under basic conditions. Dehydrogenation to glyceraldehyde was the rate-determining step, high glycerol and H_2 concentration facilitated 1,2-PG selectivity.

12.2.2 Production of 1,3-Propylene Glycol

Several reports using homogeneous catalysts to produce 1,3-PG [25] have been published. Organic solvents such as 1-methyl-pyrrolidinone and water–sulfolane mixtures, with acid additives (e.g., H_2WO_4, rhodium carbonyl acetoacetonate) were investigated. One problem of using homogeneous catalysts is the separation of the product from the catalyst. An advantage of heterogeneous catalysis is the isolation of catalyst in many cases and is conveniently accomplished. Heterogeneous catalyst studies dominate hydrogenolysis of glycerol into 1,3-PG research with the focus to increase the selectivity of 1,3-PG by developing very efficient catalysis systems.

Kurosaka et al. [26] used $Pt/WO_3/ZrO_2$ as the catalyst and the yield of 1,3-PG reached 24% in 1,3-dimethy-2-imidazolidinone at 170 °C and 8.0 MPa. Although other noble-metal catalysts were screened, Pt was the most effective. Catalyst preparation method also influenced the catalytic activity. Sequential impregnation of WO_3 and then Pt on ZrO_2 was required to provide the active sites for hydrogenolysis of glycerol. Many studies revealed that an aqueous phase favored the formation of 1,2-PG, while polar aprotic solvent favored 1,3-PG generation. While, the use of organic solvents could avoid formation of 1,2-PG, employment of polar aprotic organic solvents (e.g., NMP, DMF, and DMSO) was restricted as they are not environmentally benign. Therefore, methodologies to produce 1,3-PG without utilizing these organic solvents has elicited great attention.

An important consideration of crude glycerol from biodiesel manufacture, is the high water content. Water is a cheap green solvent hence great synergy is possible if water is selected as reaction medium. Gong et al. [27] demonstrated that protic solvents, water and ethanol favored 1,3-PG formation in selective dehydroxylation of glycerol over $Pt/WO_3/ZrO_2$ catalyst. Binary solvents showed a synergetic solvent effect on selectivity of 1,3-PG. In their further study, bifunctional catalyst consisting of $Pt/WO_3/TiO_2$ on silica was prepared to promote hydrogenolysis of glycerol in water [28]. Existence of TiO_2 species improved the dispersion of Pt metal, and WO_3 species could increase the acidity. Their results showed that water was the preferred solvent for this catalyst. Qin et al. [29] reported yields of 1,3-PG up to 32% in aqueous phase at 130 °C and 4 MPa of H_2, with $Pt/WO_3/ZrO_2$ as the catalyst. They proposed that deoxygenation of glycerol was via ionic mechanism, which includes both proton transfer and hydride transfer. In the mechanism, hydrogen molecules first undergo heterolytic cleavage on the catalyst to form protons and hydride ions. The H^+ reacts with glycerol and next follows the reaction with H^- (Fig. 12.7). High selectivity of 1,3-PG was attributed to the preferential dehydration of secondary hydroxyl group. Nakagawa et al. [30] also conducted glycerol hydrogenolysis reaction to produce 1,3-PG in aqueous media. They prepared the ReO_x modified Ir/SiO_2 catalyst and a 38% yield of 1,3-PG was attained with 81% conversion of glycerol. Through further studies, they proposed a possible reaction mechanism. One of the terminal hydroxyl groups of glycerol first gets adsorbed on ReO_x surface to form 2,3-dihydroxypropoxide; then H activated by Ir metal attacks the secondary carbon to give 3-hydroxypropoxide; the hydrolysis of 3-hydroxypropoxide releases 1,3-PG. Growth of oxidized metal clusters was sufficient to enable the preferred formation of terminal alkoxide, which was the key for obtaining high selectivity of 1,3-PG. In this study, they also demonstrated that adjacent hydroxyl group of substrate was important according to this mechanism.

A solvent-free system was also developed for glycerol hydrogenolysis by Huang et al. [31], in which direct and continuous method of converting glycerol into 1,3-PG over nonprecious bifunctional metal catalyst was demonstrated. 32.1%

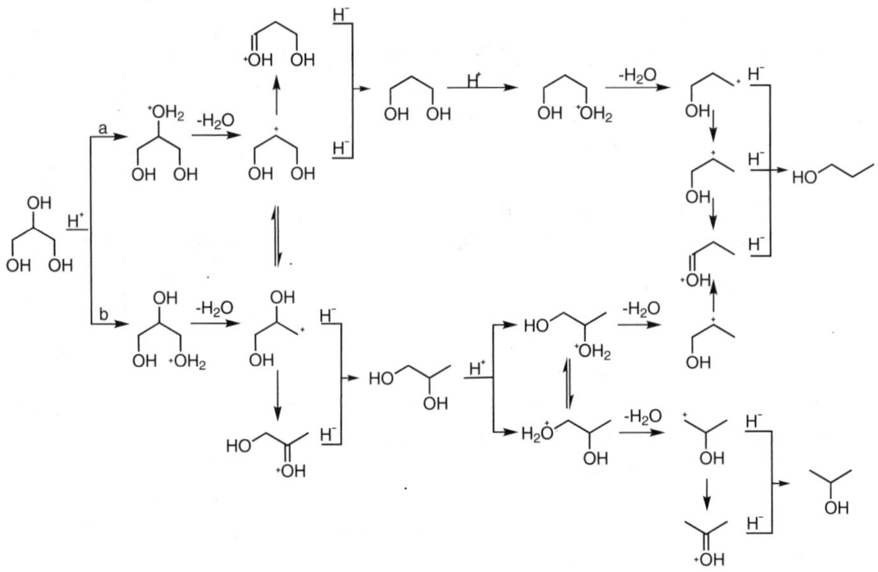

FIGURE 12.7 Proposed reaction mechanism of glycerol deoxygenation over Pt/WZ. (Adapted from Ref. [29] by permission of The Royal Society of Chemistry.)

selectivity of 1,3-PG was obtained over Cu–$H_4SiW_{12}O_{40}$(STA)/SiO_2 in vapor reaction. The STA (silicotungstic acid) provided acid sites for dehydration of glycerol to 3-hydroxypropanal and then copper metal promoted the catalytic hydrogenation of 3-hydroxypropanal to 1,3-PG. However, vaporization of glycerol at elevated temperatures in the presence of copper metal readily led to by-products via the acetol route. In addition, various cyclic acetals were also detected.

12.2.3 Selective Formation of Ethylene Glycol

EG is usually formed with low yields in glycerol hydrogenolysis due to competitive cracking between the C–C and C–O bond. In order to improve selectivity of EG, the key issue is to promote the cracking of C–C bond and inhibit cracking of C–O bond simultaneously. Ueda et al. [32] modified Ni/γ-Al_2O_3 catalyst with small amount of Pt, which promoted the production of EG in the glycerol hydrogenolysis. EG selectivity obtained was 48% with 16% glycerol conversion. Studies revealed that Pt atoms on Ni-rich Pt–Ni alloy surface can enhance the selectivity of EG and two possible explanations were given. First was that an addition of Pt promoted the retro-aldol reaction (also stated by Montassier [33]). Glycerol is dehydrogenated to glyceraldehyde, followed by the retro-aldol reaction, which lastly is hydrogenated to EG. The second explanation is the direct dissociation of C–C bond (Fig. 12.8). These two mechanisms of the EG formation from glycerol have been cited widely. However, there are very few

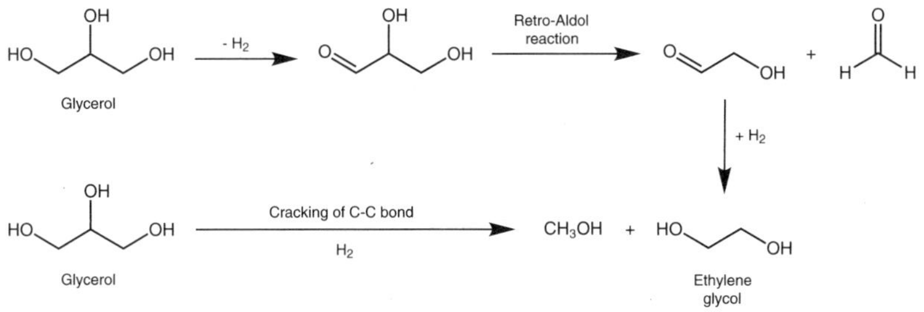

FIGURE 12.8 Proposed reaction pathways of EG formation from glycerol hydrogenolysis.

publications stating an increase in selectivity of EG. Further investigations of effective catalysts for this transformation are a priority.

12.3 CATALYTIC OXIDATION OF GLYCEROL TO VALUABLE CHEMICALS

Catalytic oxidation is regarded as an attractive technique for conversion of alcohols to corresponding aldehydes, ketones, or carboxylic acids. Glycerol is a class of alcohol which has a highly functionalized structure where oxidation can produce a series of valuable chemicals, such as glyceric acid, hydroxyacetone, tartronic acid, and keto-malonic acid (Fig. 12.9). Oxidation of primary hydroxyls of glycerol can

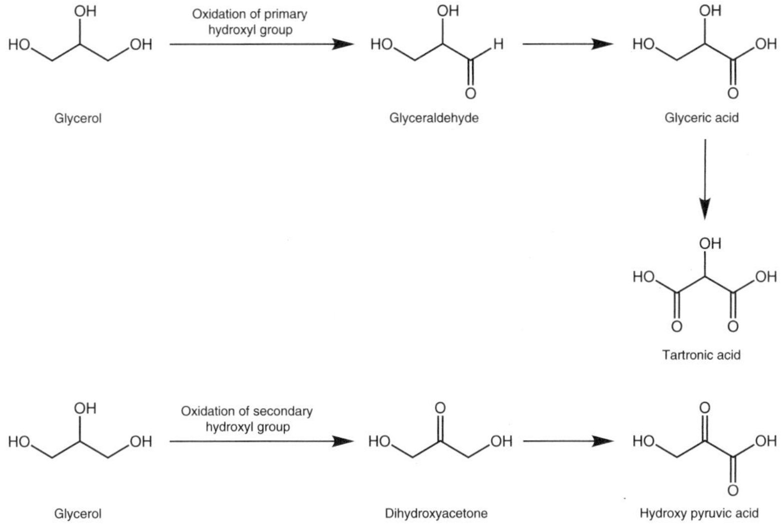

FIGURE 12.9 Possible products from glycerol oxidation.

yield glyceric acid and glyceraldehyde. Glyceric acid has many important uses, in particular for synthesis of polymers, treatment of skin disorders, an exothermic and volatile agent for packaging materials, etc. Further oxidation of glyceric acid leads to tartronic acid [5]. Oxidation of the secondary hydroxyl of glycerol can yield dihydroxyacetone (DHA) and hydroxypyruvic acid (HYPA). DHA has been used as sunless tanning agent in the cosmetics industry and can also be used as building blocks for new degradable polymers. Further oxidation of DHA leads to hydroxypyruvic acid. DHA is produced on an industrial scale by microbial oxidation of glycerol [4]. Similar to hydrogenolysis, the process for glycerol oxidation is also very complex, in which the primary and secondary hydroxyls are both susceptible to oxidation. Therefore, an efficient and selective catalyst system is required in order to obtain desirable products in high yields.

12.3.1 Oxidation of Primary Hydroxyl to Produce Glyceric Acid

Supported noble-metal catalysts including Pt, Pd, and Au catalysts are suitable for the oxidation of glycerol to glyceric acid. Pt metal catalysts favor the oxidation of primary hydroxyls over secondary hydroxyls. Selectivity of glyceric acid is upto 55% over the Pt/C catalyst [34]. Pd/C was also efficient with 70% selectivity for glyceric acid at 100% conversion [34]. Recent studies showed that Au catalysts are promising for glycerol oxidation. Since Au catalysts are considerably more resistant to oxygen poisoning than Pt- or Pd-based catalysts, they can tolerate higher oxygen partial pressures. Various supports have been studied in the preparation of Au-supported catalysts for the oxidation of glycerol. Porta et al. [35] used carbon as a support to prepare Au-based catalyst, and high selectivity of glyceric acid was obtained (>90%) at 90% conversion of glycerol. Sobczak et al. [36] explored V_2O_5, Nb_2O_5, and Ta_2O_5 as supports for Au and catalytic activities were tested for the oxidation of glycerol. Au/Nb_2O_5 catalyst prepared by gold-sol method showed high glyceric acid selectivity (42%) at 79% glycerol conversion. Villa et al. [37] prepared Au nanoparticles supported on three spinel-type $MgAl_2O_4$ supports for the selective liquid-phase oxidation of glycerol. Selectivity of glyceric acid was 60% at 50% glycerol conversion. Dimitratos et al. [38] showed that Au–Pd supported on TiO_2 prepared by a sol-immobilization method gave higher catalytic performance than impregnation method. The selectivity of glyceric acid was 64% at 97.7% glycerol conversion. In above studies, base was necessary in for glycerol oxidation, which abstracts proton from one of the primary hydroxyl groups of glycerol. Dehydrogenation was proposed as the first step in the oxidation process to yield glyceraldehyde. Of note, rapid oxidation of glyceraldehyde favors glyceric acid formation. Zope et al. [39] studied reactivity of the Au/water interface over selective oxidation of glycerol under high-P_H conditions. Through labeling experiments with $^{18}O_2$ and $H_2{}^{18}O$, they determined that the oxygen atom incorporation into the alcohol was from the hydroxide ions instead of molecular oxygen. Molecular oxygen participates in the regeneration of hydroxide ions through decomposition of peroxide intermediate.

Base can react with acid products to form the salts in case of base-catalyzed oxidation. Therefore, direct preparation of acid products in acidic conditions is required. Villa et al. [40] prepared a catalyst of Au–Pt nanoparticles supported on the zeolite H-mordenite to catalyze oxidation of glycerol in base-free conditions. Glycerol was directly and selectively converted to the glyceric acid with 81% selectivity at 100% conversion of glycerol. Role of catalyst support was to prevent the formation of H_2O_2 which could cleave the C–C bond.

12.3.2 Oxidation of Secondary Hydroxyl to Produce Dihydroxyacetone

Although Pt and Pd catalysts are suitable for glycerol oxidation, oxidation of primary hydroxyl group is preferred rather than the secondary. To increase selectivity for oxidation of secondary hydroxyl group, Kimura et al. [41] introduced p-block metals which have obvious effect. They found that after addition of bismuth to Pt catalysts, DHA selectivity greatly increased. The DHA selectivity reached as high as 80% at 80% of glycerol conversion when reaction was conducted over Pt–Bi/C catalyst in fixed bed reactor. Hu et al. [42] reported that in a semibatch reactor, DHA yield could reach 48% at 80% glycerol conversion over Pt–Bi catalyst at 80 °C and 30 psi. Painter et al. [43] used a cationic Pd catalyst to promote the oxidation of glycerol to DHA with benzoquinone or oxygen as an oxidizing agent. An impressive DHA selectivity up to 99% at 97% of glycerol conversion was attained.

Au-based catalysts were also studied in the production of DHA from glycerol. Dimirel et al. [44] used Au/C catalyst in NaOH and obtained up to 50% glycerol conversion, with 26% of DHA yield and 44% of HYPAC yield. On addition of Pt to Au/C catalyst, yield of DHA was increased slightly to 30% with the same glycerol conversion.

12.4 DEHYDRATION OF GLYCEROL TO VALUABLE INTERMEDIATES

One of the most promising routes utilizing glycerol is dehydration. Elimination of one or more water molecules from glycerol could provide many important intermediates and chemicals, such as acetol, 3-hydroxypropionaldehyde, acrolein, and acrylic acid (Fig. 12.10). Acetol can be used to produce 1,2-PG through hydrogenation reaction. This catalytic process provides another route for the production of 1,2-PG from renewable resources. Acetol can also be used as a convenient protecting group for carboxylic acids in peptide synthesis [45]. Acrolein is an important bulk chemical intermediate in the chemical and agricultural industry, which can be manufactured into, polymers, super polymers (adhesives), and detergents [1]. The commercial production method is based on propylene oxide over Bi/Mo-mixed oxide catalysts, which relies heavily on propylene obtained from petroleum. Acrylic acid is widely used in polymer dispersions, adhesives, fibers, and plastics as well as other chemical intermediates. The current route to produce acrylic acid from glycerol

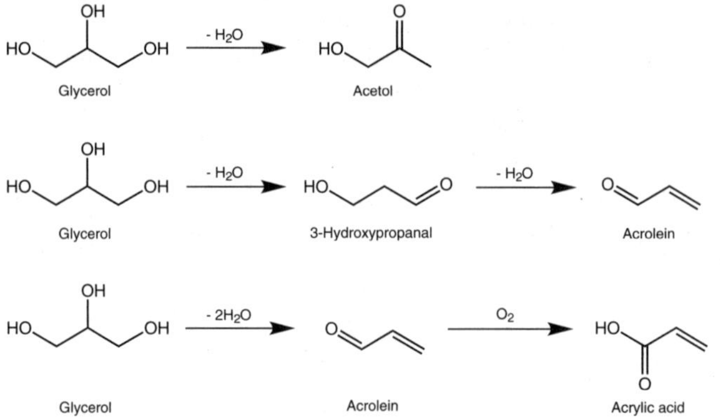

FIGURE 12.10 Dehydration of glycerol to valuable intermediates. (Reproduced from Ref. [45] with permission from The Royal Society of Chemistry © 1992.)

consists of two steps including dehydration of glycerol to acrolein and subsequent oxidation to produce acrylic acid.

12.4.1 Production of Acetol

In the elimination reaction involving one terminal hydroxyl of glycerol to acetol, Lewis acid sites and metal sites are important. Cu-based catalysts have been proposed as a good starting point for this reaction. Chin et al. [46] successfully prepared acetol via dehydration of glycerol, as the transient intermediate, eventually producing 1,2-PG. They employed reactive distillation technology to control the equilibrium. High selectivity of acetol was achieved (>90%) over copper-chromite catalyst. Niu et al. [47] have demonstrated that Cu/SiO_2 catalyst exhibited high activity and selectivity for acetol under mixed atmosphere of H_2–N_2. Glycerol conversion and selectivity for acetol was up to 98.8% and 84.6%, respectively. Employment of a mixed-gas atmosphere inhibited the deactivation of catalyst in N_2 and prevented overhydrogenation of acetol. Zhao et al. [48] utilized Cu/Al_2O_3 as a dehydration catalyst, and 86% selectivity for acetol was achieved at 220 °C and under ambient hydrogen pressure. Their study revealed production of acetol would benefit from determining the optimal loading of Cu, dehydration temperature, and H_2 atmosphere. The study also indicated that acid center was produced by Al_2O_3, and was partially generated *in situ* with dissociative adsorption process of H_2 by the CuO surface. These mechanisms resulted in increasing the catalytic dehydration performance. Sato et al. [49] used Cu/Al_2O_3 catalyst for the dehydration of glycerol. The acetol selectivity achieved 90.1% with 100% conversion of glycerol at 250 °C.

Kinage et al. [50] developed another class of catalyst for the dehydration of glycerol to acetol. They demonstrated that 5 wt.% Na-doped CeO_2 gave high selectivity (68.6%) for acetol at 10% glycerol conversion. In their study, they measured the acidic and basic sites on various Na-doped catalysts and calculated the ratio of basicity over acidity. In conclusion, authors suggest that an optimum number (2.1) of basicity/acidity ratio has lead to higher selectivity in case of Na/CeO_2. They also proposed that the formation of acetol was via dehydration of glycerol at acidic sites whereas dehydrogenation of glycerol occurs at the basic site followed by dehydration, leading to acetol. Large number of acidic or basic sites would lead to by-products. Therefore, the appropriate number of acidic and basic sites was needed to improve the selectivity of acetol.

12.4.2 Production of Acrolein

Dehydration of glycerol to acrolein can be carried out in liquid or gas phase. Homogeneous and heterogeneous acid catalysts as well as biocatalysts have been used for this reaction. A number of solid acid catalysts for the dehydration of glycerol have been reported including heteropolyacids, zeolites, and mixed-metal oxides.

The supported heteropolyacids exhibit good performance for this reaction. Alhanash et al. [51] demonstrated that the water-insoluble Cs heteropoly salt ($Cs_{2.5}H_{0.5}PW_{12}O_{40}$) (CsPW) was an active catalyst for the dehydration of glycerol to acrolein, and the catalyst exhibited 98% acrolein selectivity with full conversion of glycerol. They studied the influence of acid sites on dehydration through comparison of Brønsted acid (CsPW) with Lewis acid (Zn–Cr oxide), and also proposed reaction pathways (Fig. 12.11). Over Brønsted acid, proton transfer is not limited by steric constraint, so the secondary alcohol oxygen of glycerol is protonated. Enol 1,3-dihydroxypropene was produced through elimination of H_3O^+ and can form the aldehyde, 3-hydroxypropanal, by tautomerism. Subsequently, dehydration gave the final product, acrolein. Proton can be regenerated by interaction with H_3O^+. A difference from the Brønsted acid mechanism is the interaction of Lewis acid sites

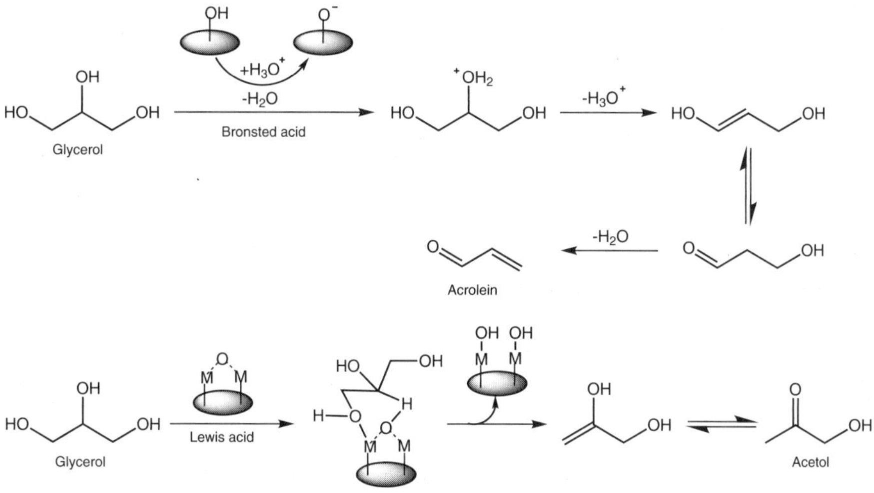

FIGURE 12.11 Possible reaction mechanism of glycerol dehydration over Brønsted acid and Lewis acid catalysts. (Reproduced from Ref. [51] with permission from Elsevier © 2010.)

with glycerol which is affected by steric constraints. Primary alcohol readily coordinates to the Lewis acid to form oxo-bridge bond with metal rather than secondary alcohol. Transfer of primary alcohol and migration of H^+ can lead to 2,3-dihydroxypropene. and hydrated active site. Through tautomerism, acetol was formed. The Lewis acidic site was regenerated by thermal dehydration of hydrated form. Katryniok et al. [52] recently prepared an acid catalyst of silicotungstic acid supported on an SBA-15 modified by zirconia grafting. Catalyst has a long lifetime performance in glycerol dehydration to acrolein. Yield of acrolein was 71% after 5 h and 69% after 24 h on stream, thus there was only a slight decrease in catalysts activity. They concluded that decrease of Brønsted acid site strength of STA by zirconia grafting has a positive impact on long-term catalytic performance. This was attributed to a reduction in carbon deposits and therefore catalyst deactivation was slowed. However, the presence of zirconia increased the number of Lewis acidic sites, which led to side reactions and decreased the selectivity of acrolein. Therefore, the overall balance of these two effects must be considered.

Zeolites-based compounds are also promising for the dehydration of glycerol to acrolein, with the exception of heteropolyacids. High catalytic performance also depends on Brønsted acid sites. Jia et al. [53] have studied the replacement of the Brønsted protons by sodium cations, via ion-exchange methodologies, which can modify the intensity of Brønsted acid sites. They used a commercial HZSM-5 catalyst with a silica/alumina ratio of 65:1 and different surface protons numbers. The catalytic performance decreased with increasing sodium ion concentration, and the best performance was obtained over sodium-free zeolites with 60% acrolein selectivity at full conversion of glycerol. This result also proved that higher concentrations of Brønsted acid sites benefited the dehydration of glycerol to acrolein. Adjusting silica/alumina ratio is another method for changing the number of Brønsted acid sites. Kim et al. [54] studied the influence of Si/Al ratio in HZSM-5 on the dehydration of glycerol. HZSM-5 catalysts with different Si/Al ratios were selected. Results indicated that low silica/alumina ratios could lead to higher number of strong acid sites through the NH_3–TPD analysis. They also studied type of the acid sites and found that lower silica/alumina ratios resulted in Lewis acid catalysis, whereas higher ratios gave Brønsted acid catalysis. The best result was obtained as 63.8% selectivity at 75.8% conversion in HZSM-5 with a silica/alumina ratio of 150:1.

In addition, some metal oxides, phosphates, and pyrophosphates have been proved as efficient catalysts for the dehydration of glycerol to acrolein. Tao et al. [55] prepared acidic binary metal oxide catalysts from SnO_2, ZrO_2, TiO_2, Al_2O_3, SiO_2, and ZnO and tested the activity in dehydration of glycerol. The highest selectivity of acrolein is obtained as 52% at 67% glycerol conversion over TiO_2–Al_2O_3 catalyst. The study revealed that presence of basic sites in binary oxide catalysts promoted the production of acetol therefore decreasing selectivity of acrolein. Small pore size also decreased the acrolein selectivity. In 2009, Liu et al. [56] reported glycerol dehydration over rare-earth pyrophosphates catalysts in gas phase. They found that pyrophosphate catalysts were active and selective toward acrolein production. Weak acidic sites formed on the surface by polyphosphate anions were proposed as the active center in this work. Among the tested catalysts, $Nd_4(P_2O_7)_3$ exhibited the best result with 80% selectivity to acrolein at 87% conversion of glycerol. Suprun et al. [57] prepared phosphate-modified TiO_2, Al_2O_3, and silica/aluminum. They found that total acidity and textural properties of catalyst could influence glycerol conversion and acrolein selectivity. Indeed, the most acidic silica-aluminum phosphate catalyst exhibited high selectivity of acrolein (72%). Deleplanque et al. [58] found iron phosphates were highly active and selective toward acrolein, the yields of which reached 80–90% after 5 h with almost 100% glycerol conversion.

12.4.3 Oxidative Dehydration of Glycerol to the Acrylic Acid and Acrolein

The oxidative dehydration of glycerol can generate acrolein, and also acrylic acid. Acrylic acid obtained in this way is potentially commercially viable. Wang et al. [59] recently reported a direct method to produce acrylic acid in one-step using vanadium–phosphate oxide (V–P–O) catalyst. In the study, molecular oxygen and glycerol were co-fed through a fixed-bed reactor. However, acrylic acid was obtained in only 5 wt% with the main product being acrolein. A reaction network starting from glycerol was proposed, in which the dehydration of glycerol becomes complex in the presence of oxygen, which possibly involves dehydration, oxidation, hydrogenation, and dehydrogenation reactions (Fig. 12.12). When protonation occurs at secondary hydroxyl group of glycerol, a water molecule and a proton

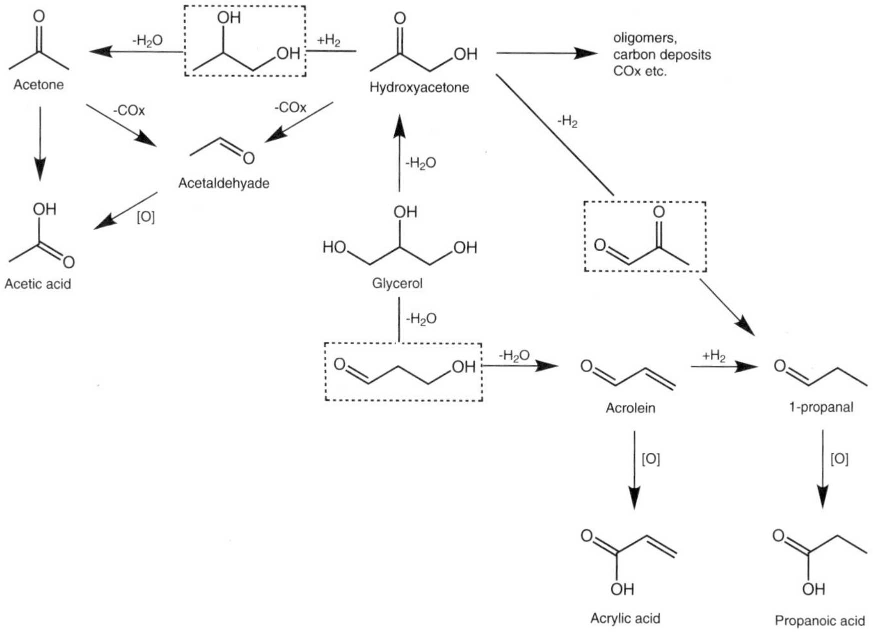

FIGURE 12.12 Proposed scheme for oxidative dehydration of glycerol. (Adapted from Ref. [59] with the permission of Elsevier © 2010.)

are eliminated from the protonated glycerol, and then 3-hydroxypropanal is produced via tautomerism of an enol. The 3-hydroxypropanal is readily dehydrated into acrolein and then oxidized to the desired product, acrylic acid, in the presence of oxygen. In contrast, when protonation proceeds at a primary hydroxyl group of glycerol, hydroxyacetone is produced through a sequence of dehydration, deprotonation, accompanied by tautomerism.

Although the yield of acrylic acid was low, these results encouraged authors to search for catalysts that are more active and selective for acrylic acid synthesis. Wang et al. [60] developed an embedded and bifunctional catalyst with FeO_x domains on the surface of $FeVO_4$ phase. The catalyst exhibited improved performance for oxidative dehydration of glycerol, with the best yield for acrylic acid reported as 14%. In the catalyst, $FeVO_4$ phase provided active sites for dehydration, and FeO_x domains provided active sites for the oxidation of acrolein to acrylic acid. The embedded structure could stabilize nanometer-sized FeO_x domains. This structure provided a suitable environment for the dehydration of glycerol and oxidation of acrolein to acrylic acid. Deleplanque et al. [58] have used molybdenum/tungsten/vanadium-based catalysts to conduct oxidative dehydration of glycerol and isolate acrylic acid. 28.7% yield of acrylic acid was obtained over MoVTeNbO catalyst with almost full conversion of glycerol. Although yields in acetic acid were also moderate (23%), catalyst deactivation was observed as well.

12.5 PRODUCTION OF FUEL AND FUEL ADDITIVES FROM GLYCEROL

Fuels are important for the industry and our high standards of life now demand a sustainable and reliable supply. With dwindling petroleum reserves and increasing concerns for the environment, developing new routes for the utilization of renewable energy sources are urgently required. Ethanol and biodiesel obtained from biomass as renewable fuels have already been developed and are covered in depth in other chapters of this book. New routes or new products are still needed to be discovered and established. The conversion of glycerol to fuels or additives can provide an alternative route for sustainable chemistry development.

12.5.1 Etherification of Glycerol to Fuel Additives

Fuel additives can enhance the performance of fuels greatly. For example, oxygenated additives can improve the combustion efficiency and reduce emission of pollutants. *Tert*-butyl ethers are one kind of oxygenated additive which bestow valuable performance enhancements. For example, methyl *tert*-butyl ether (MTBE) has excellent antidetonant and octane-improving properties as a fuel additive [4]. Therefore, the conversion of biomass to such additives is important and significant. The glycerol *tert*-butyl ether (GTBE) can be obtained through the etherification of glycerol with *iso*-butene or alcohols, which shows properties similar to MTBE in diesel and biodiesel. In the process of etherification, glycerol is converted to a mixture of mono-ethers, di-ethers, and tri-ethers (Fig. 12.13). Due to low solubility

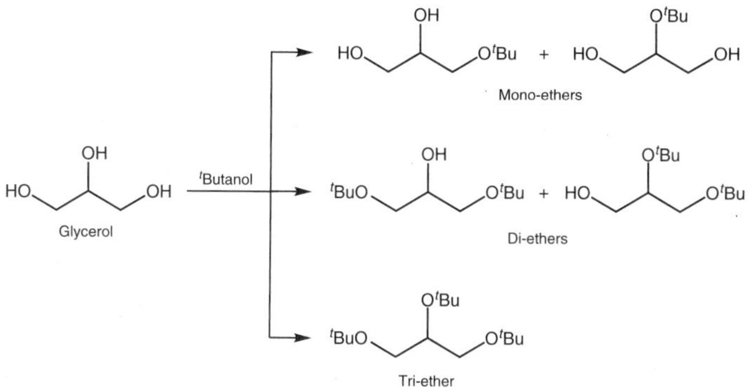

FIGURE 12.13 Etherification of glycerol to fuel additives.

of mono-*tert*-butyl ethers of glycerol in fuels, studies have focused on the etherification of glycerol to di-esters or tri-ethers.

The etherification of glycerol usually involves acid catalysis, including both homogeneous and heterogeneous acid catalysts, in which the latter are preferentially employed. Mravec's group studied etherification of glycerol with *iso*-butylene or *tert*-butyl alcohol over strong acid ion-exchange resins and large-pore size zeolites [61]. They found that zeolite catalysts have low catalytic performance, whereas the strong acid macro reticular ion-exchange resins have excellent properties. Selectivity of di-ethers and tri-ethers was greater than 92% with 100% conversion of glycerol over such ion-exchange resins. Detailed studies revealed that large pore diameter of resins promotes the reaction, whereas, zeolite catalysts have a relatively small pore size. Frusteri's group also studied the etherification of glycerol with *tert*-butyl alcohol [62], comparing the catalytic performances between the silica- supported acid catalysts and acid ion-exchange resins. The results revealed that activity of acid ion-exchange resin was higher than that of silica-supported acid catalyst. This was attributed to large pore diameter of the resins which allowed convenient access for reagent molecules to acidic sites. Lee et al. [63] conducted the etherification of glycerol with *iso*-butylene to obtain high yields of triether. With various acid catalysts, ionic liquid and heteropolyacid suppressed the formation of tri-ether in initial stage, whereas Ag^+ and Al^{3+} modified Amberlyst resin demonstrated higher yield for tri-ether formation.

12.5.2 Reforming of Glycerol to H_2 Gas or Syngas

Being a clean fuel with high energy density, H_2 is considered to be one of the most important alternative energy sources. Glycerol has a relatively high content of hydrogen; therefore, degradation of glycerol could lead to efficient H_2 or synthesis gas (mixture gas of H_2 and CO) preparation. H_2 can be directly used as gas fuel, and synthesis gas can also be used to produce fuels and chemicals through Fischer–Tropsch synthesis. Production of H_2 from glycerol is usually via reforming method, which includes steam reforming, aqueous phase reforming, and an autothermal reforming.

The method of steam reforming is studied widely and glycerol steam reforming is represented as [1]:

$$C_3H_8O_3 + 3H_2O \rightarrow 3CO_2 + 7H_2.$$

The advantage of steam reforming lies in using atmospheric pressure to improve H_2 selectivity. Noble-metal catalysts are very efficient for this reforming and the reforming temperature is usually lower than 450 °C. Shen's group studied the steam reforming of glycerol over ceria-supported metal catalysts and found that Ir/CeO_2 catalyst exhibited promising catalytic performance at 400 °C. Selectivity of H_2 was more than 85% at 100% glycerol conversion [64].

Aqueous-phase reforming could decrease reaction temperature and reduce the amount of carbon monoxide. Dumesic's group prepared Raney Ni–Sn catalyst for the aqueous-phase reforming of glycerol at 225 °C [65] and selectivity of H_2 could reach 75%. King's group conducted the same reaction over 3% Pt–3% Re/C [66] and highest hydrogen yield was obtained by the addition of KOH. Reaction pathways involved dehydrogenation and decarbonylation. These reactions were repeated until glycerol was completely converted to CO and H_2.

Schmidt's group studied the autothermal reforming of pure glycerol under aqueous solutions over Pt and Rh-based catalysts [67]. Autothermal reforming means that the process needs no external heat source, in which oxygen, steam, and glycerol are all reactants. In this kind of reforming of glycerol, H_2 selectivity was obtained as 79% at the full conversion of glycerol over Rh–Ce/Al_2O_3.

12.6 OTHER RECENT UTILIZATIONS OF GLYCEROL

As a low-carbon polyol, glycerol has considerable advantages in reacting with different substrates to produce valuable chemicals and fuels. In addition to the examples presented earlier in this chapter, other utilization methods for glycerol include esterification, condensation, and dehydrogenation reactions.

Esterification of glycerol with carboxylic acids could produce glycerol esters, including monoesters, diesters, or triesters, all of which can be used as emulsifiers in food industry, cosmetics, and pharmaceuticals, with some esters being used as valuable petrol fuel additives. Jagadeeswaraiah et al. [68] conducted esterification of glycerol with acetic acid over tungstophosphoric acid (TPA) supported on Cs-containing zirconia catalysts. Since esterification activity is related to the acidity of catalysts, Cs-containing catalysts with strong acidic sites have shown good catalytic performance. Liu et al. [69] used a series of Brønsted acidic ionic liquids as catalysts to conduct esterification of glycerol with acetic acid. In their study, they found that di-SO_3H-functionalized ILs exhibited excellent catalytic activity, and glycerol conversion was achieved as up to 95% within 30 min by using catalytic amounts of ionic liquids (only 0.1 mol% based on glycerol) (Fig. 12.14).

Apart from the oxidation of glycerol, DHA can also be obtained by dehydrogenation of glycerol. Crotti et al. [70] reported that dehydrogenation of

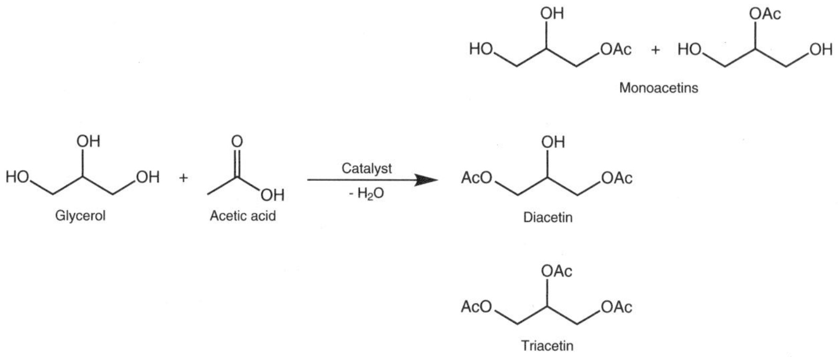

FIGURE 12.14 Esterification of glycerol with acetic acid.

glycerol was catalyzed by iridium complexes with P,N ligands, which was a base-free hydrogen-transfer reaction involving ketone, olefin, or aldehyde as hydrogen acceptor. Yield of DHA was 23% when benzaldehyde was used as hydrogen acceptor (Fig. 12.15).

CO_2 is produced as by-product and is emitted into the atmosphere during combustion of fossil resources. As a result, global climate is getting warmer, which has already become a contentious environmental issue. Therefore, utilization of CO_2 is a worthwhile goal [71]. From utilizing biomass-derived feedstocks, the conversion of glycerol with CO_2 to obtain glycerol carbonates serves as a promising sustainable route (Fig. 12.16). Glycerol carbonate is an intermediate chemical with many potential applications, which can be used to prepare new polymeric materials for the production of polycarbonates and polyurethanes, and can also be used as replacement solvents of ethylene carbonate or propylene carbonate. Direct method of glycerol carbonate production with ethylene carbonate and $scCO_2$ has been conducted under supercritical conditions [72]. Recently,

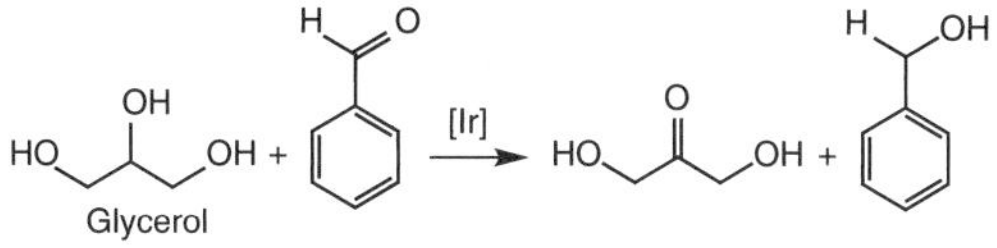

FIGURE 12.15 Hydrogen transfer reactions between glycerol and benzaldehyde over n-$Bu_2Sn(OMe)_2$ catalyst.

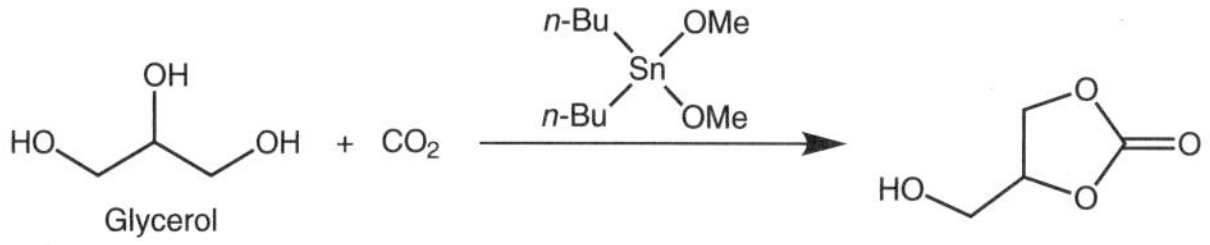

FIGURE 12.16 Glycerol reacted with CO_2 to glycerol carbonate over n-$Bu_2Sn(OMe)_2$ catalyst.

under more mild conditions direct carboxylation of glycerol with CO_2 under 5 MPa CO_2 and 450 K was realized [73]. The catalysts investigated were transition-metal alkoxides, including Sn-catalysts (n-$Bu_2Sn(OMe)_2$, n-Bu_2SnO, or $Sn(OMe)_2$). Glycerol successfully reacted with carbon dioxide to form glycerol carbonate over these catalysts.

12.7 OUTLOOK

Faced with diminishing fossil fuel reserves, developing sustainable resources is an essential issue. Biomass has attracted great attention as an important renewable resource. Due to the rapid development of technologies to realize commercially viable biodiesel, large amounts of glycerol are produced as the principle by-product. Methodologies that convert glycerol to value-added products, not only supply useful chemicals for the industry, but also enhance value, and indeed can offset cost of biodiesel production.

This chapter introduced recent studies of conversion of glycerol into valuable chemicals and fuels. Due to highly functionalized structure of glycerol, every hydroxyl group is ready to participate in reactions such as hydrogenolysis, oxidation, or dehydration. Hence an inherent problem is to control the selectivity and direct the reaction towards desired products. Therefore, a detailed study of the reactivity of each hydroxyl is necessary.

As the demand for fuels exceeds supply from nonrenewable petroleum resources, the usage of renewable biomass to produce fuels is essential. Developing efficient biodiesel synthesis is one type of sustainable route, while conversion of biodiesel-derived glycerol to fuels or additives provides an alternative. Current studies mainly focus on gas fuel production of H_2 and synthesis of additives (e.g. GTBE).

Lastly, reaction of glycerol with CO_2 can generate useful materials and also contribute to carbon capture of the greenhouse gas, CO_2. This process arguably offers an ideal application of glycerol with regard to the protection of the environment.

ACKNOWLEDGMENTS

The authors (NG, MG) wish to thank Enterprise Ireland (EI), the Irish Research Council for Science, Engineering and Technology (IRCSET), Science Foundation Ireland (SFI) and the Environmental Protection Agency (EPA) in Ireland for funding green chemistry research in Nicholas Gathergood's group.

REFERENCES

1. C. H. Zhou, J. N. Beltramini, Y. X. Fan and G. Q. Lu, *Chem. Soc. Rev.* **2008**, 37, 527–549.
2. T. Werpy and G. Petersen, Top Value Added Chemicals from Biomass: Volume I—Results of Screening for Potential Candidates from Sugars and Synthesis Gas, **2004**, Available electronically at http://www.osti.gov/bridge.

3. L. R. Morrison, *Kirk-Othmer Encyclopedia of Chemical Technology*, WileyNew York, **2001**: particular, glycerol, 4.
4. A. Behr, J. Eilting, K. Irawadi, J. Leschinski and F. Lindner, *Green Chem.* **2008**, 10, 13–30.
5. M. Pagliaro, R. Ciriminna, H. Kimura, M. Rossi and C. Della Pina, *Angew. Chem. Int. Ed.* **2007**, 46, 4434–4440.
6. H. Ma, X. Nie, J. Y. Cai, C. Chen, J. Gao, H. Miao and J. Xu, *Sci. China Chem.* **2010**, 53, 1497–1501.
7. N. Ji, T. Zhang, M. Y. Zheng, A. Q. Wang, H. Wang, X. D. Wang and J. G. G. Chen, *Angew. Chem. Int. Ed.* **2008**, 47, 8510–8513.
8. (a) T. Miyazawa, S. Koso, K. Kunimori and K. Tomishige, *Appl. Catal. A-Gen.* **2007**, 318, 244–251; (b) Y. Kusunoki, T. Miyazawa, K. Kunimori and K. Tomishige, *Catal. Commun.* **2005**, 6, 645–649; (c) T. Miyazawa, S. Koso, K. Kunimori and K. Tomishige, *Appl. Catal. A-Gen.* **2007**, 329, 30–35; (d) T. Miyazawa, Y. Kusunoki, K. Kunimori and K. Tomishige, *J. Catal.* **2006**, 240, 213–221.
9. M. Balaraju, V. Rekha, P. S. S. Prasad, B. L. A. P. Devi, R. B. N. Prasad and N. Lingaiah, *Appl. Catal. A-Gen.* **2009**, 354, 82–87.
10. S. Bolado, R. E. Trevino, M. T. Garcia-Cubero and G. Gonzalez-Benito, *Catal. Commun.* **2010**, 12, 122–126.
11. A. Alhanash, E. F. Kozhevnikova and I. V. Kozhevnikov, *Catal. Lett.* **2008**, 120, 307–311.
12. I. Gandarias, P. L. Arias, J. Requies, M. B. Guemez and J. L. G. Fierro, *Appl. Catal. B-Environ.* **2010**, 97, 248–256.
13. (a) E. P. Maris and R. J. Davis, *J. Catal.* **2007**, 249, 328–337; (b) E. P. Maris, W. C. Ketchie, M. Murayama and R. J. Davis, *J. Catal.* **2007**, 251, 281–294.
14. Z. L. Yuan, P. Wu, J. Gao, X. Y. Lu, Z. Y. Hou and X. M. Zheng, *Catal. Lett.* **2009**, 130, 261–265.
15. (a) J. Zhao, W. Q. Yu, C. Chen, H. Miao, H. Ma and J. Xu, *Catal. Lett.* **2010**, 134, 184–189; (b) W. Q. Yu, J. Xu, H. Ma, C. Chen, J. Zhao, H. Miao and Q. Song, *Catal. Commun.* **2010**, 11, 493–497; (c) W. Q. Yu, J. Zhao, H. Ma, H. Miao, Q. Song and J. Xu, *Appl. Catal. A-Gen.* **2010**, 383, 73–78.
16. M. A. Dasari, P. P. Kiatsimkul, W. R. Sutterlin and G. J. Suppes, *Appl. Catal. A-Gen.* **2005**, 281, 225–231.
17. Z. W. Huang, F. Cui, H. X. Kang, J. Chen, X. Z. Zhang and C. G. Xia, *Chem. Mater.* **2008**, 20, 5090–5099.
18. L. Y. Guo, J. X. Zhou, J. B. Mao, X. W. Guo and S. G. Zhang, *Appl. Catal. A-Gen.* **2009**, 367, 93–98.
19. J. X. Zhou, L. Y. Guo, X. W. Guo, J. B. Mao and S. G. Zhang, *Green Chem.* **2010**, 12, 1835–1843.
20. J. Chaminand, L. Djakovitch, P. Gallezot, P. Marion, C. Pinel and C. Rosier, *Green Chem.* **2004**, 6, 359–361.
21. S. Wang and H. C. Liu, *Catal. Lett.* **2007**, 117, 62–67.
22. M. Balaraju, V. Rekha, P. S. S. Prasad, R. B. N. Prasad and N. Lingaiah, *Catal. Lett.* **2008**, 126, 119–124.
23. A. Bienholz, F. Schwab and P. Claus, *Green Chem.* **2010**, 12, 290–295.
24. S. A. Wang, Y. C. Zhang and H. C. Liu, *Chem. Asian J.* **2010**, 5, 1100–1111.

25. (a) T. M. Che, US Patent Application 464 2394, **1987**; (b) E. Drent, US Patent Application 6 080 898, **2000**.
26. T. Kurosaka, H. Maruyama, I. Naribayashi and Y. Sasaki, *Catal. Commun.* **2008**, 9, 1360–1363.
27. L. F. Gong, Y. Lu, Y. J. Ding, R. H. Lin, J. W. Li, W. D. Dong, T. Wang and W. M. Chen, *Chin. J. Catal.* **2009**, 30, 1189–1191.
28. L. F. Gong, Y. Lu, Y. J. Ding, R. H. Lin, J. W. Li, W. D. Dong, T. Wang and W. M. Chen, *Appl. Catal. A-Gen.* **2010**, 390, 119–126.
29. L. Z. Qin, M. J. Song and C. L. Chen, *Green Chem.* **2010**, 12, 1466–1472.
30. Y. Nakagawa, Y. Shinmi, S. Koso and K. Tomishige, *J. Catal.* **2010**, 272, 191–194.
31. L. Huang, Y. L. Zhu, H. Y. Zheng, G. Q. Ding and Y. W. Li, *Catal. Lett.* **2009**, 131, 312–320.
32. N. Ueda, Y. Nakagawa and K. Tomishige, *Chem. Lett.* **2010**, 39, 506–507.
33. C. Montassier, J. C. Menezo, L. C. Hoang, C. Renaud and J. Barbier, *J. Mol. Catal.* **1991**, 70, 99–110.
34. R. Garcia, M. Besson and P. Gallezot, *Appl. Catal. A-Gen.* **1995**, 127, 165–176.
35. F. Porta and L. Prati, *J. Catal.* **2004**, 224, 397–403.
36. I. Sobczak, K. Jagodzinska and M. Ziolek, *Catal. Today* **2010**, 158, 121–129.
37. A. Villa, A. Gaiassi, I. Rossetti, C. L. Bianchi, K.van Benthem, G. M. Veith and L. Prati, *J. Catal.* **2010**, 275, 108–116.
38. N. Dimitratos, J. A. Lopez-Sanchez, J. M. Anthonykutty, G. Brett, A. F. Carley, R. C. Tiruvalam, A. A. Herzing, C. J. Kiely, D. W. Knight and G. J. Hutchings, *Phys. Chem. Chem. Phys.* **2009**, 11, 4952–4961.
39. B. N. Zope, D. D. Hibbitts, M. Neurock and R. J. Davis, *Science* **2010**, 330, 74–78.
40. A. Villa, G. M. Veith and L. Prati, *Angew. Chem. Int. Ed.* **2010**, 49, 4499–4502.
41. H. Kimura, *Appl. Catal. A-Gen.* **1993**, 105, 147–158.
42. W. B. Hu, D. Knight, B. Lowry and A. Varma, *Ind. Eng. Chem. Res.* **2010**, 49, 10876–10882.
43. R. M. Painter, D. M. Pearson and R. M. Waymouth, *Angew. Chem. Int. Ed.* **2010**, 49, 9456–9459.
44. S. Demirel, K. Lehnert, M. Lucas and P. Claus, *Appl. Catal. B-Environ.* **2007**, 70, 637–643.
45. B. Kundu, *Tetrahedron Lett.* **1992**, 33, 3193–3196.
46. C. W. Chin, M. A. Dasari, G. J. Suppes and W. R. Sutterlin, *AICHE J.* **2006**, 52, 3543–3548.
47. S. S. Niu, Y. L. Zhu, H. Y. Zheng, W. Zhang and Y. W. Li, *Chin. J. Catal.* **2011**, 32, 345–351.
48. J. Zhao, W. Q. Yu, D. C. Li, H. Ma, J. Gao and J. Xu, *Chin. J. Catal.* **2010**, 31, 200–204.
49. S. Sato, M. Akiyama, R. Takahashi, T. Hara, K. Inui and M. Yokota, *Appl. Catal. A-Gen.* **2008**, 347, 186–191.
50. A. K. Kinage, P. P. Upare, P. Kasinathan, Y. K. Hwang and J. S. Chang, *Catal. Commun.* **2010**, 11, 620–623.
51. A. Alhanash, E. F. Kozhevnikova and I. V. Kozhevnikov, *Appl. Catal. A-Gen.* **2010**, 378, 11–18.

52. B. Katryniok, S. Paul, M. Capron, C. Lancelot, V. Belliere-Baca, P. Rey and F. Dumeignil, *Green Chem.* **2010**, 12, 1922–1925.
53. C. J. Jia, Y. Liu, W. Schmidt, A. H. Lu and F. Schuth, *J. Catal.* **2010**, 269, 71–79.
54. Y. T. Kim, K. D. Jung and E. D. Park, *Micropor. Mesopor. Mat.* **2010**, 131, 28–36.
55. L. Z. Tao, S. H. Chai, Y. Zuo, W. T. Zheng, Y. Liang and B. Q. Xu, *Catal. Today* **2010**, 158, 310–316.
56. Q. B. Liu, Z. Zhang, Y. Du, J. Li and X. G. Yang, *Catal. Lett.* **2009**, 127, 419–428.
57. W. Suprun, M. Lutecki, T. Haber and H. Papp, *J. Mol. Catal. A-Chem.* **2009**, 309, 71–78.
58. J. Deleplanque, J. L. Dubois, J. F. Devaux and W. Ueda, *Catal. Today* **2010**, 157, 351–358.
59. (a) F. Wang, J. L. Dubois and W. Ueda, *Appl. Catal. A-Gen.* **2010**, 376, 25–32; (b) F. Wang, J. L. Dubois and W. Ueda, *J. Catal.* **2009**, 268, 260–267.
60. F. Wang, J. Xu, J. L. Dubois and W. Ueda, *ChemSusChem* **2010**, 3, 1383–1389.
61. K. Klepacova, D. Mravec and M. Bajus, *Appl. Catal. A-Gen.* **2005**, 294, 141–147.
62. F. Frusteri, F. Arena, G. Bonura, C. Cannilla, L. Spadaro and O.Di Blasi, *Appl. Catal. A-Gen.* **2009**, 367, 77–83.
63. H. J. Lee, D. Seung, K. S. Jung, H. Kim and I. N. Filimonov, *Appl. Catal. A-Gen.* **2010**, 390, 235–244.
64. B. C. Zhang, X. L. Tang, Y. Li, Y. D. Xu and W. J. Shen, *Int. J. Hydrogen Energ.* **2007**, 32, 2367–2373.
65. R. D. Cortright, R. R. Davda and J. A. Dumesic, *Nature* **2002**, 418, 964–967.
66. D. L. King, L. A. Zhang, G. Xia, A. M. Karim, D. J. Heldebrant, X. Q. Wang, T. Peterson and Y. Wang, *Appl. Catal. B-Environ.* **2010**, 99, 206–213.
67. P. J. Dauenhauer, J. R. Salge and L. D. Schmidt, *J. Catal.* **2006**, 244, 238–247.
68. K. Jagadeeswaraiah, M. Balaraju, P. S. S. Prasad and N. Lingaiah, *Appl. Catal. A-Gen.* **2010**, 386, 166–170.
69. X. Liu, H. Ma, Y. Wu, C. Wang, M. Yang, P. Yan and U. Welz-Biermann, *Green Chem.* **2011**, 13, 697–701.
70. C. Crotti, J. Kaspar and E. Farnetti, *Green Chem.* **2010**, 12, 1295–1300.
71. A. J. Hunt, E. H. K. Sin, R. Marriott and J. H. Clark, *ChemSusChem* **2010**, 3, 306–322.
72. C. Vieville, J.W. Yoo, S. Pelet and Z. Mouloungui, *Catal. Lett.* **1998**, 56, 245–247.
73. M. Aresta, A. Dibenedetto, F. Nocito and C. Pastore, *J. Mol. Catal. A-Chem.* **2006**, 257, 149–153.

CHAPTER 13

Ultrasonics for Enhanced Fluid Biofuel Production

DAVID GREWELL and MELISSA MONTALBO-LOMBOY

13.1 INTRODUCTION

In the last decade, the biofuels industry has been investigating ultrasonics as a means to improve the rates and efficiencies of biofuel production. The use of ultrasonics to enhance chemical pathways has already been widely demonstrated in a variety of fields [1]. In addition, the introduction of very high power systems (>200,000 W of mechanical energy), and new tooling designs has made the adoption of ultrasonics even more attractive. In the majority of ultrasonic applications, the pathways that result in faster and more efficient biofuel production are mechanical. That is, utilization of ultrasonics improves mass transport by enhanced mixing and increased available reaction sites through increased surface area. While some benefits have been achieved through traditional sonochemistry and the production of free radicals, these benefits are secondary and typically not significant.

This chapter provides an overview of ultrasonics and its effect in liquids, the predominate medium for biofuel production. With these fundamentals described, the use of ultrasonics in a range of biofuel production processes including energy balances, enhanced pathways, and costs is analyzed.

Before ultrasonics and biofuels are discussed, however, the important terms "high-power ultrasonics" and "low-power ultrasonics" must be defined. Although no clear distinction exists between low and high power in terms of power or power density, the difference is generally accepted as related to their respective effects on the media being treated. Low-power ultrasonics has no lasting chemical or mechanical effects, while high-power ultrasonics does have long-term effects. As the goal of this work is to convert biomass into fuels or value-added chemicals, the effects are long lasting. Thus, the term ultrasonics implies "high-powered ultrasonics."

The Role of Green Chemistry in Biomass Processing and Conversion, First Edition.
Edited by Haibo Xie and Nicholas Gathergood.

13.2 ULTRASONICS

Ultrasonics is literally defined as mechanical vibrations above the human audible range (+18–20 kHz). However, in some cases, frequencies as low as 10–15 kHz have been included because of the similar effects and ability to produce higher power levels. These vibrations have four major attributes: frequency (f), amplitude, speed (v), and wavelength (λ). Three of these attributes are related: frequency, speed, and wavelength, as defined in Equation 13.1.

$$\lambda = \frac{v}{f}. \tag{13.1}$$

The speed of sound is primarily dependent on the material through which it passes. For example, in air at 20 °C, the speed of sound is approximately 340 m s^{-1}, while in water the speed is approximately 1480 m s^{-1}. It is important to note that these estimations assume a nondispersive medium, which is relatively true for air and water. The speed of sound in dispersive media is frequency-dependent. Thus, in a dispersive media, a broad-spectrum sound (sound containing multiple frequencies) is generated from a point source which would separate into packets of common frequencies.

There are different types of vibration in various media. For example, in wire, a vibration can translate as transverse and/or longitudinal. A wave that propagates out from a source tends to do so in a spherical shape. A good example is the comparison to the waves created when a stone and a plate drop into a still pond. The stone creates waves that travel in a circular pattern. The plate, however, creates waves that are planer near the plate but, as the distance from the plate increases (several dimensions of the length of the plate), the waves will become more spherical (this is further detailed in later sections). When a wave travels along an infinitely long tube, it will continue outward, thereby forming a standing wave. If the tube is ½ a wavelength or a multiple of a ½ a wavelength long and the end of the tube terminates with a rigid wall, the reflected wave will be superimposed on the original wave, so that a standing wave is also produced. This results in areas of no displacement (nodal points) and areas of maximum displacement (antinodal points or peaks), as seen in Figure 13.1.

13.2.1 Power

Fundamentally, energy is the ability to do work. The rate of energy dissipation (W, J s^{-1}) is power. In mechanical terms, power is force multiplied by velocity. Thus, for a given load, the power is proportional to the velocity of a vibrating tool (horn), as well as a "looseness" (how well the horn and load move together) of the load to which the horn is coupled. In general, power is proportional to the stiffness of the load (the loose component), frequency, and the amplitude squared. This relationship is generally true for solid loads, assuming perfect coupling between the load and horn face. With fluid loads, the power is generally proportional to the amplitude (not the amplitude squared).

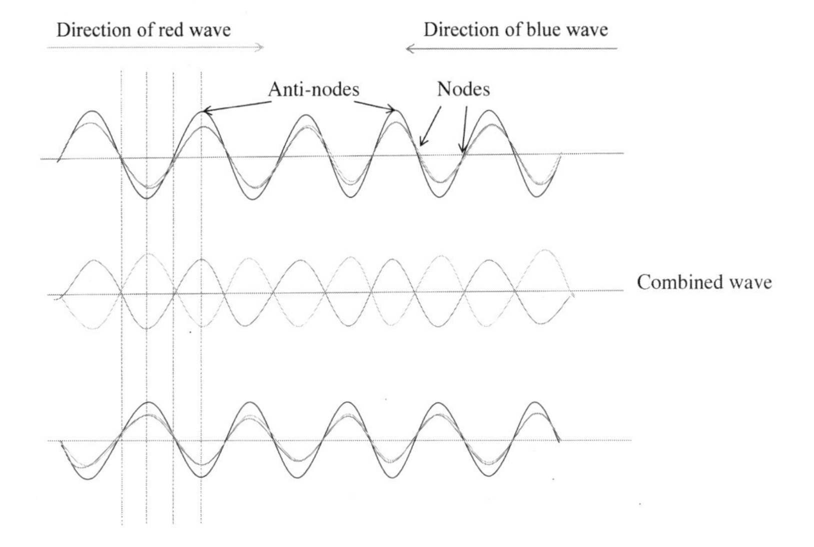

FIGURE 13.1 Illustration of standing waves.

While it would be convenient to operate at a defined power level, most power supplies simply operate at a defined amplitude. Thus, the actual dissipated power is related not only to the amplitude, but also to the load. The more the losses in the load, the greater is the dissipated load. It is important to note that for a given system, the load may vary as a function of time and temperature because the mechanical properties of the load change during treatment. However, with a liquid load, which is the most common load in the area of biofuels, the dissipation is different because shearing is a significant dissipation mechanism. In this case, the dynamic viscosity can correspond to the lossy factors.

While a common unit of power is Watts (W), two other measurements are often employed in mechanical processing. These are power density and intensity. Power density (PD) is often used in static or batch treatment of biomass (or other liquid loads) and is the average dissipated power (mechanical) over a treatment time divided by the volume of the biomass; the resulting units are $W{\cdot}L^{-1}$ (or equivalents). In addition, it can be useful to report the energy density (ED), where the mechanical energy dissipated over a given time (t) period is determined by integrating the power as a function of time and dividing this energy by the volume, so that the units are $J\,L^{-1}$. These two relationships are displayed as:

$$\mathrm{PD} = \frac{P_{\mathrm{avg}}}{\mathrm{volume}} \Rightarrow \frac{W}{L} \tag{13.2a}$$

$$\mathrm{ED} = \frac{E}{\mathrm{volume}} = \frac{\int_0^t P\,dt}{L} \Rightarrow \frac{J}{L} \tag{13.2b}$$

These relationships become particularly useful when scaling from a batch process to a continuous flow-through process that has a flow rate of q (with units such as $\mathrm{L\,s^{-1}}$). For example, it is common to optimize a process in batch processing, and then scale-up to a continuous flow process for pilot-plant demonstration or production. By using the energy density of the batch process, it is possible to estimate the optimum condition of power and flow rate. More specifically, the energy density for a continuous flow-through process is defined as:

$$\mathrm{ED} = \frac{P}{q} \Rightarrow \frac{J}{L}. \tag{13.3}$$

Thus, by measuring the dissipated power, it is possible to match the energy densities by adjusting the flow rate or amplitude accordingly. It is important to note that because of the geometric difference between a batch container and a continuous-flow container, this only provides a first-order approximation.

Acoustic intensity is the dissipated power divided by the area. For example, in the case of a horn that has a face with an area of 1 cm^2, the intensity is simply the dissipated mechanical power divided by that area. In addition, it is possible to define the intensity in terms of the mechanical impedance (stiffness, Z) of a load.

$$I = \frac{P}{A} = \frac{p^2}{Z} = \xi^2 \omega^2 Z, \tag{13.4}$$

where, P is the power, p the acoustic pressure, and ξ the particle displacement. In addition, the impedance is defined as:

$$Z = c\rho, \tag{13.5}$$

where c is the speed of sound in the fluid and ρ the fluid density.

13.3 NEAR AND FAR FIELD

As with any propagating wave from a source, attenuation and diffraction are the two major factors that define the wave field. These factors are critical in determining the design of the tooling and reaction chamber for the desired chemical reaction. If the reactants are near the source of the ultrasonics, then the field intensity is relatively uniform. The Rayleigh distance (R_0) is often used to determine the maximum distance between the source with a radius (a) and the substrate in the near-field condition. The Rayleigh distance is based on the wavelength (λ)of the sound wave in the fluid and is defined in Eq. 13.6.

$$R_0 = \frac{\pi a^2}{\lambda}. \tag{13.6}$$

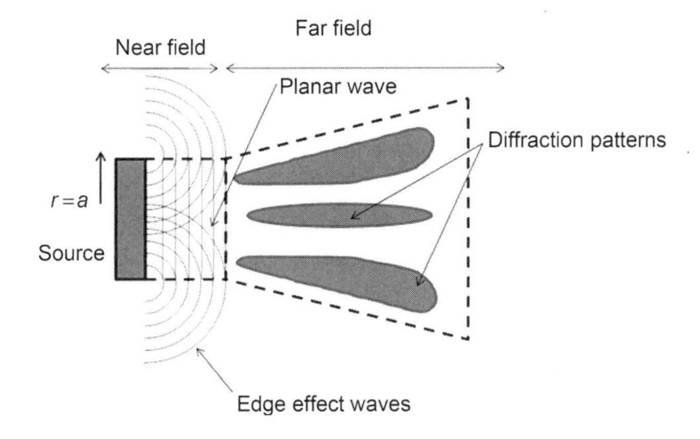

FIGURE 13.2 Illustration of near and far field.

Beyond this distance, diffraction patterns are generated which produce constructive and destructive (minimum and maximum) intensity patterns. It is important to note that diffraction can also be generated by waves reflected from the reaction chamber. The Rayleigh distance assumes that the only diffraction patterns are generated by edge effects of the ultrasonic source as seen in Figure 13.2.

Any interface with a media with different impedance (Z) will produce a reflection (R) that may result in complex diffraction patterns. The relative intensity of the reflective wave can be determined using Eq. 13.7, where Z_1 and Z_2 are the impedance of the liquid and the material of the interfacial material, such as a glass beaker or stainless reaction chamber.

$$R = \frac{(Z_2 - Z_1)}{(Z_2 + Z_1)}. \tag{13.7}$$

In a practical application, substrates are commonly treated outside the near-field condition. They still assume uniform treatment as long as the substrate is mixed, either internally by streaming (detailed in Section 13.3.3) or externally by a mixing system. In addition to diffraction, attenuation can reduce the effective distance between the source and the substrate.

13.3.1 Frequency

Ultrasonic waves are sound waves above 18–20 kHz that are above the human audible range. The operating frequency of a system greatly affects its performance. Higher frequencies (+100 kHz) tend to be more effective in generating free-radical species (detailed in Section 13.3.3.4), which can be very effective in promoting sonochemical reactions. However, if mass transfer is a limiting factor, lower frequencies can be more effective. Thus, it is common to distinguish between sonochemistry (free-radical generation) and sonomechanical systems (mass-transfer driven). While both effects are observed over a wide range of frequencies, lower frequencies (<40–60 KHz) tend to be more effective in mass transfer and higher frequencies (>100 kHz) tend to be more effective in free-radical generation. All of these effects are further detailed in Section 13.3.3. However, it is important to know the limiting factors of the pathways that need to be enhanced to optimize the production of the particular products of interest, in this case, biofuels. As will be detailed in Section 13.4, the primary limiting factor for enhanced biofuel production is mass transfer. Thus, with current technologies, lower frequencies tend to be more effective in enhancing biofuel production, particularly for large-scale production.

The frequency of an ultrasonic system defines the maximum allowable power generation for a given ultrasonic transducer (the motor). Because these motors (electric-to-mechanical energy converters) are designed to resonate at the operating frequency, their physical size is inversely proportional to frequency, thereby limiting the maximum deliverable power of an individual converter (motor). Thus, to maximize the efficiency of the converters, they are designed to resonate at the operating frequency. As a result, efficiencies greater than 95% can be achieved [2]. This contrasts with forced vibration systems (operating at nonresonance),where the efficiency can be very low (<10%) and the maximum allowable power is greatly limited. Because the physical size of the converter is inversely proportional to frequency, a ½ wavelength converter at 20 kHz is approximately 12–14 cm in length, while at 40 kHz, its length is only 6–7 cm. As the physical size decreases, and thus for a defined power (W) the power density (W cm^{-3}) increases, resulting in heating of the converter. Forced air is used to enhance the dissipation of the generated heat. This is often achieved with very high-power densities, but because the physical size of the converter defines the relative amount of surface area for a given converter, the amount of power that can be dissipated (removed) from a converter is limited. If the heat is not effectively removed, the converter will heat, thereby changing the mechanical resonance frequency, increasing the losses within the components, further decreasing the efficiency, and further increasing the rate of heating, causing thermal run away, that is, the temperature increases uncontrollably.

13.3.2 Ultrasound Generation

Overall, the effective production of ultrasonics requires three major components: power supply, converter, and tooling. Each is discussed in detail in the following sections.

13.3.2.1 Power Supply The power supply has several key functions. The first is to convert the line voltage and current to match the specifications of the converter. This means high voltage/low current for piezoelectric converters and low voltage/high current for magnetostrictive converters. In addition, the power supply must monitor the voltage and current to the converter and minimize the phase between the two, so that the system operates at the resonant frequency. While the power supply and stack assembly is an electromechanical system, the resonant frequency is primarily defined by the mechanical component (stack frequency and load). The other function of the power supply is process control. For example, in a batch mode treatment, the power supply can be set to treat the load (sample) for a predetermined length of time or to dissipate a predetermined amount of energy. The energy is measured by integrating the electrical power delivered to the converter over time. Thus, assuming the converter losses are minimal, this energy corresponds to the mechanical energy dissipated in the load. Most power supplies can also report dissipated energy using a time mode or even peak power. In many power supplies, it is possible to electrically set the amplitude of vibrations produced in the converter. While this is convenient because the stack does not have to be disassembled and the amplitude can be varied over a wide range, this type of amplitude control reduces the maximum available power that can be generated by the power supply. For example, if a 2000 W power supply is operated at 50% amplitude, its maximum available power will typically be 1000 W. Thus, it is recommended that the amplitude be primarily set by the gain in the stack (mechanical gain as detailed below) and the amplitude control of the power supply be used only for fine adjustment.

13.3.2.2 Converter There are two primary converter designs, piezoelectric and magnetostrictive. Each has its own advantages and limitations, but overall, more piezoelectric designs are used in industry, primarily because of their high efficiency and relatively low costs. For example, piezoelectric converters rely on materials that change physically in size when exposed to an electric field. In most cases, the electric field is produced by relatively high voltages (500–1000 V) across a relatively short distance (4–6 mm). This results in a high voltage gradient that allows these materials to expand and contract with relatively moderate displacements (1–3 μm). By stacking multiple actuators/discs together into a single converter and clamping them together, many microns of displacement can be generated. These single actuators are typically made from ceramics and sintered from a powder in an electric field to produce a predefined polarity, which allows the actuator to move equally in both directions. Because these converters operate at relatively high voltages (>500 V) and relatively low currents (approximately 1–10 amps), their resistive losses are minimal and they tend to operate at very high efficiencies (>95%). In addition, because the basis (polarity) is set into the material at a high temperature, typically they are thermally stable. However, because these systems are usually made from ceramic actuators in the shape of a disc, they are not tolerant of adverse loadings, such as flexural mode or high transient loads, such as impacts, that can cause the ceramic components to fail in a brittle mode. In contrast, magnetostrictive-based converters are typically made from tougher materials that can tolerate such adverse loads. In magnetostrictive designs, the actuating material is one that changes shape in a magnetic field, such as nickel. Newer materials, such as Terfenol [3], also have the additional benefit of being able to produce relatively high displacements. However, producing a magnetic field with sufficient strength to cause this displacement requires a high current that promotes resistive thermal losses. These thermal losses often have to be dissipated with heat exchangers (liquid/gas phase change types). These heat exchangers cool the converter below these materials' relatively low Curry point, where the bias can be lost and greatly reducing the effect efficiency of the system. Because of the resistive losses and losses in the cooling systems, magnetostrictive designs tend to operate with efficiencies below 50% [4].

13.3.2.3 Tooling Once the ultrasonic vibrations are generated in the converter, the vibrations are often amplified by a mechanical booster. These devices are usually made from aluminum or titanium because of their high stiffness-to-density ratio. In addition, the devices are also typically ½ wavelength long and designed to resonate at the operating frequency. The boosters are usually manufactured with discrete gains in ratios of 0:0.6, 1:1, 1:1.5, 1:2.0, and 1:2.5 and color coded to identify the gain. The ultrasonic energy is then transmitted to the horn, which delivers the energy into the work piece or load. Similar to the booster, its design may increase the amplitude through a mechanical gain, and it is often ½ or a full wavelength long and made from aluminum or titanium. However, the horn design must match the specifications of the application. In treatment of liquids in a batch mode, it is common to have a flat-face horn that simply dissipates the energy into the liquid as an expanding wave front. Other designs are submersible, which are simply sealed metal boxes that have transducers coupled to their inner walls. These boxes are then submersed in a fluid bath. It is also possible to use flat-faced horn designs for continuous treatment or with a specially designed reaction chamber.

13.3.3 Effects in Liquids

When ultrasonic energy is dissipated in a fluid at high power levels, there are a number of possible effects, including cavitation, acoustic streaming, heating, and free-radical generation. All of these can dramatically influence the media/fluid in a number of ways. For example, if solid biomass is suspended in the fluid, the cavitation can break the individual particles into smaller particles. If the fluid consists of two immiscible liquids, the cavitation and streaming can form stable emulsions. In both cases, the surface area-to-volume ratio of the various phases increases, thus increasing the number of possible reaction sites. These effects simply enhance the chemical pathways through mechanical means. In contrast, free-radical generation can have direct chemical effects on the substrate, as these species are highly reactive. However, for most industrial systems, the generation of free radicals is insignificant compared to cavitation and acoustic streaming (also called micro-streaming). These effects are detailed in the following sections.

13.3.3.1 Cavitation Many industrial applications that employ ultrasonic liquid processing rely on cavitation as the primary phenomenon. When an ultrasonic wave travels through a liquid, an alternating pressure gradient is created throughout the medium, thereby putting the liquid under varying stress as seen in Figure 13.3. In practical terms, the liquid undergoes cyclic tensile loading and compression loading. At a certain load, a bubble is nucleated and entraps dissolved gases and vapors from the liquid. On reaching a critical size, the bubble implodes, compressing all the gases and vapors and creating a local massive energy release with a shock wave or jetting as depicted in Figure 13.4. This section will detail the different

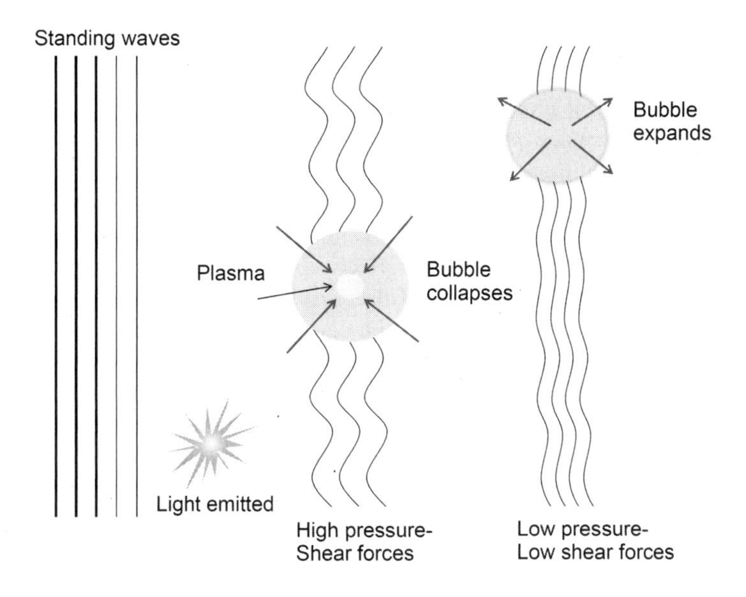

FIGURE 13.3 Illustration of acoustic cavitation [5,6].

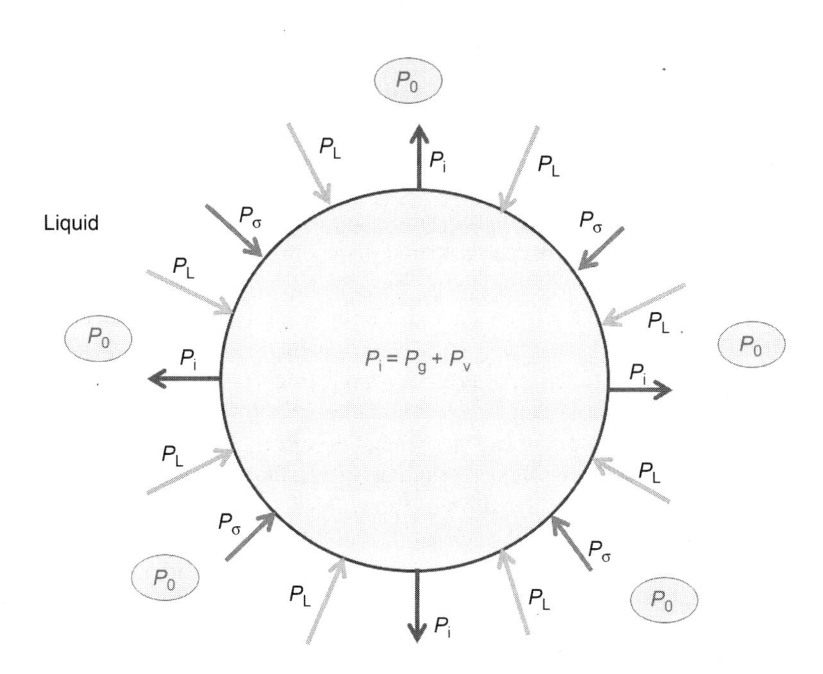

FIGURE 13.4 Schematic representation of pressures in a static gas bubble.

mechanisms of nucleation, along with the types of collapse and the effects of ultrasonics in liquids.

Cavitation by homogeneous nucleation principally occurs because of fracture/tearing of a liquid under very high stress conditions that are created in the medium because of the pressure gradient, as seen in Figures 13.1 and 13.2 [5]. Under real conditions, homogeneous nucleation is hindered by the presence of impurities, such as dust in the liquid or surface roughness near interfaces. The impurities in the liquid act as stress concentrators, reducing the effective strength of the medium which leads to nucleation at relatively low stress levels [5,7,8]. This mechanism of nucleation is called heterogeneous nucleation and is the only type of nucleation observed under practical conditions.

After nucleation, the bubble containing the dissolved gases and vapor freed from the liquid expands and contracts under the dynamic pressure gradient traveling through the liquid medium. Because of certain effects, the net diffusion of gas into the bubble increases with time because of rectified diffusion [6,8]. Rectified diffusion can be explained by two mechanisms: surface area effect and shell effect. During the compression of a bubble, the concentration of gases inside the bubble increases with respect to the medium immediate to the bubble surface. This results in a net outward diffusion of gases from the bubble into the medium. However, because the bubble is compressed, its surface area is relatively small, thereby limiting the outward diffusions. During the expansion cycle, the bubble's surface area increases, as does its volume, which causes a reduction in the concentration of gases inside the bubble. Now, inward diffusion of gases occurs and may reach relatively high levels because of the increased amount of surface area. With the cyclic pressure and mass transfer in and out of the bubble, the net mass of gases inside the bubble grows continuously until it reaches a critical size [6,8].

The second mechanism of rectified diffusion is called the shell effect. In a contracted bubble, the immediate medium at the surface of the bubble has a deficiency of gases and thus resembles a shell. This results in a net diffusion of gases away from the bubble. When the bubble expands, the relative concentration of gases is lower inside the bubble as compared to the new shell, thus resulting in a net inward diffusion of gases. Similar to the surface-area effect, the bubble continues increasing until it reaches a critical size [7]. The bubble growth continues because of rectified diffusion until it reaches a critical size, which occurs when a balance is achieved between the forces/pressures exerted by the medium and the surface tension. At that point, the internal gas pressure is broken, which causes the bubble to collapse. Figure 13.4 explains this phenomenon. Note that the bubble will collapse if the sum of the inward pressures, including P_σ (surface tension pressure), liquid pressure near the bubble P_L and hydrostatic pressure (P_O) are greater than P_i (internal gas pressure). In this case, P_i is the sum of the gas (P_g) and vapor pressure (P_v). The bubble can collapse symmetrically or asymmetrically. In the symmetrical mode, the bubble's radius reduces uniformly and implodes on itself, sending an explosive shock wave (a wave traveling faster than the speed of sound) through the medium, as seen in Figure 13.4. In the asymmetrical mode, which is caused by instability of the bubble and the surrounding medium, caving of the bubble surface occurs and an

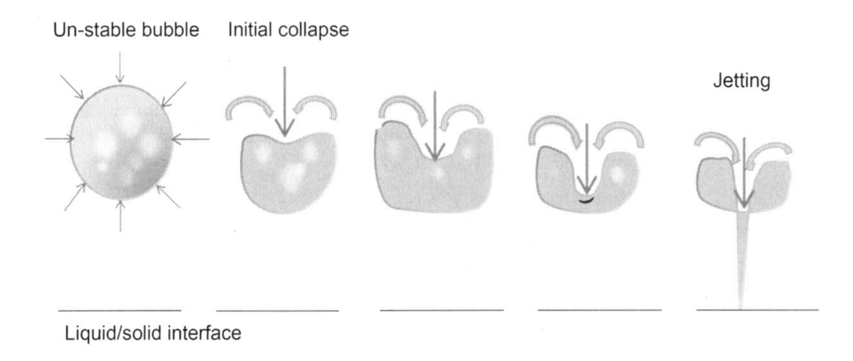

FIGURE 13.5 Asymmetric collapse of bubble via jetting [9].

intermediate donut shape is formed before collapsing by jetting, as seen in Figure 13.5. So, the symmetrical implosion creates a shock wave; asymmetrical implosion results in jetting, which is equally destructive in power.

13.3.3.2 Acoustic Streaming Acoustic streaming is another mechanism that occurs when a liquid is sonicated. Acoustic streaming [10] was first studied in 1831; it occurs at a solid/liquid interface when the solid interface experiences harmonic vibrations. The main benefit of streaming in liquid processing is mixing, which facilitates uniform distribution of ultrasound energy within the sludge mass, convection of the liquid, and distribution of any heating that occurs. Overall, there are three regions of acoustic streaming. The largest region, Eckart streaming (region I as shown in Fig. 13.6), is furthest from the vibrating tool. It has circulating currents that are defined by the shape of the container and are of the size of the wavelength of the acoustic wave in the liquid. The region near the tooling (region II) has circulating currents whose size and shape are primarily defined by the acoustic tooling. These circulations, called Rayleigh streaming, are much smaller than the wavelength of the acoustic wave in the liquid. The region nearest the tool (region III) is called Schlichting streaming region and its size is similar in size to wavelengths [11]. All three regions play a critical role in mixing of fluids.

13.3.3.3 Heating Once the ultrasonic energy is discontinued and the internal movement of streaming and cavitation are dissipated, the only lasting energy remaining in the system is heat. In fact, calorimetry has been used to measure the energy efficiency of the ultrasonic system, suggesting that nearly all the energy is eventually converted into heat [4]. Obviously, some of the energy is dissipated in particle size breakdown and sound, but the majority is heat. Thus, as further detailed in the "energy balance" Section 13.5.8, it is critical that the process/application is enhanced or accelerated enough to justify the use of ultrasonics. That is to say, one must ask the question "would it be more cost effective to use heat alone?"

In many experiments, it is desirable to minimize the temperature increase because of the heat generated by the ultrasonics. This allows the effects of the ultrasonics,

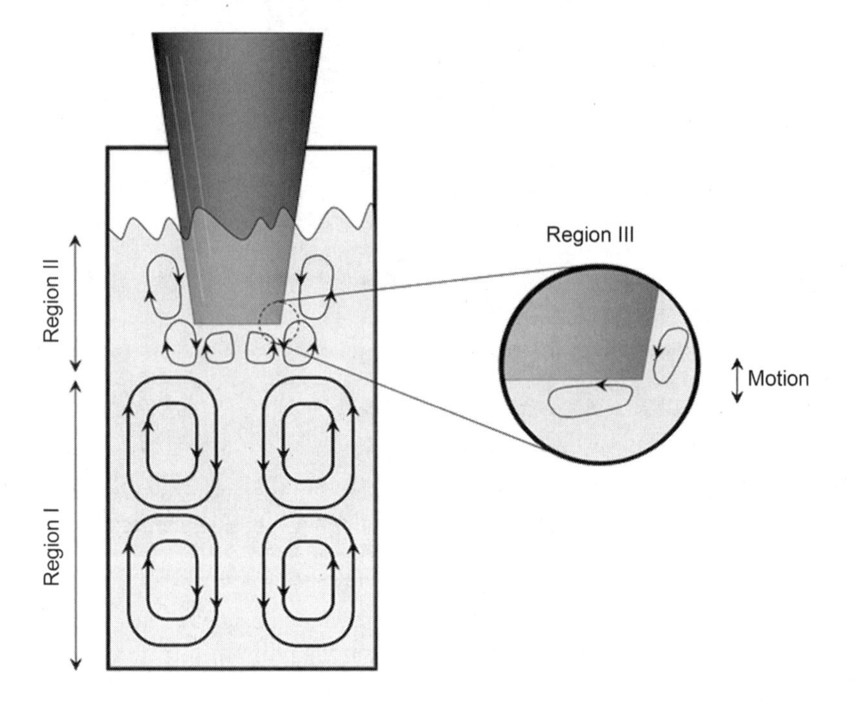

FIGURE 13.6 Regions of acoustic streaming.

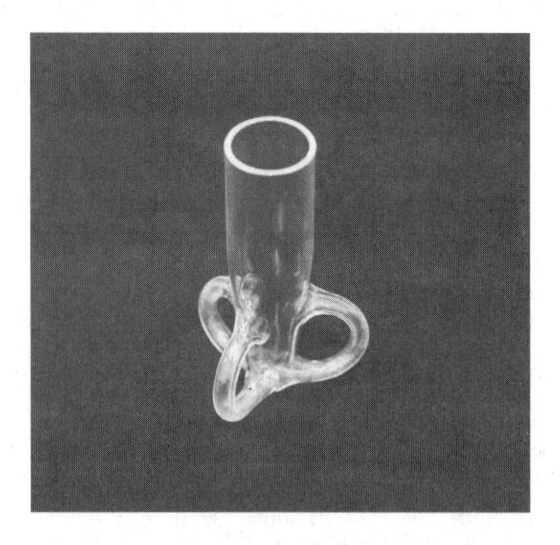

FIGURE 13.7 Photograph rosette reaction chamber.

cavitation, and streaming, to be separated from simple thermal effects. Thus, it is common to use heat exchangers to remove the heat buildup. This can be accomplished with a wide range of reaction chamber designs, but a common design is the rosette seen in Figure 13.7. The fins (three tubes) of the chamber are designed so that streaming promotes fluid flow through the fins and the entire chamber is submerged in a chilled bath, such as ice water or ethanol and dry ice, depending on the desired level of cooling.

13.3.3.4 Free-Radical Generation As previously detailed, the application of ultrasonics to a liquid can cause cavitation, as the bubble collapse induced by cavitation produces intense local heating and high pressures. This leads to extreme conditions (5000 °C and 500 atmospheres) under which the pyrolysis of water produces $H^\bullet$ and $HO^\bullet$ [1].

The number of these radicals and chemical species formed by ultrasonic treatment depend on many experimental parameters, including ultrasonic power, frequency, solution temperature, amount and type of dissolved gas, existence of radical scavengers, and ultrasonic coupling from the transducer to the reactor. Because of the higher reactivity of these species, this can greatly affect the media through direct chemical reactions. However, because free-radical generation is not a significant factor in chemical processes at lower frequencies (<50,000 Hz), where sufficient power density can be generated to enhance industrial-scale reaction, free-radical generation is insignificant compared to cavitation and streaming. More specifically, at higher frequencies, the density of cavitation bubbles that are generated is relatively high and their relative sizes are typically small. Once temperatures exceed 5000 °C, these conditions are conducive to the collapse of a large number of symmetrical bubbles. This will result in a relatively high density of free radicals. At lower frequencies (high possible power), however, the number of cavitation bubbles generated is relatively small and the maximum bubble size is relatively large. These conditions often result in asymmetrical collapse where jetting is predominant and the final temperature during collapse is relatively low.

13.3.3.5 Vapor Barrier Effects/Coupling When ultrasonics are applied to a system, ideally the motion of the horn is perfectly matched by the substrate being treated, so that the horn/part interface behaves as a solid system and the two interfaces are fully coupled. However, because of several effects, this is not always the case. For example, in the case of solid systems, where the acceleration at the horn interface can be as high as 100,000*g* (acceleration of gravity), inertial effects can prevent the part from staying coupled to the horn/tooling. In the case of liquid systems, the formation of a vapor barrier can reduce coupling of the tooling to the medium. Because cavitation is analogous to boiling and cavitation bubbles are highly dissipative, if the intensity of the sound near the horn/tooling is too high, excessive cavitation bubbles near the horn can form, thereby preventing far-field effects. That is to say, while large amounts of energy are dissipated near the horn through cavitation, there is little effect further away from the horn, preventing effective bulk treatment of the medium. This effect has been demonstrated for many years in the art of "fire walkers." These performers have learned that if the coals of a fire are hot enough and their feet are wet enough, they can quickly walk across hot coals because the heat instantaneously forms a layer of vapor (vapor barrier) between the coals and their skin. This vapor has a low thermal conductivity which acts as an insulator,

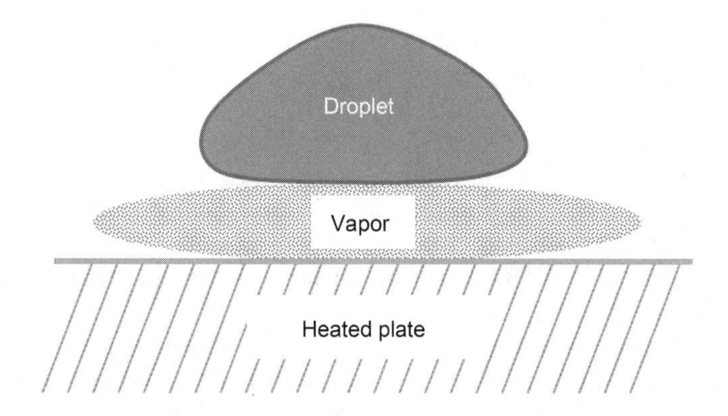

FIGURE 13.8 Illustration of vapor barrier.

preventing their feet from being burnt. The other similarity is placing a drop of water on a very hot pan; the drop bounces around the pan for several seconds before enough heat is transferred to the drop to vaporize it as shown in Figure 13.8.

13.4 BIOFUELS FEEDSTOCK AND PROCESSING

As previously mentioned, the goal of applying ultrasonics to biofuel production is to enhance the various chemical pathways that convert biomass into high-value products, such as ethanol, biodiesel, and chemical building blocks. As new biofuels and their respective processing methods are developed, there will be unforeseen chemical pathways that maybe enhanced by the use of ultrasonics. However, currently, there are a few key pathways that are critical for the conversion of biomass to biofuel. These are detailed in the following section.

13.4.1 Summary of Chemical Pathways

Ultrasonics has been applied to enhance many chemical reactions and has been shown to accelerate some reactions as much as 1000 fold [1]. In terms of biofuels, there are currently many possible reactions that can be enhanced. Many of these have been studied, and some are in production. For example, the extraction of starch from corn has been shown to be greatly enhanced by ultrasonics, as has the hydrolysis of starch into fermentable sugars [12]. In addition, the use of ultrasonics to enhance hydrolysis of lignocellulosic feedstocks has been studied. However, one of the major hurdles to overcome for lignocellulosic feedstocks is the removal of lignin. While nearly an endless number of possible applications for ultrasonics in the biofuel industry exist, some of the most promising in terms of cost and energy efficiency for various fuels are explained below.

Corn to ethanol

- Pretreatment of corn for particle-size reduction
- Accelerated gelatinization

- Accelerated hydrolysis
- Enhanced fermentation
- Energy reduction during distillation

Plant oil and tallow to biodiesel

- Enhanced oil extraction
- Oil pretreatment
- Accelerated transesterification

Lignin cellulosic to ethanol

- Removal of lignin
- Accelerated hydrolysis
- Enhanced fermentation
- Energy reduction during distillation

Algae to biodiesel

- Enhanced oil extraction
- "Milking"
- Oil pretreatment
- Accelerated transesterification

Thermal chemical

- Defouling of beds
- Accelerated conversion

13.4.2 Current Generation, Next Generation, and Advanced Fuels

As detailed in other chapters, there are a wide range of possible biomass-to-liquid fuel conversions applicable to transportation, as seen in Figure 13.9. Currently, most

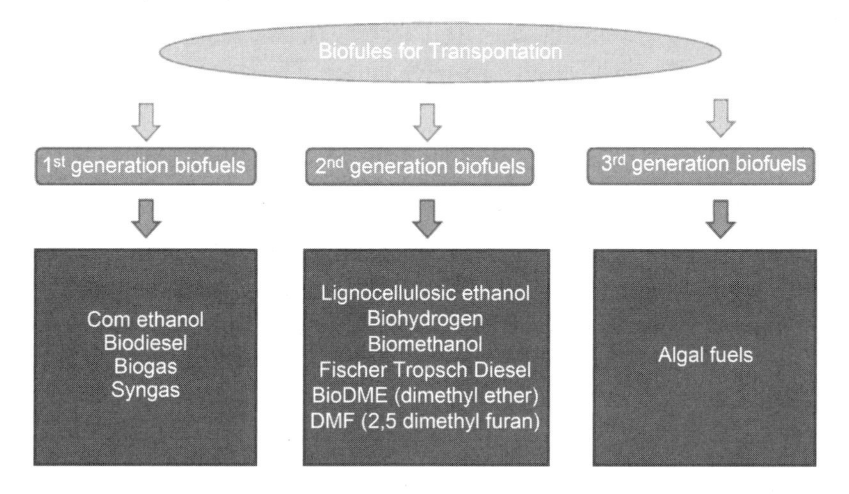

FIGURE 13.9 Schematic of possible biofuels.

biofuels are first generation and an example ethanol being derived from corn. However, pilot-plants currently produce ethanol from lignocellulosic materials, such as corn stover. This technology as well as producing fundamental chemical structures that can be used as building blocks are second generation. In the future, it is planned that fuels from algal and other sources, such as bacterial microorganisms, can be used to produce large amounts of biofuels (third generation).

13.5 ULTRASONICS AND BIOFUELS

The use of ultrasonics to enhance the production of biofuels has been proposed by many. Researchers have studied it for many years, and it has been implemented in some cases [13]. However, currently, the adoption of ultrasonics is limited. One of the main reasons is that historically the generation of large-scale ultrasonic systems has not been realized. Between 1990 and 2010, industrial systems were introduced which could meet the needs of bioprocessing plants. These systems can handle large-scale processing (i.e., +50 MG·year^{-1} of biomass), but, as with any new technology, the acceptance of these high-powered ultrasonic systems by industry has been slow because of its novelty and unforeseen risks. As more systems are implemented by industry, it is expected that the growth will become exponential.

As discussed above and detailed in the following sections, a number of already identified opportunities for the use of ultrasonics in biofuels production exist, but future processing technologies will create even more opportunities.

13.5.1 Pretreatment

13.5.1.1 Corn In the United States, corn is currently the major feedstock used in fuel ethanol production. In April 2011, fuel ethanol production in the United States was at a daily average of 886 thousand barrels [14]. Although ethanol can be made by synthesis from ethylene or by a fermentation process, the latter is commonly used at most fuel ethanol plants. Two methods are currently used to produce fuel ethanol from corn: dry milling and wet milling. While wet-milling facilities are relatively expensive, they are more efficient in producing high-value coproducts (such as zein and corn oil) from raw feedstock. In contrast, dry-grind facilities are simpler because they have less unit operations and require less capital investment.

As of 2011, there are 204 ethanol plants in the United States, with 10 plants under construction or expanding. The total U.S. production capacity is 13,507 MGY (million gallons year^{-1}) [15]. It is expected that world ethanol production will exceed 20 BGY (billion gallons year^{-1}) by 2012 [16]. Multiple strategies are being developed to increase ethanol yields, including superior enzyme cocktails, innovative unit processes, and higher fermentation productivity microorganisms. In the corn-to-ethanol process, depolymerization of starch to release fermentable sugar is the key. When enzymes are exposed to the inner regions of corn particles, they are able to depolymerize starch. Enzyme activity is therefore a function of surface area.

Ultrasonication is one of the most promising technologies that could effectively increase the available surface area for enzyme activity.

In conventional dry milling plants, the jet-cooking step is an important part of the process, as it induces starch gelatinization, aids enzymes in the breakdown of starch, and sterilizes the corn slurry in preparation for fermentation. Because of the high amount of energy required for jet-cooking, studies have shown ultrasonication can be a potential substitute. For example, in a study by Khanal et al. [17], the researchers reported enhanced saccharification of fermentable sugars, a 20-fold corn slurry particle-size reduction, and a more than 100% energy gain (discussed in Section 13.5.8) from sugar produced over the ultrasonic energy used. Ultrasonication breaks down the corn slurry, creating a higher surface-to-volume ratio and providing more effective enzyme regions to depolymerize starch to fermentable sugars. The ultrasonic effects in corn are in agreement with the findings of Nitayavardhana et al. [18] using cassava chip as the feedstock. Figure 13.10 shows the scanning electron microscopy (SEM) images comparing sonicated and unsonicated corn slurry [17]. In the figure, raw and cooked corn represent samples obtained before and after the jet-cooking process, respectively. The corn has been considerably disintegrated in both sonicated samples.

As previously noted, the two main mechanisms of ultrasonication in aqueous solution are cavitation and acoustic streaming. However, prolonged ultrasonication can also raise the temperature of the aqueous solution during pretreatment.

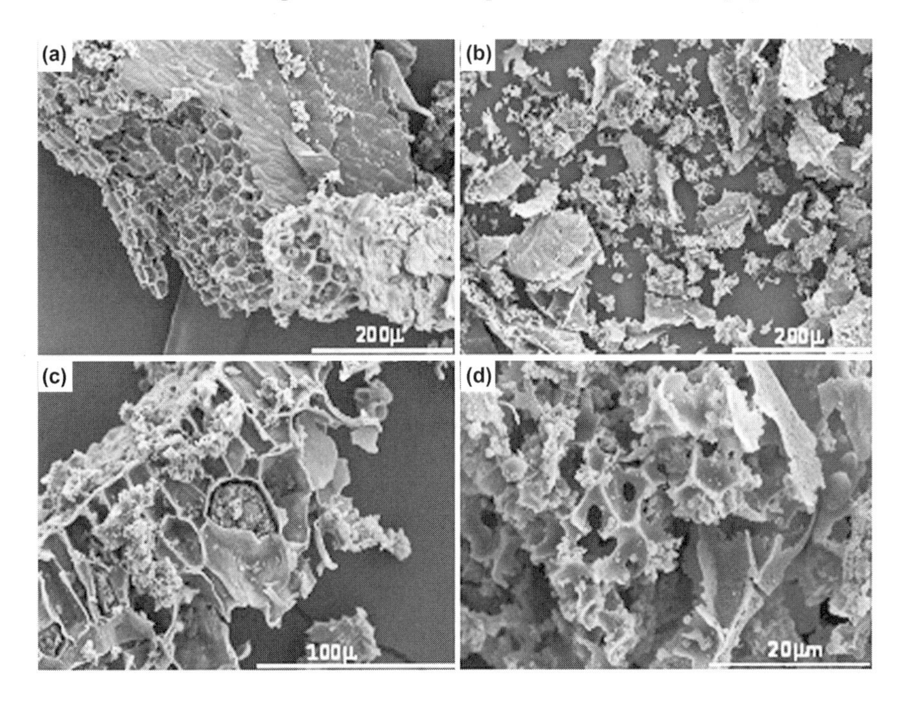

FIGURE 13.10 SEM images of sonicated and unsonicated corn slurry: (**a**) raw corn, (**b**) sonicated raw corn, (**c**) cooked corn, and (**d**) sonicated cooked corn [17].

This could cause gelatinization of corn slurry. Gelatinization is an important mechanism in starch processing. Starch gelatinization is the loss of the semi-crystalline structure or the melting of starch crystallites in the presence of water. Gelatinization can be achieved either by chemical or heat treatment, but the latter is more commonly used. It is therefore important that ultrasonics be considered as more than a heat treatment. In a study by Montalbo-Lomboy et al. [19], ultrasonication was found to induce starch swelling and gelatinization better than conventional heating. This provided evidence that the synergistic effects of heating and hydromechanical shear because of the cavitation, when coupled with acoustic streaming, caused the enhanced enzymatic hydrolysis effects and not heat treatment of corn slurry alone. However, it is noted that the heat-treatment temperatures used in the study were limited only to the temperature range obtained during ultrasonication.

To fully understand the potential of ultrasonication as a substitute for jet cooking, these two treatments be compared. In Figure 13.11, jet-cooking, ultrasonication, and untreated samples were compared, using Sugary-2 corn as the feedstock. The results of ultrasonication and jet cooking were shown to be comparable. This demonstrates that ultrasonication cannot only be as effective as jet cooking, but it may require less energy. Details will be discussed in the cost justification Section 13.5.9. Additionally, it is noted that the results represent only a 3 h saccharification period; therefore, the starch conversion did not reach 100%.

One of the challenges in implementing ultrasonic technologies in ethanol plants is its scale-up capability. A study by Montalbo-Lomboy et al. [20] presented the use of the innovative "donut" horn, which has been applied in many large-scale operations [21].

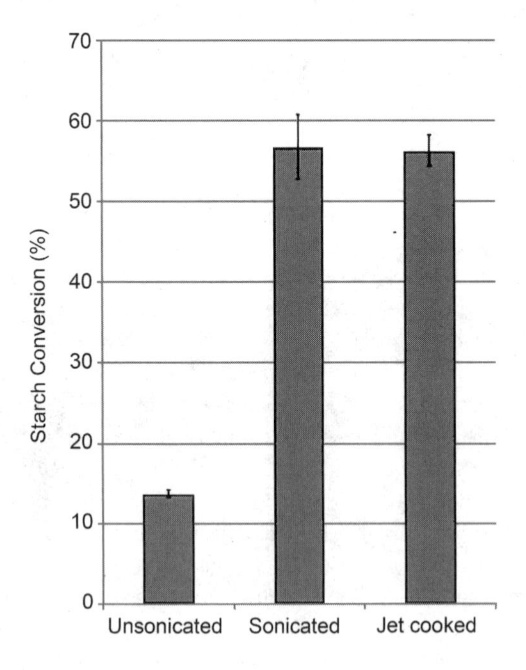

FIGURE 13.11 Comparison of starch conversion in ultrasonicated and jet cooked samples [19].

Because it can be used in a continuous flow system and can easily be retrofitted into any ethanol plant processes, it showed more promise for scale-up than many ultrasonic batch systems. The results of the study showed that while the batch ultrasonic system obtained higher sugar yields, using the "donut" horn provided higher energy efficiency compared to the typical ultrasonic batch process. It is interesting to note that the reduction of particle size because of sonication was also proportional to the dissipated ultrasonic energy regardless of the type of ultrasonic system used. The number of "donut" horns employed in the ethanol plants is therefore critical and will dictate the extent of the corn slurry particle size reduction and ultrasonics efficiency.

13.5.1.2 Lignocellulosic Biomass Most ethanol in the United States is produced from fermentable sugars derived from corn starch. However, even if all the corn in the United States and sugarcane in Brazil were used for ethanol production, the anticipated worldwide annual production of 20 billion gallons of ethanol [16] would only supply approximately 10% of U.S. transportation fuel needs [22]. In addition, there is an increasing concern about the economic and social impacts of redirecting food resources into energy production. Corn, sugarcane, and other starchy sources are not sufficient to supply the world's energy needs.

Lignocellulosic biomass is the most abundant fermentable sugar source in the world. The use of lignocellulosic feedstocks is among the leading alternatives for corn and sugarcane as fuel resources. The recalcitrance of lignocellulosic biomass, such as switchgrass, corn stover, and wood, that has served plants well in their evolutionary development with microorganisms, the environment, and insects, is a major challenge for the use of biomass as a feedstock for biochemical conversion to liquid transportation fuels. This recalcitrance requires the application of mechanical and/or chemical pretreatments prior to the enzymatic hydrolysis of the cellulose and hemicellulose fractions of lignocellulosic biomass. It has been known that direct enzymatic hydrolysis without pretreatment is generally ineffective on these feedstocks [23,24].

Lignocellulosic materials are mainly composed of celluloses, hemicelluloses, and lignin. Production of ethanol from these materials is more difficult compared to the process used for corn. In nature, the polysaccharide component in lignocelluloses protects the plant against the environment and attack from microbes. Although enzymes are currently available to convert these polysaccharides into glucose (fermentable sugar), the main challenge is to develop an energy-efficient, cost-effective, and environment-friendly method to separate the polysaccharides from their complex structure. Among the known chemical barriers to enzymatic hydrolysis are lignin, hemicelluloses, and acetyl group fractions [25,26]. Among these, lignin has been considered the major impedance [27–29]. Lignin, acting as the binder, is cross-linked into a network, with cellulose and hemicellulose making the biomass stronger and difficult to degrade. Lignin and its derivatives are toxic to microorganisms and thus considered inhibitory to saccharification and fermentation. Elimination or reduction of lignin is usually done using various pretreatment methods.

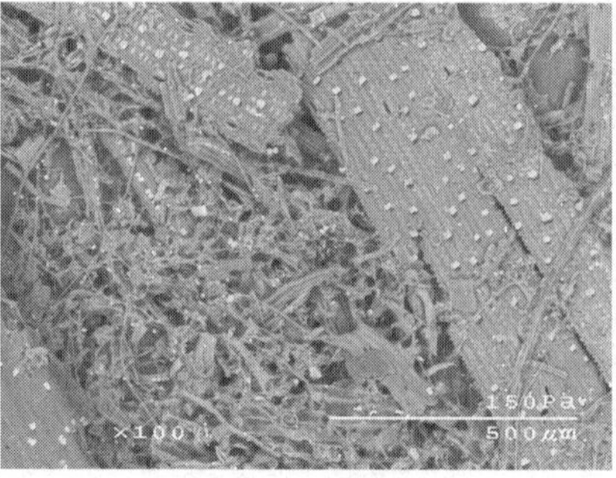

FIGURE 13.12 SEM images of untreated switchgrass (left) and treated switchgrass using ammonia-steeping ultrasonics as pretreatment (right) [31].

Historically, a number of pretreatment methods have emerged on cellulosic extraction from lignocellulosic biomass. This includes chemical, physical, and biological pretreatment [23,24,30]. Ultrasonication is one of the physical pretreatment processes studied for biofuel from biomass [31–34]. Multiple studies have employed ultrasonication to improve fermentable sugar yield or ethanol production. Most of these studies use ultrasonication as a coupled pretreatment with another pretreatment method, either biological or chemical. For example, in studies by Montalbo-Lomboy et al. [31], switchgrass was steeped in ammonium hydroxide and then later ultrasonicated to further disrupt the crystallinity in the cellulosic structure. As shown in Figure 13.12, untreated switchgrass samples had crystalline structures, while the sonicated switch grass samples had broken down into long and small strips creating higher volume-to-surface ratios for enzymatic attack. The white, star-shaped structure found in the SEM picture was silica adhering to the cellulose. The results indicated approximately a 10% increase in the cellulose conversion after 24 h of enzymatic hydrolysis compared to untreated switchgrass.

Another chemical pretreatment that can be coupled with ultrasonics that is attracting attention is the dissolution of cellulose in ionic liquid. First introduced by Graenacher [35] in 1934, it did not draw much attention until a few decades ago. Ionic liquids, considered by many as a green solvent, have been widely used in various chemistry applications. Ionic liquids are advantageous compared to other solvents because they are chemically and thermally stable, nonflammable, and have low vapor pressure [36]. Several studies have shown successful dissolution of lignocellulosic materials in ionic liquids [37–39]. Additionally, some researchers even claimed that lignocellulosic dissolution in ionic liquids had much higher enzymatic hydrolysis yields than other pretreatment methods, such as steam explosion and chemical treatment [40], thereby producing high and faster fermentable sugar yields. Despite the low vapor pressure of ionic liquids, studies have shown that ultrasonic energy has enhanced the dissolution process of ionic liquids. In a study by Mikkola et al. [41], ultrasonic-assisted cellulose dissolution occurred in 7–22 min, depending on the type of cellulosic material used. Typically, it takes

Ionic liquid treatment for 11 hours

Ionic liquid – ultrasonics treatment – 3 minutes

FIGURE 13.13 SEM images of switchgrass using ionic liquid and ionic liquid ultrasonics as pretreatment.

several hours to dissolve cellulose using ionic liquids [42]. Ultrasonication treatment has reduced this dissolution time by at least half the conventional dissolution.

Figure 13.13 shows SEM images of switchgrass treated with ionic liquid for 11 h at 130 °C and ultrasonic-assisted ionic liquid treatment for 3 min. After ionic liquid treatment alone, most of the cellulosic materials are still intact, while with the ultrasonically assisted treatment, the particles were disintegrated, thereby providing a higher surface-to-volume ratio for enzymatic reactions. Additionally, with ultrasonics, the treatment only required 3 min as compared to 11 h for ionic liquid alone.

13.5.1.3 Waste and Coproducts (Animal, Commercial, etc.) Research has shown that ultrasonics can enhance anaerobic digestion of biomass for the generation of biogas (biomethane) [43,44]. It is important to note that these gasses usually require scrubbing of CO_2 and sulfur prior to use. However, anaerobic digestion is able to use a wide range of feedstocks, many of which are coproducts from other biofuel processes or waste. For example, it has been shown that ultrasonics can increase anaerobic digestion of waste-activated sludge by several fold [45,46]. The largest known system to treat municipal waste was built by Sonix (Kidderminser, UK) in Manukau, New Zealand, a municipality with a population of 800,000. The system, which treated 290,000 $m^3\,d^{-1}$, had more than 400,000 W of power capabilities and proved to enhance digestion effectively [47]. Other locations include the Nottinghamshire region of England, Kävelinge in Sweden, Ulu Pandan in Singapore, Beenyup in Australia, Riverside, California, and Edmonton, Canada. The systems were based on the "donut" horn as detailed in Section 13.6.

Animal waste can also be pretreated with ultrasonics [48,49]. With this approach, the rate and final production of methane can be enhanced, with overall energy efficiency over 100% [50]. It has been reported that the type of animal waste can affect the optimum treatment conditions. For example, swine manure typically requires less aggressive (lower amplitudes) treatment than dairy manure. The primary driving forces for using ultrasonics in biomethane production is particle-size reduction and increased digestion. In addition to increased methane production, ultrasonics has been used to mitigate pathogens, including pyrene [51].

The primary coproduct from corn to ethanol is dry distillers grain with solubles (DDGSs). As with many other feedstocks, the reduction of particle size increased the rate and total biomethane production. In addition, under many conditions, the energy efficiency can be over 100%. Many studies have focused on various waste streams (including oranges peels [52]) as biomass for the production of liquid fuels, and it is envisioned that ultrasonics could enhance the production of biofuels from these feedstocks, as well as many others.

13.5.2 Fermentation

The core process in fuel ethanol production (using both wet and dry milling methods) is the simultaneous saccharification and fermentation of starch to sugar to ethanol. If ultrasonication is to be applied during saccharification, the effect of ultrasonics on enzymes is important. A number of notable studies have analyzed the stability of enzymes under sonication. At low acoustic power, some enzymes are not deactivated, whether immobilized or free in solution [53,54]. However, at high ultrasonic intensity, enzymes can be denatured [55,56]. In a study by Wood et al. [57], intermittent ultrasonication was found to increase ethanol yield and enhance enzymatic reaction in the simultaneous saccharification and fermentation (SSF) of mixed waste office paper. The study also showed that in the ultrasonic-assisted experiments only half the amount of enzyme used in the unsonicated SSF was required to produce a similar ethanol yield. In the study by Montalbo-Lomboy et al. [58], ultrasonication was compared to jet cooking as a pretreatment step for simultaneous saccharification and fermentation. The study showed similar starch-to-ethanol conversion rates for ultrasonication and jet cooking. However, because of the high energy demand in jet cookers, the economic analysis of the study indicates lower overall costs for ultrasonication than jet cooking. These findings suggest that ultrasonics can be an attractive and cost-effective method for reduction in the enzyme used and potential energy savings.

13.5.3 Transesterification

Although vegetable oils, such as soybean oil, have long been considered as fuel for diesel engines [59], such oils cannot be used directly in standard diesel engines because of their high molecular weight, kinematic viscosity, and poor atomization properties, as well as problems with lubrication and carbon deposition as a result of incomplete combustion [60]. These issues can be resolved by dilution, micro-emulsification, pyrolysis [61], and transesterification with methanol; the last approach is used most commonly in industry [62]. The conversion of plant oil to biodiesel fuel occurs during a transesterification process in the presence of a

Oil		Methanol		Biodiesel		Glycerin
$R-C(=O)-O-CH_2$ $R-C(=O)-O-CH$ $R-C(=O)-O-CH_2$	+	3MeOH	$\xrightarrow[\text{Heat}]{\text{Catalyst}}$	$3R-C(=O)-O-CH_3$	+	$HO-CH_2$ $HO-CH$ $HO-CH_2$

FIGURE 13.14 Transesterification reaction of oil and methanol in the presence of sodium hydroxide as catalyst.

catalyst and heat (Fig. 13.14). This process requires continuous mixing at 60 °C, which represents significant energy consumption.

When ultrasonic waves are passed through a mixture of immiscible liquids, such as vegetable oil and methanol, extremely fine emulsions can be generated. These emulsions have large interfacial areas, which provide more reaction sites for catalytic action and therefore increase the rate of the transesterification reaction. Thus, ultrasonic energy increased the reaction rate by several fold, reducing the reaction time from approximately 30 to 45 min to less than 1 min. In addition, ultrasonication of liquids produce acoustic streaming [5,10], which further promotes good mixing of reactants.

Previously, researchers have reported that ultrasonic energy can facilitate transesterfication of oils [63–65]. These studies focused on relatively low-amplitude ultrasonic generators (cleaners), which typically have amplitudes below 10 $\mu m_{p\text{-}p}$. Even at these relatively low amplitudes, enhancement of the reaction rates was reported. In addition, others reported the use of continuous flow systems, with energy densities of 11 $kJ\,L^{-1}$ ("1000 W 1.5 $gal^{-1}\cdot min^{-1}$") for enhancing transesterfication [66].

While it has been observed that the transesterification reaction time can be significantly reduced by treating the reactants with ultrasonic sound waves at room temperature [67], very high power amplitudes (+180 $\mu m_{p\text{-}p}$) can reduce transesterification time from 30 to 45 min at 60 °C to less than 20 s [68]. Figure 13.15 shows this in greater detail. Note the weight percentage of biodiesel yields obtained for all continuous, pulse mode (5 s on/25 s off) ultrasonic treatment, as well as conventional heat and stirring (commercial method). To differentiate between the treated samples and the control sample, a log/log plot was used, because without the use of a log/log plot, the extreme scales of the various plot (0–60 s for the treated samples and 0–3000 s for the control sample) made it difficult to visualize the separate plots. It is seen that the application of ultrasonics greatly reduces the reaction time, while maintaining good biodiesel yield. Thus, the reaction time can be lowered to a few seconds via either the pulse or continuous sonication mode, as compared to 45–60 min required for the commercial method.

These studies show that ultrasonics uses 35% less energy for transesterification when compared to industrial reports [69,70]. The details of these energy

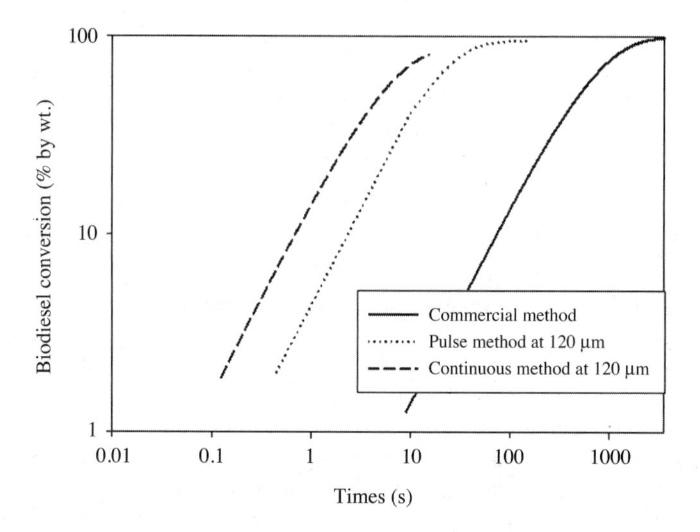

FIGURE 13.15 Comparison of biodiesel conversion (wt%) obtained by the commercial method, pulse sonication mode, and continuous sonication mode (log/log plot).

measurements are further detailed in Section 13.5.8. It is important to note that enhancement of biodiesel from oil is independent of the oil source, as similar effects have been seen with rapeseed oil [71].

13.5.4 Oil Extraction

Because many oil-generating organisms produce oil for food storage, these organisms do not tend to release this oil willingly and exhibit built-in mechanisms to retain this oil. Although methods such as centrifugation, extrusion, and crushing are typically used to extract oil from soybeans, rapeseeds, yeasts, and algae, these processes can be energy intensive and provide low yields. Ultrasonics have been found to enhance the extraction of oil from soybeans using hexane extraction and other techniques [72,73]. In addition, other studies have found that ultrasonic and microwave energy also increased oil extraction from soybeans [74] and that ultrasonics enhanced the extraction of oil and biodiesel production from microorganisms, including algae [75,76]. In addition, there have been reports of using ultrasonics to extract oil from seaweed in conjunction with microwaves [73].

13.5.5 Defoaming

Foam can often be an issue in many biofuel productions, as the final product may contain residual chemicals. Depending on costs and issues related to removing the residual chemicals, the industry may consider other techniques. One highly effective alternative is the use of ultrasonics, because plate horns, as well as other designs, can produce airborne waves that destabilize and collapse the bubbles in the foam. This can be achieved in a few seconds and does not have any residual effects [77].

13.5.6 Separation

While ultrasonics often facilitate mixing and can produce emulsions, with proper tooling design it can also be used to separate mixtures, even in aqueous solutions [78], as well as to enhance filters [79]. In this technique, forces of the ultrasound field can separate particles of different sizes, densities, or compressibility in a laminar flow system [80]. The basic concept is that the forces will drive the large, less dense particles toward the nodal plane of a half-wavelength resonant chamber more than smaller, or denser, particles. On scales of microns, the acoustic field can produce forces much greater than gravity. This effect can also be enhanced with the addition of a porous medium. When the acoustic field is applied, the interactions between the medium and the ultrasonic field entrap the smaller particles [79].

13.5.7 Emulsions

Often the production of emulsions in chemical processing is not desired, as phase separation of various compounds becomes energy intensive. However, acoustic streaming and cavitation are highly effective to produce emulsions. For example, motor oil and water can be mixed and remain stable for hours. The advantage is that the increased surface area of contact between the phases can facilitate chemical pathways.

13.5.8 Energy Balance

While there are some open questions about how energy balance relates to biofuels and ultrasonics, some points are well accepted. For example, it is well accepted that an energy balance is a ratio of the energy dissipated by ultrasonics and the additional energy recovered from the biomass. In this case, the energy dissipated (E_{in}) by the ultrasonic system is the energy from the outlet and includes all of the losses in the conversion of electrical energy into mechanical energy, such as losses in the power supply, transmission from the power supply to the converter, and losses within the converter and stack assembly. The additional energy recovered (E_{in}) from the biomass includes the chemical energy of the fuel as well as coproducts that are generated by the addition of the ultrasonic energy. Specifically, the additional products (fuels and coproducts) correspond to the difference of these products with and without the use of ultrasonics. Thus the energy balance (EB) is:

$$\mathrm{EB} = \frac{E_{out} - E_{in}}{E_{in}} (100\%). \tag{13.8}$$

To justify the use of ultrasonics in biofuel production, one would typically desire an energy balance greater than 100%. While other factors may justify the use, such as processing time, the concept of producing more energy than is used makes the justification straight forward. It is important to note that the statement "produce more energy than is used" may lead to the misconception of perpetual motion, but this is not accurate. Rather, the plants (feedstock) have stored energy from the sun and environment in chemical structures, and the use of ultrasonics only has the potential to release this energy more effectively.

13.5.9 Cost Justification

As detailed above, ultrasonication has been considered a promising technology that has the potential to replace jet cooking in ethanol production. It is therefore important that ultrasonication and jet cooking are compared in terms of cost and energy consumption. In a study by Montalbo-Lomboy et al. [58], a model was developed comparing jet cooking and ultrasonication. The model assumes a 50 MGY ethanol plant operating 350 days annually. The study limited its scope to the two pretreatment technologies only and did not consider the rest of the downstream processing of the ethanol plant. Additionally, labor cost is not incorporated into the calculation. The utility cost used for ultrasonics and jet cooking were based on electricity cost and steam cost, respectively.

The study reported that the capital cost of building an ultrasonication system cost 10 times more than a jet cooker. However, because jet cookers use steam, the utility costs offset the overall costs of the ultrasonication pretreatment. In terms of net present value, both pretreatments resulted in a favorable financial investment, but ultrasonication has a 40% higher benefit-to-cost ratio (BCR) than jet cooking. In a sensitivity analysis of the model, the equipment cost and number of ultrasonic units installed made the greatest impact on the overall cost. Doubling the utility cost for jet cooking caused the pretreatment cost to double as well. In the same study, the energy consumption of both technologies was also compared. They reported that ultrasonication has at least a 2.4 times lower energy requirement than the jet cooker.

13.6 INDUSTRIAL SYSTEMS

A number of high-power ultrasonic systems are available in the market throughout the world, and it is beyond the scope of this chapter to review all of them. However, some designs should be reviewed. One is the so called "donut" horn. This is a donut-shaped horn, manufactured by the Branson Ultrasonics Corp. (Danbury, CT) that vibrates in a radial mode; the sound is focused toward the center where the biomass/biofuel is passed. The horn is driven from a single coupling bar (see Fig. 13.16) that is driven by a transducer. This design has the advantage that it can be placed in series or parallel to increase flow rates or residence time. A single donut horn system can have a flow rate of more than 15 gal min^{-1} with a 5 kW converter.

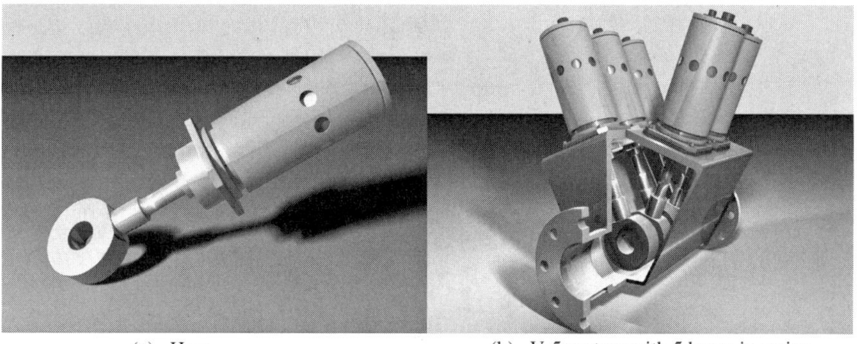

(a) Horn (b) V-5 system with 5 horns in series

FIGURE 13.16 Donut horn and converter.

Hielscher Ultrasonics GmbH (Teltow, Germany) has several systems that are targeted for the biofuels industry and, in particular, for the biodiesel industry. These systems have maximum deliverable power levels as high as 16,000 W. They can help defoul filters for cleaning of feedstock as well as promote mixing of feedstocks to produce nanoemulsions with rates as high as $4\,L\,s^{-1}$, as seen in Figure 13.17. It is important to note that Hielscher also offers smaller laboratory-scale systems.

Aurizon Ultrasonics LLC (Kimberly, WI) offers radial systems as well as "flexural" mode systems that can be used to promote turbulent mixing as seen in Figure 13.18.

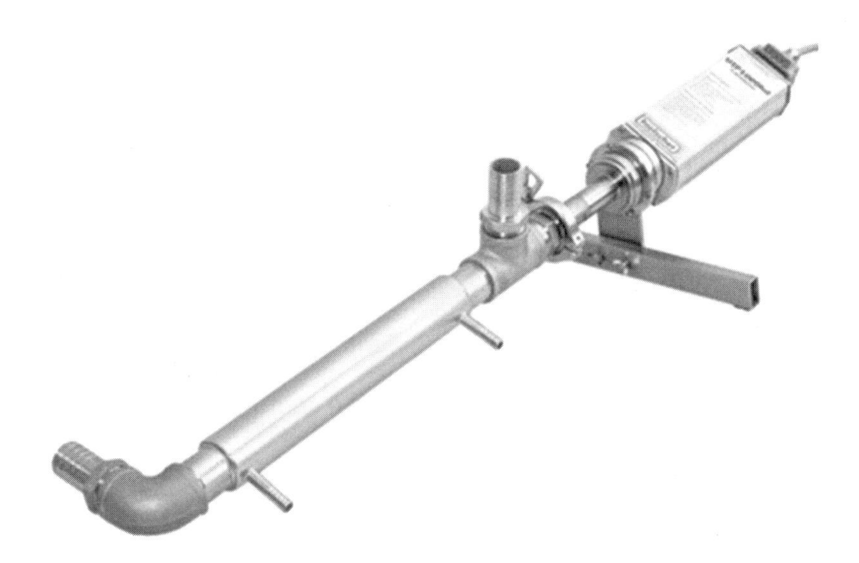

FIGURE 13.17 Hielsher Ultrasonics System (bench top 1 kW system).

FIGURE 13.18 Photograph of Aurizon's flexural horn design.

Other related technologies include multiple frequency transducers from Blackstone Ultrasonics (Jamestown, NY) that reduce the development of dead zones within reaction chambers, as well as improve mixing [81].

13.7 CONCLUSIONS/SUMMARY

It is clear that ultrasonics can enhance the conversion of biomass to biofuels in a wide range of unit operations. It can reduce particle size of biomass for increased reaction sites for hydrolysis of carbohydrates to fermentable sugars, it can promote the release of oil from various substrates and it can accelerate many chemical pathways. While the future use of ultrasonics in this industry is unclear because of the rapid changes in various technologies as well as use of new feedstocks, the biofuels industry will certainly adopt this technology. Obviously, the economics of scales will have to prove the viability of this technology in terms of costs and efficiency in order to assure success of ultrasonics in the biofuels industry. In addition, the advantages of this technology must be attractive enough to overcome industry's concerns regarding using new technologies.

REFERENCES

1. Y. Didenko and K. Suslick, *Nature* **2002**, 418, 394.
2. D. Grewell, K. Graff and A. Benatar, *Experimental Evaluation of Methods for Characterization of Power Output of High Power Ultrasonic Transducers*, 60th Annual Technical Conference for the Society of Plastic Engineers Proceedings, Society of Plastic Engineers, Brookfield, CT, **2002**.
3. A. Clark, in *Ferromagnetic Materials: Vol. 1: A Handbook on the Properties of Magnetically Ordered Substances*, (Ed. E. P. Wohlfarth), Elsevier Science B. V., Amsterdam, North-Holland, **1980**, 531.
4. D. Grewell, K. Graff and A. Benatar, *Proceedings of the 60th Annual Technical Conference for the Society of Plastic Engineers, The Society of Plastics Engineers*, Brookfield, Connecticut, **2002**.
5. A. A. Atchley and L. A. Crum, in *Ultrasound: Its Chemical, Physical, and Biological Effects* (Eds. K. Suslick), VCH Publishers, Inc., New York, NY, **1988**, 7–10.
6. A. Fin,"Nano Cavitation for Algal Oil Harvesting," can be found under http://alfin2300.blogspot.com/2010_01_01_archive.html, **2011**.
7. T. G. Leighton, *The Acoustic Bubble*, Academic Press Inc., San Diego, CA, **1997**, 3.
8. R. Baldev, P. Palanichamy and V. Rajendran, *Science and Technology of Ultrasonics*, Alpha Science International Ltd., Pangbourne RG8 8UT, UK, **2003**, 10.
9. D. Storz, "Shockwave" can be found under www.lockstockuae.com/products/_storz_-duolith_shockwave, **2011**.
10. M. Faraday, *Phil. Trans. R. Soc. London* **1831**, 121, 299.
11. K. Graff, Lecture/Class, Weld Eng 793 Individual Studies, The Ohio State University, Ohio **1989**, Unpublished.
12. S. Khanal, M. Montalbo, J.van Leeuwen, G. Srinivasan and D. Grewell, ASABE Annual International Meeting, American Society of Agricultural and Biological Engineers (ASABE), Minneapolis, Minnesota, **2007**.
13. T. Graham, *"The Use of Ultrasonic Reactors in a Small Scale Continues Biodiesel Process"*, G&M Global Enterprises, **2007**, Unpublished.
14. U.S. Energy Information Administration, "Monthly Oxygenate Report," can be found under www.eia.gov/dnav/pet/pet_pnp_oxy_dc_nus_mbblpd_m.htm, **2011**.
15. Renewable Fuels Association, "Ethanol Industry Overview," can be found under www.ethanolrfa.org/pages/statistics, **2011**.
16. Market Research Analyst, "Worlds Ethanol Production Forecast 2008–2012," can be found under www.marketresearchanalyst.com/2008/01/26/world-ethanol-production-forecast-2008-2012/, **2011**.
17. S. K. Khanal, M. Montalbo, J.van Leeuwen, G. Srinivasan and D. Grewell, *Biotechnol. Bioeng.* **2007**, 98, 978.

18. S. Nitayavardhana, S. K. Rakshit, D. Grewell, J.van Leeuwen and S. K. Khanal, *Biotechnol. Bioeng.* **2008**, 101, 487.
19. M. Montalbo-Lomboy, L. Johnson, S. K. Khanal, J.van Leeuwen and D. Grewell, *Bioresour. Technol.* **2010**, 101, 351.
20. M. Montalbo-Lomboy, S. K. Khanal, J.van Leeuwen, D. R. Raman, L. Dunn Jr. and D. Grewell, *Ultrason. Sonochem.* **2010**, 17, 939.
21. Sonico,"Sonix,™" can be found under www.sonico.net, **2011**.
22. U.S. Energy Information Administration, "Petroleum & Other Liquids," can be found under www.eia.gov/dnav/pet/pet_cons_wpsup_k_w.htm, **2011**.
23. P. Kumar, D. M. Barrett, M. J. Delwiche and P. Stroeve, *Ind. Eng. Chem. Res.* **2009**, 48, 3713.
24. G. Brodeur, E. Yau, K. Badal, J. Collier, K. B. Ramachandran and S. Ramakrishnan, *Enzyme Res.* **2011**, 2011, 1.
25. V. S. Chang, D. J. Mitchell, M. E. Himmel, B. E. Dael and Schroeder, H.A., *Appl. Biochem. Biotechnol.* **1989**, 20, 45.
26. R. Kong, C. R. Engler and E. J. Soltes, *Appl. Biochem. Biotechnol.* **1992**, 34, 23.
27. W. Schwald, H. H. Brownell and J. Saddler, *J. Wood Chem. Tech.* **1988**, 8, 543.
28. E. B. Cowling and T. K. Kirk, *Biotechnol. Bioeng. Symp.* **1976**, 6, 95.
29. J. Polcin and B. Bezuch, *Wood Sci. Technol.* **1977**, 11, 275.
30. R. C. Saxena, D. K. Adhikari and H. B. Goyal, *Renew. Sust. Energ. Rev.* **2009**, 13, 167.
31. M. Montalbo-Lomboy, G. Srinivasan, D. R. Raman, R. P. Anex and D. Grewell, ASABE Annual Meeting, American Society of Agricultural and Biological Engineers (ASABE), Minneapolis, Minnesota, **2007**.
32. Y. Zhang, E. Fu and J. Liang, *Chem. Eng. Technol.* **2008**, 31, 1510.
33. M. Imai, K. Ikari and I. Suzuki, *Biochem. Eng. J.* **2004**, 17, 79.
34. E. M. Sul'man, M. G. Sul'man and E. A. Prutenskaya, *Biocatalysis* **2011**, 3, 28.
35. C. Graenacher, US Patent 1,943,176, **1934**.
36. S. Zhu, Y. Wu, Q. Chen, Z. Yu, C. Wang, S. Jin, Y. Ding and G. Wu, *Green Chem.* **2006**, 8, 325.
37. I. Kilpelainen, H. Xie, A. King, M. Granstrom, S. Heikkinen and D. S. Argyropoulos, *J. Agric. Food Chem.* **2007**, 55, 9142.
38. R. P. Swatloski, S. K. Spear, J. D. Holbrey and R. D. Rogers, *J. Am. Chem. Soc.* **2002**, 124, 4974.
39. L. Liying and C. Hongzhang, *Chin. Sci. Bull.* **2006**, 51, 2432.
40. S. Zhu, *J. Chem. Technol. Biotechnol.* **2008**, 83, 777.
41. J-P. Mikkola, A. Kirilin, J-C Tuuf, A. Pranovich, B. Holmbom, L. M. Kustov, D. Y. Murzin and T. Salmi, *Green Chem.* **2007**, 9, 1229.
42. R. Rinaldi, R. Palkovits and F. Schuth, *Angew. Chem. Int. Ed.* **2008**, 47, 8047.
43. K. Nickel and K. U. Neis, *Ultrason. Sonochem.* **2007**, 14, 450.
44. F. Wang, Y. Wang and M. Ji, *J. Hazar. Mater.* **2005**, 123, 145.
45. H. Choi, S. W. Jeong and Y. J. Chung, *Bioresour. Technol.* **2006**, 97, 198.
46. B. Akin, S. K. Khanal, S. Sung, D. Grewell and J. van Leeuwen, *Water Sci. Technol.* **2006**, 6, 35.
47. S. Collet, T. Amato, M. Griffiths and F. Rooksby, 9th European Biosolids Conference, Wakefield, UK, **2004**.
48. C. P. Chu, D. J. Lee, B. Chang, C. S. You and J. H. Tay, *Water Res.* **2002**, 36, 2681.

49. Y. T. Chyi and R. R. Dague, *Water Environ. Res.* **1994**, 66, 670.
50. W. Wu-Haan, R. Burns, L. Moody, D. Grewell and R. Raman. *Trans. ASABE* **2010**, 53, 577.
51. T. Benabdallah El-Hadj, J. Dosta, R. Marquez-Serrano and J. Mata-Alvarez, *Water Res.* **2006**, 41, 87.
52. M. Wilkins, W. Widmer and K. Grohmann, *Process Biochem.* **2007**, 42, 1614.
53. N. Kardos and J. Luche, *Carbohydr. Res.* **2001**, 332, 115.
54. S. Barton, C. Bullock and W. Weir, *Enzyme Microb. Technol.* **1996**, 18, 190.
55. T. Mason, L. Paniwnyk and J. Lorimer, *Ultrason. Sonochem.* **1996**, 3, 253.
56. B. Özbek and K. Gen, *Process Biochem.* **2000**, 35, 1037.
57. B. E. Wood, H. C. Aldrich and L. O. Ingram, *Biotechnol. Prog.* **1997**, 13, 323.
58. M. Montalbo-Lomboy, S. Khanal, J.van Leeuwen, R. Raman and D. Grewell, *Biotechnol. Prog.* **2011**, doi: 10.1002/btpr.677.
59. G. Knothe, J.van Gerpen, J. Krahl, (eds.) *The Biodiesel Handbook*, American Oil Chemists Society Press, Champaign, IL, **2005**.
60. F. Ma and M. A. Hanna, *Bioresour. Technol.* **1999**, 70, 1.
61. M. Toda, A. Takagaki, M. Okamura, J. N. Kondo, S. Hayashi, K. Domen and M. Hara, *Nature* **2005**, 438, 178.
62. A. Srivastava and R. Prasad, *Renew. Sust. Ener. Rev.* **2000**, 4, 111.
63. C. Stavarache, M. Vinatoru, R. Nishimura and Y. Maeda, *Ultrason. Sonochem.* **2005**, 12, 367.
64. C. Stavarache, M. Vinatoru and M. Y. Maeda, *Ultrason. Sonochem.* **2006**, 13, 401.
65. A. K. Singh, S. D. Fernando and R. Hernandez, *Energy Fuels* **2007**, 21, 1161.
66. R. Kotrba,"Ultrasonic Biodiesel Processing," can be found under www.biodieselmagazine.com/articles/4202/ultrasonic-biodiesel-processing, **2011**.
67. J. A. Colucci, E. E. Borrero and F. J. Alape, *J. Am. Oil Chem. Soc.* **2005**, 82, 525.
68. P. Chand, C. Reddy, J. Verkade and D. Grewell, *Energy Fuels* **2010**, 24, 2010.
69. Energy information administration, "Steam Coal Prices for Industry," can be found under www.eia.doe.gov/emeu/international/stmforind.html, **2011**.
70. Office of industrial technologies, "Benchmark the Fuel Cost of Steam Generation," can be found under www.energystar.gov/ia/business/industry/bnch_cost.pdf, **2011**.
71. S. Gryglewicz, *Bioresour. Technol.* **1999**, 70, 249.
72. H. Li, L. Pordesimo and J. Weiss, *Food Res. Int.* **2007**, 37, 731.
73. G. Cravotto, L. Boffa, S. Mantegna, P. Perego, M. Avogadro and P. Cintas, *Ultrason. Sonochem.* **2008**, 15, 898.
74. H. Li, L. O. Pordesimo and J. Weiss, L. R. Wilhelm, *Trans. ASABE* **2004**, 47, 1187.
75. M. Koberg, M. Cohen, A. Ben-Amotz and A. Gedanken, *Bioresour. Technol.* **2011**, 102, 4265.
76. T. M. Mata, A. Martins and N. S. Caetano, *Renew. Sust. Energ. Rev.* **2010**, 14, 217.
77. J. Gallego-Juarez, UIA Annual Conference, Ultrasonics Industry Association, University of Glasgow, Scotland, **2011**.
78. G. Pangu, PhD thesis, Case Western Reserve University (USA), **2006**.
79. M. T. Grossnera, J. M. Belovich and D. L. Fekea, *Chem. Eng. Sci.* **2005**, 60, 3233.
80. M. Kumar, D. L. Feke and J. M. Belovich, *Biotechnol. Bioeng.* **2005**, 89, 129.
81. F. J. Fuchs, W. L. Puskas, "*Application of Multiple Frequency Ultrasonics*," Blackstone application notes, **2005**, Unpublished.

CHAPTER 14

Advanced Membrane Technology for Products Separation in Biorefinery

SHENGHAI LI, SUOBO ZHANG, and WEIHUI BI

14.1 INTRODUCTION

The use of renewable biological resources has been receiving increased attention due to the reliance on sometimes problematic sources of fossil fuels, the effect of nonrenewable fossil fuel combustion on the earth's climate, and the natural inexhaustible supply of biomass [1]. Biomass, mainly composed of cellulose, hemicellulose, lignin, starch, carbohydrates, glycerides, is being considered as a renewable and alternative source to the traditional fossil for producing fuels and chemicals [2].

Biomass is a very complex substance, and its transformation usually involves a series of organic reactions, such as hydrolysis, dehydration, reforming, C—C cleavage, C—O hydrogenolysis, and selective oxidation [3]. Usually, highly mixed products are obtained during the conversion of biomass, either by chemical pathways or biological pathways [4]. For example, hydrolytic products of carbohydrates, glucose ($C_6H_{12}O_6$), xylose ($C_5H_{10}O_5$), maltose, and oligosaccharides, are key intermediates of biomass conversion [5]. They can be converted by fermentation to ethanol, carbon dioxide, and a range of coproducts and also converted by chemical transformation to furfural or valuable platform chemicals, such as 5-hydroxymethylfurfural (HMF) [6]. HMF can be further transformed into levulinic acid and 2,5-disubstituted furan derivatives [7]. Moreover, through hydrogenation, HMF can be converted into biofuel molecules (e.g., 2,5-dimethylfuran and other liquid alkanes) that can be used as fuels directly [8]. The separation of targeted products from biomass is one of the greatest challenges for the foundation a bio-based chemical industry [9]. Although a variety of products from the biorefinery are potentially attainable, liquid transportation fuels in the form of ethanol (or what is now referred to as bioethanol) is rapidly gaining significance because it can be

The Role of Green Chemistry in Biomass Processing and Conversion, First Edition.
Edited by Haibo Xie and Nicholas Gathergood.

produced from fermentable sugars and can be used as fuel alternatives for gas directly [10]. Therefore, this chapter will focus on recent progress in separation technologies incorporating bioethanol as the principal product. Progress in separation technologies for butanol investigation are also included.

The common dry-milling ethanol plant consists of the following typical steps. The corn kernels are ground into flour; water and enzymes are added; and the slurry is cooked to break down the starch and promote the formation of glucose. This mixture is transferred to fermenters where the yeast converts the sugars into ethanol. The beer, which contains about 10 wt% ethanol and a high concentration of suspended solids, is directed to distillation which is carried out in two steps. In the first step, called the beer column, the stillage (the solids and a large fraction of the water) is collected at the bottom. A water–ethanol mixture containing 40–70 wt% ethanol flows to a second column, called the rectification column, where the alcohol content is increased to the maximum which can be reached by conventional distillation, 90–95 wt% (the azeotrope is 95.6 wt%). Final drying to fuel-grade ethanol(specification: water less than 1.0 wt%) is done with molecular sieves. The stillage is centrifuged to produce distiller wet grain. The liquid fraction is concentrated by evaporation to produce condensed distiller solution which is mixed with the distiller wet grain to produce dry distiller grain with solubles (DDGS) [11].

The total energy used in this process varies between 38,900 and 50,000 BTU gal^{-1} (10.8–13.9 MJ L^{-1}), Figure 14.1. representing the best available conventional technologies according to Kim and Dale (2005). The distillation and molecular

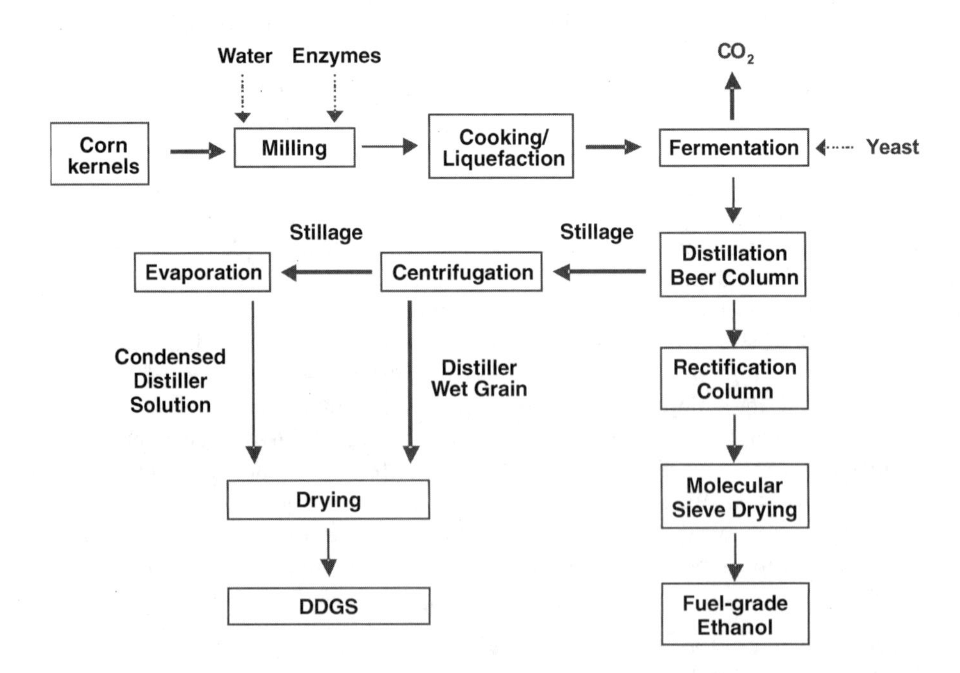

FIGURE 14.1 Conceptual process diagram for a corn-based, dry-milling ethanol plant.

TABLE 14.1 Distribution of Energy Use in a Dry Milling Ethanol Plant

Progress	Distribution (%)	
	Kim and Dale [11a]	McAloon et al. [11b]
Milling	0.8	1.0
Cooking/liquefaction	29.6	19
Fermentation	3.5	1.0
Distillation/dehydration	56.5	45
DDGS recovery	9.6	34
Total	100	100

sieve processes described above account for the largest fraction of energy used in a dry-milling plant, as shown in Table 14.1. Applying an average fraction of 50% from Table 14.1 to the total energy figures cited above, the energy needed for distillation/dehydration varies between 19,450 and 25,000 BTU gal^{-1} (5.4–6.9 MJ L^{-1}). Except the cost of raw materials, the largest cost in the production of ethanol is energy and the largest amount of energy is consumed to purify the ethanol by distillation.

To decrease energy consumption, innovative processing technologies, high separation efficiency and energy-saving process are in high demand. For a few decades, a number of novel separation processes with different features, such as membrane separation [12], foam separation [13], supercritical gas extraction [14], chromatographic separation [15], and high-gradient magnetic-separation technology [16] are researched, developed, and evaluated [17]. These new separation technologies progressed rapidly, and some have a certain scale in production separation, but most are still in the experimental research and pilot-plant scale development stage. However, the research into membrane-separation technologies has taken the prominent position in this area.

14.2 MEMBRANE-SEPARATION TECHNOLOGY

Membrane-separation technology is considered as one of the most promising high technologies from the end of the 20th century to the mid 21st century, and is a research hotspot in the world today. As a new high-tech separation technology, the application of membrane-separation technology has been expanded into many fields, including chemical, food, water processing, pharmaceutical, environmental protection, biotechnology, and energy engineering [18].

Basically, membrane separation refers to a process which uses the osmosis effect and semipermeable membrane as the barrier layer, driven by the energy, concentration, or chemical potential difference, leading to different components of the mixture being separated and purified (Fig. 14.2). The pore size in the semipermeable membrane is different, allowing some of the components to get through the membrane, while the other components are retained in the mixture, resulting in a certain degree of separation. The pressure-driven membrane processes include in

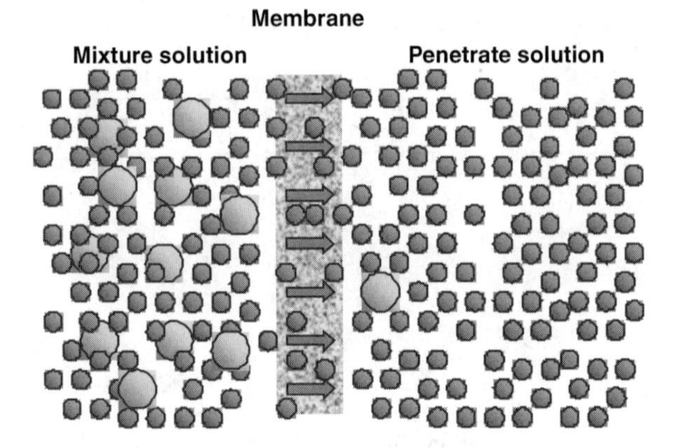

FIGURE 14.2 Schematic diagram of membrane separation.

general the microfiltration (MF), ultrafiltration (UF), nanofiltration (NF), membrane pervaporation (PV) and reverse osmosis (RO). Membranes can be divided into homogeneous membranes and composite membranes. Homogeneous membranes can be divided into symmetrical and asymmetrical membranes; homogeneous membranes have no ultra-thin active layer whereas asymmetrical membranes have. The composite membranes are the membranes typically consisting of an ultra-thin active layer formed in situ on the surface of a microporous substrate. Figure 14.3

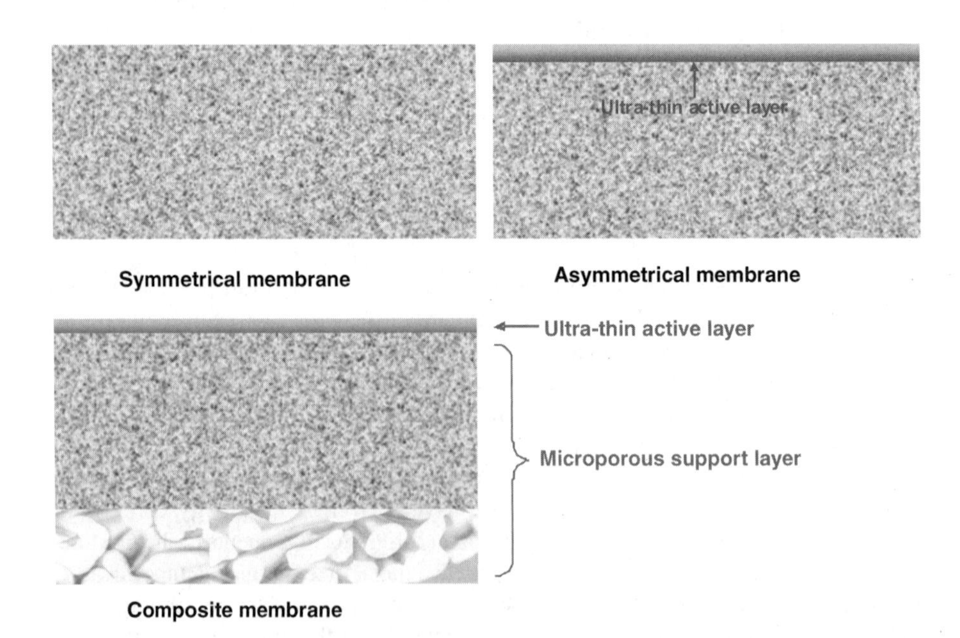

FIGURE 14.3 Schematic diagram of membrane.

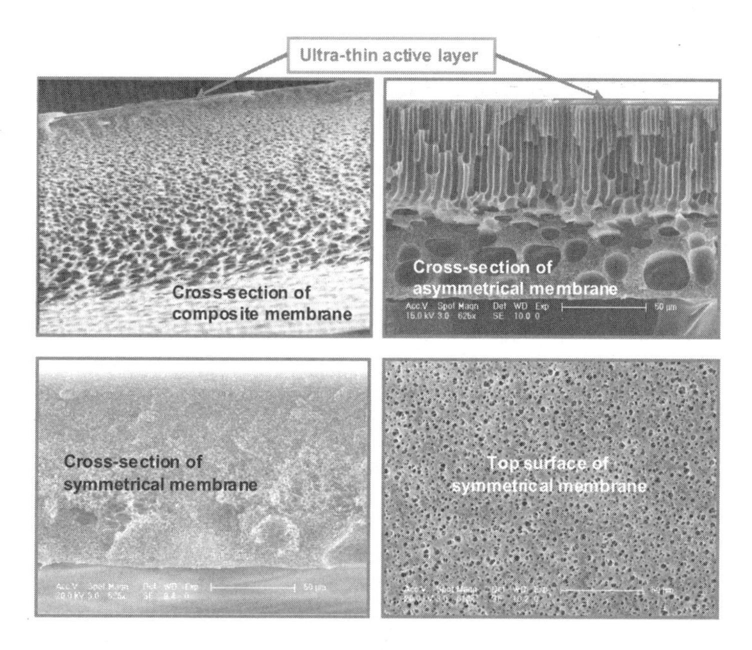

FIGURE 14.4 SEM images of three kinds of separation membranes.

represents the schematic diagram of these membranes and Figure 14.4 gives the SEM images of three kinds of separation membranes. In composite membranes, the active layer is the key component, which mainly controls the separation properties of the membrane, while the microporous support gives the necessary mechanical strength. Composite membranes have advantages over single-material asymmetric membranes in that, the top-active layer is formed in situ and hence the chemistry and performance of the top barrier layer and the bottom porous substrate can be independently modified to maximize the overall membrane performance. The separation membrane can be applied to a variety of separate fields according to the membrane performance. The separation performances of these membranes are usually expressed as the permeation flux and solute rejection, and the operation modes are classified into two types—concentration and filtration.

Lipnizki et al. reported the membrane processes that were used to separate the main products from the starch-based bioethanol-production process [19]. For the case of starch-based bioethanol production, the initial steps of the process would generally include liquification and hydrolysis, followed by fermentation. An overview of this process with potential membrane applications is given in Figure 14.5. The first potential membrane application in this concept is the purification of feedstock after hydrolysis and before fermentation. By using MF/UF it is possible to remove impurities, such as enzymes and starch residues from the glucose before the fermentation step. This includes the option to recover and recycle enzymes for the hydrolysis step. For the fermentation step, different approaches to combine membranes directly with the fermentor have been investigated. Processes such as MF,

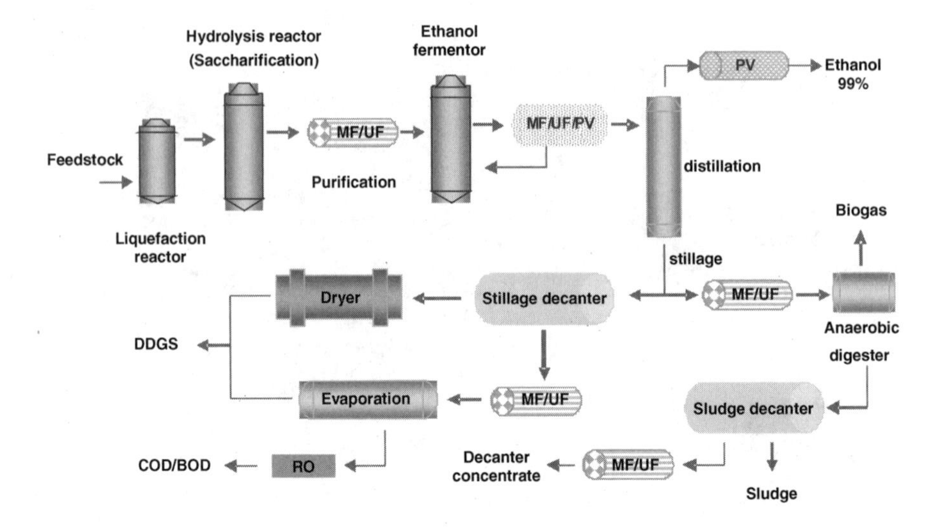

FIGURE 14.5 Overview of liquification and hydrolysis process for starch-based bioethanol production with potential membrane applications.

UF, and hydrophobic PV have been considered to be combined with the fermentation step [20]. The idea is to remove bioethanol continuously from the fermentor and thus overcome product inhibitions.

After fermentation, the preconcentrated and prepurified bioethanol stream is then passed to distillation for further concentration. One of the limitations of ethanol concentration by distillation is the azeotropic point of ethanol. To overcome this point and to achieve purities of >99 wt% (water content <1 wt%) either entrainer distillation or molecular sieves are used as conventional technologies. Alternatively, hydrophilic PV or vapor permeation (VP) which are neither limited by the azeotropic point nor require periodical regeneration can be applied.

A by-product from the distillation step is stillage, which is commonly converted to DDGS by using a decanter, and followed by evaporation and drying processes. Using this concept, it is possible to integrate RO to polish the evaporator condensate, which might contain high levels of chemical oxygen demand (COD)/biochemical oxygen demand (BOD). Alternatively, two concepts for stillage handling are under investigation [21]. The first concept is based on the classic stillage handling to produce DDGS. But apart from RO as evaporator condensate polisher, it includes a MF/UF step after the decanter and before evaporation. This MF/UF step reduces the load on the evaporation by preconcentrating the stillage. Further, the permeation from this step can be recycled as a set-back into the process. The selection of MF/UF membranes therefore influences the characteristics of the set-back. The second concept is based on the use of an anaerobic digester to convert stillage to biogas. In this concept, MF/UF can be integrated as a preconcentration step before the anaerobic digester, if the stillage has a low content of total solids. Furthermore, a combination of decanter and MF/UF can be used in the treatment of sludge from the anaerobic digester.

Apart from being integrated directly in the process, membrane processes can be used within the water cycles of the bioethanol plants. NF and RO can be applied for pretreatment of the intake water and membrane bioreactors (MBRs) can be used in the wastewater treatment.

14.3 OVERVIEW ON THE LATEST PROGRESS ON BIOPRODUCTS SEPARATION FROM BIOMASS CONVERSION PROCESS

14.3.1 Microfiltration Process

Traditional methods to separate solids from liquids are membrane processes such as MF or UF processes. MF has been widely employed and studied for the recovery of a variety of biopolymers, solid product, microorganisms, or solid impurity from fermentation broths [22] together with other common methods such as centrifugation, flocculation, which often are combined [23]. So far, a large number of MF membranes, for example, Al_2O_3 ceramics membrane [24], polytetrafluoroethylene (PTFE) [25], polysulfone membranes (atop a polypropylene-support material) [26], and commercially available Membralox® TI-70 MF membranes [27] have been investigated for the recovery of solid product.

14.3.2 Ultrafiltration Process

MF is capable of removing most fermentative microorganisms. If proteins or enzymes must be retained as well, then UF or NF may be required. For example, Knutsen and Davis have reported the use of sedimentation and UF to recover and reuse cellulase enzymes during the hydrolysis of lignocellulosic biomass [28]. Lignocellulosic particles larger than ca. 50 μm in length were first removed via sedimentation using an inclined settler. UF was then applied to retain the remaining lignocellulosic particles and the cellulose enzymes, while transmitting fermentable sugars and other small molecules. The permeating flux from the UF step for a feed consisting of 0.22 w/v% cellulase is $64 \pm 5\,L{\cdot}m^{-2}{\cdot}h^{-1}$, while that for a feed consisting of the settler overflow from a mixture 0.22 w/v% cellulase and 10 wt% lignocellulose fed to the settler is $130 \pm 20\,L{\cdot}m^{-2}{\cdot}h^{-1}$. The higher permeating flux is presumably due to the binding of a portion of the cellulase enzymes to the lignocellulosic particles during the hydrolysis and filtration process, thus preventing the enzymes from fouling the membrane. A filter-paper activity assay shows little loss in enzymatic activity throughout the combined sedimentation/UF separation process [29].

14.3.3 Membrane Fouling of MF and UF

MF and UF have been widely employed and studied for the recovery of a variety of biopolymers from fermentation broths. However, high-performance MF and UF membranes are strongly fouled by particles, proteins, polysaccharides, and humic substances. Fouling reduces performance of membranes due to blocking of some

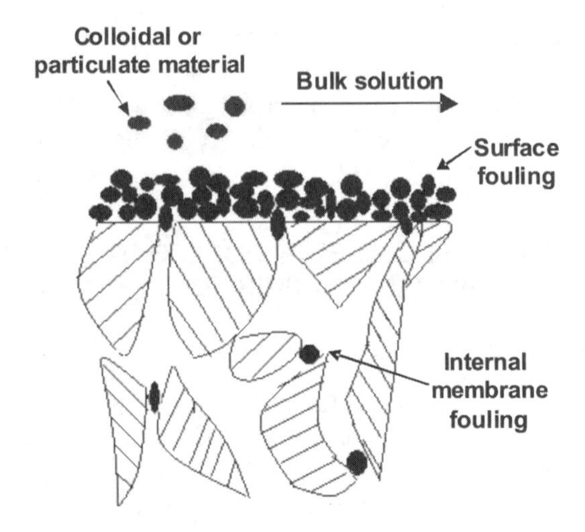

FIGURE 14.6 Schematic diagram of membrane fouling.

membrane pores (Fig 14.6) [30]. In this context, membrane fouling is a rather complex phenomenon that depends upon particle-size distributions and microbial physiology, feed concentration, temperature, pH value, ionic strength, (bio)surfactants, and separation-system hydrodynamics [31].

Poly(arylene ether sulfone)s (PES) are a class of materials which has been most widely used in the manufacture of MF or UF membranes due to its excellent mechanical, thermal, and chemical stabilities [32]. However, PES membrane is strongly hydrophobic and can easily be contaminated (Fig. 14.7). Fouling reduces productivity due to longer filtration times and shortens membrane life because of the harsh chemicals necessary for cleaning [33]. Therefore, chemical or physical modification of the separation membranes is required to improve their antifouling performance.

A novel cardo poly(arylene ether sulfone)s, bearing pendant zwitterionic carboxybetaine groups (PES-CB) (Fig. 14.8) was synthesized and used for preparation of

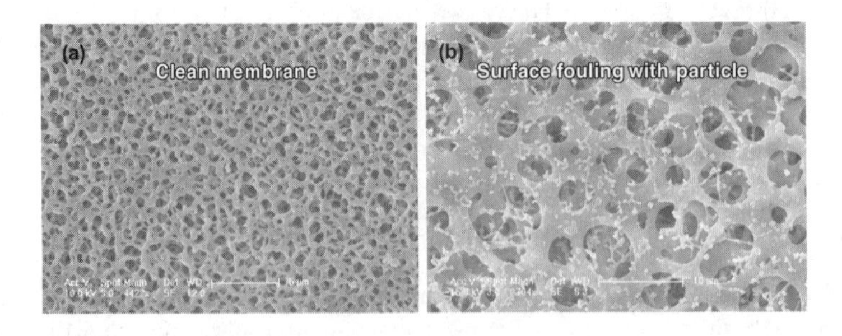

FIGURE 14.7 SEM images of PES microfiltration membrane before (**a**) and after (**b**) filtrating of solid product from fermentation broths.

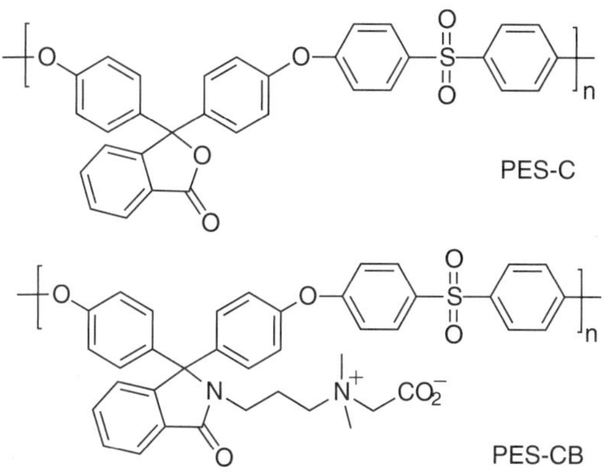

FIGURE 14.8 The chemical structures of poly(arylene ether sulfone) cardo (PES-C) and PES-CB.

asymmetric membranes (Fig. 14.9) by the phase-inversion method in our group [34]. The antifouling effect of the resultant membrane was measured under dynamic protein-adsorption experiments. The carboxybetaine poly(arylene ether sulfone) (PES-CB) membrane showed significant resistance to protein adsorption. After three cycles of lysozyme (LYZ, pI = 12.0, 14 kDa) solution filtration, a water flux recovery ratio as high as 95 wt% was achieved for the PES-CB ultrafiltration membrane, whereas only about 25 wt% of water flux recovery ratio was obtained for the conventional poly(arylene ether sulfone) cardo (PES-C) ultrafiltration membrane. This result indicated that the protein fouling in PES-CB membrane was suppressed significantly under the dynamic ultrafiltration process (Fig. 14.10).

Some recent studies have explored membrane materials challenged with complex feed solutions and determined their response to physical process parameters during MF. Choi et al. [35] used a synthetic paper mill wastewater to foul MF membranes,

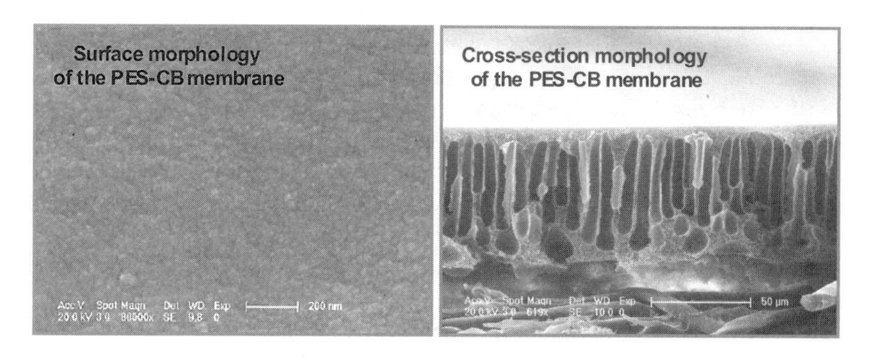

FIGURE 14.9 Surface and cross-section morphologies of the PES-CB flat sheet ultrafiltration membrane.

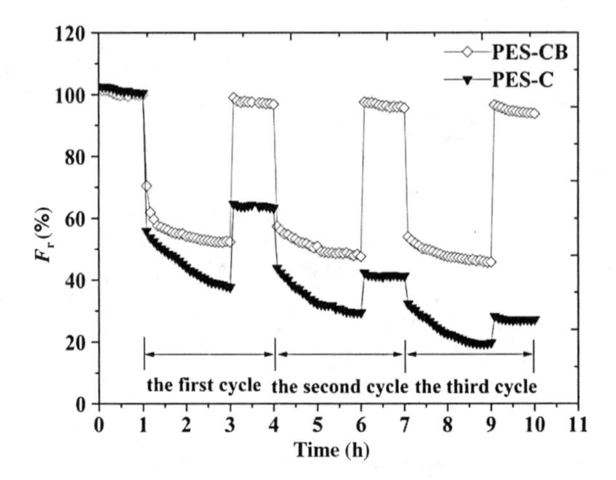

FIGURE 14.10 Time-dependent flux recovery ratio of PES-CB and PES-C ultrafiltration membranes during the LYZ solution filtration process.

and the results showed that permeate flux declined rapidly with the increase in feed concentration, decrease in both membrane pore size and tangential shear. Others also demonstrated that this phenomena can be induced by the biological properties of feed. Krstic et al. [36] investigated the effects of trans-membrane pressure, cross-flow feed velocity, biomass structure and feed composition on MF membranes that were fouled with fermentation broth specifically used to be a support of the growth of the fungus *Polyporus squamosus*. It was found that when cross-flow velocity increased, the membrane fouling mode changed from one more influenced by surface biomass associations to a mode more influenced by internal pore constrictions. Fouling rates were sensitive to the concentrations of soluble components, which were judged by conventional filtration and gravimetric bioassay. In a recent study by Choi et al. [37] the influence of cross-flow velocity on formation of fouling layers during MF and UF of biological suspensions was investigated. This study showed that the permeate flow-rate increased linearly with increasing cross-flow velocity.

Ford et al. [38] reported the application of tubular ceramic membranes with nominal pore sizes 0.5 and 0.8 μm to carry out the MF of the bio-oil. The results demonstrated the removal of the major quantity of char particles with an obvious reduction in overall ash content of the bio-oil. Results of fouling analysis obtained from longer runs of bio-oil through the membranes clearly showed that the cake-formation mechanism of fouling was predominant in this process.

Toussaint et al. [39] studied the influence of temperature on the recovery of extracellular α-agarase enzymes from fermentation broth using a polypropylene hollow-fiber filter with 0.5 μm pores. Whereas at 37 °C higher permeate flux values were recorded compared to 22 °C, the study suggested the low temperature would lead to greater overall enzyme recovery due to better molecular stability at lower temperatures.

Many other studies have strived to understand the fouling mechanisms of complex fermentation broths and different protein mixtures. Guell et al. [40] studied fouling mechanisms associated with protein mixtures and the effects of yeast-cell concentration on protein transmission through a MF membrane. This work suggested that severe membrane fouling occurred when filtering a protein mixture alone, which is evident by a decline in protein concentration in the permeate to less than 50% of its initial value in a 3 h period. However, when small numbers of yeast cells were present, either suspended in the feed or predeposited as a thin cake on the membrane surface, the protein transmission was maximized during an otherwise identical 3 h period. Kelly and Zydney et al. [41] reported that the mechanical aspects of MF protein fouling follow at least two distinct mechanisms: deposition of large protein aggregates, and the chemical attachment of proteins to other surface-associated deposits.

Bearing in mind the challenges of biotechnical separations, Persson et al. [42] investigated the transmission of protein A (molecular size: 42 kDa) through 0.2 μm nylon and 0.16 μm polyethersulfone membranes, and found that transmission of protein A was higher when the host microorganisms were removed from the feed solutions. These authors proposed that protein A was sequestered by the microorganisms presenting in the feed and/or to a dynamic biomass layer that was retained on the membrane surface. The negative effects observed by Persson and coworkers contrasts with some positive influences that the presence of certain microbial cells can have on protein separations [43].

Enhanced dynamic MF has been tested as a method for reducing fouling and concentration polarization at low trans-membrane pressure. In 2002, Brou et al. [44] reported a study that MF of yeast suspensions by using dynamic systems provided a clear enhancement of the process performance. Recently, Torras et al. [24] further developed this method, optimizing the conditions, by adjusting the transmembrane pressure and the rotational speed. Results demonstrated that the dynamic filtration can be used to reduce membrane fouling, maximizing the turbulence, and thus the shear stress over the membrane [45].

14.3.4 Membrane Materials for Alcohol Recovery by Pervaporation

14.3.4.1 Pervaporation Process PV is a process in which a liquid stream containing two or more miscible components is placed in contact with one side of a nonporous polymeric membrane or molecularly porous inorganic membrane (such as a zeolite membrane) while a vacuum or gas purge is applied to the other side (Fig. 14.11). In principle, PV is based on the solution–diffusion mechanism, which is driven by the gradient of the chemical potential between the feed and the permeate sides of the membrane. The components in the liquid stream sorb into/onto the membrane, then permeate through the membrane, and evaporate into the vapor phase. The resulting vapor, referred to as "the permeate," is then condensed. One of the components, even at low concentration in the feed, can be highly enriched in the permeate due to their different affinities to the membrane materials and different diffusion rates through the membrane.

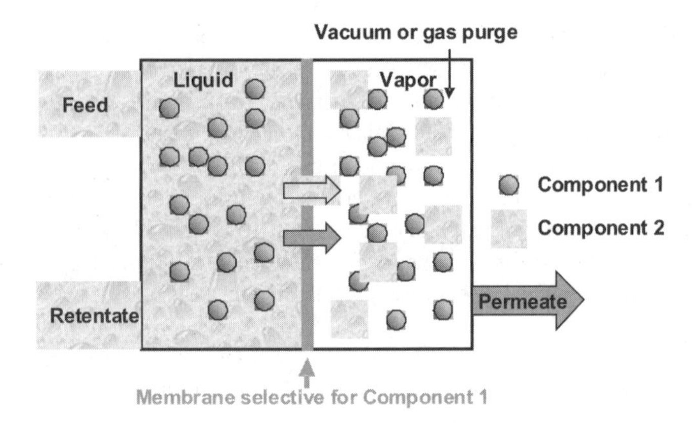

FIGURE 14.11 Schematic diagram of pervaporation process.

The properties of the membrane material determine the separation efficiency in this process. For example, if a hydrophobic membrane is applied, the membrane will preferentially permeate organic compounds (i.e., ethanol, methanol, acetone, chloroform etc.), compared to water, thus the permeate will be enriched in the organic compounds. Alternatively, if a hydrophilic membrane is applied, water will thus be enriched in the permeate and the organic compound in the feed liquid will be dehydrated. An advantage is that the general process components are the same, and only the membrane material has been changed. The dehydration of organic solvents, particularly those which form azeotropes with water (such as ethanol and isopropanol) is the main commercial use of PV today. In the dehydration application, PV competes with molecular-sieve sorption and ternary or vacuum distillation. For the production of biofuels, PV can therefore be applied to both the recovery of alcohols from water and for the dehydration of the alcohols to meet fuel dryness specifications [46].

Membrane PV can now replace the traditional azeotropic distillation, and the technology has been industrialized. The U.S. Department of Energy considers that PV can effectively recover and concentrate the organic matters in the fermentation system [47], thus PV is known as a novel technology for bio-energy development. The integration of bio-reaction and PV membrane technology can transform the traditional biomass fermentation in the preparation of high-purity ethanol by removing ethanol from the fermentation area expediently, thus not only achieving continuous production, but also improving fermentation efficiency. The process presented a number of advantages, such as, release of product inhibition, energy-saving technology (typically saving 50–75% than that of azeotropic distillation), clean production technology without introducing other components hence avoiding external contamination, easy scale-up, and straightforward integration with existing process. It is especially applied to PV membrane technology has clear economic and technical advantages in these applications [48].

The phenomenon of PV was first observed by Kober in 1917 [49], and the basic principles and potential of PV technology were established by Binning et al. in 1956

[50]. The breakthrough of PV in industrial applications was made in 1980s by GFT (Gesell-schaft für Trenntechnik, Hamburg, Germany) with the development of a series of poly(vinyl alcohol)-poly(acrylnitrile) (PVA–PAN) composite membranes for the dehydration of alcohol/water azeotropic mixtures [51]. More than 90 industrial PV units were installed worldwide between 1985 and 1996 [52]. Concurrently, around 200 European and US patents about PV were filed. As far as industrialization is concerned, the choice of PV-based hybrid systems is indispensable [53]. With current limitations of membrane materials and lack of commercially available PV membranes, PV alone cannot replace as well as compete with distillation over a wide spectrum of feed compositions. The latter is a well-established and better-understood separation process with available infrastructure on many sites. However, PV has unique and superior performance over distillation. Incorporating PV into existing distillation processes can not only lower the overall cost but also synergize the advantages of individual technologies [54].

Currently, many ethanol permeable PV membrane materials are under investigation. Ethanol permeable membrane materials including silicone polymer, fluorinated polymers, and zeolites have been studied. Polydimethylsiloxane (PDMS) membranes are widely used because of excellent hydrophobic property, low penetration resistance, good chemical stability, and thermodynamic properties. The selectivity of PDMS membrane is usually moderate and increasing the separation factor is at the expense of flux in general. In the following sections, the membrane materials design strategy based on the solution–diffusion mechanism, and the potential and the challenges that will impact membrane ability to be competitive for PV application will be reviewed and analyzed.

14.3.4.2 Alcohol–Water Separation Through Zeolite Membranes Zeolite membranes have uniform, molecular-sized pores, and they separate molecules based on differences in the molecules' adsorption and diffusion properties (Fig. 14.12). Zeolite membranes are thus well suited for separating liquid-phase mixtures by PV, and the first commercial application of zeolite membranes has been in dehydrating organic compounds [55].

Type-A zeolite membranes are almost ideally suited for organic dehydration because they are highly hydrophilic and their XRD pore diameter (0.4 nm) is smaller than that of almost all organic molecules but larger than water. The A-type membranes that have been prepared also have nonzeolite pores [56] that contain hydrophilic silanol groups [57]. These properties allow preferential permeation of water over organic compounds with separation factors that are often over 1000 and sometimes higher than 10,000. Morigami and coworkers reported that the NaA zeolite membrane had the separation factor of up to 10,000 and water flux of 2.15 $kg{\cdot}m^{-2}{\cdot}h^{-1}$ at 75 °C for removal of water in water/ethanol mixture (10/90) [58]. The application of this technology is relatively mature in Japan, especially in Yamaguchi University where Kita et al. carried out fruitful research work on this aspect [59]. In 1999, Japan promoted large-scale NaA zeolite membrane PV device to the market for the first time. The device consists of 16 sets of membrane parts, each with 125 membrane tubes (diameter of the tube is 12 mm, length is 80 cm, average pore size is 1 μm). The device

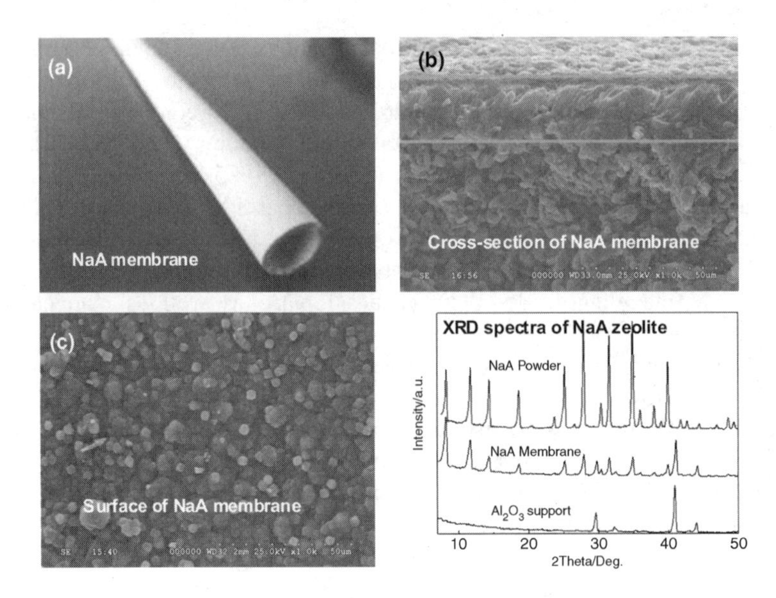

FIGURE 14.12 (**a**) Image of NaA membrane, (**b**) SEM image of cross-section of NaA membrane, (**c**) SEM image of surface of NaA membrane, and XRD spectra of NaA membrane.

has a processing capacity of 480 kg·h^{-1} with mass fraction of 90% ethanol/water mixture at 393 K. After the separation, residual water content is less than 0.2% and recovery rate of ethanol is 96%, showing a strong application future.

Okamoto and coworkers [56,60] and van den Berg et al. [61] have reported some successful zeolite membranes for water removal from organic compounds. Fluxes during ethanol dehydration tend to increase as the water concentration in the feed increases. In contrast, permeate concentrations are relatively independent of feed concentration except near 0% H_2O. Jafar et al. [62] reported that H_2O/EtOH separation factors increased as the water concentration approached zero using a NaA zeolite membrane on a ZrO_2–TiO_2 coated porous carbon tube. On the other hand, Kondo et al. [63] reported that H_2O/EtOH separation factors dramatically decreased as the H_2O feed concentration neared zero for a NaA zeolite membrane on a mullite-/alumina/cristobalite porous support. These studies also found that increasing the temperature resulted in an increase in the flux with little effect on the separation factor.

Water has also been removed from ethanol with ZSM-5, mordenite, and X-, Y-, and T-type zeolite membranes, but fluxes and separation factors are generally lower than observed for A-type zeolite membranes. This is because the pore diameters are larger for these membranes than those of A-type membranes, thus they are less size selective for water. Furthermore, they are less hydrophilic than A-type membranes because their Si/Al ratios are higher. Zeolite T membranes (OFF structure, 0.68 nm XRD pore diameter, Si/Al = 3.6) prepared by Tanaka et al. [64] and the mordenite membrane prepared by Zhang et al. [65] had higher separation factors than observed for the ZSM-5 and X- and Y-type zeolite membranes, but their fluxes and separation factors were, respectively, about 1/3 and 1/9 of those using the best A-type membranes [63]. The

X- and Y-type zeolite structures have a three-dimensional pore system of 12-member rings (MR), whereas the T-type and mordenite zeolite structures have 12-MR pores in only one direction, but have smaller 8-member ring pores in parallel with and perpendicular to their 12-MR pores. Because the membranes were randomly oriented, the 8-MR pores in the T-type and mordenite membranes might have provided alternate pathways for H_2O, while ethanol could not diffuse through. Thus, these 8-MR pores might explain why the T-type and mordenite membrane H_2O/ethanol separation factors were higher than those of the X- and Y-type membranes.

An extensive review of zeolite materials and the fundamentals of using zeolites for PV applications by Bowen et al. was recently published [66]. Silicalite-1, the most studied hydrophobic zeolite for this application, has been reported to deliver ethanol–water separation factors ranging from 12 to 72 with a typical value around 40-four to five times that of PDMS. The ethanol–water separation factor for pure silicalite-1 membranes with few nonzeolite pores larger than the zeolite pores is over 100. In addition, fluxes observed with silicalite-1 membranes meet or exceed those of the thinnest PDMS films reported. As with inorganic membranes in other applications, silicalite-1 membranes are expected to be more expensive than polymer membranes on a unit area basis. However, silicalite-1 membranes may be cost effective on a per unit ethanol basis owing to the higher separation factor and flux afforded by the silicalite-1. Efforts to produce silicalite-based modules on a larger scale are ongoing, including development of tubular silicalite-1 membranes at Bio-Nanotec Research Institute, Inc (BNRI, part of Mitsui & Co, Ltd, Tokyo, Japan) and of multichannel monolithic membranes at CeraMem Corporation (Waltham, Massachusetts USA) [67].

14.3.4.3 Alcohol–Water Separation Through Carbon Molecular Sieve Membranes To the best of our knowledge, very few investigations have been conducted on the application of carbon molecular sieve membranes (CMSMs) supported on inorganic porous support for PV. Recently, a preliminary study was reported by Dong et al. [68] where their tetramethylammonium bromide (TMAB) carbon membrane showed a high separation factor in the PV of water/ethanol and water/isopropanol mixtures at low temperature. Nonetheless, the fluxes obtained in their study are low as compared to those obtained by polymeric membranes.

A further study on the potential application of CMSMs in biofuel separation through PV was reported by Tin et al. [69] "Defect-free" carbon/ceramic composite membranes were prepared as the separation barrier for the dehydration of ethanol. The supported CMSM carbonized at 650 °C showed a great total flux of $4\,kg{\cdot}m^{-2}{\cdot}h^{-1}$ with a reasonably high separation factor of 50 at a feed temperature of 60 °C and did not swell in high-temperature operations. These carbon membranes presented outstanding separation performance and offered a viable alternative to current membranes for the separation and purification of biofuels.

14.3.4.4 Ethanol–Water Separation Through Polymeric Membranes
The current benchmark hydrophobic PV membrane material is PDMS, often referred to as "silicone rubber." PDMS is an elastomeric material that can be used to fabricate

hollow fiber, tubular, unsupported sheet, or thin-layer supported sheet membranes. [70] Several companies have manufactured thin PDMS-supported membranes over the years. At present, Membrane Technology and Research, Inc (MTR) of Menlo Park, CA is the leading supplier, manufacturing spiral wound modules out of their supported silicone rubber membranes. The reported ethanol-water separation factor with 'pure' PDMS membranes ranged from 4.4 to 10.8. The broad range in ethanol–water separation factors for PDMS membranes is typical of performance parameters reported in the literature for a given polymer and separation situation. The range in values arises from a variety of factors, including the source of the polymer starting materials (although called "PDMS," there are often differences), the method of casting the film, the cross-link density, the thickness of the selective layer, the porous support material (if any), and the test conditions.

Much effort has been expended searching for polymeric materials with better ethanol–water separation performance than PDMS. Unfortunately, seldom materials have been reported to improve upon PDMS. Enhanced performance has been observed in a few references with composite membranes based on PDMS, including PDMS impregnated in a porous PTFE support ($\alpha = 14.0$) [71]; silicone oil supported in a porous polypropylene (PP) support (α up to 12.6) [72]; and a PDMS film treated with octadecyldiethoxymethylsilane ($\alpha = 16.3$) [72].

Another material that has received significant attention is poly[1-(trimethylsilyl)-1-propyne] or "PTMSP"-a high free volume polymer displaying a permeability greater than that of PDMS [73]. The ethanol–water separation factor for PTMSP has been reported to be higher than that of PDMS, ranging from 9 to 26 [74]. Unfortunately, PTMSP membranes have, so far, proven to deliver unstable performance, with flux and selectivity declining with running time. Such changes have been attributed to the compaction of the polymer film or the sorption of foulants inside the film [75]. Recently, PTMSP has been cross-linked to yield a more physically stable material which may improve prospects for this polymer [76].

So far, hydrophilic polymeric PV membranes, for example, cellulose acetate butyrate membrane [77], PDMS-PS IPN supported membranes [78], aromatic polyetherimide membranes [79] have been investigated. Ruckenstein et al. utilized polydimethylsiloxane polystyrene interpenetrating polymer network (PDMS-PS IPN) supported membranes for PV separation of ethanol from aqueous solutions. As PS is more hydrophobic and has higher tensile strength than PDMS, the mechanical and film-forming properties of PDMS-PS are better than those of PDMS. The selectivity of these PDMS-PS membranes varied with the feed compositions. For the feed having low ethanol concentration, the membrane was more selective for ethanol, while for the feed with high ethanol concentration it was more selective for water. Schué et al. [79] investigated the sorption, diffusion, and PV of ethanol solution in homogeneous and composite aromatic polyetherimide membranes. The performance of these membranes was found to be dependent on the permeate diffusivity rather than its solubility.

14.3.4.5 Alcohol–Water Separation Through Composite or Mixed Membranes

To combine the advantages of inorganic membrane and polymeric membrane for obtaining high ratio of membrane performance/cost, recently various inorganic–polymer or polymer–polymer composite membranes, such as polystyrenesulfonate/alumina [80], polyelectrolytes multilayer [81], KA zeolite-incorporated cross-linked PVA multilayer mixed matrix membranes (MMMMs) [82], and PVA–sodium alginate (SA) blend membranes [83], have been studied for pervaporation separation of ethanol/water mixtures. It is demonstrated by Martin that the separation factor of polystyrenesulfonate/alumina composite membranes was up to 400 [80].

Dong et al. prepared a hollow-fiber composite membrane, PVA–SA blend, supported by a polysulfone (PS) hollow-fiber ultrafiltration membrane for pervaporation ethanol dehydration. With ethanol concentration at 90 wt% in the feed at 45 °C, the separation factors and permeation fluxes were 384 and 384 $g{\cdot}m^{-2}{\cdot}h^{-1}$, respectively [83].

Due to the difficulty and cost of manufacturing defect-free commercial-scale silicalite-1 membranes, several groups have investigated the potential of mixed matrix membranes consisting of silicalite-1 particles dispersed in PDMS. In fact, a membrane of this type was, until recently, available commercially from Sulzer Chemtech (Neunkirchen, Germany). The performance of these mixed matrix materials is dependent on the loading of silicalite-1, size of the particles, source of silicalite-1, and membrane casting conditions. Although some performance gains have been observed with a loading as low as 30 wt% silicalite-1 [84], loadings of 60 wt% may be needed to deliver consistently high separation factors [85]. Ikegami et al. have combined PDMS and silicalite-1 in another manner to alter properties [86]. Instead of dispersing silicalite-1 in PDMS, they coated the surface of a silicalite-1 membrane with a thin layer of PDMS to protect the silicalite-1 surface and to fill defects, such as nonzeolitic pores, in the silicalite-1 layer. PDMS-coated silicalite-1membranes delivered ethanol–water separation factors ranging from 47 to 125. Since mixed matrix membranes would be fabricated and used in much the same way as standard polymer membranes, the cost per unit area of mixed matrix membrane modules should also be similar to that of standard polymer membrane modules and, therefore, several times lower than that of inorganic membrane modules. The increased performance characterized with little cost increase has led to the interest in mixed matrix membrane materials.

Yang et al. [87] reported a nanocomposite membrane, in which an ultrathin (300 nm) homogeneous silicalite–PDMS nanocomposite membrane was fabricated on capillary support by a “Packing-filling” method. First, silicalite-1 nanocrystals were deposited onto a porous alumina capillary support using dip-coating technique (packing); second, the interspaces among the nanocrystals were filled with PDMS phase (filling). No voids between nanocrystals and PDMS phase were observed by scanning electron microscopy (SEM), suggesting good zeolite–polymer adhesion. The membrane possesses very high flux (5.0–11.2 $kg{\cdot}m^{-2}{\cdot}h^{-1}$) and good separation factor (25.0–41.6) for the pervaporative recovery of *iso*-butanol from aqueous solution (0.2–3 wt%) at 80 °C. Such properties offer great potential towards applications in fermentation–pervaporation coupled processes.

Dong et al. [88] reported a novel chitosan/titanium dioxide (CS/TiO_2) nanocomposite membranes for pervaporation test. The characterization results demonstrated that nanosized TiO_2 particles dispersed homogeneously within the CS matrix, which could be assigned to the hydrogen and titanoxane bonds formed between CS and TiO_2. Moreover, the pervaporation performance of these membranes was investigated using the separation of ethanol–water mixture as model system. Compared with CS/TiO_2 hybrid membranes prepared by blending method, most of the CS/TiO_2 nanocomposite membranes prepared by in situ sol–gel process exhibited higher permeation flux and separation factor under the identical conditions. Among all the prepared membranes, CS/TiO_2 nanocomposite membrane containing 6 wt% TiO_2 exhibited the best pervaporation performance, whose averaged permeation flux and separation factor were 0.340 kg m^{-2} h^{-1} and 196 for 90 wt% aqueous solution of ethanol at 80 °C, respectively.

Huang et al. [89] prepared a type of novel nanofiller, polyphosphazene nanotubes (PZSNTs), which was incorporated in PDMS to form nanocomposite membrane for water–ethanol separation. PZSNT can be easily prepared and can be dispersed effectively in most general solvents. The incorporation of PZSNT in PDMS membrane increased the permeability and selectivity of membrane for water–ethanol mixtures. As PZSNT content increased from 0% to 10%, permeation flux and selectivity both increased. However, as PZSNT content further increased up to 20%, permeation flux and selectivity enhancement was only slight. A decrease in PZSNT diameter also leads to an increase in permeation flux and separation factor.

14.3.4.6 Alcohol–Water Separation Through Multilayer Membranes

Tieke et al. [81] prepared multilayer membranes by alternate adsorption of cationic and anionic polyelectrolytes onto porous support membranes, and achieved highest separation capability when polyelectrolytes with high charge density such as polyetherimide (PEI) and polyvinylsulfate (PVS) were used.

In 2006, Guan et al. [90] fabricated MMMMs consisting of a selective mixed matrix membrane (MMM) [top layer, a porous poly(acrylonitrile-comethyl acrylate) [poly(AN-co-MA)] intermediate layer and a polyphenylene sulfide (PPS) nonwoven fabrics substrate]. The group found that the separation performance of the MMMM is superior to that of multi-ply homogenous membranes (MHM) containing no zeolite. In addition, a series of three-layer zeolite-embedded PVA composite membranes have been successfully fabricated using different zeolites with a loading of 20 wt%, including 3A, 4A, 5A, NaX, NaY, silicalite, and beta. Results showed that the addition of zeolite resulted in decrease in activation energies for water and ethanol, and hence increase in separation selectivity [82]. The hydrophilic PVA is chosen as the polymeric material because it is the most attractive and economical polymer material for ethanol dehydration [91].

A novel trilayer composite membrane consisting of the active layer polydimethylsiloxane (PDMS, Sylgard® 184) and dual support layers of high-porosity polyethylene (PE) and high mechanical stiffness perforated metal was investigated for the separation of 1-butanol from aqueous solution by means of pervaporation [92]. The results showed that total flux and separation factor were both increased by placing a layer of hydrophobic PE between the PDMS and the metal support. The enhancement was especially significant at low temperatures. With the feed solution of 2% 1-butanol at 37 °C, the PDMS/PE/Brass support composite membrane conferred a total flux of 132 g $m^{-2}h^{-1}$ and a separation factor of 32. With the increase of the PDMS thickness, the separation factor increased as the total flux declined. It is suggested that while the water flux remained stable, the 1-butanol flux had a linear relationship with respect to the feed concentration of 1-butanol. The overall mass-transfer coefficient for butanol was determined to be 6.9E-7 m s^{-1} using the resistance-in-series model. Using a semiempirical Sherwood number correlation, the mass transfer coefficient of 1-butanol through the liquid side boundary layer was estimated to be 25.5E-7 m s^{-1}. This was more than three times higher than the overall mass transfer coefficient, indicating that the membrane dominated the mass transfer of the pervaporation process.

Huang et al. [93] prepared a multilayer PDMS/PVDF[poly(vinylidene fluoride)] composite membrane with an alternative PDMS/PVDF/non-woven-fiber/PVDF/PDMS configuration. The porous PVDF substrate was obtained by casting PVDF solution on both sides of nonwoven fiber with immersion precipitation phase-inversion method. PDMS was then cured by phenyltrimethoxylsilane (PTMOS) and coated onto the surface of porous PVDF substrate one layer by the other to obtain multilayer PDMS/PVDF composite membrane. The multilayer composite membrane was used for ethanol recovery from aqueous solution by pervaporation, and exhibited enhanced separation performance compared with one side PDMS/PVDF composite membranes, especially in the low ethanol concentration range. The maximum separation factor of multilayer PDMS/PVDF composite membrane was obtained at 60 °C, and the total flux increased exponentially along with the increase of temperature. The composite membrane gave the best pervaporation performance with a separation factor of 15, and permeation rate of 450 g $m^{-2}h^{-1}$ with a 5 wt% ethanol concentration at 60 °C.

The layer-by-layer chitosan (CS)/polyacrylonitrile (PAN) composite membranes with high structural stability for PV dehydration were fabricated recently [94]. The structural stability of CS/PAN composite membrane for PV dehydration was improved by the introduction of mussel-adhesive-mimetic molecule, carbopol (CP). The composite membranes were simply fabricated by layer-by-layer technique, in which CP acted as an intermediate layer bridging the CS active layer and the PAN support layer. The membrane pervaporation performance was investigated by varying the molecular weight (CP981, 940, 974) and concentration of CP as well as the cross-linking degree of the CS active layer. When the concentration of CP974 was 0.5 wt%, GCCS(60)/CP(0.5)/PAN composite membrane displayed the permeation flux of 1247 g $m^{-2}h^{-1}$ and separation factor of 256 for ethanol dehydration at 353 K.

14.3.4.7 Removal of Inhibitors Fermentation broth generally contains inhibiting substances including ethanol product, flavors (phenolics), and other chemicals that are inhibitory to microorganisms and lead to apparent reduction in fermentation yield and productivity. In general, there are three major groups of inhibitors: aliphatic

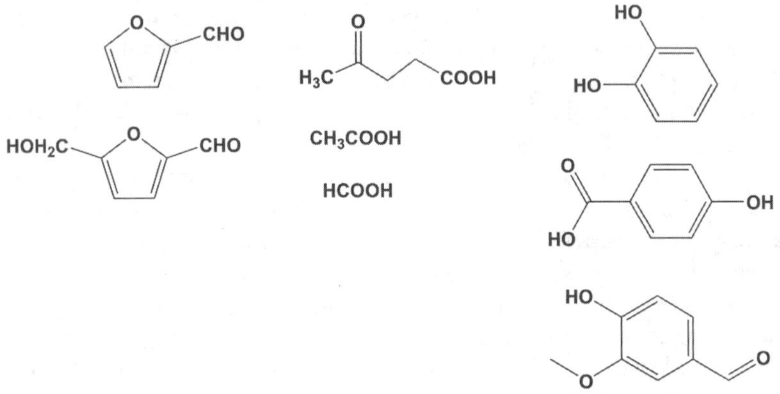

FIGURE 14.13 Molecular structure of some inhibitors.

acids (acetic, formic, and levulinic acid), furan derivatives furfural and HMF, and phenolic compounds (phenol, vanillin, *p*-hydroxybenzoic acid) (Fig. 14.13) [95].

In order to enhance the efficiency of hydrolyzate fermentation, in addition to optimization of the pretreatment and hydrolysis process for minimizing formation of the hydrolysis by-products (inhibitors), it is necessary to remove inhibitors (detoxify hydrolyzates) prior to fermentation or in situ detoxification. The detoxification can be either chemical, physical, or biological [96]. The most commonly used methods for detoxification of hydrolyzates before fermentation are: evaporation [97], solvent extraction [98], overliming with calcium hydroxide [95b,99], activated charcoal [95b,100], ion exchange resins (IER) [95b,100c], and enzymatic detoxification [101].

Evaporation is a simple way to remove acetic acid, furfural, and other volatile components, but it is difficult to remove the heavier components with higher boiling points. Extraction with solvent (e.g., ethyl acetate) is efficient in removing all the inhibitors, but it needs additional solvent (ethyl acetate) and solvent recovery for recycle use. Activated charcoal adsorption can remove the phenolic compounds, but it is not so efficient in removal of acetic acid and furfural. Overliming and IER are more efficient procedures for removal of different inhibitors from hydrolyzates, but the former leads to a large amount of gypsum. It could be concluded that IER method is the current best choice for detoxification because of its high detoxification efficiency, easy (continuous) operation, and flexible combination of different anion and cation exchangers, while the enzymatic treatment can possibly be the future choice. In addition, extractive-fermentation, membrane pervaporation-bioreactor hybrid, and VMD-bioreactor hybrid are very promising processes to remove inhibitory compounds in addition to increasing ethanol yield.

Membrane Pervaporation-Bioreactor Hybrid Fermentation coupled with PV membrane bioreactor technology is now at the laboratory stage. The key to achieve industrialization of this technology is preparing ethanol permeable PV membrane with high performance [102].

This problem can be overcome by combining fermentation with hydrophobic membrane PV for removal of the inhibitors from the fermentation broth [103]. Hence, the process can be carried out continuously and the recovered volatile organic compound (VOCs) (ethanol, acetone, butanol, 2-propanol) can be reused within other processes.

In the real application, a MF membrane is added before PV to avoid fouling of the hydrophobic membrane. Besides, the ethanol-enriched solution, that is, the permeate of the hydrophobic membrane, can be further dehydrated to produce anhydrous ethanol.

Vacuum Membrane Distillation (VMD)-Bioreactor Hybrid Membrane distillation (MD) is an appealing process suitable for separation of aqueous mixtures [104]. There are four types of MD: direct contact membrane distillation (DCMD) [105], air gap membrane distillation (AGMD) [106], sweeping gas membrane distillation (SGMD) [107] and vacuum membrane distillation (VMD) [108]. Actually, VMD is quite similar to PV, the only difference being that the separation factor here is established by vapor–liquid equilibrium of the feed solution which is not affected by the membrane used [109].

Similar to membrane PV-bioreactor hybrid, VMD-bioreactor hybrid process is also suitable for separation of ethanol and the other inhibitory compounds from fermentation broths. Till now, there has been much reported on this topic [110]. As an example, Gryta et al. produced ethanol in a membrane distillation bioreactor where porous capillary polypropylene membranes were applied for separating volatile compounds, including ethanol and other inhibitors, from the feed (broth), leading to increase in the productivity and the sugar-to-ethanol conversion rate. VMD is commercially competitive because of its high selectivity of ethanol over water, large flux, high thermal efficiency, and low energy cost [111].

14.4 CONCLUSIONS AND OUTLOOK

Nowadays, improvement in product quality and lowering the production cost are continuously required. Considering this requirement, membrane technology has many obvious advantages, which will inevitably replace the traditional inefficient separation techniques. In addition to this significant progress, it can be concluded that the use of membrane-separation technologies in biorefinery has presented many advantages over the traditional separation technologies. For example, they are relatively simple, and easy to operate, structurally compact, low cost, convenient for maintenance, and easy to be controlled automatically. Membrane-separation process can be cycled in a closed system, and thus prevent the contamination from external sources. Without any foreign chemical substances added, the permeated liquid can be recycled, thereby reducing the costs and reducing environmental pollution. All the advantages of membrane-separation technology above establish its importance with high separation efficiency in modern biochemical separation technologies, which can replace the traditional filtration, adsorption, evaporation, condensation, and other separation technologies. Bearing the advantages in mind, limitations of membrane-separation technology still needed to be determined as part of a study leading to an economic and efficient separation process for the biorefinery. Take the PV technology as an example, which is an emerging technology with significant potential to efficiently recover ethanol and other biofuels from fermentation broths. Several issues must be addressed for PV to be an economically viable for biofuel recovery: (1) increased energy efficiency origins from improved ethanol–water separation factor and heat integration/energy recovery. (2) Longer term trials with actual fermentation broths to assess membrane and module stability and fouling behavior. (3) Optimized integration of PV with fermentor which includes filtration (MF or UF) to increase cell density in fermentor and allow higher PV temperatures, as well as removal/avoidance of inhibitors. (4) Updated economic analyses of PV which provide comparisons to competing technologies on even bases at various biofuel production scales [112].

In the membrane-separation process, the concentration polarization and membrane fouling are two phenomena frequently presented, which affect the application of membrane-separation technology in the biological chemical engineering. Therefore, the research on mechanism and control technology of membrane fouling, the development of low cost, antifouling membrane materials and defect-free membranes, as well as new membrane devices design are key factors to determine whether the membrane technology will have large-scale application in the new fields of biological industries.

To play a more important role in the separation industry, membrane technology needs to address the following issues: (1) to develop the membrane with high throughput and high recognition, to improve membrane materials, membrane devices and equipments, to raise the membrane selectivity, and continue to develop functional polymer membrane materials and inorganic membrane materials; (2) to prevent membrane fouling and to develop energy-efficient decontamination technology. It involves research on membrane fouling mechanism and membrane materials with excellent antipollution performance; (3) to integrate membrane-separation techniques into existing biorefinery processes.

In conclusion, the continuing development of new membrane-separation technologies and the successful integration of them into modern biorefinery will contribute to the foundation of green and sustainable biorefinery.

REFERENCES

1. B. Kamm and M. Kamm, *Chem. Biochem. Eng.* **2004**, 18, 1.
2. (a) A. Demirbas, *Progr. Energy Combust. Sci.* **2007**, 32, 1; (b) D. L. Klass, *Biomass for Renewable Energy, Fuels, and Chemicals*, Academic Press, San Diego, **1998**; (c) D. M. Lai, L. Deng, J. Li, B. Liao, Q. X. Guo and Y. Fu, *ChemSusChem* **2011**, 4, 55; (d) A. Onda, T. Ochi and K. Yanagisawa, *Green Chem.* **2008**, 10, 1033; (e) J. Pang, A. Wang, M. Zheng and T. Zhang, *Chem. Commun.* **2010**, 46, 6935; (f) R. Rinaldi, N. Meine, J. V. Stein, R. Palkovits and F. Schüth, *ChemSusChem* **2010**, 3, 266; (g) R. Rinaldi, R. Palkovits and F. Schüth, *Angew. Chem. Int. Ed.* **2008**, 47, 8047.

3. G. W. Huber, S. Iborra and A. Corma, *Chem. Rev.* **2006**, 106, 4044.
4. (a) P. M. Dewick, *Nat. Prod. Rep.* **2002**, 19, 181; (b) Y. Li, Z. Zhao and F. Bai, *Enzyme Microb. Technol.* **2007**, 41, 312; (c) A. Mukhopadhyay, A. M. Redding, B. J. Rutherford and J. D. Keasling, *Curr. Opin. Biotechnol.* **2008**, 19, 228; (d) M. A. Rude and A. Schirmer, *Curr. Opin. Microbiol.* **2009**, 12, 274; (e) X. Zhao, C. M. Hu, S. G. Wu, H. W. Shen and Z. K. Zhao, *J. Ind. Microbiol. Biot.* **2011**, 38, 627; (f) A. Chatzifragkou, A. Makri, A. Belka, S. Bellou, M. Mavrou, M. Mastoridou, P. Mystrioti, G. Onjaro, G. Aggelis and S. Papanikolaou, *Energy* **2011**, 36, 1097.
5. R. Rinaldi and F. Schüth, *ChemSusChem* **2009**, 2, 1096.
6. (a) Z. H. Zhang, Q. A. Wang, H. B. Xie, W. J. Liu and Z. B. Zhao, *ChemSusChem* **2011**, 4, 131; (b) Z. H. Zhang and Z. B. Zhao, *Bioresour. Technol*, **2011**, 102, 3970; (c) C. Carlini, P. Patrono, A. M. R. Galletti, G. Sbrana and V. Zima, *Appl. Catal., A* **2005**, 289, 197.
7. (a) L. D. Schmidt and P. J. Dauenhauer, *Nature* **2007**, 447, 914; (b) Z. H. Zhang, K. Dong and Z. B. Zhao, *ChemSusChem* **2011**, 4, 112.
8. J. A. Melero, J. Iglesias and G. Morales, *Green Chem.* **2009**, 11, 1285.
9. C. Angels and P. Cristina, *J. Membr. Sci.* **2007**, 291, 96.
10. (a) K. L. Kadam and J. D. McMillan, *Bioresour. Technol*, **2003**, 88, 17; (b) J. Zaldivar, J. Nielsen and L. Olsson, *Appl. Microbiol. Biot.* **2001**, 56, 17; (c) J. D. McMillan, *Renew. Energ.* **1997**, 10, 295; (d) S. Kim and B. E. Dale, *Biomass Bioenerg.* **2004**, 26, 361.
11. (a) S. Kim and B. E. Dale, *Biomass Bioenerg.* **2005**, 28, 475; (b) A. McAloon, W. Yee and F. Taylor, *Production of Ethanol from Corn by the Dry Grind Process*, USDA-ARS Eastern Regional Research Center, Wyndmoor, PA, **2004**.
12. P. Sukitpaneenit and T. S. Chung, *J. Membr. Sci.* **2011**, 374, 67.
13. C. S. Jiang, Z. L. Wu, R. Li and Q. Liu, *Biochem. Eng. J.* **2011**, 55, 43.
14. C. Schacht, C. Zetzl and G. Brunner, *J. Supercrit. Fluid* **2008**, 46, 299.
15. J. Heinonen and T. Sainio, *Ind. Eng. Chem. Res.* **2010**, 49, 2907.
16. M. Heyd, M. Franzreb and S. Berensmeier, *Biotechnol. Progr.* **2011**, 27, 706.
17. H. J. Huang, S. Ramaswamy, U. W. Tschirner and B. V. Ramarao, *Sep. Purif. Technol.* **2008**, 62, 1.
18. C. J. Geankoplis and P. R. Toliver, Membrane separation processes, *Transport Processes and Separation Process Principles*, 4th ed., Prentice Hall PTR, New Jersey, **2003**.
19. F. Lipnizki, *Desalination* **2010**, 250, 1067.
20. F. Lipnizki, S. Hausmanns, G. Laufenberg, R.W. Field and B. Kunz, *Chem. Eng. Technol.* **2000**, 23, 569.
21. B. Drosg, T. Wirthensohn, G. Konrad, D. Hornbachner, C. Resch, F. Wager, C. Loderer, R. Waltenberger, R. Kirchmayr and R. Braun, *Water. Sci. Technol.* **2008**, 58, 1483.
22. (a) N. Nagata, K. Herouvis, D. Dziewulski and G. Belfort, *Biotech. Bioeng.* **1989**, 34, 447; (b) S. L. Li, K. S. Chou, J. Y. Lin, H. W. Yen and I. M. Chu, *J. Membr. Sci.* **1996**, 110, 203; (c) H. V. Adikane, R. K. Singh and S. N. Nene, *J. Membr. Sci.* **1999**, 162, 119; (d) S. M. Bailey and M. M. Meagher, *J. Membr. Sci.* **2000**, 166, 137; (e) K. Graves, G. Rozeboom, M. Heng and C. Glatz, *Biotech. Bioeng.* **2006**, 94, 346.
23. (a) N. Rossignol, L. Vandanjon, P. Jaouen and F. Quemeneur, *Aquacult. Eng.* **1999**, 20, 191; (b) N. Rossi, O. Jaouen, P. Legentilhomme and I. Petit, *Food Bioprod.* **2004**, 82,

244; (c) E. Molina Grima, E. H. Belarbi, F. G. Acien Fernandez, A. Robles Medina and Y. Chisti, *Biotechnol. Adv.* **2003**, 20, 491.

24. S. D. Rios, E. Clavero, J. Salvadó, X. Farriol and C. Torras, *Ind. Eng. Chem. Res.* **2011**, 50, 2455.
25. C. Huang, W. Jiang and C. Chen, *Water Sci. Technol.* **2004**, 50, 133.
26. T. Furukawa, K. Kokubo, K. Nakamura and K. Matsumoto, *J. Membr. Sci.* **2008**, 322, 491.
27. R. van Reis and A. Zydney, *J. Membr. Sci.* **2007**, 302, 271.
28. J. S. Knutsen and R. H. Davis, *Appl. Biochem. Biotechnol.* **2002**, 98–100, 1161.
29. W. D. Mores, J. S. Knutsen and R. H. Davis, *Appl. Biochem. Biotechnol.* **2001**, 91–93, 297.
30. (a) S. Metsämuuronen and M. Nyström, *Desalination* **2006**, 200, 290; (b) Q. Sun, Y. L. Su, X. L. Ma, Y. Q. Wang and Z. Y. Jiang, *J. Membr. Sci.* **2006**, 285, 299; (c) D. B. Mosqueda-Jimenez, R. M. Narbaitz and T. Matsuura, *J. Appl. Polym. Sci.* **2006**, 99, 2978.
31. R. S. Juang, H. L. Chen and Y. S. Chen, *J. Membr. Sci.* **2008**, 323, 193.
32. (a) Z. G. Wang, T. L. Chen and J. P. Xu, *Macromolecules* **2001**, 34, 9015; (b) K. J. Lee, J. Y. Jae, Y. S. Kang, J. Won, Y. Dai, G. P. Robertson and M. D. Guiver, *J. Membr. Sci.* **2003**, 223, 1; (c) F. Wang, M. Hickner, Y. S. Kim, T. A. Zawodzinski and J. E. McGrath, *J. Membr. Sci.* **2002**, 197, 231; (d) Z. L. Xu and F. A. Qusay, *J. Membr. Sci.* **2004**, 233, 101.
33. J. A. Koehler, M. Ulbricht and G. Belfort, *Langmuir* **1997**, 13, 4162.
34. Q. F. Zhang, S. B. Zhang, L. Dai and X. S. Chen, *J. Membr. Sci.* **2010**, 349, 217.
35. H. Choi, K. Zhang, D. D. Dionysiou, D. B. Oerther and G. A. Sorial, *Sep. Purif. Technol.* **2005**, 45, 68.
36. D. M. Krstic, S. L. Markov and M. N. Tekic, *Biochem. Eng. J.* **2001**, 9, 103.
37. H. Choi, K. Zhang, D. Dionysiou, D. Oerther and G. Sorial, *J. Membr. Sci.* **2005**, 248, 189.
38. A. Javaid, T. Ryan, G. Berg, X. Pan, T. Vispute, S. R. Bhatia, G. W. Huber and D. M. Ford, *J. Membr. Sci.* **2010**, 363, 120.
39. G. Toussaint, L. H. Ding and M. Y. Jaffrin, *Sep. Sci. Technol.* **2000**, 35, 795.
40. C. Guell, P. Czekaj and R. H. Davis, *J. Membr. Sci.* **1999**, 155, 113.
41. (a) S. T. Kelly and A. L. Zydney, *J. Membr. Sci.* **1995**, 107, 115; (b) S. T. Kelly and A. L. Zydney, *Biotech. Bioeng.* **1997**, 55, 92.
42. A. Persson, A. S. Jonsson and G. Zacchi, *Appl. Biochem. Biotechnol.* **2004**, 112, 151.
43. E. Kujundzic, A. R. Greenberg, R. Fong, B. Moore, D. Kujundzic and M. Hernandez, *J. Membr. Sci.* **2010**, 349, 44.
44. A. Brou, L. H. Ding, P. Boulnois and M. Y. Jaffrin, *J. Membr. Sci.* **2002**, 197, 269.
45. (a) C. Torras, J. Pallares, R. Garcia-Valls and M. Y. Jaffrin, *Desalination* **2009**, 235, 122.
46. P. Côté, C. Roy and N. Bernier, *Sep. Sci. Technol.* **2009**, 44, 110.
47. S. A. Amartey, P. C. J. Leung, N. Baghaei-Yazdi, D. J. Leak and B. S. Hartley, *Process Biochem.* **1999**, 34, 289.
48. K. Sriroth, K. Piyachomkwan, S. Wanlapatit and S. Nivitchanyong, *Fuel*, **2010**, 89, 1333.
49. P. A. Kober, *J. Am. Chem. Soc.* **1919**, 39, 944.
50. R. C. Binning, R. J. Lee, I. Joseph, F. Jennings and E. C. Martin, *Ind. Eng. Chem.* **1961**, 53, 45.
51. G. F. Tusel and H. E. A. Bruschke, *Desalination* **1985**, 53, 327.
52. A. Jonquieres, R. Clement, P. Lochon, J. Neel, M. Dresch and B. Chreticn, *J. Membr. Sci.* **2002**, 206, 87.

53. (a) F. Lipnizki, R. W. Field and P. K. Ten, *J. Membr. Sci.* **1999**, 153, 183; (b) S. Sridhar, B. Smitha and A. Shaik, *Sep. Purif. Rev.* **2005**, 34, 1.
54. L. Y. Jiang, Y. Wang, T. Chung, X. Y. Qiao and J. Lai, *Prog. Polym. Sci.* **2009**, 34, 1135.
55. Y. Morigami, M. Kondo, J. Abe, H. Kita and K. Okamoto, *Sep. Purif. Technol.* **2001**, 25, 251.
56. K. Okamoto, H. Kita, K. Horii, K. Tanaka and M. Kondo, *Ind. Eng. Chem. Res.* **2001**, 40, 163.
57. (a) M. A. Camblor, A. Corma, S. Iborra, S. Miquel, J. Primo and S. Valencia, *J. Catal.* **1997**, 172, 76; (b) H. Takaba, A. Koyama and S. Nakao, *J. Phys. Chem. B* **2000**, 104, 6353; (c) T. Sano, T. Kasuno, K. Takeda, S. Arazaki and Y. Kawakami, *Stud. Surf. Sci. Catal.* **1997**, 105, 1771; (d) Y. Oumi, A. Miyajima, J. Miyamoto and T. Sano, *Stud. Surf. Sci. Catal.* **2002**, 142, 1595.
58. Y. Morigami, M. Kondo, J. Abe, H. Kita and K. Okamoto, *Sep. Purif. Technol.* **2001**, 25, 251.
59. (a) H. Kita, K. Fuchida, T. Horita, H. Asamura and K. Okamoto, *Sep. Purif. Technol.* **2001**, 25, 261; (b) H. Kita, T. Inoue, H. Asamura, K. Tanaka and K. Okamoto, *Chem. Commun.* **1997**, 45; (c) K.-I. Okamoto, H. Kita, M. Kondo, N. Miyake and Y. Matsuo, US Patent 5,554,286, **1996**. (d) X. Lin, H. Kita and K. Okamoto, *Ind. Eng. Chem. Res.* **2001**, 40, 4069; (e) M. Kondo, T. Yamamura, T. Yukitake, Y. Matsuo, H. Kita and K.-I. Okamoto, *Sep. Purif. Technol.* **2003**, 32, 191; (f) X. Lin, H. Kita and K. Okamoto, *Chem. Commun.* **2000**, 19, 1889; (g) H. Kita, K. Horii, Y. Ohtoshi, K. Tanaka and K. I. Okamoto, *J. Mater. Sci. Lett.* **1995**, 14, 206.
60. K. I. Okamoto, H. Kita, M. Kondo, N. Miyake and Y. Matsuo, US Patent 5,554,286, **1996**.
61. A. W. C. van den Berg, L. Gora, J. C. Jansen, M. Makkee and T. Maschmeyer, *J. Membr. Sci.* **2003**, 224, 29.
62. J. J. Jafar, P. M. Budd and R. Hughes, *J. Membr. Sci.* **2002**, 199, 117.
63. M. Kondo, M. Komori, H. Kita and K. Okamoto, *J. Membr. Sci.* **1997**, 133, 133.
64. K. Tanaka, R. Yoshikawa, C. Ying, H. Kita and K. Okamoto, *Catal. Today*, **2001**, 67, 121.
65. Y. F. Zhang, Z. Q. Xu and Q. L. Chen, *J. Membr. Sci.* **2002**, 210, 361.
66. T. C. Bowen, R. D. Noble and J. L. Falconer, *J. Membr. Sci.* **2004**, 245, 1.
67. T. C. Bowen, H. Kalipcilar, J. L. Falconer and R. D. Noble, *J. Membr. Sci.* **2003**, 215, 235.
68. Y. R. Dong, M. Nakao, N. Nishiyama, Y. Egashira and K. Ueyama, *Sep. Purif. Technol.* **2010**, 73, 2.
69. P. S. Tin, H. Y. Lin, R. C. Ong and T. S. Chung, *Carbon*, **2011**, 49, 369.
70. L. M. Vane, *J. Chem. Technol. Biotechnol.* **2005**, 80, 603.
71. Y. Mori and T. Inaba, *Biotech. Bioeng.* **1990**, 36, 849.
72. T. Kashiwagi, K. Okabe and K. Okita, *J. Membr. Sci.* **1988**, 36, 353.
73. (a) S. L. Schmidt, M. D. Myers, S. S. Kelley, J. D. McMillan and N. Padukone, *Appl. Biochem. Biotechnol.* **1997**, 63–65, 469; (b) T. Masuda and E. Isobe, T. Higashimura, *J. Am. Chem. Soc.* **1983**, 105, 7473; (c) I. Pinnau and L. G. Toy, *J. Membr. Sci.* **1996**, 116, 199; (d) V. V. Volkov, V. S. Khotimsky, E. G. Litvinova, A. G. Fadeev, Y. A. Selinskaya, N. A. Plate, J. D. McMillan and S. S. Kelley, *Polym. Mater. Sci. Eng.* **1997**, 77, 339; (e) Y. Ichiraku, S. A. Stern and T. Nakagawa, *J. Membr. Sci.* **1987**, 34, 5; (f) V. V. Volkov,

A. G. Fadeev, V. S. Khotimsky, E. G. Litvinova, Y. A. Selinskaya, J. D. McMillan and S. S. Kelley, *J. Appl. Polym. Sci.* **2004**, 91, 2271.

74. A. G. Fadeev, Y. A. Selinskaya, S. S. Kelley, M. M. Meagher, E. G. Litvinova, V. S. Khotimsky and V. V. Volkov, *J. Membr. Sci.* **2001**, 186, 205.
75. A. G. Fadeev, S. S. Kelley, J. D. McMillan, Y. A. Selinskaya, V. S. Khotimsky and V. V. Volkov, *J. Membr. Sci.* **2003**, 214, 229.
76. (a) C. J. Ruud, J. Jia and G. L. Baker, *Macromolecules*, **2000**, 33, 8184; (b) J. Jia and G. L. Baker, *J. Appl. Polym. Sci.: Part B*, **1998**, 36, 959.
77. W. S Wu, W. S. Lau, W. W. Y. Tangaiah and G. P. Sourirajan, *J. Coll. Interf. Sci.* **1993**, 160, 502.
78. E. Ruckenstein and L. Liang, *J. Membr. Sci.* **1996**, 114, 227.
79. H. Qariouh, R. Schué, F. Schué and C. Bailly, *Polym. Int.* **1999**, 48, 171.
80. C. R. Martin, P. Aranda and W. J. Chen, *J. Membr. Sci.* **1995**, 107, 199.
81. B. Tieke and L. Krasemann, *Chem. Eng. Technol.* **2000**, 23, 211.
82. Z. Huang, H. M. Guan, W. L. Tan, X. Y. Qiao and S. Kulprathipanja, *J. Membr. Sci.* **2006**, 276, 260.
83. Y. Q. Dong, L. Zhang, J. N. Shen, M. Y. Song and H. L. Chen, *Desalination*, **2006**, 193, 202.
84. I. F. J. Vankelecom, E. Scheppers, R. Heus and J. B. Uytterhoeven, *J. Phys. Chem.* **1994**, 98, 12390.
85. (a) W. J. Groot, M. R. Kraayenbrink, R. G. J. M. van der Lans, K. Ch. and A. M. Luyben, *Bioproc. Eng.* **1993**, 8, 189; (b) M. M. Meagher, N. Qureshi and R. W. Hutkins, US Patent 5,755,967, **1998**.
86. (a) T. Ikegami, D. Kitamoto, H. Negishi, K. Haraya, H. Matsuda, Y. Nitanai, N. Koura, T. Sano and H. Yanagishita, *J. Chem. Technol. Biotechnol.* **2003**, 78, 1006; (b) H. Matsuda, H. Yanagishita, H. Negishi, D. Kitamoto, T. Ikegami, K. Haraya, K. Nakane, Y. Idemoto, N. Koura and T. Sano, *J. Membr. Sci.* **2002**, 210, 433.
87. X. L. Liu, Y. S. Li, Y. Liu, G. Zhu, J. Liu and W. Yang, *J. Membr. Sci.* **2011**, 369, 228.
88. D. Yang, J. Li, Z. Jiang, L. Lu and X. Chen, *Chem. Eng. Sci.* **2009**, 64, 3130.
89. Y. W. Huang, P. Zhang, J. Fu, Y. Zhou, X. Huang and X. Tang, *J. Membr. Sci.* **2009**, 339, 85.
90. H. M. Guan, T. S. Chung, Z. Huang and M. L. Chng, S. Kulprathipanja, *J. Membr. Sci.* **2006**, 268, 113.
91. H. Ohya, K. Matsumoto, Y. Negishi, T. Hino and H. S. Choi, *J. Membr. Sci.* **1992**, 68, 141.
92. S. Y. Li, R. J. Srivastava and R. S. Parnas, *J. Membr. Sci.* **2010**, 363, 287.
93. X. Zhan, J. D. Li and J. Q. Huang, *Appl. Biochem. Biotech*, **2010**, 160, 632.
94. J. Ma, M. H. Zhang, H. Wu, X. Yin, J. Chen and Z. Jiang, *J. Membr. Sci.* **2010**, 348, 150.
95. (a) E. Palmqvist and B. Hahn-Hägerdal, *Bioresour. Technol.* **2000**, 74, 25; (b) A. K. Chandel, R. K. Kapoor, A. Singh and R. C. Kuhad, *Bioresour. Technol.* **2007**, 98, 1947.
96. E. Palmqvist and B. Hahn-Hägerdal, *Bioresour. Technol.* **2000**, 74, 17.
97. A. Converti, J. M. Dominguez, P. Perego, S. S. Silva and M. Zilli, *Chem. Eng. Technol.* **2000**, 23, 1013.
98. (a) J. M. Cruz, J. M. Dominguez, H. Dominguez and J. C. Parajo, *Food Chem.* **1999**, 67, 147; (b) J. Gonzalez, J. M. Cruz, H. Dominguez and J. C. Parajo, *Food Chem.* **2004**, 84,

243; (c) J. J. Wilson, L. Deschatelets and N. K. Nishikawa, *Appl. Microbiol. Biotechnol.* **1989**, 31, 592.

99. M. Cantarella, L. Cantarella, A. Gallifuoco, A. Spera and F. Alfani, *Process Biochem.* **2004**, 39, 1533.
100. (a) J. C. Parajo, H. Dominguez and J. M. Dominguez, *Bioresour. Technol.* **1996**, 57, 179; (b) L. Canilha, J. B. de, A. e Silva and A. I. N. Solenzal, *Process Biochem.* **2004**, 39, 1909; (c) M. L. M. Villarreal, A. M. R. Prata, M. G. A. Felipe, J. B. Almeida and E. Silva, *Enzyme Microb. Technol.* **2006**, 40, 17.
101. (a) L. J. Jönsson, E. Palmqvist, N. O. Nilvebrant and B. Hahn-Hägerdal, *Appl. Microbiol. Biotechnol.* **1998**, 49, 691; (b) T. Gutiérrez, L. O. Ingram and J. F. Preston, *J. Biotechnol.* **2006**, 121, 154.
102. (a) P. Kaewkannetra, N. Chutinate, S. Moonamart, T. Kamsan and T. Y. Chiu, *Desalination* **2011**, 271, 88; (b) T. Ikegami, D. Kitamoto, H. Negishi, K. Haraya, H. Matsuda, Y. Nitanai, N. Koura, T. Sano and H. Yanagishita, *J. Chem. Technol. Biot.* **2003**, 78, 1006; (c) V. V. Volkov, A. G. Fadeev, V. S. Khotimsky, E. G. Litvinova, Y. A. Selinskaya, J. D. McMillan and S. S. Kelley, *J. Appl. Polym. Sci.* **2004**, 91, 2271; (d) T. A. C. Oliveira, J. T. Scarpello and A. G. Livingston, *J. Membr. Sci.* **2002**, 195, 75; (e) F. Lipnizki, S. Hausmanns, G. Laufenberg, R. Field and B. Kunz, *Chem. Eng. Technol.* **2000**, 23, 569; (f) M. W. Reij, J. T. F. Keurentjes and S. Hartmans, *J. Biotechnol.* **1998**, 59, 155; (g) K. Brindle and T. Stephenson, *Biotechnol. Bioeng.* **1996**, 49, 601; (h) K. Kargupta, S. Datta and S. K. Sanyal, *Biochem. Eng. J.* **1998**, 1, 31.
103. (a) F. Lipnizki, S. Hausmanns, G. Laufenberg, R. Field and B. Kunz, *Chem. Ing. Tech.* **1998**, 70, 1587; (b) F. Lipnizki and R. W. Field, *Sep. Sci. Technol.* **2001**, 36, 3311; (c) K. L. Wasewar and V. G. Pangarkar, *Chem. Biochem. Eng. Q* **2006**, 20, 135.
104. (a) G. Lewandowicz, W. Bialas, B. Marczewski and D. Szymanowska, *J. Membr. Sci.* **2011**, 375, 212; (b) Y. Y. Lu and J. H. Chen, *Ind. Eng. Chem. Res.* **2011**, 50, 7345; (c) M. Khayet, *Adv. Coll. Interf.* **2011**, 164, 56.
105. H. J. Hwang, K. He, S. Gray, J. H. Zhang and I. S. Moon, *J. Membr. Sci.* **2011**, 371, 90.
106. (a) K. He, H. J. Hwang and I. S. Moon, *Kor. J. Chem. Eng.* **2011**, 28, 770; (b) J. Kim, D. S. Chang and Y. Y. Choi, *Ind. Eng. Chem. Res.* **2009**, 48, 5431.
107. (a) K. Charfi, M. Khayet and M. J. Safi, *Desalination* **2010**, 259, 84; (b) M. Khayet, P. Godino and J. I. Mengual, *J. Membr. Sci.* **2000**, 170, 243.
108. (a) N. Tang, H. J. Zhang and W. Wang, *Desalination* **2011**, 274, 120; (b) J. P. Mericq, S. Laborie and C. Cabassud, *Water. Res.* **2010**, 44, 5260.
109. C. Gostoli and G. C. Sarti, *J. Membr. Sci.* **1989**, 41, 211.
110. (a) J. Bausa and W. Marquardt, *Ind. Eng. Chem. Res.* **2000**, 39, 1658; (b) M. Gryta, A. W. Morawski and M. Tomaszewska, *Catal. Today*, **2000**, 56, 159; (c) M. A. Izquierdo-Gil and G. Jonsson, *J. Membr. Sci.* **2003**, 214, 113; (d) J. Phattaranawik, A. G. Fane, A. C. S. Pasquier and W. Bing, *Desalination* **2008**, 223, 386; (e) J. Phattaranawik, A. G. Fane, A. C. S. Pasquier, W. Bing and F. S. Wong, *Chem. Ing. Tech.* **2009**, 32, 38; (f) T. H. Khaing, J. F. Li, Y. Z. Li, N. Wai and F. S. Wong, *Sep. Purif. Technol.* **2010**, 74, 138; (g) T. H. Khaing, J. F. Li, Y. Z. Li, N. Wai and F. S. Wong, *Sep. Purif. Technol.* **2010**, 74, 138.
111. Z. Lei, B. Chen and Z. Ding, *Special Distillation Processes*, 1st ed., Elsevier, Amsterdam, **2005**.
112. L. M. Vane, *J. Chem. Technol. Biotechnol.* **2005**, 80, 622.

CHAPTER 15

Assessment of the Ecotoxicological and Environmental Effects of Biorefineries

KERSTIN BLUHM, SEBASTIAN HEGER, MATTHEW T. AGLER, SIBYLLE MALETZ, ANDREAS SCHÄFFER, THOMAS-BENJAMIN SEILER, LARGUS T. ANGENENT, and HENNER HOLLERT

15.1 INTRODUCTION

Biofuels derived from organic materials, such as starch, vegetable and animal fats, or (ligno)cellulose, have found increased importance and interest for the transport sector in many countries during the past few years. They are considered to be renewable alternatives with the benefits of a reduced dependence on fossil fuels and a potential to slow down the effect of global climate change due to decreased greenhouse gas (GHG) emissions from the transport sector. These positive potentials have led to a considerable proliferation of biofuel production.

Biofuel raw materials can be obtained from food crops and traditional domestic production processes. These production processes demand large acreage to satisfy yields, and thus compete with food-crop production. Therefore, conflicting land-use interests and potential adverse effects on the environment accompany the rise in bioenergy demand. From an environmental point of view, the intensification of farming for higher yields may increase the pollution of terrestrial and aquatic ecosystems with fertilizers and pesticides. Furthermore, land clearings to increase the available cultivation area pose severe problems to ecosystem stability, biodiversity, and soil quality. In addition, the replacement of (rain)forests, savannas, or grasslands by cropland for the production of biofuels can cause an increase in GHG emission instead of lowering them [1,2]. Therefore, sustainability criteria for biodiesel production concerning the conservation of rain forests and their endemic species have to be considered. This might be especially important for Europeans with respect to the ambitious EU goal of reaching a 10% share of biofuels for the overall fuel consumption in the transport sector by 2020 [3]. In Germany, for example, arable

The Role of Green Chemistry in Biomass Processing and Conversion, First Edition.
Edited by Haibo Xie and Nicholas Gathergood.

land for the cultivation of energy crops is not sufficiently available to achieve this objective, and thus substantial biomass imports from non-EU countries will be required [4]. This emphasizes the necessity to establish global guidelines and a worldwide solution to conflicting land use if large-scale adverse effects of bioenergy use are to be avoided.

Alternatively, biomass for biofuel production can be obtained from organic waste products (e.g., wheat straw, poplar twigs, and corn stover), which are environmentally more sustainable, and are not competing with food production. Researchers are already studying the conversion of wooden or dry lignocelluloses, as well as green parts of plants, which are not used as comestible goods, to platform chemicals (sugar, carboxylates, and syngas [i.e., synthesis gas]) and further into biofuel or biofuel candidates. Biofuels that are produced from these organic wastes would prevent a competition with food production. Further improvements in environmental performance of waste products are related to agricultural productivity enhancement that lead to higher biomass production per acre and the optimization of feedstock processing towards lower energy inputs, which results in a better energy balance and lower GHG emissions. Besides waste products, microalgae are another example of a potential alternative feedstock as they permit the use of nonarable land as well as nonpotable water and do not compete with food-crop cultures (cell carbon is fixed from the atmosphere) [5–7]. In addition to research on suitable biomass and improvements of the environmental performance, approaches to optimize engine combustion of biofuels and combinations of renewable fuel components to enhance engine performance and reduce emissions have been reported [8].

Altogether, the advancement of biofuels is being directed by concerns regarding the potentially negative environmental impacts and the sustainability of renewable fuels. These concerns are mainly based on GHG emissions and energy balances of biofuels in regards to life cycle assessments (LCAs) compared to the fossil fuels they are supposed to replace. In this context, comprehensive LCA approaches are a helpful tool in elucidating the true potential for GHG savings when energy-consuming steps in the production process and emissions associated with logistics are taken into account. However, alongside efforts to investigate and verify the advantages of biofuels in terms of sustainability and reduction of emissions, further potential impacts to the environment require consideration. Unintended environmental consequences can result from a technological application aimed to improve a different environmental issue, which is, for example, the case of the fuel additive methyl tertiary butyl ether (MTBE). MTBE has been used to reduce pollutants from automobile fuel combustion and later attracted attention due to potential inhalation health effects and the contamination of water resources [9]. Consequently, the authors demanded the evaluation of potential environmental impacts in a comprehensive manner.

A literature research by Bluhm et al. [10] revealed an increase in research activities on biofuels particularly in the last 4 years. Figure 15.1 emphasizes this finding, by displaying that research on biofuels is a fast-growing sector with the potential to outpace the research on fossil fuels within the next couple of years. In contrast to the overall research on biofuels, comparatively few literature sources are

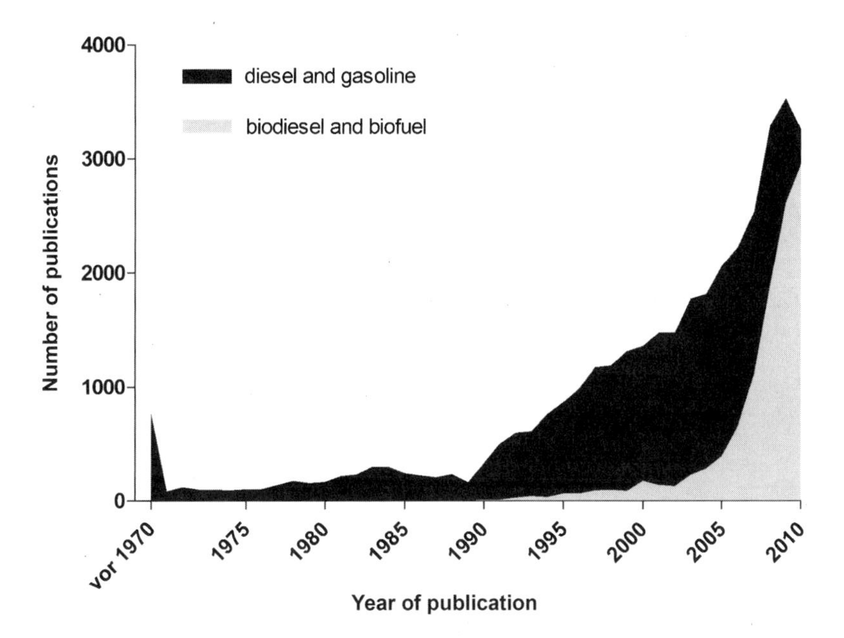

FIGURE 15.1 Development of the numbers of scientific publications on fossil fuel and biofuel between 1945 and November 2010 listed in Thompson ISI – Web of Science extended. Search terms for fossil fuels were "diesel∗" and "gasoline∗," for biofuels the terms "biofuel∗" and "biodiesel∗" were used. Publications matching both search terms for either fossil fuels or biofuels were considered only once.

available on their potential adverse consequences to the environment (Fig. 15.2). In fact, yearly publications discussing the (eco)toxicological potentials of biofuels (some of these studies only mentioned toxic potential without any further investigation) were less than half the number found for (eco)toxicological approaches with fossil fuels (Figs. 15.1 and 15.2). After excluding all studies not using or referring to bioassays for the evaluation of ecotoxicological potentials of biofuels, less than 22% remain (57 publications). Thus, there is an obvious lack of information for specific (eco)toxicological effects, for example, mutagenicity, teratogenicity, endocrine disruption, or dioxin-like activity compared to fossil fuels (Fig. 15.3).

15.2 BACKGROUND

15.2.1 Overview of Ecotoxicological Studies on Biofuels

Only few studies with an explicit focus on the effect of biofuel emissions are available (Fig. 15.3). Most of these studies investigated mutagenic and genotoxic hazards of exhaust emissions from biofuels and biofuel blends in comparison to fossil fuels [e.g., [11–20]]. Data from these studies were, however, conflicting. In some of these papers, it was shown that biofuel emissions can reduce the mutagenic

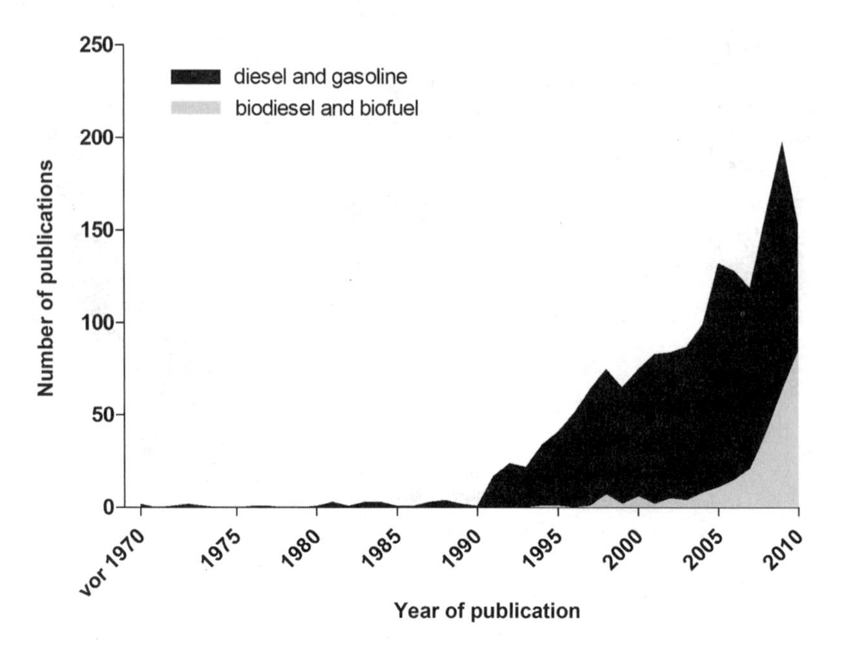

FIGURE 15.2 Development of the numbers of scientific publications on (eco)toxicological investigations of fossil fuels and biofuels between 1945 and November 2010. Combinations of "tox∗" or "ecotox∗" and "diesel∗"or "gasoline∗" as well as "tox∗" or "ecotox∗" and "biodiesel∗" or "biofuel∗" were choosen as search terms in a literature search using ISI Web of Science extended. Publications found repeatedly for search terms regarding (eco)tox and fossil fuels were considered only once. The same applies for the search terms with respect to (eco)tox and biofuels.

potential when compared to exhaust of fossil fuels [11–13,15]. In contrast, some other papers reported either no differences or an increase in mutagenic activity of biofuel or blend emission extracts [14,16,17,21]. However, the finding that not only emission particles from fossil fuels but also those from biofuels contain directly- and indirectly-acting mutagens seems crucial with respect to new developments in the field of biofuels. Further studies on biofuels and vegetable oils, which can be used as biofuels, were conducted to determine the effects on aquatic and soil organisms [22–27]. Results from these studies indicate that biofuels can pose a substantial risk to aquatic organisms [22,23]. Investigations on soil microorganisms, however, revealed a much higher toxic potential for petrodiesel [27] and no toxic potential for a biofuel (meeting the requirements of E DIN 51606) at concentrations up to 12% w/w [24]. Additionally, the study by Li et al. [26] about the impacts of vegetable-oil-contaminated sediments on soil microorganisms and the amphipod *Hyalella azteca* reported that the toxicity may be limited. However, they recommend additional studies to verify their results with respect to natural conditions. In case of *in vivo* investigations of mammals, clinical, histopathological, and biochemical effects were revealed for oral exposure to biodiesel and inhalation of biodiesel exhaust [28–30].

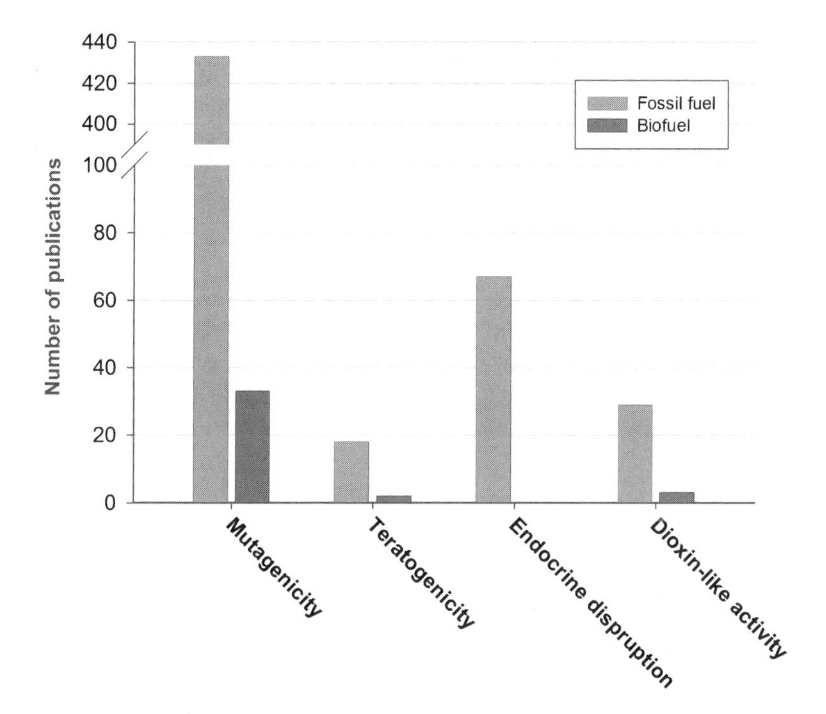

FIGURE 15.3 Number of scientific publications on investigations of fossil fuels and biofuels, respectively, using bioassays and published between 1945 and 2010 (November). The studies were selected from all publications found with the search criteria described for Figure 15.2. Publications subsumed under mutagenicity also include articles on genotoxicity, carcinogenicity, and DNA damage; publications subsumed under endocrine disruption also include articles on reproductive effects.

Based on their results, Brito et al. [28] concluded that particulate matter (PM) from biodiesel soybean ethyl esters causes equal or stronger toxic effects than diesel, due to promoted cardiovascular alterations as well as pulmonary and systemic inflammation. In contrast, the studies by Poon et al. [29,30] revealed that fossil diesel fuels caused more histopathological and biochemical effects than the biofuels tested. Overall, results of these studies indicate oxidative stress after treatment with both fossil fuels and biofuels. In addition to fuels themselves and biofuel emissions, waste or by-products should be investigated because they are potential sources of problematic substances that have to be disposed. An example of the formation of a waste product with adverse effects is given by the biodiesel production process using castor bean seeds. During this process, a toxic and allergenic castor bean waste is formed that must be detoxified before it can further be used or discarded without the risk of damaging the environment [31].

Based on the results presented, potential adverse effects to the environment or human health due to the release of biofuels via combustion, accidental spills or other inadvertent discharges during transportation, storage, or use cannot be ruled out. But to date comparatively few (eco)toxicological investigations of biofuels have been conducted and the data do not allow a definite conclusion on the potential effects on the risks for the environment and human health. Therefore, additional experimental studies covering investigations of substances and compositions generated at different stages of a biofuel-production process or during biofuel usage are required for a comprehensive risk assessment.

In the following sections, four common methods for the pretreatment of lignocellulosic biomass in a biorefinery are presented along with potentially toxic by-products. This is followed by the presentation of a proof-of-concept study implemented to evaluate the applicability of dilute acid pretreated biomass samples for (eco)toxicological investigations. This study provided one of the first quantified indications regarding the toxic effectiveness of by-products generated in biofuel refineries.

15.2.2 Biomass Pretreatment

Lignocellulosic biomass is a resilient matrix of cellulose fibers intertwined with hemicellulose and lignin (Fig. 15.4). Currently, several different platforms convert lignocellulosic biomass to precursors of biofuels and biochemicals [32,33]. The differences between these platforms are essentially the method of biomass conversion and the resulting precursor chemicals (e.g., sugars, carboxylates, and syngas) because the subsequent conversion of the precursor chemicals into finished bioproducts is not platform specific. Here, we describe three platforms: (1) *the sugar platform* consists of enzymatic conversion of biomass to sugar monomers that can be further converted into fuel by, for example, yeast or bacterial fermentation; (2) *the carboxylate platform* relies on the activity of hydrolytic enzymes produced in-situ by undefined mixed cultures to generate carboxylates that can be converted into fuels biologically or chemically; and (3) *the syngas platform* relies on thermochemical conversion of biomass (e.g., pyrolysis) to generate syngas that can be further fermented or synthesized into fuels.

In a biorefinery, several treatment steps are operating in parallel or in sequence with the goal to maximize product value from crude lignocellulosic feedstocks. In general, biomass storage, handling, and pretreatment steps are followed by a platform biomass-conversion step, a fuel production step, and fuel separation, purification, and storage steps [34]. Each of the platform-conversion steps has special requirements for biomass pretreatment. For example, chemical pretreatment allows enzymatic or biological hydrolysis (with either isolated enzymes or whole microbes in the sugar and carboxylate platforms, respectively) to penetrate the lignocellulose matrix and to free monomers on a relatively short timescale for further conversion [35]. More specifically, to increase hydrolysis rates, chemical pretreatment achieves one or more of the following, it: (i) increases accessible surface area to hydrolytic enzymes; (ii) frees fermentable carbon by disrupting the lignocelluose matrix; and (iii) decreases crystallinity of cellulose [35]. Many types

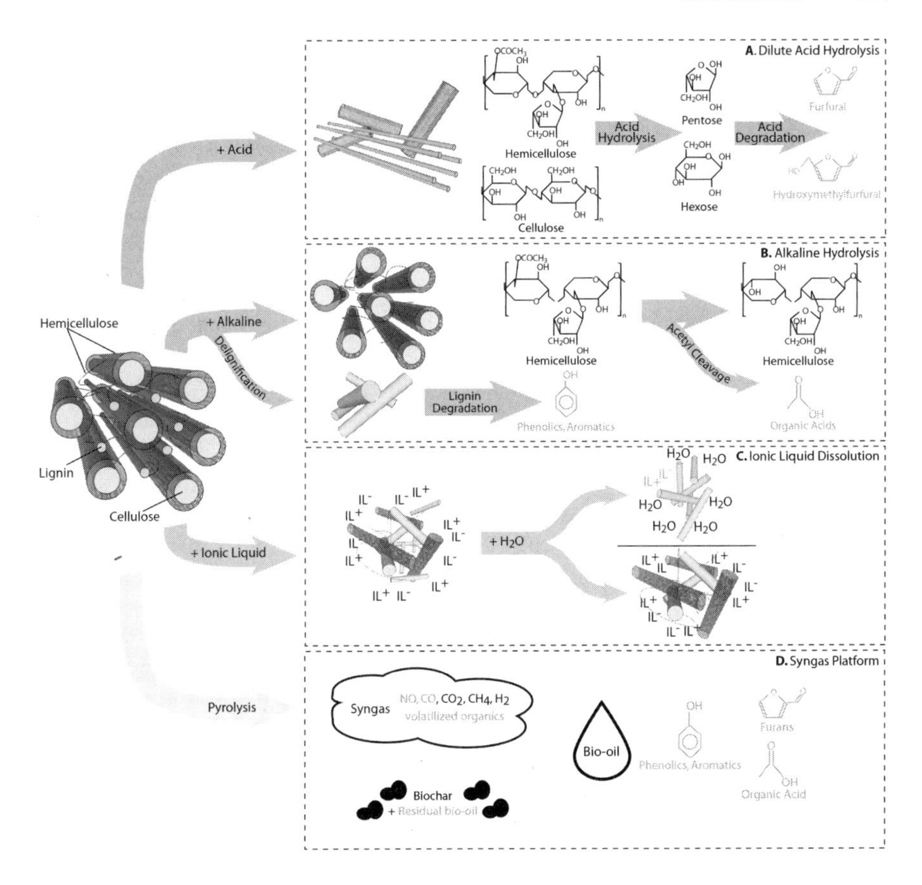

FIGURE 15.4 Production of toxic compounds (red) in three common chemical pretreatment steps and in the syngas platform step (e.g., pyrolysis). (**a**) Dilute acid pretreatment depolymerizes cellulose and hemicellulose and then hydrolyzes the backbone to sugar monomers that are further converted to furans. (**b**) Alkaline pretreatment delignifies biomass, forming phenolics and aromatics, and cleaves organic acid substitutions from hemicellulose. (**c**) Ionic liquids dissolve lignocellulose, and the noncrystalline cellulose is then dissolved into water, which may contain residual ionic liquids. (**d**) Pyrolysis converts lignocellulose to syngas (containing CO, NO, and volatilized organics), bio-oil (containing organic compounds), and biochar (with residual tar and bio-oil).

and variations of chemical pretreatments have been developed. Here, we discuss three commonly studied chemical pretreatment methods: dilute acid hydrolysis; alkaline hydrolysis; and ionic liquid dissolution. Each chemical pretreatment results in compounds that are toxic to fermentation organisms [36–38]. The compounds, including organic acids, furans, phenolics, and ionic liquids, are produced as by-products of hydrolysis or are residual pretreatment chemicals. In pyrolysis kilns, much higher operating temperatures may circumvent the need for chemical pre-treatment all together, but physical treatment to decrease feedstock particle size will still be necessary. The advantage of not generating toxic by-products from chemical pretreatment, however, may be overshadowed because pyrolysis itself generates toxic compounds, such as phenolic-rich bio-oils, that could inhibit downstream bioprocesses or leave the biorefinery and enter the environment. Thus, to minimize formation and negative effects of by-products and maximize the efficiency of biomass treatment, an understanding of the mechanisms of toxic by-product formation in biorefineries is required.

15.2.2.1 Dilute Acid Pretreatment Dilute acid pretreatment is an effective method designed to condition lignocellulosic biomass for hydrolysis, and relatively simple and inexpensive reaction conditions add to its appeal. The most commonly used acid that is added to the biomass is sulfuric acid in concentrations ranging from 0.5 to 1.5% v/v in reactors operated in the range of 120–180 °C [39]. Dilute acid pretreatment employs several mechanisms that improve yields for downstream processes. Initially, the biomass particle size is reduced, increasing the surface area available for enzymes. Next, cleavage of organic acid substitutions from hemicellulose allows better access for hemicellulases (e.g., xylanase), and acid hydrolysis of the cellulose or hemicellulose backbone releases hexose and pentose monomers (Fig. 15.4a). Hot-water pretreatment of biomass is also commonly used and its mechanism is essentially the same as dilute acid pretreatment, albeit at lower rates, because acid hydrolysis only occurs due to accumulated protons from cleavage of hemicellulosic organic acid substitutions [35]. The primary toxic compounds of concern due to dilute acid pretreatment are furfural and hydroxymethyl furfural, which inhibit both enzymatic hydrolysis and fermentation, and aromatic compounds (from lignin). These furans result from degradation of hexoses and pentoses under acidic conditions (Fig. 15.4a). In addition, organic acids, primarily acetic acid, derived from hemicellulose substitutions, also accumulate, and their unionized form are especially toxic to microbes at the low pH of the dilute acid hydrolysate. For dilute acid pretreatment, formation of toxic compounds is correlated to high temperatures, high acid loading, and long reaction times, while moderate reaction conditions limit their formation [39].

15.2.2.2 Alkaline Pretreatment Alkaline pretreatment is useful in many biorefinery situations because when high-temperature pretreatment systems are impractical, contacting biomass with alkalies for long periods can attain satisfactory treatment. In fact, efficient treatment time is dependent on temperature: reactions of the order of weeks are necessary for room temperature, while minutes to hours suffice at elevated temperatures (~120 °C). In addition, chemicals for alkaline pretreatment, such as $Ca(OH)_2$ or CaO, have a relatively low cost and are widely available [40]. The alkaline pretreatment employs mechanisms to remove lignin that is a diverse polymer of phenolic and aromatic compounds and the most resilient of the lignocellulose polymers to enzymatic or biological hydrolysis (Fig. 15.4b). A common variation on alkaline pretreatment improves lignin-removal mechanisms by adding oxygen to increase lignin degradation [41]. Cellulose and hemicellulose are well protected by lignin, and its removal greatly increases enzyme access to the sugar polymers. Release of the phenolic and aromatic monomers that make up lignin must be considered, however, since these compounds are highly toxic. For example, a wide variety of lignin-derived phenolic compounds, such as vanillic and syringic acids, were produced and accumulated during alkaline treatment of wheat straw with oxygen [42]. In addition, similar to dilute acid pretreatment, alkaline pretreatment improves hemicellulase action by removing organic-acid substitutions from hemicellulose (Fig. 15.4b). The resulting organic acids from alkaline pretreatment, however, are expected to be less toxic than those produced during dilute acid pretreatment because the high hydrolysate pH ensures that the acids are in the ionized form. Of course, this advantage would disappear if the fermentation broth becomes acidified.

15.2.2.3 Ionic Liquid Pretreatment Ionic liquids (ILs) are a class of salts that are liquid at room temperature and have very low vapor pressures. These chemicals have unique properties dependent on their structure, and have attracted researchers seeking to exploit their novelty for a wide range of applications [43]. For example, the ability to dissolve lignocellulosic biomass has prompted a flurry of activity developing ILs as a novel biomass pretreatment option [44]. Wood or other lignocellulosic biomass dissolves due to interactions between the IL and lignin, hemicellulose, and cellulose. The result is a nearly complete loss of the protective structure of the lignocellulosic biomass (Fig. 15.4c). Cellulose can subsequently be recovered by the addition of a solvent, such as water, which interrupts bonds between cellulose and the IL, causing it to precipitate in its noncrystalline form (Fig. 15.4c). Lignin, on the other hand, stays dissolved in solution, potentially being recovered for other uses [45,46] (Fig. 15.4c). IL pretreatment represents a physical change of the lignocellulose matrix rather than a chemical one, and to our knowledge, available literature does not report the formation of toxic compounds by IL pretreatment. Studies, however, report that certain ILs may be toxic to microbes [47,48]. Because some ILs partly dissolve in water [43], residual ILs in recovered cellulose may cause detrimental effects to biofuel production processes (Fig. 15.4c).

15.2.2.4 Syngas Platform (Pyrolysis) The syngas platform step destroys the lignocellulose matrix chemically by heating in the presence of low or no oxygen (i.e., pyrolysis). It is considered advantageous, especially for resilient woody biomass, because high temperatures and partial combustion completely degrade the lignocellulose structure. Depending on reaction conditions, pyrolysis produces variable fractions of synthesis gas (syngas: H_2, CO_2, CO, CH_4), liquid bio-oil, and solid biochar [49] (Fig. 15.4d). Downstream fermentation for production of biofuels and biochemicals has been performed on the gaseous fraction (syngas) and the liquid fraction (bio-oil) [50,51]. The main toxic component in syngas is CO formed by incomplete combustion. Therefore, fermentation of syngas to, for example, ethanol, requires selection of microbial strains, such as the carboxydotrophic strains *Clostridium carboxidivorans* P7 and *Clostridium ljungdahlii* ATCC 49587, which are known to utilize and remove CO at low concentrations [50] (Fig. 15.4d). Syngas may also contain toxic levels of NO or organics, such as tars, ash, ethylene, ethane, and acetylene, which have all been shown to negatively affect syngas fermentation by *C. carboxidivorans* P7 [52,53] (Fig. 15.4d). Bio-oil contains high levels of toxic organic compounds formed during pyrolysis, including acids, aldehydes, and furans derived from carbohydrates, and phenolic and aromatic compounds derived from lignin [54] (Fig. 15.4d). Bio-oil has been treated to remove toxic compounds and utilized as a fermentation substrate in the production of ethanol by yeast [51]. Solid fractions (biochar) are useful as soil additives, and thus are commonly released into the environment [55,56]. It is, therefore, pertinent to understand how toxic components should be limited because biochar may contain some tar or residual liquid fractions with environmentally damaging toxicants similar to those described for bio-oil. Due to the ubiquitous formation of toxic compounds in pyrolysis and the use of its products in fermentation and environmental applications, negative effects of the toxicants are an important consideration.

The formation of toxic compounds due to pretreatment methods is of high relevance, especially regarding the inhibition of downstream bioprocesses. Besides adverse effects on microbial cultures used for fermentation processes, potential ecotoxicological effects are also relevant considering the possibility of a release of products from pretreatment processes into the environment. Therefore, a proof-of-concept study was performed to examine the applicability of dilute acid pretreated biomass samples for (eco)toxicological investigations. The results obtained from these samples provide indications regarding potential effects on organisms and are discussed in the following section.

15.2.3 Ecotoxicological Investigations of Dilute Acid Pretreated Samples

The ecotoxicological hazard potential of a sample can be investigated using different approaches, for example, chemical analyses, application of bioassays, or analyses of the structure of ecological communities, but also combinations thereof. Chemical analyses, such as gas chromatography (GC) or high performance liquid chromatography (HPLC), are used to determine chemical contaminants. Bioassays can be used to determine the toxicity of a sample or to provide information about the responses of ecological communities to chemical stressors. Integrated approaches combine several lines of evidence, such as the Sediment Quality Triad (SQT), which proposes the combination of biological and chemical analyses as part of a comprehensive investigation of sediment toxicity [57,58]. With respect to biofuel-related samples, such as pretreated substrates for biorefinery processes, only a few studies examined the (eco)toxicological potential, and thus little is known about potential toxicants. Therefore, a battery of biotests instead of a chemical analysis was applied to reveal a first indication of the hazard potentials originating from pretreated biomass samples. Chemical analyses would allow the identification and quantification of many substances of interest in a given sample. However, although the detection and identification of every single compound in a complex sample would theoretically be conceivable, it is not practicable from an economical point of view. Consequently, chemical analyses require an idea of the substances that might be of interest, which in case of samples from biorefinery processes is difficult to conceive because information regarding this purpose is scarcely available. Another reason for the initial decision against a chemical analysis was that the detection of hazardous substances does not allow precise prediction regarding adverse effects of complex samples on organisms or the environment, even if the toxic potential of any single compound would be known. In this context, several factors need to be considered. Effects, for example, are organism specific and depend on the bioavailability of a given substance. Additionally, interactions between different chemicals, such as additive, synergistic, or antagonistic effects, could significantly increase or decrease the effects of a sample [59,60], but these combined effects are not detectable by chemical analysis.

While biological test systems are not capable of identifying single contaminants, they can be used to determine the actual adverse effects of a pure chemical or a complex sample on organisms or cell-based *in vitro* systems [61]. Complex samples applicable for investigations in bioassays are environmental samples, for example, sediment samples as required for the SQT, but also a multitude of other kind of samples, such as organic waste sludge [62] or pharmaceuticals and personal care products [63]. In this respect, bioassays enable researchers to reveal the combined effects of all compounds in the given sample, even if the active compounds or their metabolites are unknown or unidentified. Metabolites formed during biotransformation, for example, often show increased bioactivity. Thus, their detection is as important as the detection of the parent substances [64]. A further advantage of biotests compared to chemical analyses is their ability to assess the bioavailability of toxic compounds [65]. However, the knowledge obtained by a single biotest would be very limited. Therefore, to gain a broad range of knowledge, and a more comprehensive understanding of the ecotoxicological hazardous potential, a test battery, using different test systems, can be applied [66,67]. Such a test battery consists of organisms from different trophic levels, as well as degrees of complexity and uses a combination of acute and mechanism-specific biotests [68,69], thus, representing the intricacy of ecosystems. The application of such biotest batteries is required, for instance, as part of the authorization of chemicals by the Registration, Evaluation, Authorisation and Restriction of Chemicals (REACH) framework of the European Union [70].

In the following, we present a proof-of-concept study to evaluate the applicability of several bioassays for investigations of biofuel-related samples, specifically from biofuel-production processes. In this particular study, a battery of acute and mechanism-specific *in vitro* biotests was used to determine the ecotoxicological relevance of samples obtained from a two-step *n*-butyrate and *n*-butanol fermentation process (a novel carboxylate platform technology [71]) and a common pretreatment step that is used for the sugar platform. Biological analysis focused on different pretreated and untreated corn-fiber substrate samples (corn fiber is the fractionated outer layers from a corn kernel and a waste product from the corn-to-ethanol

TABLE 15.1 Pretreatment Conditions of the Substrate Samples

Material (Reactor)	Solution	Biomass Concentration (w/w)	Temperature (°C)	Treatment Time (min)
Corn fiber (R1-S)	0.5% v/v H_2SO_4	15%	160	20
Corn fiber 4 (R4-S)	–	15%	–	–
Corn stover (CS)	0.7% v/v H_2SO_4	10%	180	10

industry), on a complementary sample of effluent after biochemical conversion of pretreated corn fiber, and a pretreated corn stover (corn stover is left-over plant material after removal of corn kernels during harvesting). We investigated whether toxic compounds are formed as a result of the pretreatment and, furthermore, whether this is affected by the intensity of the pretreatment method. Toxic compounds could inhibit microbial cultures that are grown in down-stream bioprocesses, and thus interfere with further fermentation processes (second step). Therefore, knowledge on the change of toxicity due to the pretreatment method is of enormous interest from an economical point of view, as well as with respect to the disposal of wastes. Details of the sample pretreatment are shown in Table 15.1.

The influence of the intensity of the pretreatment method will be shown through a comparison of the toxicity of the corn-fiber substrate R1-S pretreated with dilute acid and the untreated corn-fiber sample R4-S. Furthermore, the impact of the undefined-mixed-culture fermentation process (using a very diverse community of microbes) on the removal of toxic compounds was investigated by comparing the dilute acid substrate sample R1-S to the complementary effluent sample R1-E, taken from a semicontinuous thermophilic anaerobic reactor operating at a 25-day hydraulic retention time. For classification of each samples' harmful potential, one sample made from corn stover (CS) with a relatively higher lignin content was used as a reference. The pretreatment for this reference sample was even more stringent as for the previous mentioned corn-fiber sample R1-S, and both aspects would be important for a likely higher toxic impact of the pretreated corn stover compared to the pretreated corn fiber.

In detail, cytotoxicity was assessed using the Neutral red retention assay (NR assay) [72] and the 3-[4,5-dimethyltiazol-2-yl]-2,5-diphenyl tetrazolium bromide (MTT) assay [73] using rainbow trout (*Oncorhynchus mykiss*) liver cells (RTL-W1) [74]. This cell culture was also used to detect Ah-receptor agonist and EROD activity by means of the 7-ethoxyresorufin-*O*-deethylase (EROD) assay [75]. In addition, mutagenicity, teratoganicity, and endocrine activity were examined. Mutagenicity was investigated by using the Ames fluctuation assay with *Salmonella typhimurium* [76,77], teratogenicity by means of the embryotoxicity test using embryos of *Danio rerio* [78,79], and estrogenic activity based on the Yeast Estrogenic Screen (YES) assay [80,81].

Cytotoxicity, EROD activity, mutagenicity, embryotoxicity, and endocrine activity represent important endpoints in many ecotoxicological studies [82,83]. Therefore, they were used in this study to reveal the ecotoxicological relevance of a biofuel pretreatment method. Application of these biotests allows determination of both acute toxicity (cytotoxicity, mortality) and mechanism-specific effects (EROD activity, mutagenicity). The mechanism of each test will be discussed briefly in the subsequent paragraphs.

15.2.3.1 Cytotoxicity Both cytotoxicity assays investigate the acute cytotoxicity of cells exposed to a given sample. The NR assay quantifies the number of viable cells related to the accumulation of the dye neutral red in their lysosomes [84,85], whereas the MTT assay quantifies the number of living cells through an analysis of their mitochondrial metabolic function [73]. More precisely, the NR assay represents a very sensitive indicator for membrane integrity [72], whereas the MTT assay quantifies inhibition of mitochondrial succinate dehydrogenase, and thus mitochondrial metabolic function [86]. A significant correlation of the NR assay as a sensitive indicator for lysosomal membrane integrity and the MTT activity had been reported by Hollert et al. [82], with the NR assay being slightly more sensitive [82,86].

For each sample, the concentration resulting in 50% reduction of cell viability in the NR assay (NR_{50}) and the EC_{50} (corresponding concentration in the MTT assay) were calculated. These values can be used as a criterion for the cytotoxic potential of each sample. For its calculation, the lowest NR_{50}/EC_{50}-value – representing the highest effectiveness – was set equal to 1. In both biotests, this sample was the corn stover sample CS. The NR_{50}/EC_{50}-values determined for each replicate of each sample were divided by this lowest value to determine the ratio to the sample with the highest cytotoxic potential. Furthermore, the mean value for each sample was calculated, revealing the cytotoxic potential (in a range from 0 to 1) where higher values indicate a stronger cytotoxic potential. This approach allows a comparison between the results of the two different bioassays.

Additionally, statistical analyses (Student's t-test following square-root transformation) were performed to detect significant differences between the samples tested in the same bioassay and between the results of the same sample obtained with the two different bioassays.

15.2.3.2 Dioxin-Like Activity In contrast to acute cytotoxicity assays, the EROD assay represents a mechanism-specific, sublethal test for identification of cytochrome-P450-activity. The cytochrome-P450 monooxygenase system plays a major role in the phase-I metabolism of xenobiotica of many lipophilic organic substances in various organisms [86,87]. The EROD assay investigates the induction of the CYP1A-biotransformation system by dioxin-like substances [88,89]. One of the enzyme activities of the CYP1A–enzyme complex is that for 7-ethoxyresorufin-*O*-deethylase (EROD) [90]. The induction pathway is mediated by the aryl hydrocarbon receptor (AhR). Activity of the EROD enzyme is determined by degradation of the artificial substrate 7-ethoxyresorufin to resorufin [83]. Dioxin-like compounds can be characterized as hydrophobic and aromatic compounds with a planar structure. This structure, even if it is only part of the whole molecule, can fit the binding sites of the intracellular AhR, which acts in the cell nucleus as a transcription

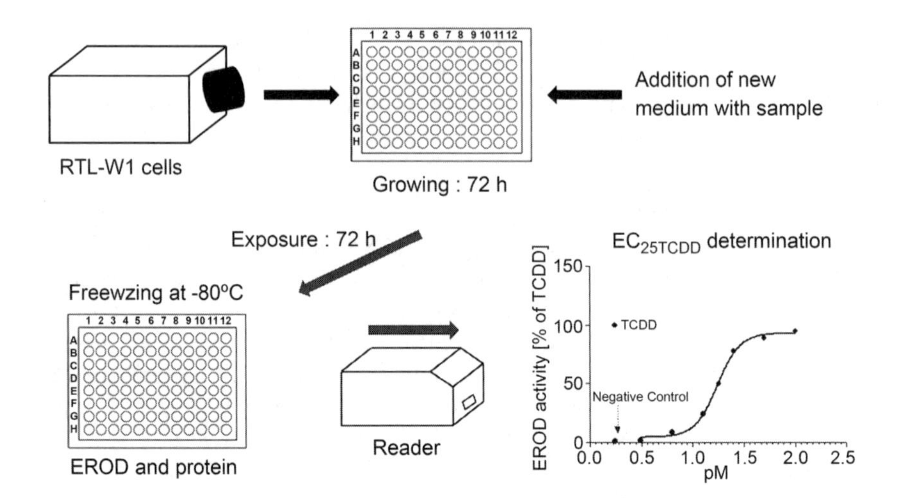

FIGURE 15.5 Scheme of the performance of the EROD assay. EROD induction in the permanent cell line RTL-W1 (rainbow trout liver) as a biomarker for dioxin-like activity.

factor for CYP1A, and thus EROD enzyme induction [88,89,91]. A scheme of the test method is given in Figure 15.5.

15.2.3.3 Mutagenicity The Ames fluctuation assay, which is a modification of the traditional Ames plate incorporation assay developed by Ames et al. [92], was performed according to Reifferscheid et al. [76] and the draft for the International Standard ISO/CD 11350 (cf. Fig. 15.6). This assay detects the mutagenic potential of chemicals and environmental samples. The used tester strain TA100 of *S. typhimurium* carries a variation in the histidine-coding region depriving the ability to synthesize histidine (histidine auxotroph). Therefore, the bacteria are unable to grow in a histidine-free medium. The test principle is based on the hypothesis that exposure to mutagenic agents leads to reverse mutations in the histidine-coding region. The modern microplate fluctuation protocol produces similar results as the plate incorporation assay and is a promising tool to test environmental samples for their mutagenic activity. The test-organism *S. typhimurium* is a Gram-negative, rod-shaped bacterium with an average generation time of 20–30 min. For the Ames assay, several bacterial strains with different mutations and modifications are available. The genotype of the tester strain used for the Ames fluctuation assay is summarized in Table 15.2.

Some chemical substances are biologically inactive unless they are metabolized to active forms. Bacteria lack this metabolic capability. Therefore, to mimic the metabolic conversion of a substance as it would occur in vertebrates a mammalian metabolic activation system (rat liver homogenate, S9-fraction) can be added [95]. The biotransformation enzymes contained in the S9 supplementation are able to (bio) activate a great number of substances, and thus to elucidate the mutagenic potential

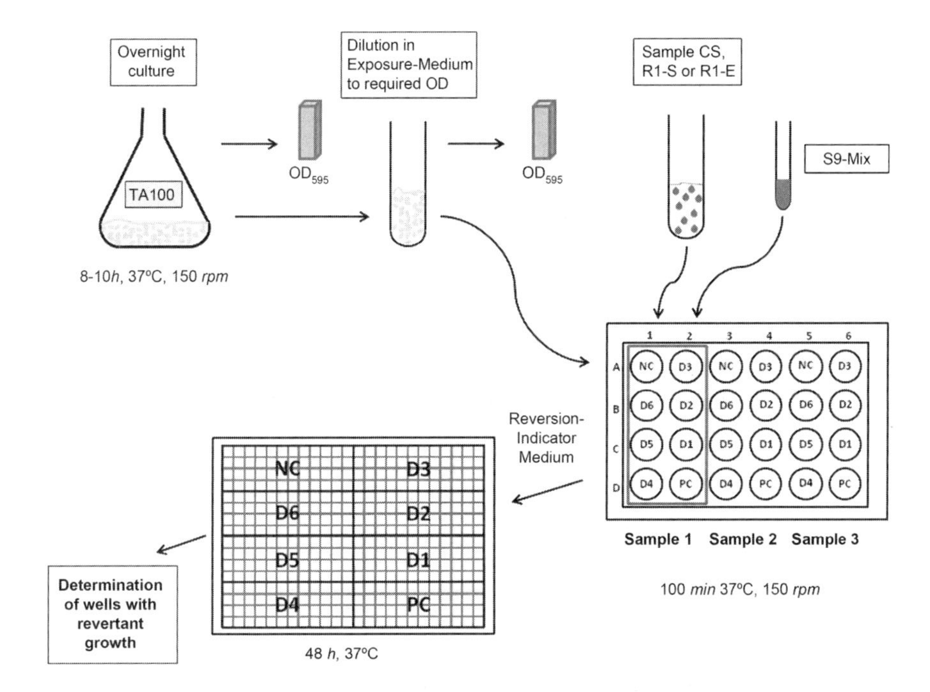

FIGURE 15.6 Scheme of the performance of the Ames fluctuation assay. The cell density of a bacterial overnight culture had to be adjusted with exposure medium to ensure comparable test results. The bacteria were exposed to six sample dilutions (D1–D6), a negative control (NC), and a positive control (PC) with or without the addition of S9-mix (S9, NADP, glucose-6-phosphate, calcium chloride, phosphate buffer, and magnesium chloride) in 24-well plates. After an incubation time of 100 min, reversion-indicator medium (pH indicator medium) was added and the solution transferred to a 384-well plate. These plates were analyzed after 48 h of incubation.

of substances showing indirect mutagenicity. Statistical evaluation was conducted using the software Toxrat (Toxrat Solutions GmbH, Alsdorf, Germany). For testing of normal distribution and variance homogeneity, the Shapiro Wilk's test and the Levene's test, respectively, were performed. All data sets passed these checks and

TABLE 15.2 Genotype of the *S. typhimurium* Strain TA100 [According to refs. [92–94]]

Mutation/Deletion	Genetic Modification or Mutation Site
*uvr*B	A deletion through the *uvr*B region eliminates the DNA excision repair system
rfa	The mutation leads to a partial loss of lipopolysaccharide coating the bacterial surface, making the bacteria more permeable and the strains nonpathogenic
amp^+	The ampicillin resistance is applied to check if the plasmid pKM-101 is present, the plasmid primarily ensures a higher sensitivity of the test
hisG46	Base pair mutation in the histidine operon of TA100

were tested for significant differences to the negative control by using the Williams' test, a multiple comparison test. Additionally, an induction factor was calculated for each concentration by dividing the mean of positive wells obtained for a particular sample concentration by the mean of positive wells of the negative control.

15.2.3.4 Embryotoxicity The fish embryo toxicity test with the zebrafish *D. rerio* represents an acute lethality and teratogenicity *in vitro* bioassay. *D. rerio* (Hamilton-Buchanan 1922) is a cyprinid with a well-characterized embryonic development. The transparency of eggs and embryos allows observation under the microscope [78,96], and the excellent knowledge of the developmental stages allows a continuous examination of embryos exposed to a given sample. After 48 h of exposure, teratogenic effects are recorded and mortality is determined by means of effects considered lethal for developing embryos based on standardized mortality criteria [78,97–99]. These include the coagulation of embryos, a lack of heart function, blood circulation, somite formation, or a nondetachment of the tail. A test was considered valid if the mortality of the negative control did not exceed 10% and the positive control induced effects in more than 20% of the embryos. Samples were considered embryotoxic, if the mortality induced by the investigated sample is higher 10%. With respect to the data obtained (e.g., identical results in three replicates for CS) and the low number of replicates (R1-S, R1-E and R4-S; $n = 2$), possibilities for the application of adequate statistical analyses were not given. Therefore, normal distribution and equality of variances were assumed and significant differences between the samples calculated by means of the t-test.

15.2.3.5 Endocrine Activity An increasing number of xenoestrogens, such as many pesticides and plasticizers, have been found to bind to the human estrogen receptor α (ERα) [100]. Furthermore, xenoestrogenic substances found in aquatic ecosystems interfere with the natural animal hormone systems. These potentials can lead to adverse effects from animal abnormalities in wildlife up to human reproductive disruption [100,101]. Reported effects are, for instance, reduction of penis size of the American alligator [102], the decreasing male sperm quality and density [103], the increasing risk of testicular and breast cancer [104–106], or even strong events on a population level of fish, as recently shown by Kidd et al. [107], using whole-lake experiments with the synthetic estrogen ethynylestradiol.

A useful tool for the determination of estrogenic activity in a given sample is the YES assay, which is based on a genetically engineered yeast strain of *Saccharomyces cerevisiae* [80]. These yeast cells carry the human ERα gene, as well as an expression plasmid containing estrogen responsive elements operating the *lacZ* reporter gene, which encodes the enzyme β-galactosidase (Fig. 15.7). Expression of this enzyme is triggered by estrogen-like substances binding to the ERα. As a consequence, the amount of synthesized β-galactosidase is directly related to the concentration of estrogen-like substances. Addition of the artificial substrate chlorophenol-red-β-galactopyranoside (CPRG; yellow) allows a quantification of the enzyme induction. The enzyme converts CPRG to chlorophenol red (red) that can be determined photometrically.

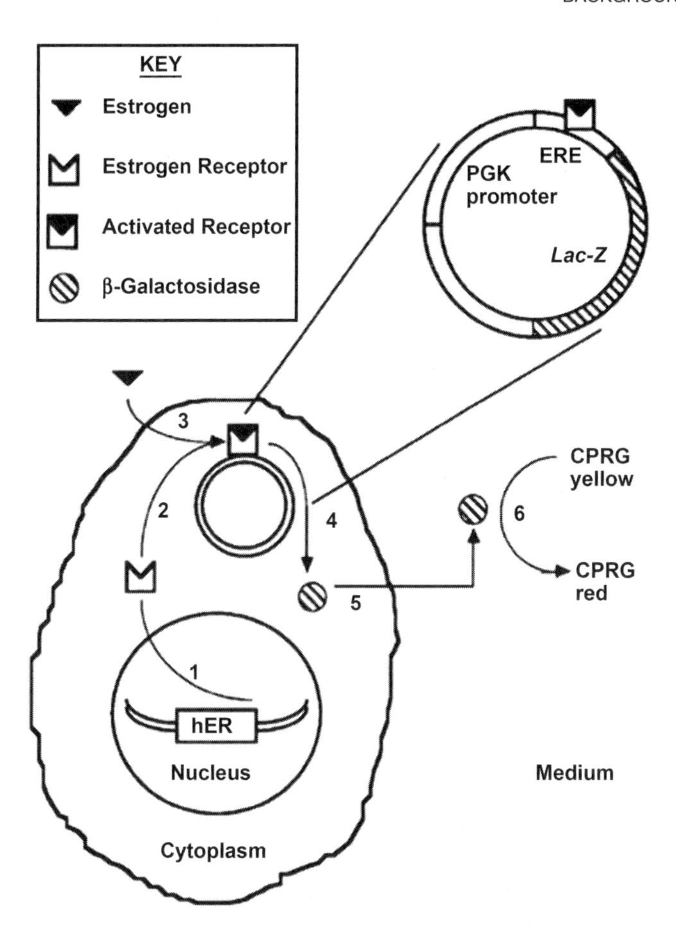

FIGURE 15.7 Schematic representation of the estrogen-inducible expression system in yeast. The human estrogen receptor gene is integrated into the main genome and is expressed (1) in a form capable of binding to estrogen response elements (ERE) within a hybrid promoter on the expression plasmid (2). Activation of the receptor (3), by binding of a ligand, causes expression of the reporter gene *Lac-Z* (4) which produces the enzyme b-galactosidase. This enzyme is secreted into the medium (5) and metabolizes the chromogenic substrate CPRG (normally yellow) into a red product (6), which can be measured by absorbance (picture published by Ref. [80]). Reproduced by permission of Wiley.

To enhance sensitivity and reduce required time, the LYES assay was used in this study. This test represents a modified version of the original YES assay by involving the enzyme lyticase for a digestion step [108]. This test approach had already been applied successfully in other studies such as by Wagner and Oehlmann [81].

For the evaluation of endocrine activities, the estradiol equivalent quotient (EEQ) was calculated by means of a 17β-estradiol (E2) standard curve according to Wagner and Oehlmann [81]. E2 is known as a very strong natural estrogen [100]. Statistical analyses were not performed because only two replicates were available.

TABLE 15.3 Highest Concentrations Used in each Bioassay in Percent [%] of the Undiluted Sample

	CS (%)	R1-S (%)	R1-E (%)	R4-S (%)
NR-/MTT-assay	50	50	50	50
EROD assay	5	3.5	29.6	25
Ames fluctuation assay	80	80	80	–
Fish embryotoxicity assay	1	1	1	1
YES assay	20.8	62.5	0.4	62.5

Different concentrations of the samples were applied in each biotest due to the experimental design or cytotoxic effects for the test organisms. Table 15.3 gives an overview of the highest concentrations used.

15.3 RESULTS AND DISCUSSION

15.3.1 Cytotoxicity

Cytotoxicity assays are often used to screen the toxic potential of a sample. In this study, the NR assay and the MTT assay were performed with the samples CS, R1-S, R4-S, and R1-E. The results are presented in Figure 15.8.

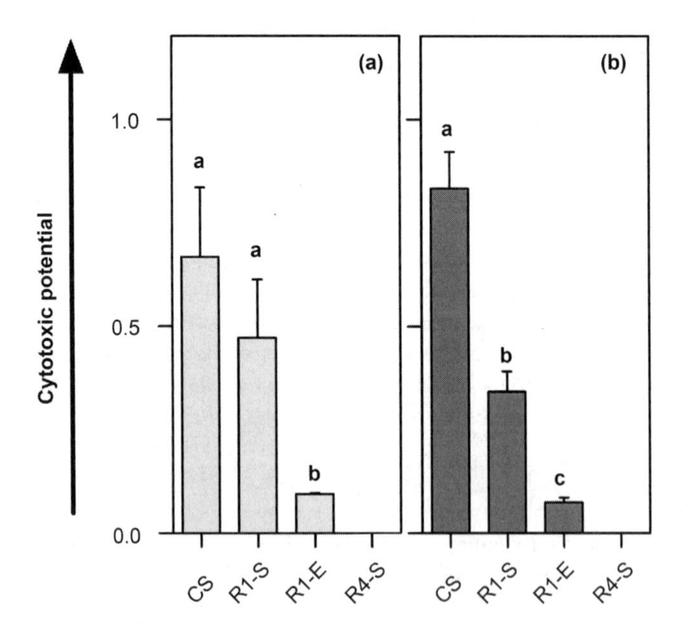

FIGURE 15.8 Cytotoxic potentials of each sample determined using NR assay (**a**) and MTT assay (**b**), respectively. Data are given as means (bars) and standard errors of the means (error bars) of the cytotoxic potential calculated by setting the lowest NR_{50}/EC_{50}-value obtained by the NR/MTT assay equal to 1 and dividing each value by this NR_{50}/EC_{50}-value. No NR_{50}/EC_{50}-values could be determined for sample R4-S; $n = 3$.

The sample CS showed the highest cytotoxic potential in the MTT assay (Fig. 15.8b), followed by the sample R1-S ($p = 0.007$). According to the results of the NR assay, the mean values obtained for these samples indicate a similar result but no significant difference was calculated due to data that spread out over a larger range of values (Fig. 15.8a). With regard to the other samples tested, R1-E showed a comparably low cytotoxic potential in both assays, whereas sample R4-S revealed no cytotoxic effects at all. Based on the statistical calculations, significant differences could be found between CS and R1-E, as well as between R1-S and R1-E (for both assays; Fig. 15.8). Furthermore, a general comparison of both assays revealed no significant differences between the cytotoxicity data generated for each sample. Therefore, the results indicate a confirmation of the findings by Hollert et al. [82] that revealed a strong correlation of both cytotoxicity assays.

The cytotoxic potential is influenced by the intensity of the pretreatment method and the composition of the lignocellulosic material. The strong cytotoxic potentials of CS and R1-S, and also the differences between both samples, can, therefore, only be partly explained by the pretreatment methods: the dilute acid pretreatment for sample CS (0.7% H_2SO_4 at 180 °C) was more stringent than that for R1-S (0.5% H_2SO_4 at 160 °C). Dilute acid pretreatment of lignocellulosic material, such as corn fiber or corn stover, is known to release a high number of by-products, such as furans and phenols, partly based on the composition of the lignocellulosic materials. When the lignin concentrations are higher, for example, in corn stover, a higher concentration of phenolic compounds is anticipated, resulting in possible differences in cytotoxicity. Besides other by-products, furan derivates and phenolic compounds were both reported to be toxic to microorganisms used for fermentation [36,109], as well as to fish cells [91]. Since the untreated substrate sample R4-S showed no cytotoxic potential at all, this is a strong indication for increased toxic potential due to dilute acid pretreatments.

After the undefined-mixed-culture fermentation process (the first-step *n*-butyrate fermentation process) the cytotoxic potential of the complementary effluent sample R1-E was reduced. This was presumably the result of the microbial degradation during fermentation, which would have reduced the concentrations of by-products because of the activity of many microbial species with different metabolic traits. Although both cytotoxicity assays revealed a significant reduction of the cytotoxic potential, apparent cytotoxic effects are obtained with both tests (Fig. 15.8a and b). We have an explanation for this result because at the same time that toxic by-products were degraded by the metabolic-rich community of microbes, the carboxylate products, such as acetate and *n*-butyrate, were accumulated. Indeed, parallel tests on the cytotoxic potential of acetate and *n*-butyrate showed an increasing cytotoxicity level with increasing carboxylate concentrations (data not shown). In particular, *n*-butyrate was found to be toxic for RTL-W1 cells at the pH levels of the biological sample.

15.3.2 Dioxin-Like Activity

Following the determination of the cytotoxic potential, dioxin-like activity was investigated. Results of the NR assay were used to determine the highest applicable concentration for the EROD assay to avoid masking cytotoxic effects. EROD induction is a widely used indicator for exposure to dioxin-like compounds. Thus, investigation of the EROD-inducing potential represents an important biological/bioanalytical endpoint in many ecotoxicological studies, such as investigations of contaminated industrial sites [88,110]. In this study, none of the samples caused any dioxin-like activity. Formation of dioxin-like compounds depends strongly on the kind of pretreatment. A study by Tame et al. [111] revealed that lignocellulosic material forms polychlorinated dibenzo-*p*-dioxins and dibenzo-furans (PCDD/F) during combustion. Thus, various pretreatment methods that are relying on application of heat, such as pyrolysis, are more likely to generate this PCDDs or PCDFs. The amount of lignin in the substrate is also important. Although carbohydrates tend to form PCDD or PCDF even when the sample contains no lignin at all, the PCDD and PCDF formation increases significantly in the presence of lignin [111]. As a consequence, an investigation of dioxin-like compounds should be considered for lignocellulosic biofuel samples pretreated with pyrolysis.

15.3.3 Mutagenicity

Mutagenicity was investigated for the samples CS, R1-S, and R1-E using the *S. typhimurium* strain TA100 with and without S9 supplementation. For these samples, the corresponding number of positive wells (wells with revertant growth) is shown in Figure 15.9. The number of positive wells allows an assessment of the mutagenic potential for each sample.

Each of the investigated samples revealed a mutagenic potential, with the samples R1-S and R1-E showing higher or equal numbers of positive wells than CS, both with and without the addition of a metabolic activation system. These samples R1-S and R1-E are considered mutagenic when tested without S9 and applied in concentrations higher than 12.5%. If tested with S9 metabolic activation, concentrations of R1-S higher than 12.5% induced significant mutagenic effects, whereas the sample R1-E had to be applied in concentrations higher than 25%. These results indicate that higher concentrations of R1-E compared to R1-S have to be applied to organisms with a metabolic activation system to cause a mutagenic effect. However, for the sample R1-E a higher induction factor was calculated (23.1) in comparison to the sample R1-S (11.5) with respect to the highest concentration tested. Therefore, if applied in higher concentrations the mutation induction seems to be increased after the fermentation process. The sample CS, on the other hand, showed a lower number of positive wells and a lower induction factor (5.5) at the highest concentration tested. Moreover, a mutagenic effectiveness was not detectable below a concentration of 50% (without metabolic activation) or 25% (when tested with the addition of S9). Therefore, the stringency of pretreatment does not seem to increase the mutagenic potential.

However, there are several issues related to the interpretation of these results especially with respect to the samples investigated. Regarding the test evaluation, a reverse mutation is indicated by a color change from purple to yellow due to an acidification of the medium. The acidification is indicated by a pH-indicator dye.

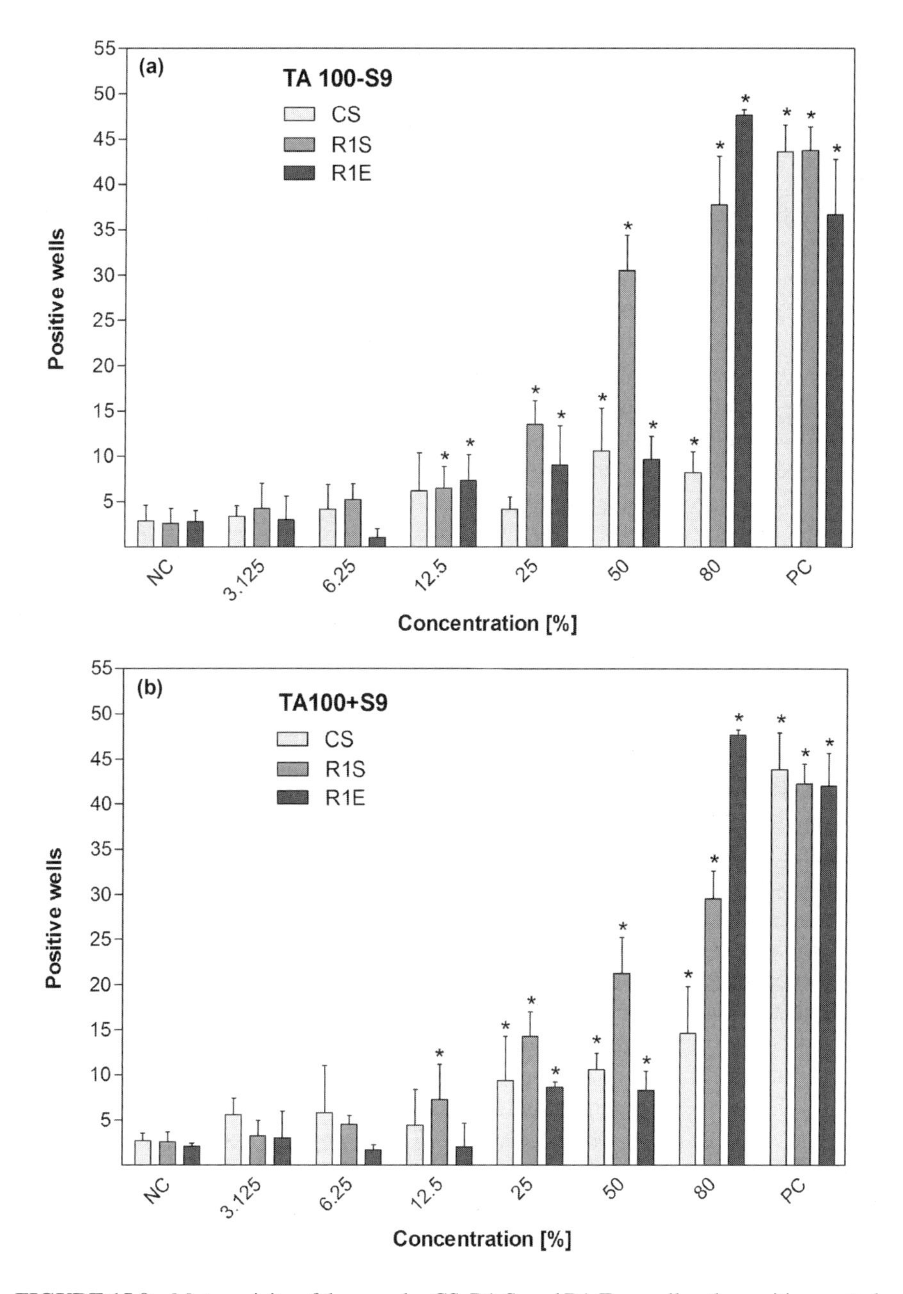

FIGURE 15.9 Mutagenicity of the samples CS, R1-S, and R1-E as well as the positive control (PC) as determined by the Ames fluctuation assay using the *Salmonella typhimurium* tester strain TA100. Data are given as means (bars) and standard deviation (error bars) of the number of positive wells with (**b**) and without (**a**) S9 supplementation. PC for tests with S9: 2-aminoanthracene, PC for tests without S9: nitrofurantoin (TA100). $n_{CS} = 5$; $n_{R1\text{-}S} = 4$; $n_{R1\text{-}E} = 3$.

Samples with a low pH might, therefore, have an influence on the result. However, the pH value of all samples tested was adjusted before testing. Hence, an effect due to a low pH value can be excluded. Another issue is related to a substance (histidine) that can lead to a positive result for the tested sample even if no substance with a mutagenic potential is present. For the determination of reverse mutations, a histidine-free medium is used in order to only allow revertants to grow. The samples tested, however, contained concentrations of proteins up to $5\,g\,L^{-1}$. This might be a problem due to the release of histidine which is a potential by-product of hydrolysis [36] and could lead to false-positive results. If histidine would occur in sufficient quantities to influence the results and would be solely responsible for the increase of positive wells, the number of these should be identically independent from a metabolic activation or a tester strain. However, this does not apply with respect to the results obtained. As an example, the findings for the sample R1-S when investigated with or without metabolic activation were compared by means of the Student's t-test. For the concentrations of 50% and 80% a significant difference between both test approaches was found. Furthermore, a lower number of positive wells and lower induction factors were found for the same sample (R1-S), identical concentrations and no metabolic activation when using the tester strain TA98 compared to TA100 (data not shown). These results are strong indications against an effect solely obtained by histidine possibly contained in the samples.

Nevertheless, the potential impact of histidine impedes an incontestable interpretation. Therefore, mutagenicity/genotoxicity should be verified with a biotest not depending on a histidine free sample, for example, the umu test, the Comet assay, or a micronucleolus test. However, to the best of our knowledge, no studies on this topic are available.

15.3.4 Fish Embryo Toxicity

The fish embryo toxicity test using *D. rerio* embryos revealed similar patterns of the ecotoxicological effects as the cytotoxicity tests. The results are shown in Figure 15.10.

Highest mortality was determined for the dilute acid pretreated samples. Again, sample CS induced stronger effects than the sample R1-S. For the concentrations of 1%, significant differences between the untreated corn-fiber sample R4-S and the samples CS and R1-S were found ($p \leq 0.05$). These results were in very good concordance to the results obtained by the MTT cytotoxicity assay. In detail, highest mortality rates in each of these bioassays were observed for the samples CS and R1-S, with CS being slightly more toxic. R4-S, on the other hand, showed a decreased toxicity compared to the dilute acid pretreated samples. Correlation between embryo toxicity assays and cytotoxicity assays had been reported before, with the embryo toxicity assays being more sensitive than the cell-based cytotoxicity assays [98]. Important for the evaluation of these data is that this biotest was performed using fish embryos, and thus more complex organisms compared to the EROD or NR assay, which used a permanent cell line.

Furthermore, the toxicity of phenolic compounds on fish embryos was investigated by Lange et al. [98] and Schulte and Nagel [112]. The observed effects on *D.*

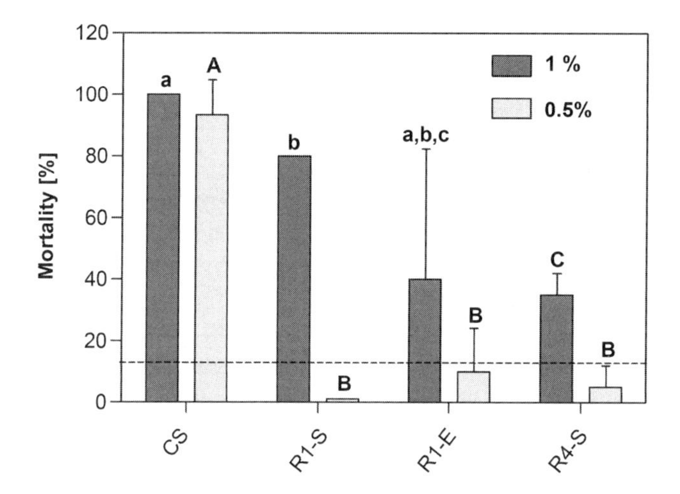

FIGURE 15.10 Embryo toxicity of the four samples in the fish embryo toxicity test with *Danio rerio*. Data are given as means (bars) and standard deviations (error bars). Presented are the mortalities in percent [%] for concentrations of 1% and 0.5% after 48 h of exposure. Small letters denote significant differences ($p < 0.05$) between samples tested at 1% and capital letters denote significant differences ($p < 0.05$) between samples tested at 0.5%. A sample is considered to be embryotoxic when exceeding 10% mortality. $n_{(CS)} = 3$; $n_{(R1\text{-}S,\ R1\text{-}E,\ R4\text{-}S)} = 2$.

rerio embryos may, thus, again be related to the formation of phenolic by-products during the dilute acid pretreatment [109]. Nevertheless, these observations are only applicable to the highest investigated concentration of 1%. With the exception of CS, which caused nearly 100% mortality at a concentration of 0.5 %, no lethal effects could be found at lower concentrations. From this finding, a steeper dose–response relationship for R1-S and R4-S compared to CS can be deducted, which would result in a lower EC_{50} value and corresponding higher toxicity potential for the sample CS. To verify this finding, further concentrations have to be tested and the number of replicates increased.

However, embryos exposed to lower concentrations of sample R1-S and likewise of sample CS, showed a considerable lack of pigmentation (cf, Fig. 15.11b). Although this effect is not considered lethal, it definitely represents an embryotoxic impact [113].

15.3.5 Endocrine Activity

Interpretation of the endocrine activity was complicated by several issues. Due to strong cytotoxic effects caused by the corn-fiber effluent sample R1-E when tested at maximum concentrations (62.5%), which is possibly caused by carboxylate product accumulation, the highest test concentrations of this sample had to be reduced to a concentration that showed no reduction in yeast numbers as determined by an OD measurement. The highest test concentration of the corn stover sample CS was also reduced, but not to the same extent as R1-E. The highest concentrations tested were

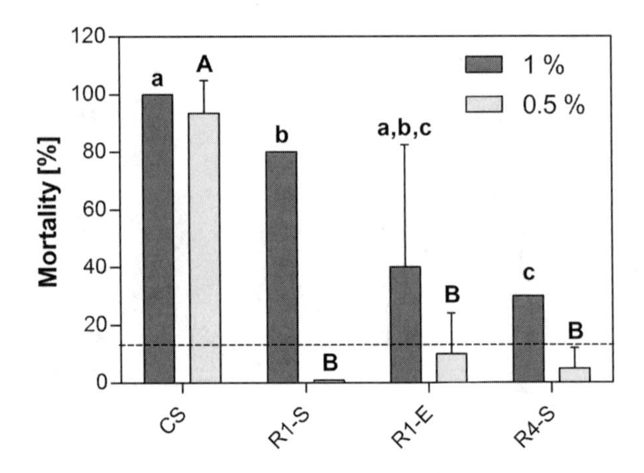

FIGURE 15.11 Embryos of *D. rerio* after 48 h of exposure to a negative control (**a**) and a sample inhibiting the development of normal pigmentation patterns (**b**).

20.8% of the undiluted sample for CS and 0.4% for the sample R1-E. However, due to the strong reduction of the highest test concentration for R1-E, the amount of potential estrogenic by-products in the sample was presumably diluted to the background activity.

The data presented in Figure 15.12 show the highest induction of the control, each sample, as well as the standard substance 17β-estradiol (E2). The induction is given as the estrogenic equivalent quotient (EEQ) in ng L^{-1} [81]. As shown in Figure 15.12, the EEQs of the samples range from 3.6 to 8.4 ng L^{-1}. Since the EEQ calculated for the negative control was 4.6 ng L^{-1}, the samples R1-E with an EEQ of 3.6 ng L^{-1} and R4-S with an EEQ of 5.4 ng L^{-1} were not considered to induce endocrine activity. Thus, the only samples that indicate slightly increased endocrine activities were CS (8.3 ng L^{-1}) and R1-S (7.8 ng L^{-1}). However, this assumption could not be confirmed by a statistical analysis.

With regard to the acute toxic effect of the sample CS on the yeast cells at higher concentrations, this sample was applied in lower concentrations than the samples R1-S and R4-S. Because CS dilute acid pretreatment was more stringent than any corn-fiber pretreatment while, at the same time, the lignin levels of corn stover were higher than corn fiber, this cytotoxicity was expected. However, some tolerance of yeast strains, such as *S. cerevisiae* for a dilute acid pretreatment with corn stover has been shown [114]. Considering the observation of no acute toxic effects on yeast by the dilute acid treated corn-fiber sample R1-S, the lethal impact of CS on yeast cells could be because of a higher concentration of aromatic compounds due to a more prevalent lignin concentration in CS.

The apparent lack of any endocrine activity for sample R1-E was unexpected. The inoculum for the reactors originated from sheep rumen and an anaerobic digester. It is well known that anaerobic treatment of organic plant material by intestinal bacteria can increase estrogenic potential of this material in bioreactors [115,116]. Although the reactor temperature was 55 °C compared to 37 °C for

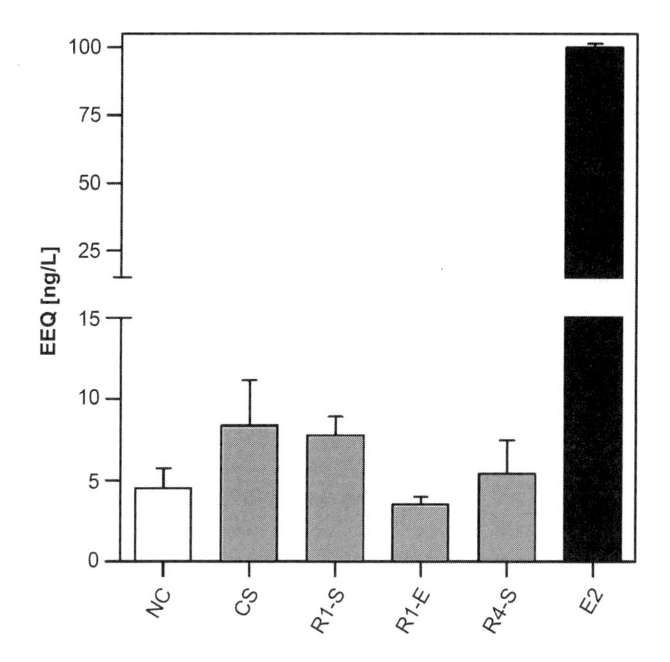

FIGURE 15.12 Highest induction of each sample in comparison to 1 pM of the E2 standard. Data are given as means (bars) and standard errors of the means (error bars) of the estradiol equivalent quotient (EEQ) in $ng\,L^{-1}$. The highest test concentrations for the samples are: CS = 20.8%, R1-S & R4-S = 62.5%, R1-E = 0.4%. $n_{all} = 2$.

intestines, these bioreactors operated analogously to intestinal processes (fermenting biomass to organic acids). Accordingly, an increase of the estrogenic activity of the corn-fiber sample during fermentation was expected, which would have resulted in the detection of a higher estrogenic potential for the effluent sample R1-E compared to the complementary substrate sample R1-S. A possible explanation for the apparent lack of any endocrine activity would be the strong cytotoxicity caused by the sample. These strong cytotoxic effects could have superimposed the putative endocrine activity. However, due to the low number of replicates ($n = 2$) these previous discussed results could not be verified by means of statistical analyses. Further replicates were considered but not performed, mainly due to extremely strong cytotoxic effects caused by sample R1-E. The cause for the high cytotoxicity of this sample is not completely understood. Toxic effects due to the main intermediates acetate and *n*-butyrate or due to bacterial toxins are possible [117,118], but were not investigated in this study. These observations collide with the results from the cytotoxicity assays, which revealed a much stronger cytotoxic potential for the samples R1-S and CS and a weaker cytotoxic potential for the effluent sample R1-E. However, large differences in the cytotoxic potential as determined by NR assay and MTT assay compared to the YES assay are conceivable due to physiological and morphological differences between RTL-W1 cells and *S. cerevisiae*. Yeasts, for example, have a cell wall, whereas RTL-W1 cells, like all cells of vertebrates, have only a cell membrane. Toxicity is likely to be influenced by structural differences as well as by different cellular responses, such as apoptosis or detoxification mechanisms.

Overall, the results obtained from the YES-assay are ambiguous. Although an increased endocrine activity was indicated by the assay for two samples, the concentrations applied for substrate and effluent samples are not comparable. Whereas the endocrine activity of sample R1-S was investigated starting at 62.5%, the highest concentration of the complementary effluent sample R1-E was 0.4%. The lack of estrogenic activity, especially regarding the effluent sample R1-E, was unexpected as discussed above. However, a higher estrogenic potential for this sample might still be conceivable but not detectable in the YES-assay due to high sample dilution. Thus, the YES assay might not be an appropriate test for this kind of samples. Consequently, the investigation of the endocrine activity of these samples should be continued using more sensitive assays, such as the ER CALUX® assay [119]. Additionally, effect-directed analysis could be applied for the investigated samples. In earlier studies, for example [88], masking toxicity of crude sediment extracts was reduced using fractionation techniques to evaluate the mechanism-specific toxicity of a given strong toxic sample.

With respect to the results presented and other available data, it seems to be appropriate to select certain endpoints separately for each biomass pretreatment method. An overview of the bioassays discussed, their applicability as well as preliminary estimations regarding their relevance for the testing of differently pretreated samples are presented in Table 15.4. We recommend assays for the investigation of cytotoxic potentials for samples from biomass pretreatment in general due to their utilization as a screening assay for further sublethal ecotoxicological endpoints, such as EROD induction. Taking into account the results presented by Tame et al. [111] and Zhao et al. [120], the EROD assay should be at least considered for investigations of ionic liquids and samples from pyrolysis. With regard to the dilute acid and the alkaline pretreatment, the implementation of the

TABLE 15.4 Applicability and Preliminary Estimation of the Relevance of Bioassays with Regard to Biomass Pretreatment Methods

Pretreatment	Cytotoxicity	EROD Activity	Embryotoxicity	Mutagenicity	Endocrine Activity
Dilute acid	+	+/–	+	+[a]	+[a]
Alkaline	+	+/–	n.a.	n.a.	+
Ionic liquids[b]	+	+	n.a.	n.a.	n.a.
Pyrolysis[c]	+	+	n.a.	n.a.	n.a.

+ = investigations highly recommended

+/– = applicable but appeared to be of less relevance with regard to the initial results.

[a]applicable but, with regard to the extracts used in this study, other test systems than the Ames fluctuation assay and the YES assay, respectively, might be more suitable; n.a. = no data available.

[b]Ref. [120].

[c]Ref. [111].

EROD assay is generally possible but, based on our results, seems to be of little relevance. Other assays, such as the embryotoxicity and the mutagenicity assay investigations, were only performed with dilute acid pretreated samples, and we are aware of no available literature on their application to other pretreatments. Concerning the results of the dilute acid pretreatment, the mutagenic and embryotoxic effects demonstrate a great importance for corresponding samples to be investigated in assays revealing effects on the embryotoxicity and mutagenicity. To make recommendations for the application of embryotoxicity and mutagenicity assays regarding samples from alkaline pretreatment, ionic liquids or pyrolysis, initial investigations are needed (these have to be assayed as well). This also holds true with regard to endocrine activity. However, with respect to samples from alkaline pretreatment, endocrine activities were indicated by first results (unpublished data), and thus tests regarding this endpoint should be considered. It has to be emphasized that all of these suggestions represent only a small fraction of all known (eco)toxicologically relevant endpoints. Nevertheless, they illustrate the need for more investigations of these biofuel-related samples.

In summary, ecotoxicological investigations of biomass samples after dilute acid pretreatment revealed cytotoxic and embryotoxic effects, no significant effects in the YES assay but indications for endocrine as well as mutagenic potentials. Dioxin-like activities were not detected. The dilute acid pretreatment of the substrates CS and R1-S were found to increase the cytotoxic potential for RTL-W1 cells and to decrease the survival rate of *D. rerio* embryos, compared to the effluent sample R1-E and the untreated substrate sample R4-S. This indicates a positive correlation with intensified pretreatments. In contrast, higher mutagenic potentials were revealed for the samples R1-S and R1-E than for CS, but the substances that caused these effects still remain unknown. Thus, there is a strong need for further investigations of the mutagenic potential. With respect to endocrine activity, however, no significant effects could be recorded, albeit indications for a low increase of endocrine potentials were found. However, further studies are highly recommended to decide whether the YES assay is the best-suited test system for this (eco)toxicological endpoint and to allow a better estimation of a reproduction toxicological potential risk for the environment and human health. Induction of the phase-I metabolism of xenobiotics, on the contrary, was not detected for any sample. Obviously, EROD-inducing compounds were not formed, neither by means of dilute acid pretreatment nor by the undefined mixed cultures during fermentation; and might not be of high relevance with respect to the samples tested. Nevertheless, depending on the pretreatment method the risk of increased dioxin-like activity has to be considered differently.

The biotest results together with the data available from literature strongly propose further ecotoxicological investigations with regard to biofuel-related samples, in order to identify any possible harm for the environment and human health. The respective biotests and endpoints to be studied should be chosen according to the kind of biofuel, the pretreatment method of the substrate, and the production process. However, further investigations are indispensable to define such decision criteria.

15.3.6 Implications for a Sustainable Production and Use of Biofuels

Knowledge on the potential hazards of biofuel-related samples is important to minimize adverse effects on the environment and/or human health. Ecotoxicological investigations provide important information regarding such effects and can therefore serve as an important tool for a sustainable development of biofuels. Beside biofuels themselves, relevant samples are also biofuel combustion products and their waste or by-products from the production processes, including biomass pretreatments and the conversion process.

At first, samples that induce certain adverse effects and, thus, permit conclusions regarding disadvantageous steps in the production process need to be identified using a multifaceted biotest battery. By chemical analyses of the samples the hazardous compounds causing adverse effects could subsequently be identified. Based on the test results, adaptations of a pretreatment method or changes in the production processes would then be carried out with the aim to minimize the hazard potential of a biofuel. A successful implementation of changes in the production process or the need for further improvement could be revealed through verification with further biotesting of samples from the adapted process.

Ecotoxicological investigations of biofuels and associated processes should be initiated early already accompanying the biofuel development process. This would allow to identify those biofuels and related production processes that are as environmentally compatible as possible and therefore avoid unintended release of a fuel with high hazard potentials. Changes in the production process as a result of ecotoxicological testing may, therefore, improve the ecological profile of biofuels.

15.4 CONCLUSIONS

Information on ecotoxicological effects of biofuels and biofuel emissions are limited but potential adverse effects to the environment cannot be ruled out. Furthermore, waste or by-products formed during pretreatment or fermentation may be sources of problematic substances, but even less data is available on this topic. Recently conducted investigations on the (eco)toxicological effectiveness of dilute-acid pretreated biofuel substrates and an effluent sample by means of a biotest battery address this topic. Test results revealed no dioxin-like activities but cytotoxic and embryotoxic effects, as well as indications for mutagenic effects and an increase of endocrine activities. Cytotoxic and embryotoxic effects varied depending on the pretreatment method. Samples pretreated with dilute acid (CS, R1-S) increased the cytotoxic potential and adverse effects on the survival rate of *D. rerio* embryos. The fermentation process, on the other hand, revealed a potential to remove cytotoxicity-inducing compounds. Mutagenic potentials were found for the samples CS, R1-S, and R1-E by using the Ames fluctuation assay and were discussed in the view of possible impacts due to the composition of the samples; but the substances behind these effect potentials still need to be identified. Nevertheless, the findings demonstrate that pretreatment can lead to adverse effects on organisms. As a consequence, the ecotoxicological impacts of biofuels, biofuel combustion products, and their waste or by-products from the production processes, including biomass pretreatments and the conversion process, have to be evaluated before a biofuel is deployed. Early identification of toxic intermediates, by-products, or wastes should be taken into consideration to allow production-process optimization regarding an improved ecological profile of biofuels. Further knowledge on adverse effects would help optimize the production process and could serve as an early warning system to prevent the formation of toxic by-products or restrain the release of dangerous substances to the environment.

ACKNOWLEDGMENTS

This work was partly performed as part of the Cluster of Excellence "Tailor-Made Fuels from Biomass," which is funded by the Excellence Initiative by the German federal and state government to promote science and research at German universities.

Further funding for this project was provided by a seed funds grant of the RWTH University of Aachen by the German Excellence Initiative of the DFG. Moreover, we thank the U.S. Department of Agriculture for the support of Largus T. Angenent and Matthew T. Agler through a NIFA grant with contract number 2007-35504-05381.

The authors thank Loren B. Iten and Bruce S. Dien (USDA Agricultural Research Service, Peoria, IL, USA) for their work on the biomass pretreatment.

REFERENCES

1. J. Fargione, J. Hill, D. Tilman, S. Polasky and P. Hawthorne, *Science* **2008**, 319, 1235.
2. T. Searchinger, R. Heimlich, R. A. Houghton, F. X. Dong, A. Elobeid, J. Fabiosa, S. Tokgoz, D. Hayes and T. H. Yu, *Science* **2008**, 319, 1238.
3. Commission of the European Communities, COM (**2006**) 848, 2006.
4. M. Faulstich and K. B. Greiff, *Umweltwiss. Schadst. Forsch.* **2008**, 20, 171.
5. L. Gouveia and A. C. Oliveira, *J. Ind. Microbiol. Biotechnol.* **2009**, 36, 269.
6. P. M. Schenk, S. R. Thomas-Hall, E. Stephens, U. C. Marx, J. H. Mussgnug, C. Posten, O. Kruse and B. Hankamer, *Bioenerg. Res.* **2008**, 1, 20.
7. V. Patil, K. Q. Tran and H. R. Giselrod, *Int. J. Mol. Sci.* **2008**, 9, 1188.
8. The Cluster of Excellence "Tailor made fuels from biomass" (TMFB), RWTH Aachen University, www.fuelcenter.rwth-aachen.de, [accessed **2011**].
9. J. M. Davis and V. M. Thomas, *Ann. NY Acad. Sci.* **2006**, 1076, 498.
10. K. Bluhm, S. Heger, T.-B. Seiler, A. V. Hallare, A. Schäffer and H. Hollert, *Energy Environ. Sci.* **2012**, 5, 7381–7392.
11. J. Bünger, J. Krahl, H. U. Franke, A. Munack and E. Hallier, *Mutat. Res. Gen. Tox. En.* **1998**, 415, 13.
12. J. Bünger, J. Krahl, K. Baum, O. Schroder, M. Muller, G. Westphal, P. Ruhnau, T. G. Schulz and E. Hallier, *Arch. Toxicol.* **2000**, 74, 490.

13. J. Bünger, M. M. Muller, J. Krahl, K. Baum, A. Weigel, E. Hallier and T. G. Schulz, *Mutagenesis* **2000**, 15, 391.
14. J. Bünger, J. Krahl, A. Munack, Y. Ruschel, O. Schroder, B. Emmert, G. Westphal, M. Muller, E. Hallier and T. Bruning, *Arch. Toxicol.* **2007**, 81, 599.
15. N. Y. Kado and P. A. Kuzmicky, NREL/SR-510-31463, Department of Environmental Toxicology, University of California, Davis, **2003**.
16. J. Krahl, A. Munack, N. Grope, Y. Ruschel, O. Schroder and J. Bunger, *Clean Soil Air Water* **2007**, 35, 417.
17. J. Krahl, G. Knothe, A. Munack, Y. Ruschel, O. Schroder, E. Hallier, G. Westphal and J. Bunger, *Fuel* **2009**, 88, 1064.
18. C. L. Song, Y. C. Zhou, R. J. Huang, Y. Q. Wang, Q. F. Huang, G. Lu and K. M. Liu, *J. Hazard. Mater.* **2007**, 149, 355.
19. M. L. Gagnon and P. A. White, *Environ. Mol. Mutagen.* **2008**, 49, 564.
20. J. L. Mauderly, *Plant Oils as Fuels: Present State of Science and Future Developments* (Eds. N. Martini, J. S. Schell), Springer, Berlin/Germany, **1998**, 92.
21. L. Turrio-Baldassarri, C. L. Battistelli, L. Conti, R. Crebelli, B. De Berardis, A. L. Iamiceli, M. Gambino and S. Iannaccone, *Sci. Total Environ.* **2004**, 327, 147.
22. B. P. Hollebone, B. Fieldhouse and M. Landriault, in *Proceedings of the International Oil Spill Conference* **2008**, 929.
23. N. Khan, M. A. Warith and G. Luk, *J. Air Waste Manage.* **2007**, 57, 286.
24. A. Lapinskiene, P. Martinkus and V. Rebzdaite, *Environ. Pollut.* **2006**, 142, 432.
25. Y. Y. Liu, T. C. Lin, Y. J. Wang and W. L. Ho, *J. Environ. Sci. Health Part A Toxic-Hazard. Subst. Environ. Eng.* **2008**, 43, 1735.
26. Z. K. Li, K. Lee, S. E. Cobanli, T. King, B. A. Wrenn, K. G. Doe, P. M. Jackman and A. D. Venosa, *Environ. Toxicol.* **2007**, 22, 1.
27. P. Gateau, F. van Dievoet, V. Bouillon, G. Vermeersch, S. Claude and F. Staat, *J. Am. Oil Chem. Soc.* **2005**, 12, 308.
28. J. M. Brito, L. Belotti, A. C. Toledo, L. Antonangelo, F. S. Silva, D. S. Alvim, P. A. Andre, P. H. N. Saldiva and D. Rivero, *Toxicol. Sci.* **2010**, 116, 67.
29. R. Poon, I. Chu, V. E. Valli, L. Graham, A. Yagminas, B. Hollebone, G. Rideout and M. Fingas, *Food Chem. Toxicol.* **2007**, 45, 1830.
30. R. Poon, V. E. Valli, M. Rigden, G. Rideout and G. Pelletier, *Food Chem. Toxicol.* **2009**. 47, 1416–1424.
31. M. G. Godoy, M. L. E. Gutarra, F. M. Maciel, S. P. Felix, J. V. Bevilaqua, O. L. T. Machado and D. M. G. Freire, *Enzyme Microb. Technol.* **2009**, 44, 317.
32. NREL, Biomass Research – What is a Biorefinery?, http://www.nrel.gov/biomass/biorefinery.html, [accessed **2009**].
33. M. T. Agler, B. A. Wrenn, S. H. Zinder and L. T. Angenent, *Trends Biotechnol.* **2011**, 29 (2), 70–78.
34. S. Fernando, S. Adhikari, C. Chandrapal and N. Murali, *Energy Fuels* **2006**, 20, 1727.
35. N. Mosier, C. Wyman, B. Dale, R. Elander, Y. Y. Lee, M. Holtzapple and M. Ladisch, *Bioresour. Technol.* **2005**, 96, 673.
36. V. V. Zverlov, O. Berezina, G. A. Velikodvorskaya and W. H. Schwarz, *Appl. Microbiol. Biotechnol.* **2006**, 71, 587.

37. P. T. Pienkos and M. Zhang, *Cellulose* **2009**, 16, 743.
38. S. A. Gangu, L. R. Weatherley and A. M. Scurto, *Curr. Org. Chem.* **2009**, 13, 1242.
39. H. Noureddini and J. Byun, *Bioresour. Technol.* **2010**, 101, 1060.
40. S. Kim and M. T. Holtzapple, *Bioresour. Technol.* **2005**, 96, 1994.
41. V. Chang, M. Nagwani, C.-H. Kim and M. Holtzapple, *Appl. Biochem. Biotechnol.* **2001**, 94, 1.
42. J. M. Lawther and R. Sun, *Ind. Crop. Prod.* **1996**, 5, 87.
43. T. Welton, *Chem. Rev.* **1999**, 99, 2071.
44. M. Zavrel, D. Bross, M. Funke, J. Büchs and A. C. Spiess, *Bioresour. Technol.* **2009**, 100, 2580.
45. S. Singh, B. A. Simmons and K. P. Vogel, *Biotechnol. Bioeng.* **2009**, 104, 68.
46. J. Zakzeski, P. C. A. Bruijnincx, A. L. Jongerius and B. M. Weckhuysen, *Chem. Rev.* **2010**, 110, 3552.
47. M. Matsumoto, K. Mochiduki, K. Fukunishi and K. Kondo, *Sep. Purif. Technol.* **2004**, 40, 97.
48. J. Ranke, K. M^lter, F. Stock, U. Bottin-Weber, J. Poczobutt, J. Hoffmann, B. Ondruschka, J. Filser and B. Jastorff, *Ecotoxicol. Environ. Saf.* **2004**, 58, 396.
49. S. Yaman, *Energ. Convers. Manage.* **2004**, 45, 651.
50. H. Younesi, G. Najafpour and A. R. Mohamed, *Biochem. Eng. J.* **2005**, 27, 110.
51. E. M. Prosen, D. Radlein, J. Piskorz, D. S. Scott and R. L. Legge, *Biotechnol. Bioeng.* **1993**, 42, 538.
52. A. Ahmed, B. G. Cateni, R. L. Huhnke and R. S. Lewis, *Biomass Bioenerg.* **2006**, 30, 665.
53. A. Ahmed and R. S. Lewis, *Biotechnol. Bioeng.* **2007**, 97, 1080.
54. K. Sipilä, E. Kuoppala, L. Fagernäs and A. Oasmaa, *Biomass Bioenerg.* **1998**, 14, 103.
55. J. Lehmann, J. Gaunt and M. Rondon, *Mitigation Adaptation Strategies Global Change* **2006**, 11, 395.
56. J. L. Gaunt and J. Lehmann, *Environ. Sci. Technol.* **2008**, 42, 4152.
57. P. M. Chapman, *Sci. Total Environ.* **1990**, 97–98, 815.
58. P. M. Chapman and H. Hollert, *J. Soils Sediments* **2006**, 6, 4.
59. C. Hoffmann, D. Sales and N. Christofi, *Int. Microbiol.* **2003**, 6, 41.
60. A. S. Bhat and A. A. Ahangar, *Toxicol. Mech. Methods* **2007**, 17, 441.
61. K. Fent, *Ökotoxikologie*, Thieme Verlag, Stuttgart, **2007**.
62. L. K. Gustavsson, N. Klee, H. Olsman, H. Hollert and M. Engwall, *Environ. Sci. Pollut. Res. Int.* **2004**, 11, 379.
63. M. E. DeLorenzo and J. Fleming, *Arch. Environ. Contam. Toxicol.* **2008**, 54, 203.
64. E. Testai, *Endocrine Disrupters and Carcinogenic Risk Assessment,* Vol. 340 (Eds. L. Chyczewski, J. Niklinski, E. Pluygers), I O S Press, Amsterdam, **2002**, 255.
65. A. V. Weisbrod, J. Sahi, H. Segner, M. O. James, J. Nichols, I. Schultz, S. Erhardt, C. Cowan-Ellsberry, M. Bonnell and B. Hoeger, *Environ. Toxicol. Chem.* **2009**, 28, 86.
66. H. Hollert, S. Heise, S. Pudenz, R. Bruggemann, W. Ahlf and T. Braunbeck, *Ecotoxicology* **2002**, 11, 311.
67. S. Keiter, A. Rastall, T. Kosmehl, K. Wurm, L. Erdinger, T. Braunbeck and H. Hollert, *Environ. Sci. Pollut. Res. Int.* **2006**, 13, 308.

68. M. Narracci, R. A. Cavallo, M. I. Acquaviva, E. Prato and F. Biandolino, *Environ. Monit. Assess.* **2009**, 148, 307.
69. L. Mariani, D. De Pascale, O. Faraponova, A. Tornambe, A. Sarni, S. Giuliani, G. Ruggiero, F. Onorati and E. Magaletti, *Environ. Toxicol.* **2006**, 21, 373.
70. Regulation (EC) No 1907/2006 of the European Parliament and of the Council, **2006**.
71. L. T. Angenent and B. A. Wrenn, *Bioenergy* (Eds. C. S. Harwood, A. L. Demain, J. D. Wall), ASM Press, Washington, DC, **2008**, 179.
72. H. Babich and E. Borenfreund, *Toxicol. In Vitro* **1992**, 6, 493.
73. T. Mosmann, *J. Immunol. Methods* **1983**, 65, 55.
74. L. E. J. Lee, J. H. Clemons, D. G. Bechtel, S. J. Caldwell, K. B. Han, M. Pasitschniakarts, D. D. Mosser and N. C. Bols, *Cell Biol. Toxicol.* **1993**, 9, 279.
75. A. Behrens, K. Schirmer, N. C. Bols and H. Segner, *9th International Symposium on Responses of Marine Organisms to Pollutants (PRIMO 9)* Elsevier Sci Ltd., Bergen, Norway, **1997**, 369.
76. G. Reifferscheid, C. Arndt and C. Schmid, *Environ. Mol. Mutagen.* **2005**, 46, 126.
77. International Organization for Standardization, ISO/CD 11350 (draft), **2010**.
78. R. Nagel, *Altex Altern. Tierexp.* **2002**, 19, 38.
79. T. Braunbeck, M. Bottcher, H. Hollert, T. Kosmehl, E. Lammer, E. Leist, M. Rudolf and N. Seitz, *Altex Altern. Tierexp.* **2005**, 22, 87.
80. E. J. Routledge and J. P. Sumpter, *Environ. Toxicol. Chem.* **1996**, 15, 241.
81. M. Wagner and J. Oehlmann, *Environ. Sci. Pollut. Res. Int.* **2009**, 16, 278.
82. H. Hollert, M. Durr, L. Erdinger and T. Braunbeck, *Environ. Toxicol. Chem.* **2000**, 19, 528.
83. H. Hollert, M. Durr, H. Olsman, K. Halldin, B. Van Bavel, W. Brack, M. Tysklind, M. Engwall and T. Braunbeck, *Ecotoxicology* **2002**, 11, 323.
84. E. Borenfreund and J. A. Puerner, *Toxicol. Lett.* **1985**, 24, 119–124.
85. E. Borenfreund and C. Shopsis, *Xenobiotica* **1985**, 15, 705.
86. K. Fent and J. Hunn, *8th International Symposium on Pollutant Responses in Marine Organisms (PRIMO 9)* Elsevier Sci Ltd, Pacific Grove, CA, **1995**, 377.
87. K. Fent and J. J. Stegeman, *Aquat. Toxicol.* **1993**, 24, 219.
88. W. Brack and K. Schirmer, *Environ. Sci. Technol.* **2003**, 37, 3062.
89. K. Hilscherova, M. Machala, K. Kannan, A. L. Blankenship and J. P. Giesy, *Environ. Sci. Pollut. Res. Int.* **2000**, 7, 159.
90. K. Schirmer, S. Bopp, S. Russold and P. Popp, *Grundwasser – Zeitschrift der Fachsektion Hydrogeologie* **2004**, 1, 33.
91. J. J. Whyte, R. E. Jung, C. J. Schmitt and D. E. Tillitt, *Crit. Rev. Toxicol.* **2000**, 30, 347.
92. B. N. Ames, F. D. Lee and W. E. Durston, *Proc. Natl. Acad. Sci. USA* **1973**, 70, 782.
93. J. McCann, N. E. Spingarn, J. Kobori and B. N. Ames, *Proc. Natl. Acad. Sci. USA* **1975**, 72, 979.
94. B. N. Ames, J. McCann and E. Yamasaki, *Mutat. Res.* **1975**, 31, 347.
95. B. N. Ames, W. E. Durston, E. Yamasaki and F. D. Lee, *Proc. Natl. Acad. Sci. USA* **1973**, 70, 2281.
96. H. W. Laale, *J. Fish Biol.* **1977**, 10, 121.

97. U. Ensenbach, *Fresenius Environ. Bull.* **1998**, 7, 531.
98. M. Lange, W. Gebauer, J. Markl and R. Nagel, *Chemosphere* **1995**, 30, 2087.
99. H. Hollert, S. Keiter, N. König, M. Rudolf, M. Ulrich and T. Braunbeck, *J. Soils Sediments* **2003**, 3, 197.
100. P. De Boever, W. Demare, E. Vanderperren, K. Cooreman, P. Bossier and W. Verstraete, *Environ. Health Perspect.* **2001**, 109, 691.
101. S. F. Arnold, M. K. Robinson, A. C. Notides, L. J. Guillette and J. A. McLachlan, *Environ. Health Perspect.* **1996**, 104, 544.
102. L. J. Guillette, D. B. Pickford, D. A. Crain, A. A. Rooney and H. F. Percival, *Gen. Comp. Endocrinol.* **1996**, 101, 32.
103. E. Carlsen, A. Giwercman, N. Keiding and N. E. Skakkebaek, *Br. Med. J.* **1992**, 305, 609.
104. E. Carlsen, A. Giwercman, N. Keiding and N. E. Skakkebaek, *Environ. Health Perspect.* **1995**, 103, 137.
105. T. M. Crisp, E. D. Clegg, R. L. Cooper, W. P. Wood, D. G. Anderson, K. P. Baetcke, J. L. Hoffmann, M. S. Morrow, D. J. Rodier, J. E. Schaeffer, L. W. Touart, M. G. Zeeman and Y. M. Patel, *Environ. Health Perspect.* **1998**, 106, 11.
106. E. Dewailly, S. Dodin, R. Verreault, P. Ayotte, L. Sauve, J. Morin and J. Brisson, *J. Natl. Cancer Inst.* **1994**, 86, 232.
107. K. A. Kidd, P. J. Blanchfield, K. H. Mills, V. P. Palace, R. E. Evans, J. M. Lazorchak and R. W. Flick, *Proc. Natl. Acad. Sci. USA* **2007**, 104, 8897.
108. T. Schultis and J. W. Metzger, *Chemosphere* **2004**, 57, 1649.
109. M. J. Lopez, N. N. Nichols, B. S. Dien, J. Moreno and R. J. Bothast, *Appl. Microbiol. Biotechnol.* **2004**, 64, 125.
110. W. Brack, H. Segner, M. Moder and G. Schuurmann, *Environ. Toxicol. Chem.* **2000**, 19, 2493.
111. N. W. Tame, B. Z. Dlugogorski and E. M. Kennedy, *P. Combust. Inst.* **2009**, 32, 665.
112. C. Schulte and R. Nagel, *ATLA-Altern. Lab. Anim.* **1994**, 22, 12.
113. E. J. van den Brandhof and M. Montforts, *Ecotoxicol. Environ. Saf.* **2010**, 73, 1862.
114. C. Fischer, D. Clein-Marcuschamer and G. Stephanopoulos, *Metab. Eng.* **2008**, 10, 295.
115. S. Possemiers, A. Heyerick, V. Robbens, D. De Keukeleire and W. Verstraete, *J. Agric. Food Chem.* **2005**, 53, 6281.
116. L. H. Xie, E. M. Ahn, T. Akao, A. A. M. Abdel-Hafez, N. Nakamura and M. Hattori, *Chem. Pharm. Bull.* **2003**, 51, 378.
117. G. Rodrigues and C. Pais, *Food Technol. Biotechnol.* **2000**, 38, 27.
118. F. M. Al-Jasass, S. M. Al-Eid and S. H. H. Ali, *J. Food Agric. Env.* **2010**, 8, 314.
119. J. Legler, C. E. van den Brink, A. Brouwer, A. J. Murk, P. T. van der Saag, A. D. Vethaak and P. van der Burg, *Toxicol. Sci.* **1999**, 48, 55.
120. D. B. Zhao, Y. C. Liao and Z. D. Zhang, *Clean-Soil Air Water* **2007**, 35, 42.

INDEX

The Role of Green Chemistry in Biomass Processing and Conversion, First Edition.
Edited by Haibo Xie and Nicholas Gathergood.